11th Nanosafe International Conference on Health and Safety Issues Related to Nanomaterials for a Socially Responsible Approach (NANOSAFE 2016)

Journal of Physics: Conference Series Volume 838

Grenoble, France
7 – 10 November 2016

ISBN: 978-1-5108-4184-0
ISSN: 1742-6588

Printed from e-media with permission by:

Curran Associates, Inc.
57 Morehouse Lane
Red Hook, NY 12571

Some format issues inherent in the e-media version may also appear in this print version.

Printed by Curran Associates, Inc. (2017)

For permission requests, please contact the Institute of Physics at the address below.

Institute of Physics
Dirac House, Temple Back
Bristol BS1 6BE UK

Phone: 44 1 17 929 7481
Fax: 44 1 17 920 0979

techtracking@iop.org

Additional copies of this publication are available from:

Curran Associates, Inc.
57 Morehouse Lane
Red Hook, NY 12571 USA
Phone: 845-758-0400
Fax: 845-758-2633
Email: curran@proceedings.com
Web: www.proceedings.com

Table of contents

Volume 838

5th Nanosafe International Conference on Health and Safety issues related to nanomaterials for a socially responsible approach (NANOSAFE 2016)

7–10 November 2016, Grenoble, France

Accepted papers received: 2 May 2017
Published online: 1 June 2017

Preface

Papers

Measurement and characterization of nano-objects

Exposure

Safer by design nanomaterials and process

Risk Management

From nanoproducts to waste

Toxicology

Safe use of nano objects for medicine applications

Regulation/Standardization

IOP Conf. Series: Journal of Physics: Conf. Series **838** (2017) 011001 doi:10.1088/1742-6596/838/1/011001

Nanosafe 2016 - Preface

This volume contains a collection of contributions presented at the 5th **Nanosafe International Conference** on Health and Safety issues related to nanomaterials for a socially responsible approach (NANOSAFE 2016) held in Grenoble, France, from 7th to 10th November 2016.

Nano objects represent a powerful "enabling technology" leading to revolutionary breakthroughs in many different areas vital for humanity including medicine, energy, environment, etc. and also preserving the rare mineral resources by rendering mater more efficient.
For one of the first time in the science history, risks have been taken into account since the very beginning of the manufactured nanomaterials and Nanosafety is now considered as a specific new scientific area, gaining in importance and maturity each days thanks to our dynamic community spread all over the world.

Following the successful outcome of the four past international conferences on Safe Production and Use of Nanomaterials: Nanosafe 2008, 2010, 2012 and 2014, **the Platform Nano Safety "PNS"** welcomed the attendees to Minatec, Grenoble, for this fifth edition with some of the most famous specialists in the field.
In 2016, the subtitle of the conference had been slightly changed to "**Health and Safety Issues Related to Nanomaterials**" in order to welcome inboard two new topics: Urban Nanoparticles and some aspect of Nanomedicine, in addition to the usual issues addressed in previous Nanosafe conferences such as Detection and Characterization, Expology, Release from Nano-enabled Products, Safer by Design Nanomaterials and Process, Risk Management, Nanoproducts to waste, Toxicology, Environmental Interactions, Regulation and Standardization and Nano Responsible Development.
Furthermore, three 3 round tables had been organized in order to promote friendly discussions between attendees: Nano-Responsible Development, Urban Particles Mitigation: What is Reasonably Possible, Nanomedecine: Benefice/Risk.

In this 5th edition, there were more than 350 registered participants from 29 different countries including 170 oral presentations and 91 posters covering the whole Nanosafety issues in 6 main sessions, satellite workshops and round tables.

This fifth edition had been a great success and the Local Organizing Committee want to acknowledge all the attendees for their presence and the quality of their presentations.

The Nanosafe 2016 organizers

François TARDIF

Jean-François DAMLENCOURT

Philippe CHARLETY

Gaelle CHARLIER

IOP Conf. Series: Journal of Physics: Conf. Series **838** (2017) 011001 doi:10.1088/1742-6596/838/1/011001

Committees

CHAIR: **François TARDIF** (CEA/PNS, FR), **Jean-François DAMLENCOURT** (CEA/PNS, FR) **& Frédéric SCHUSTER** (CEA, FR)

CO-CHAIR: **Georgios KATALAGARIANAKIS** (EC, BE)

INTERNATIONAL SCIENTIFIC COMMITTEE

Christof ASBACH (IUTA, DE)
Daniel BERNARD (CEA/PNS, FR)
Jorge BOCZKOWSKI (INSERM, FR)
Patrick BOISSEAU (CEA/PNS, FR)
Jean-Yves BOTTERO (CEA/PNS, FR)
Marie CARRIERE (CEA, FR)
Flemming CASSEE (RIVM, NL)
Alexandre CECCALDI (ETPN, FR)
Daren CHEN (University of Virginia, USA)
Eric DRAIS (INRS, FR)
Claude EMOND (University of Montreal, CA)
Steffi FRIEDRICHS (OECD, FR)
Benjamin GILBERT (Berkeley, USA)
Peter HOET (Ku Leuven, BE)
Sophie LANONE (INSERM, FR)
Simona MURA (Universite Paris Sud, FR)
Fabrice NESSLANY (Institut Pasteur, FR)
Bernd NOWACK (EMPA, CH)
Günter OBERDORSTER (U. Rochester, USA)
David PUI (University of Minnesota, USA)
Jérôme ROSE (CNRS-CEREGE, FR)
Myriam RICAUD (INRS, FR)
Vicki STONE (HWU, GB)
Raphael de THOURY (Nanobadge, Alcen, FR)
Mark WIESNER (Duke University, USA)
Wendel WOHLLEBEN (BASF, DE)

LOCAL ORGANIZING COMMITTEE - THE NANOSAFETY PLATFORM (PNS)

Gaelle CHARLIER (CEA/PNS, FR)

Philippe CHARLETY (CEA/PNS, FR)
Pascal CONCHE (CEA/MINATEC, FR)
Jean-François DAMLENCOURT (CEA/PNS, FR)
Julie DUBOIS (CEA/PNS, FR)
Vanessa GAULTIER (CEA/PNS, FR)
François TARDIF (CEA/PNS, FR)
Florian VUONG (CEA/MINATEC, FR)

THANKS TO THE SUPPORT RECEIVED FROM OUR VOLUNTEER TEAM:

S. Derrough, M. Aru, D. Boutry, S. Artous, C. Ducros, C. Philippot, S. Jacquinot, L. Givelet, S. Clavaguera, C. Brouard, A. Guiot, V. Bartolomei, D. Locatelli, S. Motellier, C. Desvergne, E. Zimmermann

IOP Conf. Series: Journal of Physics: Conf. Series **838** (2017) 011001 doi:10.1088/1742-6596/838/1/011001

The Nanosafety Platform

Fully dedicated to Health and Safety issues related to nanomaterials

Located at CEA Grenoble, France

Well-defined assignments and advanced specific tools

One of the main PNS goals is to contribute to the responsible and competitive development of nanomaterials. The PNS benefits from the support of CEA (*The French Alternative Energies and Atomic Energy Commission* - 16 000 researchers) in terms of fundamental research, safety technologies and nanomaterials pilot lines in actual use. It also relies on "equipments of excellence" for detection and characterization (Cryo Tomo TEM for complex media).

A wide range of R&D and services available straightforward through a one stop shop to help industrials, unique in France and Europe.

PNS R&D activities focuses on 6 main topics:

-Technologies and methods for advanced characterization & detection
-Safer-by-Design materials and processes
-Toxicology and ecotoxicology through partnerships
-Industrial hygiene
-Lifecycle management and dissemination in the environment
-Urban particles

PNS provides a wide range of services suitable to industry:

-Consulting and auditing on industrial health & safety issues
-Measurements at workplace
-Characterization
-Release testing in use conditions (accelerated aging / abrasion resistance, etc) in order to reduce consumers exposure to nanoparticles
-Nanoparticle toxicity evaluation
-Training (catalog of standard training courses and development of customized, company specific training courses) for operators and prevention workers

Thanks to its integrated R&D and services activities, to its independence of analysis and to its deep commitment to confidentiality, PNS provides a unique and efficient response to industrial needs in nanosafety.

Visit us at www.nanosafety-platform.com

IOP Conf. Series: Journal of Physics: Conf. Series **838** (2017) 011001 doi:10.1088/1742-6596/838/1/011001

Nanosafe 2016 – Gallery

IOP Conf. Series: Journal of Physics: Conf. Series **838** (2017) 011001 doi:10.1088/1742-6596/838/1/011001

Peer review statement

All papers published in this volume of *Journal of Physics: Conference Series* have been peer reviewed through processes administered by the proceedings Editors. Reviews were conducted by expert referees to the professional and scientific standards expected of a proceedings journal published by IOP Publishing.

IOP Conf. Series: Journal of Physics: Conf. Series **838** (2017) 012001 doi:10.1088/1742-6596/838/1/012001

Performance study of portable devices for the real-time measurement of airborne particle number concentration and size (distribution)

Sébastien Bau[1], Raphaël Payet[1], Olivier Witschger[1] and Elżbieta Jankowska[2]

[1]Institut National de Recherche et de Sécurité (INRS), Laboratoire de Métrologie des Aérosols, Rue du Morvan, CS 60027, F 54519 Vandoeuvre Cedex, France;

[2]Central Institute for Labour Protection (CIOP-PIB), Department of Chemical, Aerosol and Biological Hazards, PL 00-701 Warsaw, Poland

E-mail: sebastien.bau@inrs.fr

Abstract. The aim of this experimental study was to investigate the performance of both portable and transportable devices devoted to the real-time measurement of airborne particle number concentration and size (distribution). Electrical mobility spectrometers (SMPS, FMPS, Nanoscan) as well as diffusion chargers (DiSCmini, Nanotracer) were studied. Both monodisperse and polydisperse aerosols were produced within the CAIMAN facility to challenge the instruments. The monodisperse test aerosols were selected in the 15-400 nm diameter range using a differential mobility analyser (DMA), and presented number concentrations of between 6.10^2 and 2.10^5 cm^{-3}. The polydisperse test aerosols presented modal diameters of between 8 and 270 nm and number concentrations between 4.10^3 to 10^6 cm^{-3}. The behavior of the different devices is expressed as (1) the ratio of the reported diameter to the reference diameter, and (2) the ratio of the reported number concentration to the reference concentration. These results are displayed as boxplots to better represent the statistical distribution of the experimental results. For the group of electrical mobility spectrometers, a good agreement between SMPS and FMPS and the reference was demonstrated. A slight tendency for the Nanoscan to underestimate particle size distribution for particles above around 100 nm was observed. The data reported for the group of diffusion chargers demonstrate that all, except the Nanotracer, show a tendency to underestimate particle diameter, by a factor around -40% to -10%. In the case of particle concentration, larger deviations were observed.

1. Introduction

Nanomaterials have been increasingly developed and used in many technology and industry sectors over the last 20 years, and increasing numbers of workers are thus likely exposed to airborne nanoparticles [1]. The parameters that should be assessed in order to characterize airborne nanomaterials are still being debated, and implementation of a multi-metric approach has recently been suggested [2-4].

In addition to chemical composition, airborne particle number concentration as well as particle size are among the parameters of interest [5] as they allow determination of the nanoparticle quantity and indicate the region of the respiratory tract where inhaled nanoparticles will be deposited and potentially interact. Methodologies to assess occupational inhalation exposure to airborne particles during production, handling and use of manufactured nanomaterials have recently been proposed [6-12] and tested in various workplace environments [13-18]. In particular, these strategies emphasize the utility of real-time instruments.

Investigating the performances of instruments devoted to the measurement of airborne nanoparticles is crucial prior to their use in workplaces, e.g. [19, 20]. Both "gold-standard" instruments, such as electrical mobility analyzers (e.g., Scanning Mobility Particle Sizer – SMPS, Fast Mobility Particle Sizer – FMPS, Nanoscan – NS), and portable and battery-operated diffusion chargers, designed for occupational hygiene/workplace exposure monitoring, are studied in this work.

It is also expected that this work will feed into the standardization work currently underway in Working Group 3 of CEN/TC137 (Assessment of workplace exposure to chemical and biological agents - Particulate matter) [21, 22].

2. Materials and methods

2.1. Test aerosols

Both monodisperse and polydisperse test aerosols, consisting of metals (carbon, titanium, silver), metal alloy (constantan), salts (sodium chloride, cesium chloride) and organic compounds (DEHS – DiEthylHexylSebacate), were produced to challenge the instruments. All aerosols were produced within the CAIMAN facility developed at INRS [23]. The monodisperse aerosols consisted of originally polydisperse aerosols, which were subsequently DMA-selected. The range of particle (modal) diameter and number concentration for both monodisperse and polydisperse test aerosols are presented in Table 1.

Table 1. Range of modal diameters and number concentrations investigated.

	Range of (modal) diameter (nm)	Range of number concentration (cm^{-3})
monodisperse	15 – 400	$6.10^2 – 2.10^5$
polydisperse	8 – 270	$4.10^3 – 1.10^6$

Examples of monodisperse and polydisperse number size distributions of the test aerosols are shown in Figures 1 and 2.

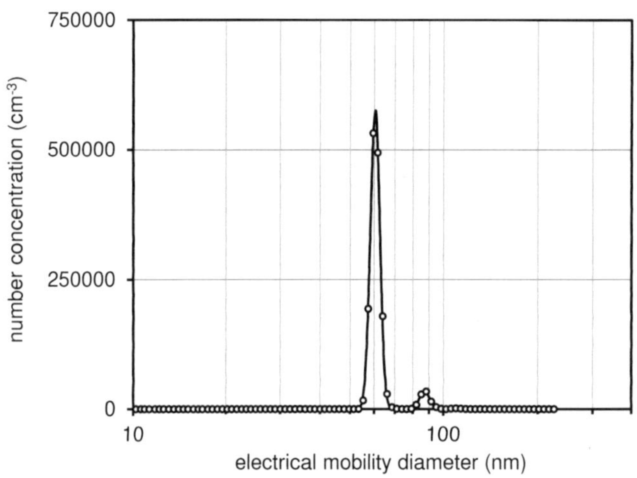

Figure 1. Example of monodisperse number size distribution (DMA-selected electrical mobility diameter 60 nm, Ag particles produced within CAIMAN [23]).

Figure 1 highlights the presence of multiple-charge particles (equivalent electrical mobility diameter of approximately 90 nm). However, in the remainder of this paper, only the DMA-selected particle diameter will be considered, the proportion of multiple-charge particles being below about 5 % of the total number and thus considered negligible.

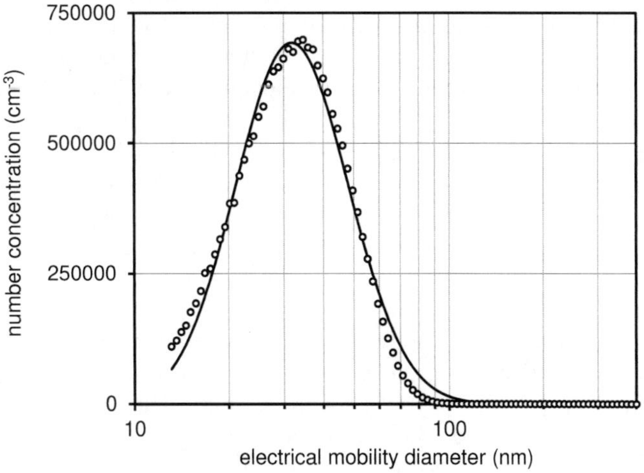

Figure 2. Example of polydisperse number size distribution (Al particles produced within the CAIMAN facility [23]).

2.2. Reference instruments

The reference particle number concentration ($C_{N,\mathrm{ref}}$) was provided by a Condensation Particle Counter (CPC; Grimm 5.401, d_{50} = 4.5 nm, $Q_{aerosol}$ = 1.5 L min^{-1}), while the reference number size distribution was determined by a Scanning Mobility Particle Sizer (SMPS; Grimm, composed of a Vienna Type Differential Mobility Analyzer — DMA — and a CPC 5.403, $Q_{aerosol}$ = 0.3 L min^{-1}, Q_{sheath} = 3 L min^{-1}). A lognormal model was fitted by a least squares method to obtain the modal diameter of the distribution (e.g., Figure 2), considered as the reference diameter (d_{ref}) in polydisperse mode. Both reference devices are "gold-standard" instruments and were calibrated prior to the measurement campaigns.

In monodisperse mode, the DMA-selected particle electrical mobility diameter was considered as the reference diameter (d_{ref}).

2.3. Instruments under study

2.3.1. Electrical mobility spectrometers.
Although not ideally suited to the monitoring of aerosols in workplaces due to their low time-resolution, lack of field-portability, complexity of use and high cost [24, 25], SMPS as well as FMPS were examined in this study. Despite their drawbacks in the field, these devices are research-grade instruments, which make it is possible to accurately measure aerosol parameters for laboratory studies [26]. Only a single specimen of each instrument was studied. The SMPS was from TSI (DMA 3081, CPC model 3787, $Q_{aerosol}$ = 0.6 L.min^{-1}, Q_{sheath} = 6 L.min^{-1}).

In addition, a new portable and battery-operated version of the SMPS, commercialized by TSI and named Nanoscan, allows airborne particle number size distribution to be measured in one minute over a size range of 10 to 420 nm. Three specimens of Nanoscan TSI were investigated in this work.

For all these devices, measured number size distributions were adjusted by means of a monomodal lognormal model; the modal diameter of the fitted distribution is considered in the remainder of this paper.

2.3.2. Diffusion chargers

Real-time portable and battery-operated devices based on particle diffusion charging and sequential electrical measurement were also investigated in this study. The DiSCmini (Matter Aerosol AG, Switzerland), developed by Fierz *et al.* [27], is a standalone real-time handheld instrument that provides the airborne nanoparticle number concentration (C_N, 10^3 – 10^6 cm^{-3}) and average diameter (d, 20 - 300 nm) with an accuracy of ± 30% according to the manufacturer [28, 29]. To avoid artifacts due to coarse particles, the DiSCmini is equipped with an inlet separator with a cutoff diameter of 700 nm. This device has previously been studied in laboratory conditions [30-33] as well as in field conditions [34-36]. In this study, ten specimens of DiSCmini were challenged by the different test aerosols.

The NanoTracer diffusion charger developed by Philips Aerasense [37] and licensed to Oxility (Eindhoven, Netherlands) has similar characteristics to the DiSCmini, and reports the average diameter and number concentration of airborne particles. The NanoTracer has been used in combination with GPS tracking, for example, to assess the contribution of different activities to personal exposure [38, 39]. One specimen of NanoTracer was investigated in this work.

2.4. Methods

Experiments consisted of measuring in parallel the aerosols produced during a 5- to 30-minute period when the test aerosol is stable and when a sufficient number of scans can be measured by means of (1) the reference instruments and (2) the instruments under study. Due to the large number of devices investigated, experiments were replicated to allow all instruments to be challenged by a sufficient number of aerosols.

The data treatment procedure was as follows:

- For each test aerosol, the average number concentration $\overline{C_N}$ and the corresponding standard deviation $\sigma(C_N)$ were calculated for all instruments.
- Data points presenting a coefficient of variation above 5 % were disregarded, i.e. when

$$\frac{\sigma(C_N)}{\overline{C_N}} > 5\ \%$$

- For each of the remaining data points, the following ratios were calculated:
    - the ratio R_C between the concentration measured by the instrument under study ($\overline{C_N}$) and the corresponding reference concentration ($\overline{C_{N,\text{ref}}}$):

$$R_C = \frac{\overline{C_N}}{\overline{C_{N,ref}}}$$

    - the ratio R_d between the modal (or average) diameter provided by the instrument under study ($\bar{d}$) and the corresponding reference diameter (d_{ref}):

$$R_d = \frac{\bar{d}}{d_{ref}}$$

- This calculation was performed for all test aerosols, and the data were presented as boxplots. In these graphs, the box surrounding the median value corresponds to the 1st and 3rd quartiles, while the error bars represent the 95% confidence interval.

3. Results

The following subsections present the experimental results obtained for both the electrical mobility spectrometers and the diffusion chargers. The latter results, expressed as ratios, are displayed as boxplots, i.e. the closer to unity the ratio, the better the agreement between the instrument under study and the reference.

3.1. Electrical mobility spectrometers

Figure 3 presents the experimental results obtained for electrical mobility spectrometers in terms of diameter ratio (R_d, top), and number concentration ratio (R_C, bottom), relative to the reference.

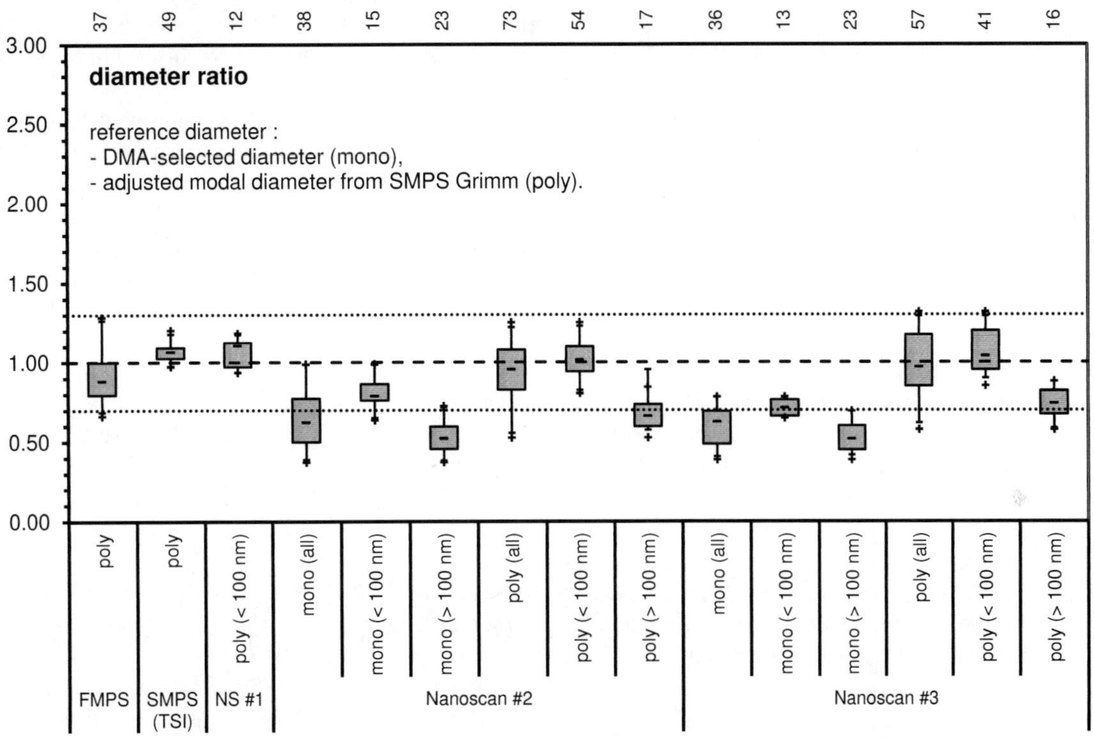

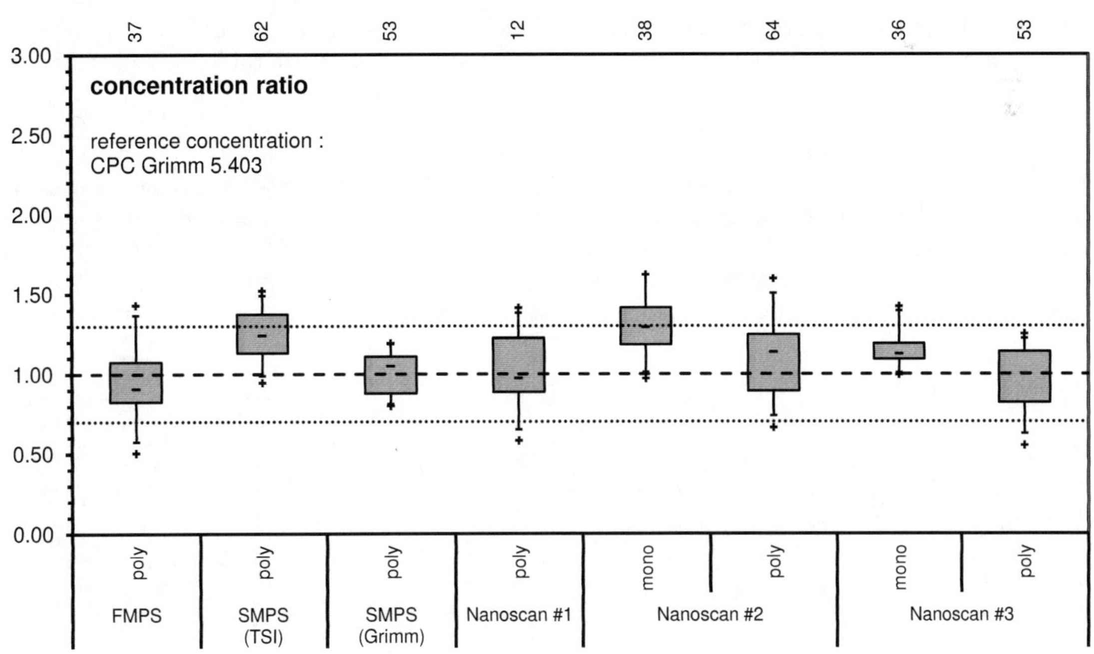

Figure 3. Boxplot of the diameter ratios (top) and number concentration ratios (bottom) found for the electrical mobility spectrometers (NS: Nanoscan).

The diameter ratios in Figure 3 show that the FMPS and SMPS are in close agreement with the reference particle diameter, within ± 30% in most cases. Nanoscans #2 and #3 underestimated particle diameter when challenged by aerosols composed of particles above 100 nm, by approximately -30% to -50%. On the contrary, all Nanoscans under investigation reported size distributions in agreement with the reference within ± 30% when measuring both mono- and polydisperse aerosols with (modal) diameters below 100 nm.

In terms of number concentration, all of the instruments investigated were in agreement with the reference concentrations, within ± 30 %, except the Nanoscan #2. It is important to remember that for all of these devices, the reported number concentration is derived from a calculation internal to the software that takes into account the electrical state of charge of the particles as well as the transfer function through the DMA. Based on physical assumptions, such inversion computations can lead to under- or overestimation of number concentrations. Larger discrepancies have previously been reported elsewhere, e.g. [40, 41].

3.2. Diffusion chargers

Figure 4 presents the experimental results obtained for various diffusion chargers in terms of diameter ratio (top), and number concentration ratio (bottom), relative to the reference. Because of their operating principles, the two ratios will be considered in parallel. Indeed, particle size and concentration are strongly dependent on each other due to the charging law, e.g. [27, 37].

3.2.1. DiSCmini

Ten specimens of DiSCmini were investigated in this study. According to Figure 4, they all behave similarly; the devices report particle sizes smaller than the reference, leading to an overestimation of the corresponding total number concentration.

More particularly, median diameter ratios of between 0.60 and 0.89 (respective relative discrepancies between -40% and -11%) were observed, the test aerosols being either mono or polydisperse. The number concentrations reported by the DiSCmini specimens were found to be somewhat higher than the reference, with the exception of model "DM CIOP 5", whose range of median ratios of 1.17 to 2.44 is not in line with the manufacturer's specifications of ± 30% (dotted lines).

3.2.2. Case of Nanotracer

Only one Nanotracer specimen was studied. Thus, the results reported here are less robust than those obtained for the DiSCmini. Nevertheless, Figure 4 suggests that this Nanotracer slightly overestimates particle diameter, by a factor of 15% on average. On the other hand, this results in an underestimation of particle number concentration, by about -30%. Investigations with multiple specimens of Nanotracer are required to better assess their behavior with respect to the reference instrument.

IOP Conf. Series: Journal of Physics: Conf. Series **838** (2017) 012001 doi:10.1088/1742-6596/838/1/012001

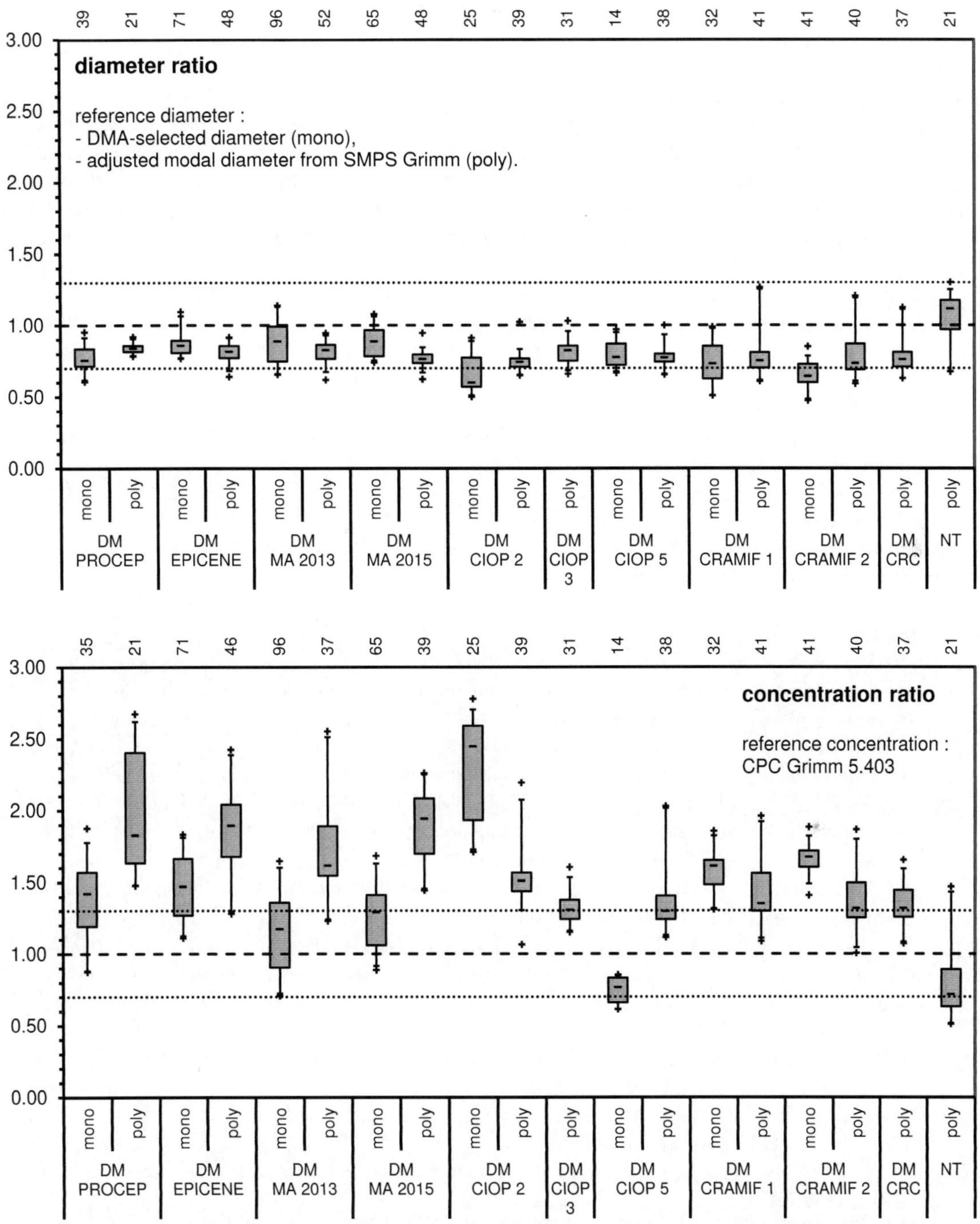

Figure 4. Boxplot of the diameter ratios (top) and concentration ratios (bottom) determined/measured by the diffusion chargers (DM: DiSCmini, NT: Nanotracer).

4. Conclusion

This study focused on the performances of real-time instruments devoted to the measurement of airborne particles. Gold-standard devices such as electrical mobility spectrometers (transportable and portable instruments) as well as personal portable diffusion chargers were investigated. All instruments were challenged by a set of monodisperse and polydisperse test aerosols covering a wide range of (modal) diameters and number concentrations. The values for these two parameters reported by these devices were compared to reference instruments, such as SMPS for the number size distribution and CPC for the total number concentration.

The experimental results obtained for the group of electrical mobility spectrometers highlight the good agreement between SMPS and FMPS and the reference. A slight tendency for the Nanoscan to underestimate particle size distribution when the particles were larger than around 100 nm was observed, and was probably due to the decrease in selectivity of the radial DMA used in the Nanoscan.

The data reported for the group of diffusion chargers demonstrate that all except the Nanotracer show a tendency to underestimate particle diameter. For the 10 DiSCminis studied, acceptable discrepancies of around -40% to -10% were observed. In the case of particle concentration, larger deviations were observed. However, in our opinion, this is probably not important when the relative concentrations provided by, for example, two DiSCmini operated in parallel are being used to conclude whether or not a given activity leads to a significant nanoparticle release. Nevertheless, this device is sensitive enough to be used as a nanoparticle monitor, e.g. in workplaces where nanomaterials are handled or produced, provided that the aerosol being measured is not composed only of particles larger than the upper limit of 300 nm.

Acknowledgements
The authors would like to thank Catherine Hedouin-Langlet of CRAMIF (Paris) and Bertrand Honnert of INRS for providing their DiSCminis during the test campaigns.

References
[1] SAVOLAINEN K., PYLKKÄNEN P., NORPPA H., FALCK G., LINDBERG H., TUOMI T., *et al.* - Nanotechnologies, engineered nanomaterials and occupational health and safety – A review. *Safety Sci.*, 2011, **48,** 957-963.
[2] SHEPARD M. N. & BRENNER S. - An occupational exposure assessment for engineered nanoparticles used in semiconductor fabrication. *Ann. Occup. Hyg.*, 2013, **58,** 251-265.
[3] GORDON S. C., BUTALA J. H., CARTER J. M., ELDER A., GORDON T., GRAY G., *et al.* - Workshop report: Strategies for setting occupational exposure limits for engineered nanomaterials. *Regul. Toxicol. Pharm.*, 2014, **68,** 305-311.
[4] O'SHAUGHNESSY P. T. - Occupational health risk to nanoparticulate exposure. *Environ. Sci.: Processes Impacts*, 2013, **15,** 49-62.
[5] WANG J., ASBACH C., FISSAN H., HÜLSER T., KUHLBUSCH T. A. J., THOMPSON D., *et al.* - How can nanobiotechnology oversight science and industry: examples from environmental, health, and safety studies of nanoparticles (nano-EHS). *J. Nanopart. Res.*, 2011, **13,** 1373-1387.
[6] BROUWER D., VAN DUUREN-STUURMAN B., BERGES M., JANKOWSKA E., BARD D., & MARK D. - From workplace air measurement results towards estimates of exposure? Development of a strategy to assess exposure to manufactured nano-objects. *J. Nanopart. Res.*, 2009, **11,** 1867-1881.
[7] BROUWER D. - Exposure to manufactured nanoparticles in different workplaces. *Toxicology*, 2010, **269,** 120-127.
[8] KUHLBUSCH T., ASBACH C., FISSAN H., GÖHLER D., & STINTZ M. - Nanoparticle exposure at nanotechnology workplaces: A review. *Part. Fiber Toxicol.*, 2011, **8,** 22.
[9] RAMACHANDRAN G., OSTRAAT M., EVANS D. E., METHNER M. M., O'SHAUGHNESSY P., D'ARCY J., *et al.* - A strategy for assessing workplace exposures to nanomaterials. *Journal of Occupational and Environmental Hygiene*, 2011, **8,** 673-685.
[10] WITSCHGER O., LEBIHAN O., REYNIER M., DURAND C., MARCHETTO A., ZIMMERMANN E., *et al.* - Préconisations en matière de caractérisation des potentiels

d'émission et d'exposition professionnelle aux aérosols lors d'opérations mettant en oeuvre des nanomatériaux. *Hygiène et Sécurité au Travail*, 2012, **226,** 41-55.

[11] METHNER M., HODSON L., & GERACI C. - Nanoparticle emission assessment technique (NEAT) for the identification and measurement of potential inhalation exposure to engineered nanomaterials – part A. *Journal of Occupational and Environmental Hygiene*, 2010, **7,** 127-132.

[12] OSTRAAT M. L., THORNBURG J. W., & MALLOY Q. G. J. - Measurement strategies of airborne nanomaterials. *Environ. Eng. Sci.*, 2013, **30,** 126-132.

[13] METHNER M., HODSON L., DAMES A., & GERACI C. - Nanoparticle emission assessment technique (NEAT) for the identification and measurement of potential inhalation exposure to engineered nanomaterials – part B: results from 12 field studies. *Journal of Occupational and Environmental Hygiene*, 2010, **7,** 163-176.

[14] METHNER M., BEAUCHAM C., CRAWFORD C., HODSON L., & GERACI C. - Field application of the Nanoparticle Emission Assessment Technique (NEAT): task-based air monitoring during the processing of Engineered Nanomaterials (ENM) at four facilities. *Journal of Occupational and Environmental Hygiene*, 2012, **9,** 543-555.

[15] RAYNOR P. C., INGRAHAM CEBULA J., SPANGENBERGER J. S., OLSON B. A., DASCH J. M., & D'ARCY J. B. - Assessing potential nanoparticle release during nanocomposite shredding using direct-reading instruments. *Journal of Occupational and Environmental Hygiene*, 2012, **9,** 1-13.

[16] KOIVISTO A. J., LYYRÄNEN J., AUVINEN A., VANHALA E., HÄMERI K., TUOMI T., *et al.* - Industrial worker exposure to airborne particles during the packing of pigment and nanoscale titanium dioxide. *Inhal. Toxicol.*, 2012, **24,** 839-849.

[17] PETERS T. M., HEITBRINK W. A., EVANS D. E., SLAVIN T. J., & MAYNARD A. D. - The mapping of fine and ultrafine particle concentrations in an engine machining and assembly facility. *Ann. Occup. Hyg.*, 2006, **50,** 249-257.

[18] HEITBRINK W. A., LO L.-M., & DUNN K. H. - Exposure Controls for Nanomaterials at Three Manufacturing Sites. *J. Occup. Environ. Health*, 2014**,** in press, DOI: 10.1080/15459624.15452014.15930559.

[19] BUONANNO G., JAYARATNE R. E., MORAWSKA L., & STABILE L. - Metrological Performances of a Diffusion Charger Particle Counter for Personal Monitoring. *Aerosol Air Qual. Res.*, 2014, **14,** 156-167.

[20] VOSBURGH D. J. H., KU B. K., & PETERS T. M. - Evaluation of a Diffusion Charger for measuring aerosols in a workplace. *Ann. Occup. Hyg.*, 2014, **58,** 424-436.

[21] CEN - prEN 17058, Workplace exposure — Assessment of inhalation exposure to nano-objects and their agglomerates and aggregates, 2016.

[22] CEN - prEN 16966, Workplace exposure – Metrics to be used for the measurements of exposure to inhaled nanoparticles (nano-objects and nanostructured materials) such as mass concentration, number concentration and surface area concentration, 2016.

[23] JACOBY J., BAU S., & WITSCHGER O. - CAIMAN: a versatile facility to produce aerosols of nanoparticles. *Journal of Physics: Conference Series*, 2011, **304,** 012014.

[24] LESKINEN J., JOUTSENSAARI J., LYYRÄNEN J., KOIVISTO J., RUUSUNEN J., JÄRVELÄ M., *et al.* - Comparison of nanoparticle measurement instruments for occupational health applications. *J. Nanopart. Res.*, 2012, **14,** 718-733.

[25] STABILE L., CAUDA E., MARINI S., & BUONANNO G. - Metrological assessment of a portable analyzer for monitoring the particle size distribution of ultrafine particles. *Ann. Occup. Hyg.*, 2014, **58,** 860-876.

[26] WITSCHGER O. - Monitoring nanoaerosols and occupational exposure, in *Nanoethics and nanotoxicology*, eds. Houdy, P., Lahmani, M., & Marano, F. (Springer Verlag, Berlin Heidelberg), pp. 163-199.

[27] FIERZ M., HOULE C., STEIGMEIER P., & BURTSCHER H. - Design, calibration, and field performance of a Miniature Diffusion Size Classifier. *Aerosol Science and Technology*, 2011, **45,** 1-10.

[28] MATTER AEROSOL - *DiSCmini instruction manual Rev. 1.10*, 2012.

[29] TESTO - http://testo-partikel.de/files/DiSCmini_EN_compressed.pdf (2016-11-02)

[30] BAU S., ZIMMERMANN B., PAYET R., & WITSCHGER O. - Laboratory study of the performance of the miniature Diffusion Size Classifier (DiSCmini) for various aerosols in the 15-400 nm range. *Environmental Science: Processes and Impacts*, 2015, **17,** 261-269.

[31] ASBACH C., KAMINSKI H., VON BARANY D., KUHLBUSCH T. A. J., MONZ C., DZIUROWITZ N., *et al.* - Comparability of portable nanoparticle exposure monitors. *Ann. Occup. Hyg.*, 2012, **56,** 606-621.

[32] MILLS J. B. - Evaluation of the DiSCmini personal aerosol monitor for submicrometer sodium chloride and metal aerosols, 2013, University of Iowa.

[33] MILLS J. B., PARK J. H., & PETERS T. M. - Comparison of the DiSCmini aerosol monitor to a handheld condensation particle counter and a scanning mobility particle sizer for submicrometer sodium chloride and metal aerosols. *Journal of Occupational and Environmental Hygiene*, 2013, **10,** 250-258.

[34] MEIER R., CLARK K., & RIEDIKER M. - Comparative testing of a Miniature Diffusion Size Classifier to assess airborne ultrafine particles under field conditions. *Aerosol Science and Technology*, 2013, **47,** 22-28.

[35] KIM B., KIM H., & YU I. J. - Assessment of nanoparticle exposure in nanosilica handling process: including characteristics of nanoparticles leaking from a vacuum cleaner. *Ind. Health*, 2014, **52,** 152-162.

[36] OH S., KIM B., & KIM H. - Comparison of nanoparticle exposures between fumed and sol-gel nano-silica manufacturing facilities. *Ind. Health*, 2014, **52,** 190-198.

[37] MARRA J., VOETZ M., & KIESLING H. J. - Monitor for detecting and assessing exposure to airborne nanoparticles. *J. Nanopart. Res.*, 2010, **12,** 21-37.

[38] BUONANNO G., STABILE L., & MORAWSKA L. - Personal exposure to ultrafine particles: the influence of time-activity patterns. *Science of the Total Environment*, 2014, **468-469,** 903-907.

[39] BEKÖ G., KJELDSEN B. U., OLSEN Y., SCHIPPERIJN J., WIERZBICKA A., KAROTTKI D. G., *et al.* - Contribution of various microenvironments to the daily personal exposure to ultrafine particles: Personal monitoring coupled with GPS tracking. *Atmos. Environ.*, 2015, **110,** 122-129.

[40] FRANK B. P., SALTIEL S., HOGREFE O., GRYGAS J., & LALA G. G. - Determination of mean particle size using the electrical aerosol detector and the condensation particle counter: comparison with the scanning mobility particle sizer. *Journal of Aerosol Science*, 2008, **39,** 19-29.

[41] LIU P. S. K. & DESHLER T. - Causes of concentration differences between a scanning mobility particle sizer and a condensation particle counter. *Aerosol Science and Technology*, 2003, **37,** 916-923.

IOP Conf. Series: Journal of Physics: Conf. Series **838** (2017) 012002 doi:10.1088/1742-6596/838/1/012002

Performance study of various Condensation Particle Counters (CPCs): development of a methodology based on steady-state airborne DEHS particles and application to a series of handheld and stationary CPCs

Sébastien Bau, André Toussaint, Raphaël Payet and Olivier Witschger

Institut National de Recherche et de Sécurité (INRS), Laboratoire de Métrologie des Aérosols, Rue du Morvan, CS 60027, 54519 Vandoeuvre Cedex, France

E-mail: sebastien.bau@inrs.fr

Abstract. Strategies for measuring occupational exposure to aerosols composed of nanoparticles and/or ultrafine particles highlight the use of techniques for determining airborne-particle number concentration as well as number size distribution. The objective of the present work was to set up a system for conducting laboratory verification campaigns of condensation particle counters (CPCs). Providing intercomparison data as well as calibrating and checking CPCs are among the key elements in ensuring reliable laboratory or field measurement campaigns. For this purpose, the reproducible aerosol source "Calibration Tool", initially developed by the Fraunhofer ITEM, was acquired by the Laboratory of Aerosol Metrology at INRS. As a first part of this study, a detailed characterization of the Calibration Tool developed at the laboratory is the subject of the parametric study presented here. The complete installation is named the "DCC" for "Device for Counter Check". Used in combination with a reference counter, the DCC can now be used for routine laboratory measurements. Unlike that used for primary calibration of a CPC, the proposed protocol allows a wide range of number concentrations and particle sizes to be investigated and reproduced. The second part of this work involves comparison of the number concentrations measured by several models of CPC in parallel at the exit of a flow splitter, with respect to a reference.

1. Introduction

Atmospheric, indoor and workplace air naturally contains particles below 1 µm in size [1, 2]. In workplace atmospheres, these submicron particles may originate from a variety of sources, including combustion processes, diesel exhaust, thermal spraying of metals, or energetic processes such as welding and grinding [3-7].

Nanomaterials offer applications in fields as wide-ranging as health, food and agriculture, energy, materials, and transport. However, their unique properties have raised questions about their potential related health effects. Due to the rising use of nanomaterials, the number of exposed workers in research laboratories and public or private industry is probably increasing at all stages of the lifecycle of these products – from synthesis, use and ageing [8], through to disposal [9, 10], waste treatment, and maintenance [11, 12] – as well as in accidental exposure situations [13, 14]. Assessing inhalation exposure to airborne nanoparticles therefore constitutes an important challenge [15], and providing robust data related to the performances of instruments devoted to the measurement of submicron particles is crucial prior to their wider use in occupational hygiene.

To complete conventional sampling approaches, which provide information on the average exposure of workers, different strategies for measuring occupational exposure to airborne nanoparticles, highlighting the use of time-resolved techniques, have been developed [16-20]. In addition to their chemical composition, airborne particles can be characterized by their number concentration as well as their size distribution. These characteristics are among the parameters of interest when aerosol toxicology is sought [21]. When possible, a multi-metric approach is advised [22-24].

1.1. Context and outline of the study

This work focuses on the condensation particle counter (CPC), an instrument devoted to the measurement of submicron particle number concentrations. First developed over a century ago [25, 26], these instruments are based on the optical detection of particles artificially grown through the condensation of an adsorbed vapor on their surface. The devices record the number concentration of airborne particles per unit of air volume. Because of their field portability, CPCs are considered reliable devices for workplace air measurement.

In order to ensure adequate data interpretation, the performances of CPCs need to be well described. In most cases, calibrations performed by independent companies (except CPC producers and distributors) are limited to a concentration range of between approximately 100 cm^{-3} and 10 000 cm^{-3}. This range is narrow compared to the levels of aerosol number concentrations that can be measured in workplace atmospheres.

In parallel, occupational exposure limits (OEL) expressed in terms of number concentration of airborne nanoparticles have been proposed in Germany by the IFA [27] and in the United Kingdom by the BSI [28]. At present, limit values based on particle numbers already exist for clean rooms [29] as well as for diesel exhaust, for example in the framework of the Euro V label. Furthermore, number concentration is the most frequent characteristic used for airborne nanoparticle monitoring, task emission classification, and for evaluating the performance of protective equipment against nanoparticles.

Given these various issues, a methodology for checking CPCs is essential to ensure reliable laboratory or field measurement campaigns.

1.2. Objective

To date, there is no standard that allows the accuracy of a CPC to be verified [30]. *Calibration* of a CPC requires a combination of a differential mobility analyzer (DMA) and an aerosol electrometer [31], which are expensive devices.

This work aims were to develop, validate and test a methodology for *checking* CPCs at the INRS Laboratory of Aerosol Metrology. *In the first stage of this study*, the aerosol to be measured was characterized by different CPCs in terms of size distribution, total concentration, time stability, reproducibility, etc. *In the second stage*, the test protocol and data treatment procedure were developed and validated. Finally, *in the third stage of this work*, the complete protocol was carried out using several different types of CPCs (both handheld and stationary models). A freshly calibrated CPC was used simultaneously as a reference measurement in compliance with standard prEN 16897 [30].

2. Test aerosol characterization

2.1. Aerosol source: The Calibration Tool

Controlled concentrations of aerosols were generated using the Calibration Tool developed by Koch *et al.* [32, 33]. This generator produces DEHS droplets by evaporation/condensation processes, which are then fed into a continuously stirred tank reactor, as shown in Figure 1.

The feed aerosol is formed in a turbulent coaxial jet inside the reactor. A hot saturated stream of an organic vapor is fed through a nozzle and generates nuclei in the inner jet by homogeneous nucleation when the inner jet is mixed with the surrounding cold clean air flow. These particles grow further through condensation and coagulation. The product aerosol in the tank reaches a stationary state that is spatially and temporally homogeneous (self-preserved distribution) after around one hour of operation. This results in a very stable and reproducible production of particles.

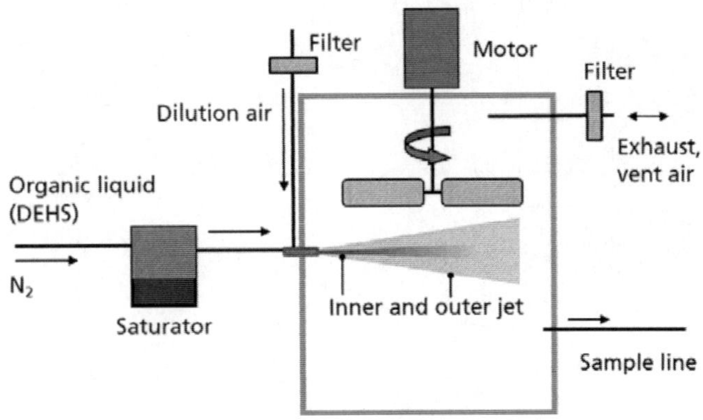

Figure 1. Schematic diagram of the Calibration Tool (courtesy of W. Koch).

As an example, Figure 2 presents the number size distribution of the aerosol produced, measured over 6 days after 5 hours of continuous operation. It can be concluded from Figure 2 that the mode of the distribution is around 60 nm, with less than 7 % variability observed over the 6 days. Furthermore, the total number concentration is around 2.10^7 cm^{-3}, associated with a variation coefficient of less than 3 %.

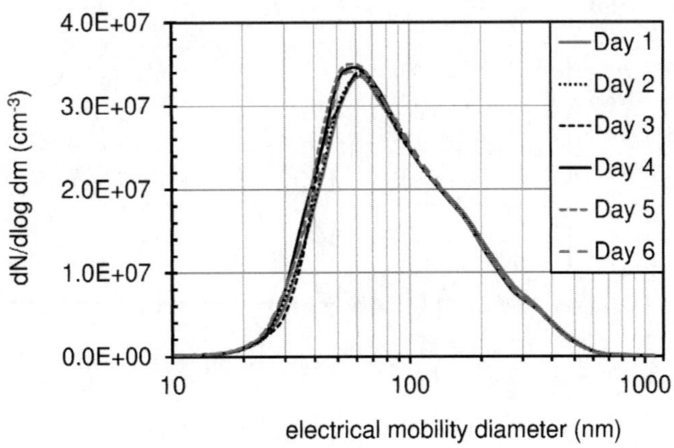

Figure 2. Number size distribution of the aerosols produced within the Calibration Tool over 6 days.

2.2. The Device for Counter Check (DCC)

The DCC consists of a generator (the Calibration Tool), a specific sampling and dilution line, and a reference CPC, as shown in Figure 3. In terms of the technical configuration, a manual valve was placed directly in the aerosol line, which is connected to a HEPA-filtered air inlet and a homogenization chamber. Thus, the aerosol concentration at the outlet can be easily controlled by turning the valve.

Because of the dilution and coalescence processes that occur between the sampling from the reactor and measurement at the CPC, the number size distribution as well as the total concentration of the aerosols are different to those presented in Figure 2; the modal diameter of the size distribution of the aerosols ranges from 220 nm to 140 nm for total concentrations of 2.10^4 to 2.10^5 cm^{-3} (Figure 4).

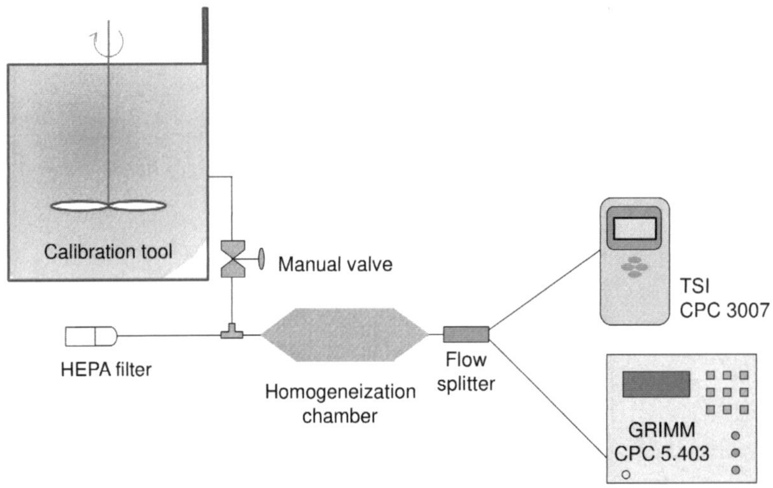

Figure 3. Schematic diagram of the Device for Counter Check (DCC).

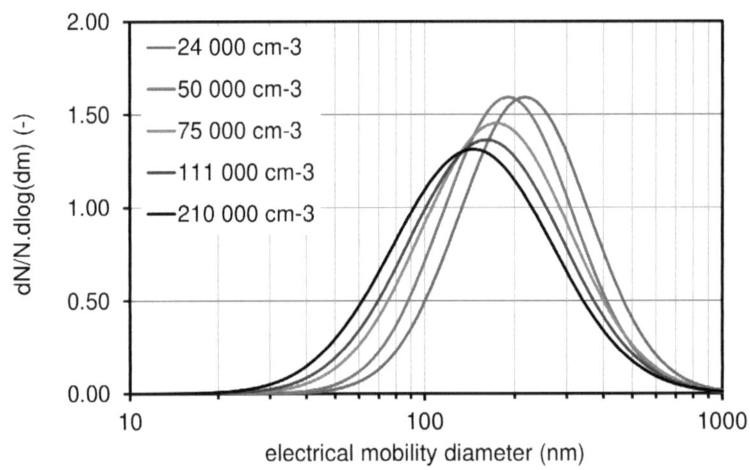

Figure 4. Number size distribution of the aerosols for different total number concentrations.

3. Testing methodology and data treatment

The following steps must be rigorously adhered to when a CPC checking campaign is conducted:

- Turn on the Calibration Tool and wait for one hour for the aerosol to stabilize.
- Turn on the CPC under study and the reference CPC.
- Adjust the dilution ratio using the manual valve so that it fits within the concentration range for all CPCs.
- Once the concentration is stable, perform a measurement for 10 minutes.
- Modify the dilution ratio to cover the widest concentration range (10^2 to 10^5 cm^{-3}).

The data treatment procedure involves the following steps:

- For each concentration level, calculate the average concentration $\bar{X}$ and the corresponding standard deviation σ for all CPCs (those being studied as well as the reference CPC):

$$\bar{X} = \frac{\sum_{i=1}^{n} x_i}{n}$$

$$\sigma = \sqrt{\frac{\sum_{i=1}^{n} (x_i - \bar{X})^2}{n}}$$

where n is the number of points and x_i is the number concentration measured at time i.

- Disregard any data points presenting a coefficient of variation of more than 5%, i.e. when:

$$\frac{\sigma}{\bar{X}} > 5\,\%$$

- For each of the remaining data points, calculate the ratio R between the concentration measured by one of CPCs under study ($\bar{X}$) and the corresponding reference concentration ($\overline{X_{\text{ref}}}$):

$$R = \frac{\bar{X}}{\overline{X_{\text{ref}}}}$$

- Perform this calculation for all concentration levels tested, and present the data as boxplots. In these graphs, the box surrounding the median value corresponds to the 1st and 3rd quartiles, while the error bars represent the 95% confidence interval.

4. Implementation of the methodology on various CPCs and discussion

The reference CPC used in this study was a butanol-based stationary CPC, Grimm model 5.403, operated in high flow mode (1.5 L.min^{-1}). The instrument was calibrated prior to the measurements. The levels of concentration investigated ranged from 10^3 to 10^5 cm^{-3}, thus corresponding to the specified range of concentration for all of the CPCs tested.

4.1. Handheld CPCs

The experimental results for the different handheld CPCs are presented in Figure 5.

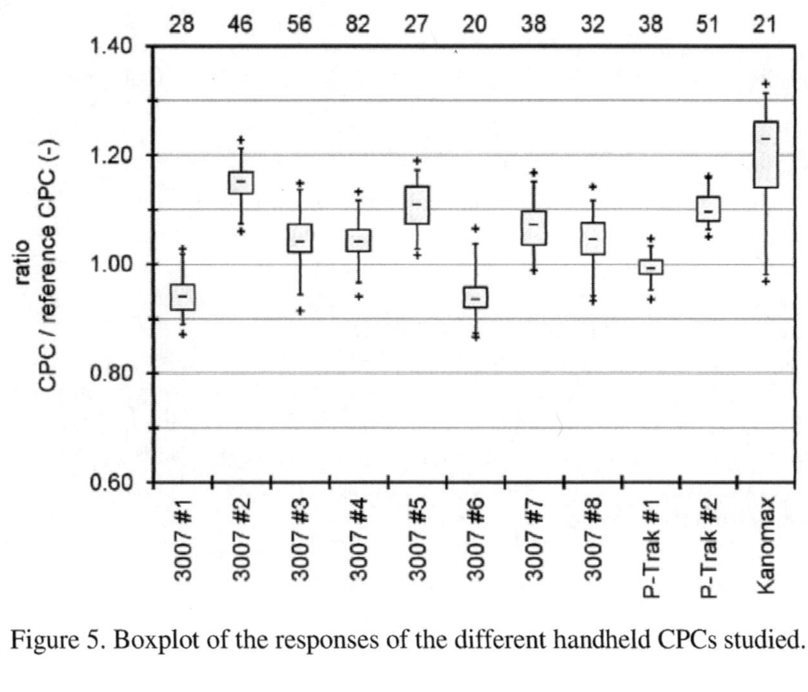

Figure 5. Boxplot of the responses of the different handheld CPCs studied.

As seen in Figure 5, all CPCs were found to lie within ± 25% deviation from the reference, except for the Kanomax CPC, which also presented the largest span, from 0.97 to 1.33. Furthermore, it can be observed from Figure 5 that different specimens of the same CPC model (TSI 3007) can differ significantly, e.g. the response of CPC 3007 #2 and 3007 #6 varies by up to 25%.

4.2. Stationary CPCs
The experimental results for stationary CPCs are presented in Figure 6.

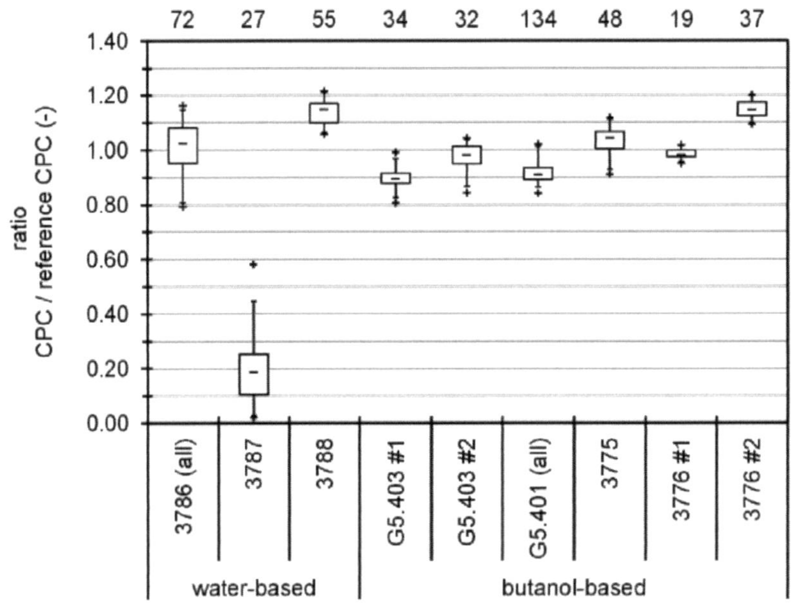

Figure 6. Boxplot of the responses of the different stationary CPCs studied.

Figure 6 shows that all of the CPCs studied were found to lie within ± 25 % of the reference, except model TSI 3787, which is known to be highly sensitive to particle hydrophobicity. Nevertheless, large spans, in particular the magnitude of the 95% confidence interval compared to the 50% confidence interval, are observed in the experimental data from CPC models TSI 3786 and Grimm 5.401.

4.3. Discussion
The compared behaviors of CPCs need to be considered when two specimens of the same model of CPC are used in workplace situations. Indeed, strategies for exposure assessment to airborne nanomaterials [16-20] advise the parallel use of a CPC near the source and another positioned far-field. Combined with results from intercomparison studies, the measured concentrations should be further interpreted by means of statistical analysis [34-38].

Although the median ratios lie within the tolerated uncertainty, significant differences between different models of CPC may lead to wrong conclusions if the devices are used in parallel. In the case of handheld CPC model TSI 3007, placing 3007 #2 near the source and 3007 #6 far-field may result in the conclusion that airborne particles are emitted by the process (as the concentration measured by CPC 3007 #2 near the source is higher than that measured far-field by 3007 #6) even in the absence of additional particles from the expected aerosol source.

The example of CPC model TSI 3786 is detailed in Figure 7, which presents the results of the concentrations measured by the CPC and the corresponding reference concentration, as well as the ratio (top, right axis) calculated for each situation.

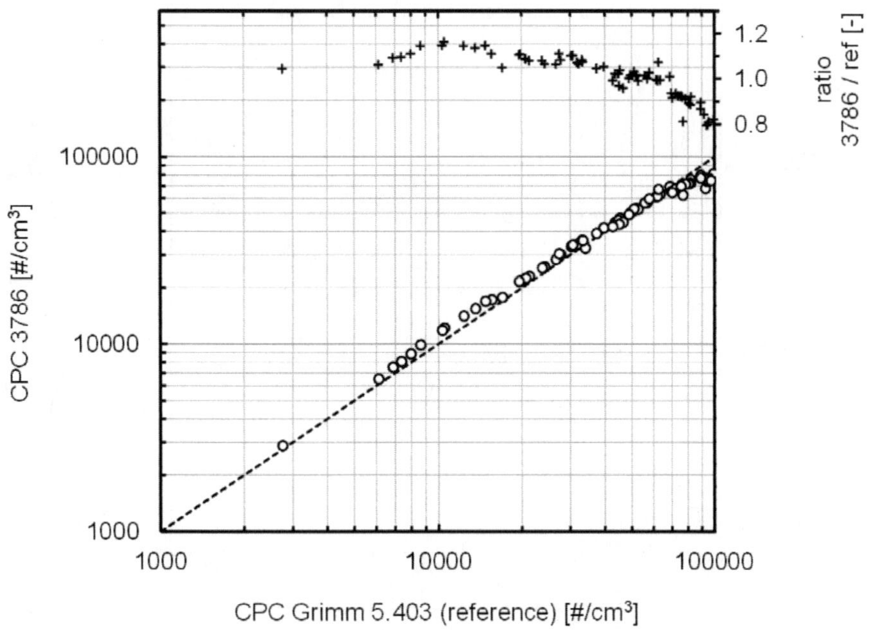

Figure 7. Response of CPC model 3786 (absolute concentration and concentration relative to the reference).

It is clear from Figure 7 that particle concentration has an effect on the response of CPC model TSI 3786. In particular, the ratio seems quite constant at around 1.1 for total number concentrations below approximately 35 000 cm^{-3}. However, at higher concentrations, the CPC under investigation starts to underestimate the number concentration of the aerosol.

This decrease in the number concentration reported by CPC model TSI 3786 might be related to unsaturated nuclei within the saturation chamber due to particle hydrophobicity. In contrast, CPC model TSI 3788 is not affected by particle hydrophobicity, as can be seen in Figure 6. This can be explained by differences in saturation / condensation temperatures, as well as in sheath air and aerosol flows within the instruments.

The case of CPC model Grimm 5.401 is described in Figure 8. This case is interesting because it clearly highlights a change in the concentration ratio from 1.0 for concentrations below 10^4 cm^{-3} to 0.9 for concentrations above 10^4 cm^{-3}. The 10^4 cm^{-3} limit corresponds to the switch from single count mode to photometric mode.

As a consequence, restricting TSI model 3786 response to number concentrations below 35 000 cm^{-3} and considering two different intervals ($< 10^4$ cm^{-3} and $> 10^4$ cm^{-3}) for the Grimm model 5.401 results in the boxplots presented in Figure 9. Compared to Figure 6, these results now show smaller confidence intervals, suggesting that the restricted intervals of concentration considered are compliant.

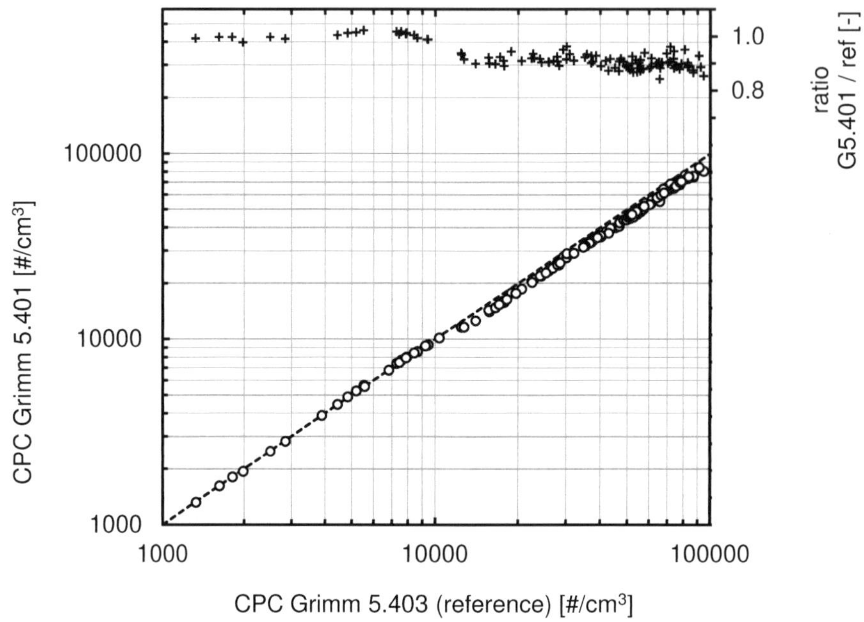

Figure 8. Response of CPC model 5.401 (absolute concentration and concentration ratio compared to the reference).

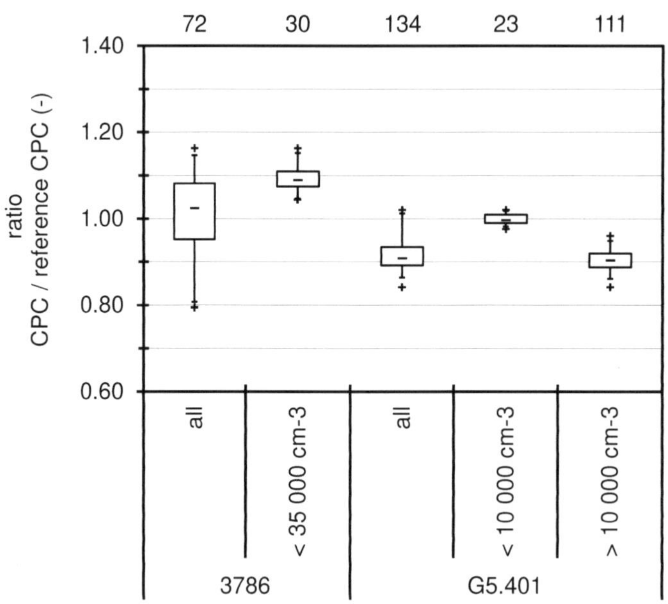

Figure 9. Detailed boxplot of the responses of CPCs 3786 and 5.401.

IOP Conf. Series: Journal of Physics: Conf. Series **838** (2017) 012002 doi:10.1088/1742-6596/838/1/012002

5. Conclusion

Strategies for assessing occupational exposure to airborne nanomaterials highlight the use of time-resolved techniques to determine, among others, number concentration and size distribution. The objective of this work focused on development of a methodology to check the accuracy of Condensation Particle Counters (CPCs) prior to laboratory or field measurement campaigns.

For this purpose, the Calibration Tool initially designed by ITEM Fraunhofer was acquired by the laboratory of Aerosol Metrology at INRS. Combined with an in-house dilution line and a reference CPC, the Device for Counter Check (DCC) can now be used under routine conditions according to a specific protocol.

This work must now be pursued for a wider range of CPCs in order to establish a map of their performance. Once checked, an inter-comparison campaign using various CPCs must be carried out on a wide variety of aerosols. This step will help to better define their scope of use and to investigate their performance when challenged by aerosols representative of those encountered in the air in workplaces.

References

[1] JOHN W. - Size distribution characteristics of aerosols, in *Aerosol measurement: Principles, techniques and applications*, eds. Kulkarni, P., Baron, P. A., & Willeke, K. (John Wiley & Sons, Inc., Hoboken, New Jersey), pp. 41-54.

[2] HINDS W. C. - *Aerosol Technology. Properties, behavior and measurement of airborne particles (2nd Edition)*, 1999**,** New York: John Wiley & Sons, 483 p.

[3] BÉMER D., RÉGNIER R., SUBRA I., SUTTER B., LECLER M. T., & MORELE Y. - Ultrafine particles emitted by flame and electric arc guns for thermal spraying of metals. *Ann. Occup. Hyg.*, 2010, **54,** 607-614.

[4] XU Z., LI X., GUAN C., & HUANG Z. - Effects of injection timing on exhaust particle size and nanostructure on a diesel engine at different loads. *Journal of Aerosol Science*, 2014, **76,** 28-38.

[5] ZIMMERMAN N., GODRI POLLITT K. J., JEONG C.-H., WANG J. M., JUNG T., COOPER J. M., *et al.* - Comparison of three nanoparticle sizing instruments: The influence of particle morphology. *Atmos. Environ.*, 2014, **86,** 140-147.

[6] KOIVISTO A. J., HUSSEIN T., NIEMELÄ R., TUOMI T., & HÄMERI K. - Impact of particle emissions of new laser printers on modeled office room. *Atmos. Environ.*, 2010, **44,** 2140-2146.

[7] PARK J. H., MUNDUNKOTUWA I. A., KIM J. S., STANAM A., THORNE P. S., GRASSIAN V. H., *et al.* - Physicochemical characterization of simulated welding fumes from a spark discharge system. *Aerosol Science and Technology*, 2014, **48,** 768-776.

[8] JIANG L., KONDO A., SHIGETA M., ENDOH S., UEJIMA M., OGURA I., *et al.* - Evaluation of Particles Released from Single-wall Carbon Nanotube/Polymer Composites with or Without Thermal Aging by an Accelerated Abrasion Test. *Journal of Occupational and Environmental Hygiene*, 2014, **11,** 658-664.

[9] SAVOLAINEN K., PYLKKÄNEN P., NORPPA H., FALCK G., LINDBERG H., TUOMI T., *et al.* - Nanotechnologies, engineered nanomaterials and occupational health and safety – A review. *Safety Sci.*, 2011, **48,** 957-963.

[10] RAYNOR P. C., INGRAHAM CEBULA J., SPANGENBERGER J. S., OLSON B. A., DASCH J. M., & D'ARCY J. B. - Assessing potential nanoparticle release during nanocomposite shredding using direct-reading instruments. *Journal of Occupational and Environmental Hygiene*, 2012, **9,** 1-13.

[11] NILSSON P., ISAXON C., ERIKSSON A., MESSING M., LUDVIGSSON L., RISSLER J., *et al.* - Nano-objects emitted during maintenance of common particle generators: direct chemical characterization with aerosol mass spectrometry and implications for risk assessments. *J. Nanopart. Res.*, 2013, **15,** 2052.

[12] ZIMMERMANN E., DERROUGH S., LOCATELLI D., DURAND C., FROMAGET J. L., LEFRANC E., *et al.* - Results of potential exposure assessments during the maintenance and cleanout of deposition equipment. *J. Nanopart. Res.*, 2012, **14,** 1209.

[13] MOTZKUS C., CHIVAS-JOLY C., GUILLAUME E., DUCOURTIEUX S., SARAGOZA L., LESENECHAL D., *et al.* - Aerosols emitted by the combustion of polymers containing nanoparticles. *J. Nanopart. Res.*, 2012, **14,** 687.

[14] CLEMENTE A., PILAR LOBERA M., BALAS F., & SANTAMARIA J. - Development of a self-cleaning dispersion and exposure chamber: application to the monitoring of simulated accidents involving the generation of airborne nanoparticles. *J. Hazard. Mater.*, 2014, **280,** 226-234.

[15] MAYNARD A. D., AITKEN R. J., BUTZ T., COLVIN V., DONALDSON K., OBERDÖRSTER G., *et al.* - Safe handling of nanotechnology. *Nature*, 2006, **444,** 267-269.

[16] WITSCHGER O., LE BIHAN O., REYNIER M., DURAND C., MARCHETTO A., ZIMMERMANN E., *et al.* - Préconisations en matière de caractérisation des potentiels d'émission et d'exposition professionnelle aux aérosols lors d'opérations mettant en oeuvre des nanomatériaux. *Hygiène et Sécurité au Travail*, 2012, **226,** 41-55.

[17] METHNER M., HODSON L., & GERACI C. - Nanoparticle emission assessment technique (NEAT) for the identification and measurement of potential inhalation exposure to engineered nanomaterials – part A. *Journal of Occupational and Environmental Hygiene*, 2010, **7,** 127-132.

[18] RAMACHANDRAN G., OSTRAAT M., EVANS D. E., METHNER M. M., O'SHAUGHNESSY P., D'ARCY J., *et al.* - A strategy for assessing workplace exposures to nanomaterials. *Journal of Occupational and Environmental Hygiene*, 2011, **8,** 673-685.

[19] BROUWER D., VAN DUUREN-STUURMAN B., BERGES M., JANKOWSKA E., BARD D., & MARK D. - From workplace air measurement results towards estimates of exposure? Development of a strategy to assess exposure to manufactured nano-objects. *J. Nanopart. Res.*, 2009, **11,** 1867-1881.

[20] OSTRAAT M. L., THORNBURG J. W., & MALLOY Q. G. J. - Measurement strategies of airborne nanomaterials. *Environ. Eng. Sci.*, 2013, **30,** 126-132.

[21] WANG J., ASBACH C., FISSAN H., HÜLSER T., KUHLBUSCH T. A. J., THOMPSON D., *et al.* - How can nanobiotechnology oversight science and industry: examples from environmental, health, and safety studies of nanoparticles (nano-EHS). *J. Nanopart. Res.*, 2011, **13,** 1373-1387.

[22] SHEPARD M. N. & BRENNER S. - An occupational exposure assessment for engineered nanoparticles used in semiconductor fabrication. *Ann. Occup. Hyg.*, 2013, **58,** 251-265.

[23] GORDON S. C., BUTALA J. H., CARTER J. M., ELDER A., GORDON T., GRAY G., *et al.* - Workshop report: Strategies for setting occupational exposure limits for engineered nanomaterials. *Regul. Toxicol. Pharm.*, 2014, **68,** 305-311.

[24] O'SHAUGHNESSY P. T. - Occupational health risk to nanoparticulate exposure. *Environ. Sci.: Processes Impacts*, 2013, **15,** 49-62.

[25] AITKEN J. - On the Number of Dust Particles in the Atmosphere. *Nature*, 1888, **37,** 428-430.

[26] MAYNARD A. D. - Learning from the past. *Nature Nanotechnology*, 2015, **10,** 482-483.

[27] IFA - Criteria for assessment of the effectiveness of protective measures, 2009. http://www.dguv.de/ifa/en/fac/nanopartikel/beurteilungsmassstaebe/index.jsp (avril 2013).

[28] BSI - Nanotechnologies – Part 2: Guide to safe handling and disposal of manufactured nanomaterials, PD 6699-2:2007, 2007.

[29] ISO - Cleanrooms and associated controlled environments - Part 1: Classification of air cleanliness, ISO 14644-1, 1999.

[30] CEN - Workplace exposure - Characterization of ultrafine aerosols / nanoaerosols - Determination of number concentration using condensation particle counters, prEN 16897, 2015.

[31] ISO - Aerosol particle number concentration - Calibration of condensation particle counters, ISO 27891, 2015.

[32] KOCH W., LÖDDING H., & POHLMANN G. - A reference aerosol generator based on Brownian coagulation in a continuously fed well stirred tank reactor. *Journal of Aerosol Science*, 2012, **49,** 1-8.

[33] KOCH W., POHLMANN G., & SCHWARZ K. - A reference number concentration generator for ultrafine aerosols based on Brownian coagulation. *Journal of Aerosol Science*, 2008, **39,** 150-155.

[34] CLERC F., NJIKI-MENGA G. H., & WITSCHGER O. - Exploratory study on a statistical method to analyse time resolved data obtained during nanomaterial exposure measurements. *Journal of Physics: Conference Series*, 2013, **429,** 012003.

[35] KLEIN ENTINK R. H., FRANSMAN W., & BROUWER D. H. - How to statistically analyze nano exposure measurement results: Using an ARIMA time series approach. *J. Nanopart. Res.*, 2011, **13,** 6991-7004.

[36] MCGARRY P., MORAWSKA L., KNIBBS L. D., & MORRIS H. - Excursion guidance criteria to guide control of peak emission and exposure to airborne engineered particles. *Journal of Occupational and Environmental Hygiene*, 2013, **10,** 640-651.

[37] PARK J. I., RAMACHANDRAN G., RAYNOR P. C., EBERLY E. B., & OLSON G. - Comparing exposure zones by different exposure metrics using statistical parameters: contrast and precision. *Annals of Occupational Hygiene*, 2010, **55,** 1-14.

[38] BEKKER C., KUIJPERS E., BROUWER D. H., VERMEULEN R., & FRANSMAN W. - Occupational exposure to nano-objects and their agglomerates and aggregates across various life cycle stages; a broad-scale exposure study. *Annals of Occupational Hygiene*, 2015, **59,** 681-704.

In vitro human digestion test to monitor the dissolution of silver nanoparticles

P Bove[1], **M A Malvindi**[2] **and S Sabella**[1]
[1] Istituto Italiano di Tecnologia, Drug Discovery and Development Department, Via Morego, 30 – 16136 Genova, Italy
[2] Istituto Italiano di Tecnologia, Center for Bio-Molecular Nanotechnologies@UniLe, Via Barsanti, 73010 Arnesano (Lecce), Italy

E-mail: stefania.sabella@iit.it

Abstract. Nanotechnology is a scientific revolution that the food industry has experienced over the last years. Widely employed as food additives and/or food contact materials in consumer products, silver nanoparticles are an example of this innovation. However, their increasing use makes also likely the human ingestion, thus requiring a proper risk analysis. In this framework, a comprehensive characterization of biotransformation of silver nanoparticles in biological fluids is fundamental for the regulatory needs. Herein, we aimed at studying the dissolution behaviour of silver nanoparticles using an *in vitro* test, which simulates the human oral ingestion of NPs during their passage through the gastrointestinal tract. The nanoparticle suspensions were characterized in the different digestion phases using several techniques to follow the changes of key physical properties (*e.g.*, size, surface charge and plasmon peak) and to quantify the biotransformed products arisen by the process, as for example free silver ions.

1. Introduction

Due to their proved antimicrobial activity and to their enhanced spectroscopic properties, silver nanoparticles (AgNPs) are highly employed by industries. They are widely used as health supplements in food industry, as transformed materials in food packaging and appear incorporated in personal care products, inks, electronics, and textile [1]. This increasing industrial production makes very likely the oral exposure of human populations, (*e.g.,* consumers, workers, children) both by voluntary or involuntary actions, thus requiring an urgent identification of the health risk profile [2, 3]. Research, regulatory and policy communities extensively debate this argument [4, 5, 6], highlighting that the quantitative description of NP fate along with the biotransformed species, in real exposure conditions, are related to the concepts of bioavailability/bioaccessibility, which are required by risk assessment.

Given the importance of these parameters, the development of an *in vitro* test using realistic conditions and in standardized manners is key to enable extraction of quantitative information. A classic example of dissolution test is applied in pharmaceutical or food industries. The test provides information on bioavailability (fraction released and adsorbed in the blood) and bioaccessibility (fraction released by a product) of a drug or a food contaminant. In this framework, we evaluated the feasibility of an *in vitro* dissolution test (previously employed for the quantification of bioavailability of food additives [7]) to quantify the dissolution of AgNPs in conditions likely occurring in the human oral ingestion. For the dissolution measurement, we applied UF/ICP-AES, DLS, zeta potential, UV-Vis to characterize size, surface charge and plasmon peak.

IOP Conf. Series: Journal of Physics: Conf. Series **838** (2017) 012003 doi:10.1088/1742-6596/838/1/012003

2. Methods

An *in vitro* dynamic model developed by Versantvoort *et al.* was applied here to digest AgNPs [7]. Briefly, artificial juices were prepared by mixing salt solutions, organic compounds and proteins at pH conditions similar to the human digestive compartments. Suspensions of AgNPs were incubated with the different juices, added in a temporal sequence that simulates the transit of food bolus along the oro-gastrointestinal (OGI) tract.

As reference materials, we used silver nanoparticles referred to as NM300k, obtained from the repository list of the European Commission Joint Research Centre (JRC) [8]. Dynamic light scattering (DLS), zeta potential and Transmission Electron Microscopy (TEM) were employed to evaluate size, morphology and surface charge of NM300k in dispersion medium and synthetic digestive juices. UV-Vis spectrophotometer was used to monitor the plasmon peak of NM300k during the digestion steps. Dissolution was measured by Ultrafiltration (UF) coupled to Inductively Coupled Plasma – Adsorption Emission Spectroscopy (ICP-AES).

3. Results and discussion

Figure 1 reports TEM images and the % of dissolution of NM300k, which are produced during the *in vitro* digestion test. Consistent with the NM datasheet, the size of NM300k is about 20 nm (Figure 1A) in either the ctrl and in the saliva compartment, although a small population of agglomerates and low level of dissolution are monitored in the saliva (Figure1B). Along the passage in stomach simulating conditions, TEM analysis evidences a higher degree of particle dissolution, also showing the presence of a population of NPs with size below 5 nm (together with some aggregates). This indicates that most of the particles dissolved. UF/ICP-AES measures a value of about 20% of free ions. This value is not in contrast with the TEM data, as it only refers to the fraction of free ions able to pass though the filter. The remaining fraction is more likely bound to matrix so that it remains into the pellet. In the intestine, the dissolved fraction appeared further reduced to about 2%. Moreover, some aggregates along with nanosized particles of about 20 nm were observed. An in depth characterization of these nano-based salts are currently on going in our laboratories.

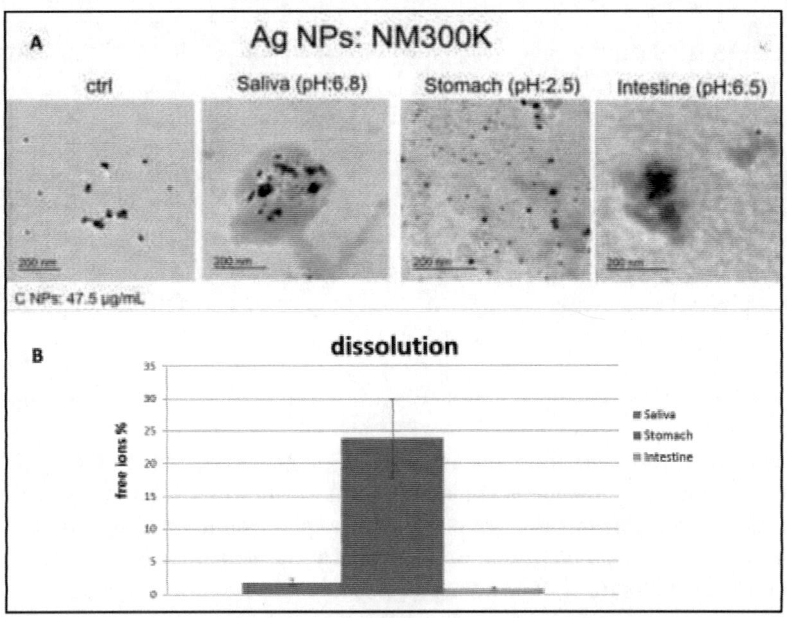

Figure 1. NM300k digestion by *in vitro* dissolution test. (A) TEM images of NM300k in simulated digestive fluids. (B) Percentage of free silver ions by *in vitro* digestion assay.

These data were confirmed by DLS and UV-Vis characterization (Figure 2). DLS spectra reported a small aggregation in the saliva with a peak size distribution ranging from 14 ± 3 nm to 120 ± 4 nm. This is in line with UV-Vis spectra that confirm the presence of the typical absorption peak at 412 nm for silver nanoparticles, along with a broad band in the red region (which is a typical indicator of agglomerates). In the stomach and in the intestine, DLS detected only big agglomerates whereas the UV-Vis spectra displayed the absence of the peak at 412 nm, further confirming the NP dissolution. Interestingly, by zeta pot analyses, we found out that the nominal surface charge for NM300k progressively changed, mostly acquiring the charge features of the media components (*e.g.*, proteins). We also observed that in the saliva and intestine compartments, NM300k acquired a negative charge (around -35mV), whereas, in the stomach, the surface charge approached quite neutral values (in line with literature data) [8]. Note that, since these NPs partially dissolve in the stomach compartment, it could be possible that the reported zeta pot values in the digestive juices refer to mixed aggregates of biological/nanostructures present in the solutions, which do not relate to the primary NPs. This point deserves future investigations and clarifications.

AgNPs NM330k

Hydrodinamic diameter		ctrl	saliva	stomach	intestine
Peak 1		40 ± 3 nm	14 ± 3 nm	150 ± 6 nm	-
Peak 2		-	120 ± 4 nm	740 ± 17 nm	700 ± 72 nm
UV-Vis peak		ctrl	saliva	stomach	intestine
Peak 1		412 nm	412 nm	-	-
Peak 2		450 nm	550 – 650 nm	-	-
Surface charge	nominal	ctrl	saliva	stomach	intestine
mV	-11.0	-24.4	-35.0	-3.0	-32.0

Figure 2. DLS, UV-Vis adsorption and zeta potential values of NM300k in the different digestive matrices.

4. Conclusions

In conclusions, we performed analytical studies to investigate AgNP fate in synthetic biological fluid, particularly focusing on tracing the changes of size, surface charges and quantifying the dissolution process using a dissolution test. The obtained results show that NM300k did not maintain their primary properties along their passage through the OGI tract due to an almost complete dissolution. In particular, NPs begin to partially agglomerate in the salivary compartment and quite totally dissolve in the stomach compartment. Once they arrive in the intestine compartment, only a small fraction of free ions is quantified to about 2%. Changes in UV-Vis adsorption and surface charge are also monitored at similar conditions.

Overall, these data inform us that, during the digestion process, complex mixtures of bio-transformed products (including different organic, inorganic and ion molecular species) occur. These are the actual species responsible of effective dose, cellular permeability and uptake *in vitro*/*in vivo*. In the future, studies on test standardization and characterization of free and bound to matrix ions will provide benefits for the development of a test useful for the nanoregulation needs.

5. References

[1] Gaillet S and Rouanet J M 2015 *Food Chem. Toxicol.* **77**
[2] Stone V *et al* 2014 *Part. Fibre Toxicol.* **11** 9
[3] Scott-Fordsmand J J *et al* 2014 *Nano Today* **9** 5

IOP Conf. Series: Journal of Physics: Conf. Series **838** (2017) 012003 doi:10.1088/1742-6596/838/1/012003

[4] OECD (Organisation for Economic Co-operation and Development) 2009 Enviroment directorate joint meeting of the chemicals committee and the working party on chemicals, pesticides and biotechnology
[5] ECHA (European Chemicals Agency) 2012 Guidance on information requirements and chemical safety assessment, Appendix R7-1 Recommendations for nanomaterials applicable to Chapter R7a - Endpoint specific guidance
[6] ECHA (European Chemicals Agency) 2015 Guidance on Information Requirements and Chemical Safety Assessment, Chapter R.7a: Endpoint specific guidance
[7] Versantvoort C H M *et al* 2005 *Food Chem. Toxicol.* **43** 1
[8] Klein C L *et al* 2011 NM-series of representative manufactured nanomaterials. NM-300 silver. Characterisation, stability, homogeneity
[9] Roger K M *et al* 2012 *Sci. Total. Environ.* **420**.

Physico-chemical separation process of nanoparticles in cosmetic formulations

R. R. Retamal Marín[1], F. Babick[1], M. Stintz[1]

1. - Research Group Mechanical Process Engineering, Institute of Process Engineering and Environmental Technology, Technische Universität Dresden, Münchner Platz 3, D-01062 Dresden, Germany.
E-mail: rodrigo.retamal@tu-dresden.de

Abstract. Understanding the world of nanoparticles, especially their interactions with the environment, begins with their correct detection and successive quantification. To achieve this purpose, one needs to perform correctly developed standard operating procedures (SOPs). Furthermore, the study of nanoparticles frequently requires their characterisation in complex media (e.g. in cosmetic formulations). In this study, a set of sample preparation procedures for the detection and extraction of NMs in emulsion-based formulations is proposed and their performance for model and real-life products is discussed. A separation or extraction of lipid phases is achieved by means of organic solvents. The polarity of the lipid phases is decisive for selecting an optimum solvent. The use of the Hansen Solubility Parameters (HSP) may clearly support this decision.

Keywords: nanomaterials (NMs), Hansen Solubility Parameters (HSP), cosmetic formulations

1. Introduction

Nanomaterials (NMs) play an important role in the design of new products, since they allow for controlled variation of different product properties. This holds particularly true for cosmetic formulations and food products, where NM additives are employed as stabilisers, pigments, UV absorber, thickeners, flow enhancing agent and even more [1, 2]. Furthermore, there is potential concern on the health effects of NMs, because of which legal authorities request reliable methods for the characterisation of NMs in such products [3, 4]. Further motivations for appropriate characterisation techniques are process control and quality assurance.

This study proposes a set of sample preparation procedures for the detection of NMs in emulsion-based formulations and discusses their performance for model and real-life products [5]. One of these procedures is the extraction of lipid phases, which facilitates the characterisation of different NM polarities (e.g. hydrophobic or hydrophilic).

2. Materials and Methods

2.1. *Methodological procedures*
The present study focuses on SiO_2 based on its use in a wide range of consumer products – including cosmetics (hydrated silica) and foods (E551) – in form of powder, dilute liquid suspensions, emulsions or pastes [6]. Experiments were conducted for complex formulations, which were composed of aqueous solutions, dispersed lipid phases (24 wt.-%), hydrophobic fumed SiO_2 (10 wt.-%; BET 200 m^2/g) and typical additives used in cosmetics (e.g. glycerol as thickening agent).
The qualitative analysis of particle size distribution (PSD) was conducted with scanning electron microscopy (SEM) – Gemini DSM 980 (Zeiss/Germany). The quantitative analysis of PSD was determined with standard analytical techniques (laser diffraction LD) – HELOS BR (Sympatec/Germany), dynamic light scattering (DLS) – HPPS (Malvern/UK), which are of low cost (compared to ICP-MS) and can quickly provide highly accurate results. However, when applied to specific particles in complex media they require appropriate preparation techniques and adopted measurement procedures [7]. Further problems arise since the dispersed phases typically cover a wide range of particle sizes (a few nanometres up to even millimetres).

2.2. *Determination of size distribution*

The selected methods for the determination of particle size distribution (PSD) are based on different measurands [8], i.e. equivalent diameters, and also on different types of quantity (e.g. DLS: intensity weighted) [9]. Furthermore, different size ranges are covered, each of which is smaller than the whole range from primary particle size to maximum agglomerate size. The detection, extraction and subsequent characterisation of NMs in multi-disperse systems as emulsions involve metrological difficulties [10]. This study discusses the development of preparation techniques for the adequate characterisation of silica particles in complex media. The characterisation of NMs in cosmetic formulations (e.g. Pickering emulsion) requires the appropriate development of SOPs for the preparation and extraction of NMs in dispersion systems. The structure of these SOPs can generally be classified into five steps, namely (1) wetting/suspending (polarity (hydrophobic and hydrophilic)), (2) dispersion (e.g. ultrasonic dispersion (cup horn, sonotrode and cleaning bath)), (3) sampling (mass/volume balance), (4) sample division/preparation (extraction, sedimentation/centrifugation), and finally (5) measurement/interpretation (e.g. particle size distribution (PSD)). The dispersion experiments can be quantitatively classified according to volume specific energy input.

3. Results – Characterisation of Cosmetic Emulsions

3.1. *Evolution of separation and extraction process*
The characterisation of NMs in food and cosmetic emulsions is challenged by the presence of different liquid phases and the huge variety of solutes including rheological additives and amphiphilic molecules [11]. The different inorganic and organic solutes (e.g. electrolytes, sugar, emulsifiers, proteins) can significantly affect the interfacial properties of all particles (e.g. by adsorption/desorption) and thus influence the behaviour of the NMs in the product (dispersibility in aqueous and lipid phases, adsorption at fluid-fluid interfaces or on the surface of large particle).
A separation or extraction of lipid phases is achieved by means of organic solvents. The polarity of the lipid phases is decisive for selecting an optimum solvent. It can be shown that the use of the Hansen Solubility Parameters (HSP) may clearly support this decision [12, 13, 14, 15]. The principal idea consists in the separation of the oil droplets from the aqueous phase by means of bi-continuous mixtures of water with a lipophilic (i.e. non-polar) solvent (see Figure 1).

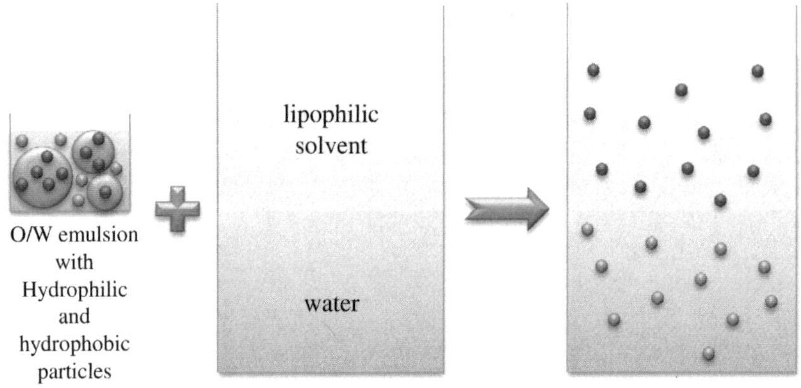

Figure 1. Proposal idea of the extraction process of solid particles of emulsion by means of bi-continuous mixtures of water with a lipophilic solvent

In this study, hydrophobic SiO_2 was separated from oil droplets by dissolving the lipid in n-heptane and subsequent phase separation by centrifugation (see Figure 2). An emulsion sample of 1 g was added to 49.5 g water and 49.5 g heptane. The mixtures were then agitated by magnetic stirring for 20 hours. A preliminary test showed that this procedure leads to a stable mixture with slow phases separation under gravity. In order to work efficiently, the separation of selected bi-continuous mixtures of heptane and water was accelerated by centrifugation (see Figure 2).

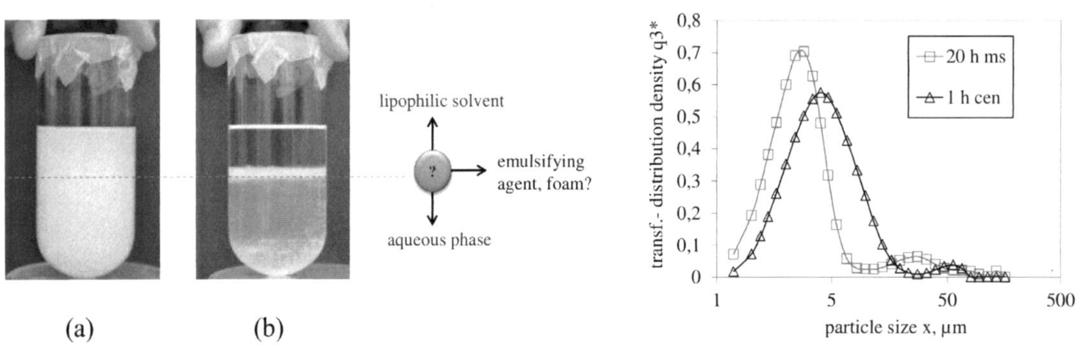

Figure 2. Left: Hydrophobic fumed SiO_2 in water and n-heptane (mixing ratio 1:1); (a) after 20 h magnetic stirring (ms), (b) after 1 hour centrifugation (cen) at 2600 g; Right: Change of size distribution (determined with LD) of a SiO_2 cosmetic emulsion within water phase

After centrifugation one can observe three phases: the organic solvent, liquid foam with a high portion of emulsifying agent, and water. The solvent phase is transparent but contains solid nanoparticles as could be demonstrated with DLS and SEM analysis (see Figure 3).

The DLS indicates the existence of particles with the same size range as the SiO_2 aggregates added to the emulsion. Moreover, SEM images impressively show the fractal morphology of particles in the n-heptane phase, which confirms that these particles present the hydrophobic fumed SiO_2. This proves the affinity of hydrophobic SiO_2 in non-polar solution and supports the validity of the separation method.

IOP Conf. Series: Journal of Physics: Conf. Series **838** (2017) 012004 doi:10.1088/1742-6596/838/1/012004

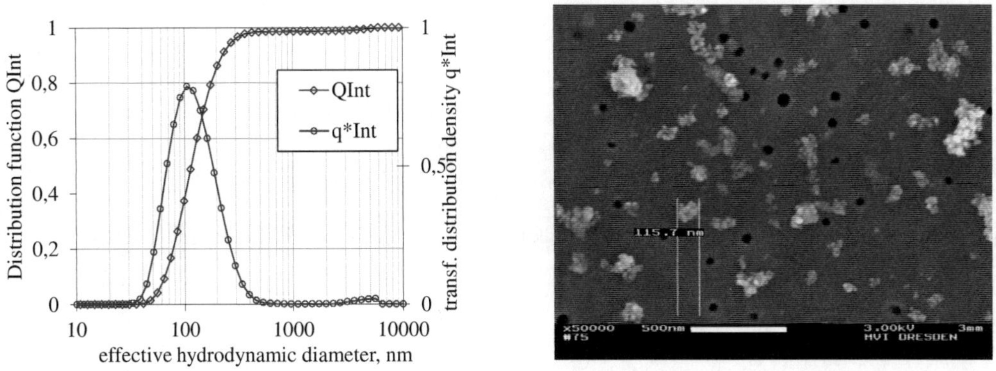

Figure 3. Left: intensity weighted distribution for SiO_2 in n-heptane phase; Right: SEM image with high magnification (finest particles found); sample carrier: track-etch-membrane with 50 nm pore

4. Summary and Conclusion

The detection of nanoparticulate material in complex media is not trivial. Common characterisation methods are focused particle systems with just one dispersed phase, preferably with low or moderate polydispersity. Therefore, it is very important, in such complex media, to implement adequate SOPs and to achieve objective interpretations of the analysis (PSD, equivalent diameters and also different types of quantity). This investigation verified an alternative procedure to separate particles with regard to the polarity of solvent and solute in complex cosmetic emulsions. This procedure is applicable to primary emulsion (e.g. O/W or W/O) or even secondary or multiple-emulsion (e.g. O/W/O or W/O/W) and it can be employed for hydrophilic as well as hydrophobic silica particles.

References

[1] Raj S., Jose S., Sumod U.S., Sabitha M. 2014 *J. Pharm. Bioall. Sci.* **4** 183-193
[2] Epstein H. A., Kielbassa A. 2013 *Bio-Nanotechnology: A Revolution in Food, Biomedical and Health Sciences* **1** 414-423
[3] Singh P., Nanda A. 2012 *Toxicol. Environ. Chem.* **94** 1467–1479
[4] Lorenz C., Von Goetz N., Schreringer M., Wormuth M. and Hungerbrühler K. 2011 *Nanotoxicology* **5** 12-29
[5] Gianeti M. D., Wagemaker T. A. L.,Seixas V. C. and Maia Campos M.B 2012 *Curr. Nanosci.* **8** 526-534
[6] He X. and Hwang H. 2016 *J. Food Drug Anal.* **24** 671-681
[7] Mu L. and Sprando R. L. 2010 *Pharm. Res.* **27** 1746-1749
[8] Babick F., Schießl K. and Stintz M. 2012 *Part. Part. Syst. Charact.* **29** 104-115
[9] Babick F. 2016 *Particle Technology Series* **20** 37-43
[10] Babick F and Ullmann C. 2016 *Powder Technol.* **301** 503-510
[11] Contado C., Ravani L. and Passarella M. 2013 *Analytica Chimica Acta* **788** 183-192
[12] Hansen C. 2007 *Solubility Parameters - An Introduction. Hansen Solubility Parameters* **2** 1-24
[13] Hansen C. 2014 *Prog Org Coat* **51** 77-84
[14] Gårdebjer S., Andersson M., Engström J., Restorp P., Persson M. and Larsson A. 2016 *Polym. Chem.* **7** 1756-1764
[15] Wang B. and Shi B. 2010 *J. Macromol Sci. Phy.* **49** 383-391

IOP Conf. Series: Journal of Physics: Conf. Series **838** (2017) 012005 doi:10.1088/1742-6596/838/1/012005

Dustiness of 14 carbon nanotubes using the vortex shaker method

Claire Dazon[1], Olivier Witschger[1], Sébastien Bau[1], Raphaël Payet[1], Karine Beugnon[2], Geneviève Petit[2], Thibaut Garin[2] and Laurent Martinon[2]

[1]Laboratoire de Métrologie des Aérosols, Institut National de Recherche et de Sécurité (INRS), Rue du Morvan, CS 60027, F 54519 Vandœuvre Cedex, France
[2]Laboratoire Amiante Fibres Particules (LAFP), Service Parisien de Santé Environnementale, Mairie de Paris / DASES, F-75013 Paris, France

E-mail: olivier.witschger@inrs.fr

Abstract. The handling of carbon nanotube (CNT) powders is a plausible scenario during the course of the CNT life-cycle. However, related exposure data remain limited. In this context, information about the dustiness of CNT is therefore of great interest, for example for control banding or exposure modelling. Here, we investigate the dustiness of fourteen CNT powders using the Vortex Shaker (VS) method. The central component of the VS method is a stainless steel cylindrical tube, continuously shaken in a circular orbital motion, in which a small volume (0.5 cm^3) of the powder to be tested is placed. All samples were obtained through the NANoREG Nanomaterials Information and Web-Order system. The test procedure that we have developed is based on four principal components: (i) a respirable cyclone for gravimetric sampling, (ii) a CPC as a reference instrument for number concentration measurement, (iii) an MPS for collection of particles for EM observations/analysis, and (iv) an ELPI for size-resolved aerosol measurement. In this paper, the data were evaluated using two parameters: (i) the mass-based dustiness index in the respirable fraction; and (ii) the number-based dustiness index in the respirable fraction. The results indicate that the method leads to relatively accurate mass- and number-based dustiness indices. The indices obtained span wide ranges, of 2 and 3 orders of magnitude variation for mass and number respectively, suggesting a corresponding significant difference in terms of potential exposure. EM observations reveal that airborne CNTs are mostly released as bundles of different shapes ranging from a few tens of nanometers up to tens of micrometers in size.

1. Introduction

Dustiness is a generic term used to define the ability of a powdered material (e.g., loose, granulated, or pelletized powders) to generate an aerosol (airborne particles) during handling. The dustiness of a powder is an important determinant for worker exposure and should be considered during the design and operation of many industrial or research processes [1]. For some years now, dustiness has been a required input parameter in control banding (CB) tools used to evaluate and control the risk of exposure to nanomaterials [2, 3]. Its use is also increasing in risk assessment, for example for the carbon nanotube group [4].

In response to the particular requirements associated with testing and measuring the dustiness of nanomaterials, a number of recent studies have aimed either to modify the two methods (the rotating drum and the continuous drop) described in the EN15051-1 European standard [5] or to develop new approaches [6]. One such approach is the Vortex Shaker (VS) method described in this work. Our motivations for developing this method were based on three main issues simulating different (including worst-case) workplace scenarios to those addressed by the EN15051-1 methods [5]; performing dustiness experiments with small amounts of nanomaterials that are either potentially very toxic and/or costly; and developing smaller set-ups that can be used in ventilated enclosures or fume hoods, thus better protecting operators in charge of experiments.

The expected widespread use of carbon nanotube (CNT) composites in industrial and commercial applications across industry calls for an assessment of their possible release and of potential worker exposure [7]. The handling of CNT powders is plausible throughout the entire life-cycle of CNTs, but in spite of evidence that shows that workers are exposed to CNTs [8], exposure data obtained in workplaces are unfortunately scarce. Consequently, there is a real interest in developing dustiness test methods in order to produce data for CNTs and rank them according to their ability to generate aerosols.

The current consensus in the literature is that particle mass on its own is not an adequate metric for evaluating aerosol exposure in situations where nanomaterials are handled. Instead, a multi-metric approach is now recommended, especially for carbon nanotubes [9]. A multi-metric approach includes other aerosol properties, including airborne particle number concentration, size distribution and morphology. In addition, the usefulness of real-time data for providing information about particle emission kinetics is also now well-established because new procedures for data-analysis are being developed [6].

Given this context, the VS method was established here in order to provide relevant dustiness indices as well as information about size distribution and morphology. The dustiness of fourteen CNTs was investigated. Only results in terms of mass- or number-based dustiness indices are reported here; particle emission kinetics and size distribution of the released aerosols will be presented in a separate paper.

2. Material and methods

The vortex shaker (VS) method is a recent method in terms of standard dustiness testing, however several configurations of this method have been proposed over time [6, 10]. The technique was derived from an original concept that was developed as part of a field study devoted to evaluating the release of aerosols during the handling of single-walled carbon nanotubes (SWCNT) [11].

2.1. Vortex shaker design and aerosol measurement

The central component of the experimental set-up consists of a specially designed stainless steel cylindrical tube (inner diameter of 31 mm, volume of 102 cm^3) with a conical bottom that is continuously shaken in a circular orbital motion (displacement amplitude: 4 mm; angular frequency: 30 Hz), into which a small volume (0.5 cm^3) of the test sample is placed (Figure 1).

HEPA-filtered air is passed through the cylindrical tube at a flow-rate Q_{VS} of 8.4 l/min in order to transfer the emitted aerosol inside the tube to the sampling and measurement section. The aerosol emitted during the vibration is drawn from the outlet tube to the real-time instruments and sampling devices through an initial flow splitter, which divides the aerosol flow into two flows directed toward two identical respirable cyclones (GK2.69, SKC) operating at Q_C = 4.2 l/min. One of these respirable cyclones is equipped with a cassette containing a pre-weighed filter for gravimetric analysis. The sampled particle mass is then used to determine the respirable mass dustiness index. The second respirable cyclone acts as a particle selector for the real-time instruments and the TEM grid sampler positioned downstream. Due to the short sampling time (~10 s), the TEM grid sampler requires a by-pass system (equipped with a 25-mm filter cassette) in order to maintain a constant flow through the respirable selector. To prevent exposure of the operator during tests as well as during the disassembly

IOP Conf. Series: Journal of Physics: Conf. Series **838** (2017) 012005 doi:10.1088/1742-6596/838/1/012005

and cleaning sequences between each test, the entire test bench is housed within an approved ventilated enclosure that has been specially designed for the safe handling of powders (LEV systems, Safetech).

Because (i) the entire range of particle sizes corresponding to the respirable size fraction should be assessed for the risk of exposure by inhalation [12], (ii) a single, rather than two, size-resolved instrument is preferred, and (iii) the size-resolved instrument should have a high time resolution (~1 s), the Electrical Low Pressure Impactor (ELPI™, Dekati Ltd) was selected for the size-resolved instrument. As a reference instrument for counting particles, a condensation particle counter (CPC Model 3007, TSI Inc.) was used.

To collect airborne particles for subsequent observation and analysis by electron microscopy, a specific TEM grid sampler (MPS, Ecomesure) equipped with Holey Carbon Film on 400 Mesh Copper Grids (S147-A, Agar Scientific) was used. The electron microscopy observations were conducted at the Laboratoire Amiante Fibres Particules (LAFP) with a transmission electron microscope (120 kV JEM-1400, Jeol) equipped with a CCD camera (ES500 Erlangshen ES500, Gatan Inc.). The JEM-1400 is also equipped with an EDS microanalysis system (Oxford Instruments). Prior to each observation, a size calibration at the magnification used for EM observation of the samples was performed with certified polystyrene latex spheres of 0.88 µm diameter. The collected particles was not counted and sized in this study as our intention was only to characterize the samples from a qualitative and global point of view.

Carbon impregnated conductive flexible tubing was used to connect the different parts of the set-up as well as the instruments.

In compliance with the requirements of EN15051-1 [5], the air in the experimental set-up was controlled for humidity (RH = 50.9 ± 0.6 %) and monitored for temperature (22.1 ± 0.3 °C) for the duration of the experiments.

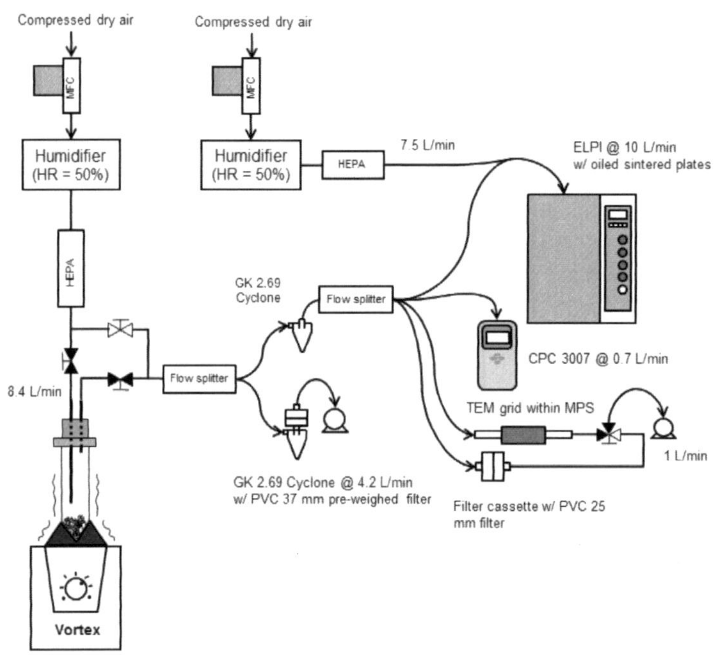

Figure 1: Experimental set-up of the vortex shaker method for determining the number-based and mass-based dustiness indices, characterizing the particle size distribution of the aerosol emitted, and collecting airborne particles for subsequent EM observations.

IOP Conf. Series: Journal of Physics: Conf. Series **838** (2017) 012005 doi:10.1088/1742-6596/838/1/012005

The test samples were weighed with a XP205 analytical balance (10 µg readability, Mettler Toledo), and the 37-mm filters used for the respirable cyclone were weighed with a MX5 microbalance (1 µg readability, Mettler Toledo). The filters used in this study are PVC membrane filters with a pore size of 5 µm (GLA 5000, SKC Inc.). This type of filter was chosen because of its low tare weight and low moisture pickup for gravimetric stability.

2.2. Test protocol

The test samples were prepared by pouring 0.5 cm^3 of powder into an Eppendorf® microtube. After filling, the tubes were weighed to the nearest 10 µg. The samples were then conditioned for at least 24 h prior to the dustiness test in a laboratory-made humidity-controlled chamber at 50 ± 2 % RH.

After cleaning the entire test bench, and/or, if necessary, changing the tubing and connections, and having equipped the cyclone with a pre-weighed filter for gravimetric analysis, the airflows were checked using a flow calibrator. The cleanliness of the air through the test bench was then assessed from CPC measurements. The by-pass line was then opened and the main line closed in order to fill the pre-conditioned 0.5 cm^3 test sample into the cylindrical tube. The mass of the test sample M_0 was obtained from the difference between the mass of the filled microtube and mass of the microtube after emptying.

In parallel, the real-time instruments started recording the background level, over about 180 s. Finally, at the same time, the incoming air flow was switched on in the main line and the agitation was turned on. The test sequence comprised 60 s of agitation, with the measurement continuing for another 540 s, thus giving a total of 600 s for the sampling (respirable cyclone) and real-time measurement (CPC and ELPI). Sampling with the MPS was carried out for around 10 s after the peak in concentration.

After each test, the respirable sampling filter was gently removed from the cassette and gravimetric analysis was then performed according to the laboratory protocol in order to obtain the mass Δm_f of the collected particles. This mass was used to determine the mass-based dustiness index (see section 2.3.

Real-time instrument data files from the CPC and the ELPI™ were saved and the data were subsequently analysed using specific calculation tools developed in the laboratory in order to determine the number-based dustiness index, emission rate and size distribution (see section 2.3.

During cleaning of the test bench and measurement instruments, the operator wore a powered respirator, protective gloves with sleeves and a cotton laboratory coat.

2.3. Data processing

In this work, the data for the vortex shaker method were evaluated using two parameters derived from the respirable filter sampling and CPC measurements. As mentioned in the introduction, the emission rate and size distribution data will be presented in a separate paper.

The mass-based dustiness index in the respirable fraction ($DI_{M,R}$), reported in milligrams of aerosol per kilogram of powder, was calculated by dividing the mass collected by the respirable sampling filter (in milligrams), by the mass M_0 of the powder placed in the cylindrical tube (in milligrams) :

$$DI_{M,R} = \frac{\Delta m_f}{M_0} \cdot \frac{Q_{VS}}{Q_C} \cdot 10^6$$

The number-based dustiness index in the respirable fraction ($DI_{N,R}$), given in particles per milligram of powder, was calculated by dividing the number of particles emitted over the duration of the vibration period (60 s) (in particles) by the mass M_0 of the powder placed in the cylindrical tube (in milligrams) :

$$DI_{N,R} = \frac{1}{M_0} \cdot \sum_{0}^{T} C_{CPC}(t) \cdot Q_{VS} \cdot \Delta t_{CPC} \cdot \frac{10^3}{60}$$

where $C_{CPC}(t)$ is the number concentration recorded by the CPC at time t, in particles per cm^3, and Δt_{CPC} is the time step of the CPC (1 s in this work).

2.4. Carbon nanotubes

Fourteen CNTs were evaluated. These candidate CNTs were supplied by the European Commission's Joint Research Centre (JRC), who have established a repository of Representative Test Materials (RTMs), which hosts industrially manufactured nanomaterials that are distributed worldwide for the safety testing of nanomaterials [13]. The CNTs were received in vials, each containing about 100 mg of CNTs. One or more vials were used for each experiment, depending on the bulk density of the sample. Given the subsampling procedure from a single batch by the JRC, the vials provided can be considered identical.

The different test CNTs, their main physical properties and the basic characteristics of the test samples are given in Table 1, below.

Table 1: Designation and main properties of the test CNTs and test samples.

CNT Designation*			JRC Data**			Test Sample	
Type*	NM	MW or SW	Diameter (nm)	Length (µm)	Surface area (m²/g)	Mass (mg)	Bulk density (g/cm³)
Core	400	MW	13.6	0.846	254	29.7	0.06
	401		64.2	4.048	18	4.8	0.01
	46000a	SW	n/a	n/a	n/a	95.6	0.19
Alternative	402	MW	12.7	2.2	226	47.1	0.09
	403		12	0.62	189	70.0	0.14
	40001a		15	50	>233	24.3	0.05
	40002a		15	50	>233	27.5	0.05
	40003a		20-30	10-30	>110	48.7	0.10
	40004a		<8	10-30	>500	198.3	0.40
	40005a		8-15	50	>233	31.1	0.06
	40006a		20-30	10-30	>110	63.3	0.13
	40007a		8	30	>500	129.9	0.26
	40008a		8-15	50	>233	24.9	0.05
	40009a		20-30	20-30	N.A	44.3	0.09

*Type according to the NanoREG project and identification code used by JRC.

Only one of the fourteen CNTs was a single-walled CNT (SWCNT); the other thirteen were multi-walled CNTs (MWCNT). According to the JRC data, the CNTs cover a wide range of diameters, lengths and surface areas.

The mass of the test samples ranged from about 200 mg down to 5 mg. These masses correspond to the volume of 0.5 cm^3 used as the reference volume in the VS method. The large variations in mass are due to the differences in the bulk density of the CNTs; the minimum and maximum bulk densities obtained were 0.01 and 0.4 g/cm^3, respectively. It is interesting to note that these values are 5 to 235 times lower than the theoretical graphite solid density of 2.27 g/cm^3.

Results and discussion

A total of 42 experiments were conducted on the fourteen CNTs using the VS method (i.e. three replicates of each CNT were tested).

Figure 2 presents the mass-based ($DI_{M,R}$) and number-based ($DI_{N,R}$) dustiness indices (average values and 95 % confidence intervals) of the respirable fraction. The limit of detection of the mass-based dustiness index (LOD($DI_{M,R}$)) is also shown. The LOD($DI_{M,R}$) corresponds to the ratio of the gravimetric detection limit to the mass of the test sample, thus the LOD($DI_{M,R}$) was determined for each experiment. The limit of detection associated with the gravimetric analysis was obtained from the reproducibility of the blank PVC membrane weights as measured before and after their assembly within a 37 mm sampling cassette used with the respirable cyclone. The repeat weight measurements took place over a period of about 1 month and were performed on six series of five PVC membranes from the same batch. The gravimetric limit of detection finally obtained was 17 µg. Assuming a mass M_0 of 100 mg, this corresponds to a LOD($DI_{M,R}$) of 340 mg/kg.

A first observation concerns the low scatter of the results obtained for each CNT. The median coefficient of variation (CV) of the $DI_{N,R}$ is 10 %, with minimum and maximum values of 1 % and 29 %, respectively. This CV is the same as that of the $DI_{M,R}$ (10 %) but the range of the latter is a little wider, with minimum and maximum values of 4 % and 63 %, respectively. Overall, these results demonstrate that the protocol developed in this study yields accurate dustiness indices.

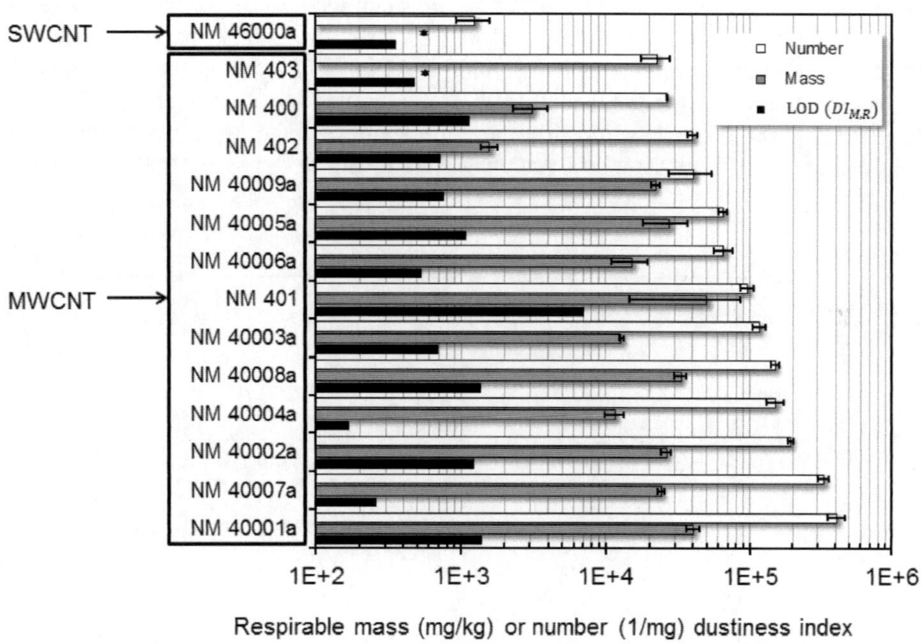

Figure 2: Respirable mass dustiness indices (in mg/kg) and respirable number dustiness indices (in 1/mg) for the fourteen CNTs, obtained using the VS method. Average values are presented with their 95% confidence interval. Black bars correspond to the LOD($DI_{M,R}$). *: mass dustiness indices were below the LOD($DI_{M,R}$).

A second observation concerns the values of the dustiness indices obtained using the VS method. The $DI_{N,R}$ values of the CNTs vary by three orders of magnitude, with a median value ~8.1 10^4 1/mg, while the $DI_{M,R}$ values span a narrower range, covering two orders of magnitude variation, with a median value of ~2.3 10^4 mg/kg. If the SWCNT is excluded, a narrower range of $DI_{N,R}$ values is observed the $DI_{N,R}$ values of only the MWCNT vary by one order of magnitude, with a median value ~9.6 10^5 1/mg. However, exclusion of the SWCNT does not change the result for the $DI_{M,R}$ as the index obtained for the SWCNT was below the LOD($DI_{M,R}$). From these results, it seems clear that the particle releases are considerably lower for the SWCNT than for the MWCNT, suggesting that there is a corresponding significant difference in terms of potential exposure, all other things being equal. Concerning the$DI_{M,R}$, the situation is somewhat different as the SWCNT is characterized by a mass dustiness index below the LOD($DI_{M,R}$). Only one of the MWCNT samples (NM 403) had a value below the LOD($DI_{M,R}$). The ratio $DI_{M,R}$/ LOD($DI_{M,R}$) varies widely among the other twelve MWCNTs, ranging from about 2 to 92.

Although a number of experimental studies have been carried out on the dustiness of carbon nanotubes [6], only two produced dustiness data that can be compared to the results of this work [14, 15]. The data available for comparison are shown in Figure 3. In one of the studies, the dustiness test method was somewhat different to the VS method, based instead on the Venturi dustiness testing device initially developed for pharmaceutical powders [16]. In the other study, the dustiness method was also based on the use of a vortex shaker, but a different configuration and measurement protocol were used [15]. Moreover, the trade names of the JRC CNT samples cannot be determined (for the second study), though the names are given for the three CNTs used in the other study [14]. For these reasons, our interpretation of the comparison remains very limited. Of the dustiness values presented in Figure 3, only two are categorized as SWCNT. Overall, the dustiness values vary by a little more than two orders of magnitude, from about ~1.6 10^3 mg/kg up to about ~3.2 10^5 mg/kg. Interestingly, only one link can be made on NM 402. The difference between the two $DI_{M,R}$ determined for NM 402 is quite important: the value obtained in this work is ~20 times lower than the value obtained in the previous study [15]. As stated earlier, this is essentially due to differences in the test protocol. In the earlier study, bronze micro-beads were added to the test samples in the cylindrical tube in order to promote de-agglomeration and release of airborne particles during the agitation. Moreover, the test sequence involved 3600 s of agitation and sampling, and thus was very different from the protocol used in this work.

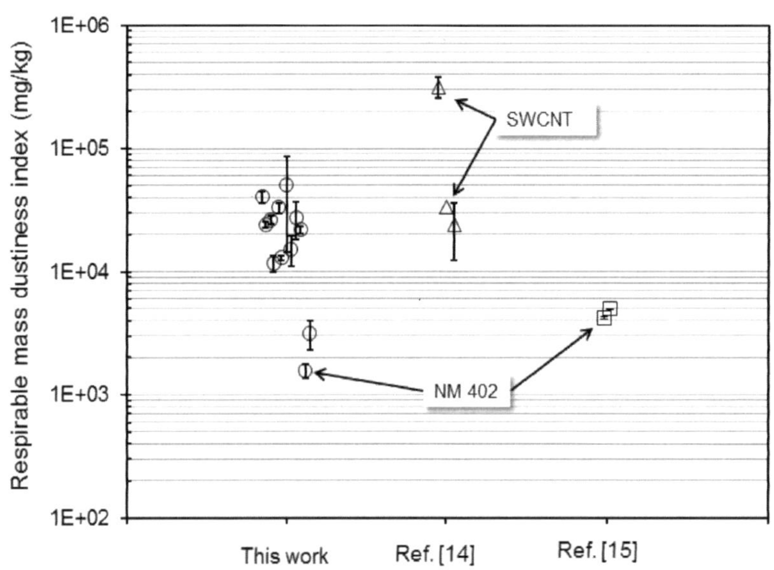

Figure 3: Comparison of the respirable mass-based dustiness indices (in mg/kg) obtained in this work with available published indices from [14, 15].

The last observation concerns the ranking of the different CNTs. In Figure 3, the dustiness data are ranked according to $DI_{N,R}$, from the highest (NM 40001a) to the lowest (NM 46000a). The lowest

IOP Conf. Series: Journal of Physics: Conf. Series **838** (2017) 012005 doi:10.1088/1742-6596/838/1/012005

$DI_{N,R}$ was obtained on the SWCNT. It can be observed that the ranking order differs according to which of the dustiness metrics, mass- or number-based, is considered. Moreover, there appears to be no correlation between the two dustiness metrics. This suggests that the choice of a dustiness metric will have significant implications for exposure modeling or when comparing powders. This supports our initial decision to develop a dustiness method following a multi-metric approach.

Figures 4, 5 and 6 present transmission electron microscopy (TEM) images of collected airborne CNTs (MWCNT 401, SWCNT 46000a, MWCNT 40002a) at different magnifications. MWCNT 401 was selected for TEM analysis because of its singular morphology compare to the others CNTs. SWCNT 46000a was selected because it was the only SWCNT included in the study. MWCNT 40002a was chosen as it is representative of the other CNTs.

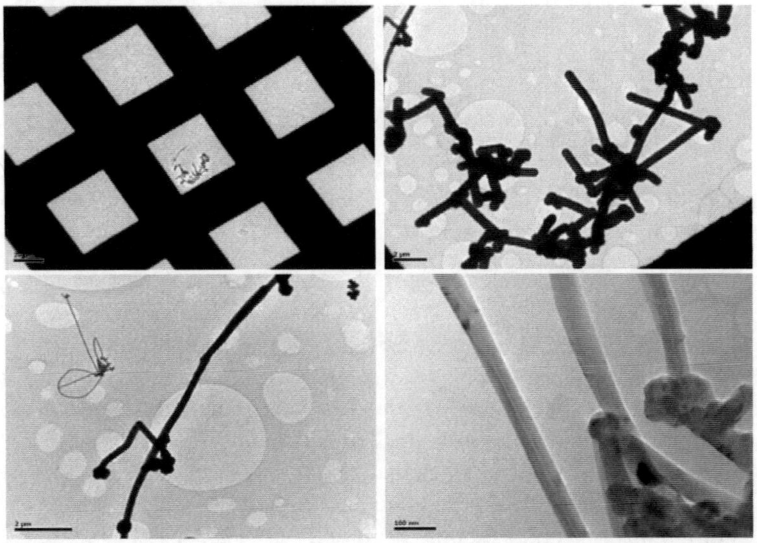

Figure 4: TEM images of airborne CNTs sampled during the MWCNT 401 experiments. Sampling duration of 10 s with the MPS equipped with 400 Mesh Holey Carbon Film TEM grids. Images © LAFP.

IOP Conf. Series: Journal of Physics: Conf. Series **838** (2017) 012005 doi:10.1088/1742-6596/838/1/012005

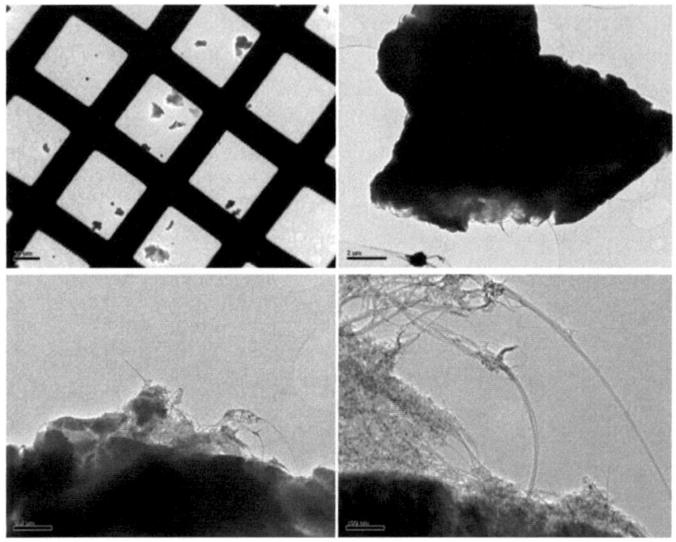

Figure 5: TEM images of airborne CNTs sampled during the SWCNT 46000a experiments. Sampling duration of 10 s with the MPS equipped with 400 Mesh Holey Carbon Film TEM grids. Images © LAFP.

The image on the top left shows the structure of the square mesh grid (openings of 40 x 40 μm^2) and the particles that were collected. The images demonstrate that the protocol used here makes it possible to obtain TEM grids that are not overloaded, i.e. the number of particles per unit of area is low enough to allow straightforward observation.

The images also illustrate that the particles that comprise the released aerosols vary greatly in terms of the size and morphology of the CNT aggregate structures. Only MWCNT 401 exhibited any individual fibers in the images (Fig. 4). Some of these fibers appear to be knotted, as shown in Figure 4. The remainder of the CNT images all show nanotubes with different diameters and aspect ratios, tied or wrapped up together in bundles (Figs. 5 and 6). The bundles, of variable density, exhibit different shapes and range from a few tens of nm to several tens of micrometers in size. Several nanotubes, either straight or curved, typically protrude from these bundles. Closer examination of the bundles reveals that some of the nanotubes have metal catalyst particles attached to them. EDS analysis indicates that these particles are essentially Fe.

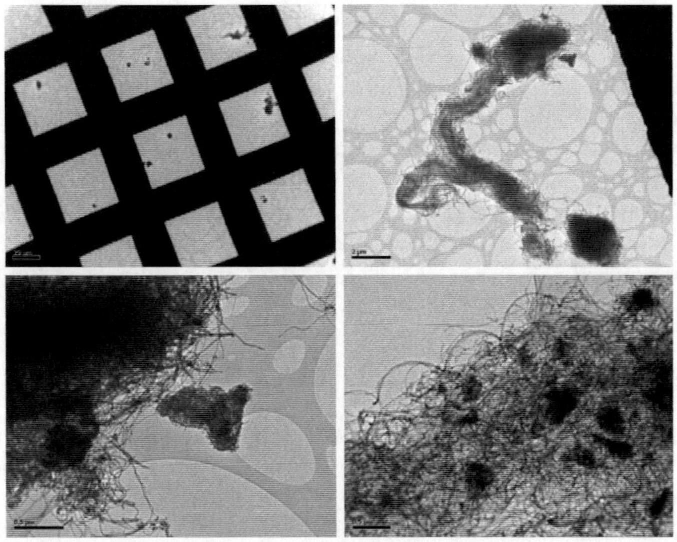

Figure 6: TEM images of airborne CNTs sampled during the MWCNT 40002a experiments. Sampling duration of 10 s with the MPS equipped with 400 Mesh Holey Carbon Film TEM grids. Images made © LAFP.

Conclusion

In this work, the VS method was developed following a multi-metric approach, in order to obtain relevant dustiness indices as well as information about size distribution and morphology. The dustiness of fourteen industrially manufactured CNTs provided by JRC was then investigated. The results indicate that the method provides relatively accurate mass- and number-based dustiness indices. The dustiness indices cover a wide range, of 2 and 3 orders of magnitude variation for mass and number metrics, respectively, suggesting a corresponding significant difference in terms of the potential for exposure. When the CNTs are ranked in order of dustiness, the order is different depending on which of the two metrics is considered, and there appears to be no correlation between the two indices. Electron microscopy observations show that CNTs are mostly released as bundles of different shapes, ranging from a few tens of nanometers up to several tens of micrometers in size. Individual nanotubes were observed to protrude from these bundles. The particle emission kinetics and size distributions of the released aerosols will be presented in a separate paper.

The few existing studies on the dustiness of powders (including CNTs) show that there is still a need for further work on the harmonization of methods and the processing of data. Moreover, efforts should be made to improve our understanding of the factors that control the dustiness of powders. These different elements are important as requests for relevant dustiness data are likely to increase in the future, particularly for risk assessments using exposure modeling or control banding tools.

Acknowledgements

The research leading to these results was partially funded by the European Union Seventh Framework Programme (FP7/2007-2013) under the project NANoREG (A common European approach to the regulatory testing of nano-materials), grant agreement 310584.

References

[1] LIDEN G. - Dustiness Testing of Materials Handled at Workplaces. *Annals of Occupational Hygiene*, 2006, **50,** 437-439.

[2] SÁNCHEZ JIMÉNEZ A., VARET J., POLAND C., FERN G. J., HANKIN S. M., & VAN TONGEREN M. - A comparison of control banding tools for nanomaterials. *Journal of Occupational and Environmental Hygiene*, 2016, **13,** 936-949.

[3] LIGUORI B., HANSEN S. F., BAUN A., & JENSEN K. A. - Control banding tools for occupational exposure assessment of nanomaterials — Ready for use in a regulatory context? *NanoImpact*, 2016, **2,** 1-17.

[4] NAKANISHI J., MORIMOTO Y., OGURA I., KOBAYASHI N., NAYA M., EMA M., *et al.* - Risk Assessment of the Carbon Nanotube Group. *Risk Analysis*, 2015, **35,** 1940-1956.

[5] CEN - Workplace exposure - Measurement of the dustiness of bulk materials - Part 1: Requirements and choice of test methods, EN 15051-1, November 2013.

[6] JENSEN K., LEVIN M., & WITSCHGER O. - Methods for Testing Dustiness, in *Nanomaterial Characterization – An Introduction*, ed. Tantra, R. (Wiley & Sons), pp. 209-230.

[7] NOWACK B., DAVID R. M., FISSAN H., MORRIS H., SHATKIN J. A., STINTZ M., *et al.* - Potential release scenarios for carbon nanotubes used in composites. *Environment International*, 2013, **59,** 1-11.

[8] DEBIA M., BAKHIYI B., OSTIGUY C., VERBEEK J. H., BROUWER D. H., & MURASHOV V. - A Systematic Review of Reported Exposure to Engineered Nanomaterials. *Annals of Occupational Hygiene*, 2016, **60,** 916-935.

[9] OBERDÖRSTER G., CASTRANOVA V., ASGHARIAN B., & SAYRE P. - Inhalation Exposure to Carbon Nanotubes (CNT) and Carbon Nanofibers (CNF): Methodology and Dosimetry. *Journal of Toxicology and Environmental Health, Part B*, 2015, **18,** 121-212.

[10] ISO - Nanomaterials - Quantification of nano-object release from powders by generation of aerosols, TS 12025, November 2011.

[11] BARON P. A., MAYNARD A. D., & FOLEY M. - Evaluation of aerosol release during the handling of unrefined single walled carbon nanotube material, 2002,

[12] WITSCHGER O. - Monitoring Nanoaerosols and Occupational Exposure., in *Nanoethics and Nanotoxicology*, eds. Houdy, P., Lahmani, M., & Marano, F. (Springer Berlin Heidelberg), pp. 163 - 199.

[13] TOTARO S., COTOGNO G., RASMUSSEN K., PIANELLA F., RONCAGLIA M., OLSSON H., *et al.* - The JRC Nanomaterials Repository: A unique facility providing representative test materials for nanoEHS research. *Regulatory Toxicology and Pharmacology*, 2016, **81,** 334-340.

[14] EVANS D. E., TURKEVICH L. A., ROETTGERS C. T., DEYE G. J., & BARON P. A. - Dustiness of Fine and Nanoscale Powders. *Annals of Occupational Hygiene*, 2012.

[15] JRC - Multi-walled Carbon Nanotubes,
NM-400, NM-401, NM-402, NM-403:
Characterisation and Physico-Chemical Properties, 2014,

[16] BOUNDY M., LEITH D., & POLTON T. - Method to Evaluate the Dustiness of Pharmaceutical Powders. *Annals of Occupational Hygiene*, 2006, **50,** 453-458.

[1] LIDEN G. - Dustiness Testing of Materials Handled at Workplaces. *Annals of Occupational Hygiene*, 2006, **50,** 437-439.

[2] SÁNCHEZ JIMÉNEZ A., VARET J., POLAND C., FERN G. J., HANKIN S. M., & VAN TONGEREN M. - A comparison of control banding tools for nanomaterials. *Journal of Occupational and Environmental Hygiene*, 2016, **13,** 936-949.

[3] LIGUORI B., HANSEN S. F., BAUN A., & JENSEN K. A. - Control banding tools for occupational exposure assessment of nanomaterials — Ready for use in a regulatory context? *NanoImpact*, 2016, **2,** 1-17.

[4] NAKANISHI J., MORIMOTO Y., OGURA I., KOBAYASHI N., NAYA M., EMA M., *et al.* - Risk Assessment of the Carbon Nanotube Group. *Risk Analysis*, 2015, **35,** 1940-1956.

[5] CEN - Workplace exposure - Measurement of the dustiness of bulk materials - Part 1: Requirements and choice of test methods, EN 15051-1, November 2013.

[6] JENSEN K., LEVIN M., & WITSCHGER O. - Methods for Testing Dustiness, in *Nanomaterial Characterization – An Introduction*, ed. Tantra, R. (Wiley & Sons), pp. 209-230.

[7] NOWACK B., DAVID R. M., FISSAN H., MORRIS H., SHATKIN J. A., STINTZ M., *et al.* - Potential release scenarios for carbon nanotubes used in composites. *Environment International*, 2013, **59,** 1-11.

[8] DEBIA M., BAKHIYI B., OSTIGUY C., VERBEEK J. H., BROUWER D. H., & MURASHOV V. - A Systematic Review of Reported Exposure to Engineered Nanomaterials. *Annals of Occupational Hygiene*, 2016, **60,** 916-935.

[9] OBERDÖRSTER G., CASTRANOVA V., ASGHARIAN B., & SAYRE P. - Inhalation Exposure to Carbon Nanotubes (CNT) and Carbon Nanofibers (CNF): Methodology and Dosimetry. *Journal of Toxicology and Environmental Health, Part B*, 2015, **18,** 121-212.

[10] ISO - Nanomaterials - Quantification of nano-object release from powders by generation of aerosols, TS 12025, November 2011.

[11] BARON P. A., MAYNARD A. D., & FOLEY M. - Evaluation of aerosol release during the handling of unrefined single walled carbon nanotube material. NIOSH DART-02-191, 2002, 22 pages.

[12] WITSCHGER O. - Monitoring Nanoaerosols and Occupational Exposure., in *Nanoethics and Nanotoxicology*, eds. Houdy, P., Lahmani, M., & Marano, F. (Springer Berlin Heidelberg), pp. 163 - 199.

[13] TOTARO S., COTOGNO G., RASMUSSEN K., PIANELLA F., RONCAGLIA M., OLSSON H., *et al.* - The JRC Nanomaterials Repository: A unique facility providing representative test materials for nanoEHS research. *Regulatory Toxicology and Pharmacology*, 2016, **81,** 334-340.

[14] EVANS D. E., TURKEVICH L. A., ROETTGERS C. T., DEYE G. J., & BARON P. A. - Dustiness of Fine and Nanoscale Powders. *Annals of Occupational Hygiene*, 2012.

[15] JRC (2014) *Multi-walled Carbon Nanotubes, NM-400, NM-401, NM-402, NM-403: Characterisation and Physico-Chemical Properties* (European Commission, Luxembourg: Publications Office of the European Union).

[16] BOUNDY M., LEITH D., & POLTON T. - Method to Evaluate the Dustiness of Pharmaceutical Powders. *Annals of Occupational Hygiene*, 2006, **50,** 453-458.

IOP Conf. Series: Journal of Physics: Conf. Series **838** (2017) 012006 doi:10.1088/1742-6596/838/1/012006

Assessment of personal exposure to airborne nanomaterials: Evaluation of a novel sampler

Bertrand Faure, Hélène Dozol, Christophe Brouard, Arnaud Guiot and Simon Clavaguera

NanoSafety Platform, Commissariat à l'Energie Atomique et aux Energies Alternatives (CEA), Univ. Grenoble Alpes, Grenoble, 38054, FRANCE

Email: simon.clavaguera@cea.fr

Abstract. A novel sampler, the NANOBADGE, has been developed to assess personal exposure to nano-objects, agglomerates and aggregates (NOAA) at the workplace. The NANOBADGE collects particles on filters subsequently analyzed by X-Ray Fluorescence spectroscopy (XRF), which provides a mass-based quantification with chemical selectivity. The NANOBADGE was benchmarked against a scanning mobility particle sizer (SMPS) and a DiSCmini by carrying out simultaneous measurements on test aerosols of ZnO or TiO_2 for particle sizes between 20 and 400 nm for which the DiSCmini has its highest accuracy. The effective density and shape of the NOAA present in the test aerosols were determined experimentally to compare number-based data obtained with the SMPS and the DiSCmini with mass-based data obtained with the NANOBADGE. The agreement between the SMPS and the NANOBADGE sampler was within ± 25 % on all test aerosols. The converted DiSCmini data matched the SMPS and sampler data for polydisperse aerosols in the specified size range as long as the DiSCmini assumptions meet the aerosol characteristics (i.e. lognormal size distribution with a given geometric standard deviation $\sigma g = 1.9$). The detection limits of the NANOBADGE sampler were in the order of tens of nanograms per filter, which is low enough to reliably detect exposure levels below the recommended exposure limit (REL) of the National Institute for Occupational Safety and Health (NIOSH) and the Institut National de Recherche et de Sécurité (INRS) for ultrafine ZnO and TiO_2 even for short-term exposure situations.

1. Introduction

The manufacturing and use of engineered nano-objects, agglomerates and aggregates (NOAA) has increased rapidly over the recent years. NOAA are now mass-produced, used for a very wide range of applications and incorporated into mainstream consumer products [1]. Besides the obvious benefits these materials bring, their potential adverse health and environmental effects, particularly caused by inhalation, are raising growing concerns [2-4]. The intrinsic ability of NOAA to reach and deposit in the deep alveolar region of the lungs suggests that their toxicity and fate may differ significantly depending on the particle size, shape and chemical composition [5, 6]. Although the likelihood for exposure to NOAA by inhalation is higher for workers producing or handling these products at the industrial scale, end-users are also potentially exposed and the airborne release of NOAA in the environment cannot be excluded [7-9]. Since the hazards and the environmental impact are unknown, precautionary safety measures in the nanomaterial industry should ideally involve a continuous tracking of the personal exposure to airborne NOAA [10].

There is no scientific agreement on the best metric for exposure assessment between number, surface area or mass concentration [11]. With the exception of fibers which are counted by number, mass concentration remains the most widespread metric [12, 13]. Most published studies on inhalation exposure used stationary measurement equipment [14, 15]. Although several personal samplers and monitors became available lately [16-20], the number of studies on their comparability and accuracy is still rather low [21-24]. Within the framework of the nanoIndEx project, we propose to evaluate herein the performance of the NANOBADGE personal sampler.

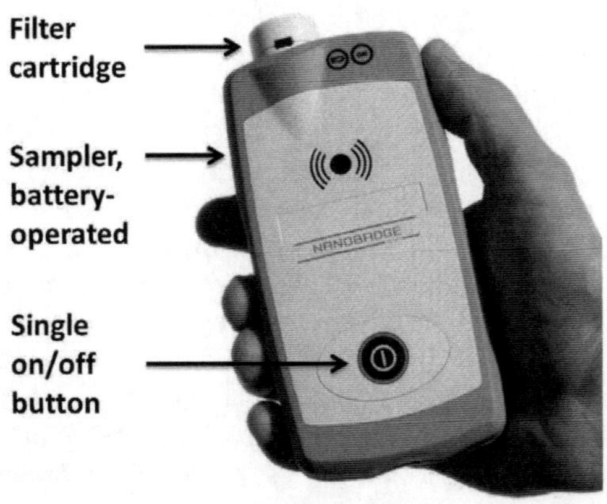

Figure 1. Picture of the NANOBADGE personal sampler equipped with a filter cartridge.

The NANOBADGE sampler is a battery-operated portable device which collects airborne particles in the breathing zone of the worker, defined as a 30-cm hemisphere around mouth and nose [25]. The device has been designed to minimize the burden in terms of weight and noise and is operated with a single button, as shown in Figure 1 [26]. The sampler is connected to filtration cartridges, whose filters are then analyzed offline by X-Ray Fluorescence spectroscopy (XRF). This highly-sensitive technique yields the elemental composition of the collected particles with sensitivity in the order of a few tens of nanograms per filter. Moreover it is a non-destructive analytical technique, meaning that the same sample can be characterized with other techniques such as scanning electron microscopy [5]. The sampler can be equipped with a pre-separator with a cut-off diameter of 4 µm to remove coarse particles.

The sampler was compared to a stationary instrument, the scanning mobility particle sizer (SMPS), and to a personal monitor, the DiSCmini [17, 24, 27]. A procedure has been developed to compare the number-based metrics of the SMPS and the DiSCmini to the mass-based value obtained with the sampler. For that purpose, aerosols with particle sizes within 20-400 nm were used since the comparability of personal monitors and SMPS on the mean particle size and number concentration are respectively within ± 30% and ± 50% in this specified size range [28]. The performance and the ability of the NANOBADGE to perform personal exposure measurements to airborne nanomaterials will be discussed.

2. Materials and methods

The DiSCmini is equipped with an impactor with a cut-off at around 700 nm to remove coarse particles which cannot be sized correctly by the monitor [17]. In order to test the instruments under the same conditions, during all tests with monodisperse and polydisperse aerosols in the 20-400 nm size range the tested NANOBADGE and DiSCmini were run without the use of a pre-separator. This is not detrimental for the performance of the instruments as the modal diameter of the test aerosols was much smaller than the cut-off of their impactors. The NANOBADGE sampler (2013 version operated with a flowrate of 0.6 L/min) was equipped with polycarbonate filters (Whatman® Nuclepore Track-Etched Membranes, pore size 0.4 µm) for the collection of particles. The filters were analyzed by XRF, providing a cumulative mass-based quantification of the chemical elements present on the filters. Thus, the sampler provided the mass concentration in the personal breathing zone averaged over the total sampling time.

The XRF analysis was performed on a Rigaku Nanohunter XRF spectrometer operating at 50 kV, 0.80 mA, equipped with copper and molybdenum anodes. The former anode was chosen to detect and to quantify titanium, and the latter for zinc. The Z-offset (vertical position of the upper surface of the sample with respect to the excitation beam) has been optimized for every measurement in order to get fluorescence signal from the filter only, based on the intensity of the element which was considered (Ti or Zn). An acquisition time of 200 s and an incident angle of 0.75° were used. The fluorescence intensity of the element has been normalized by the area of the inelastic scattering peak (16.5-18 keV) for the molybdenum anode and by the area of the inelastic and elastic scattering peak (7.71-8.40 keV) for the copper anode [29-31]. The quantification by mass of the elements deposited on the filters required that the XRF spectrometer is calibrated, which was done using a previously reported methodology [32]. In short, sets of filters of increasing particle loading were generated by sampling aerosols of ZnO or TiO_2. The filters were then analyzed by XRF, followed by dissolution of the particles for elemental quantification by Inductively Coupled Plasma Mass Spectrometry (ICP-MS). The detailed procedure is described in the Supplementary Material. The plot of the normalized XRF intensity versus the mass determined by ICP-MS lead to the calibration curves for Ti and Zn, used to convert the X-ray fluorescence intensity to mass. For comparison of the mass with the SMPS and DiSCmini data, we assumed that the elements detected were present as oxides.

The limit of detection (LOD) was estimated for ZnO and TiO_2 from the measured XRF intensity on 3 blank filters. For each filter, the minimum detectable intensity I_{LOD} was calculated from the mean normalized intensity $< I >$ of the element considered and the standard deviation σ of 5 independent measurements [33]:

$$I_{LOD} =< I > +4.65\sigma \quad (1)$$

This minimum detectable intensity was converted to a minimum detectable mass using the calibration curve.

Simultaneous samplings with the SMPS, the DiSCmini and the NANOBADGE sampler were carried out using the experimental setup shown in Scheme 1. The test aerosols of ZnO and TiO_2 were generated by atomization of aqueous dispersions (200 mg/L and 100 mg/L in water for ZnO and TiO_2, respectively). The dispersions were prepared by dilution of concentrated stock dispersions with ultrapure water (Milli-Q, Millipore, Billerica, USA): a commercial ZnO dispersion (Evonik "VP Disp ZnO 20 DW", 35 wt% in water) and a TiO_2 dispersion (TiO_2 anatase, Nanostructured & Amorphous Materials Inc., 5-30 nm APS, 15 wt% in water). A TSI aerosol generator (TSI model 3076) was used to produce the aerosols, dried with a silica gel diffusion dryer (TSI model 3062). Polydisperse aerosols were introduced through an aerosol neutralizer (^{85}Kr source, 74 MBq) into a vertical chamber (ca. 120 cm high, inner diameter 29 cm) in a laminar flow (30 L/min, Re=20) of HEPA (high-efficiency particulate arrestance)-filtered dilution air. In order to study monodisperse test aerosols, a differential mobility analyzer (DMA - TSI long model 3081, sheath air 5 L/min, aerosol flowrate 1 L/min) was added between the atomizer and the vertical chamber. The NANOBADGE sampler and the DiSCmini monitor were installed inside the tunnel, along with a thin-walled sampling probe connected to a SMPS (TSI model 3080 with a DMA from TSI, model 3081, sheath air

3 L/min). The uniformity in space of the particle number concentrations was checked in the sampling area of the chamber (see Supplementary Material).

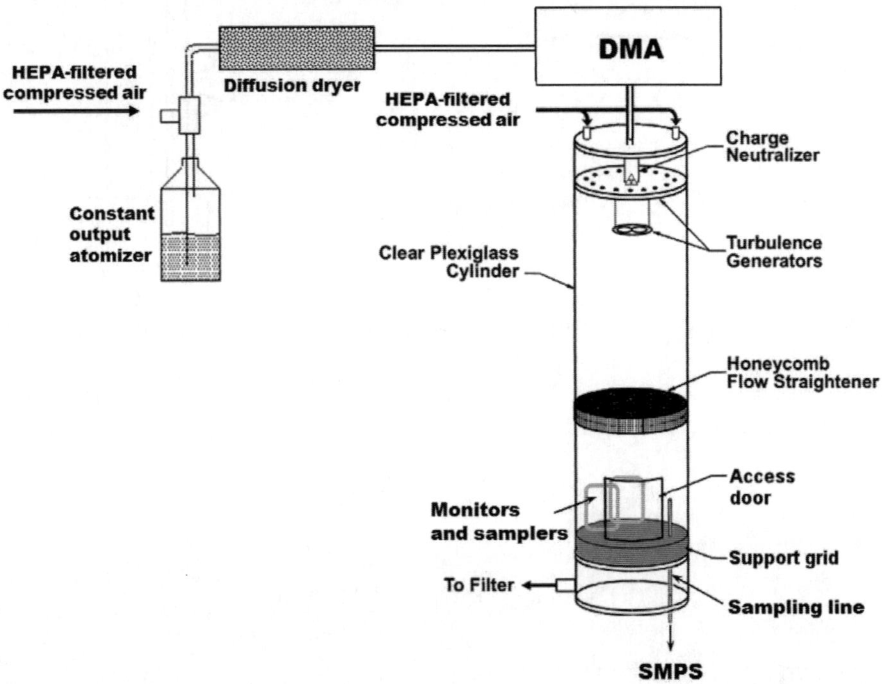

Scheme 1. Experimental setup used for the generation and simultaneous sampling and monitoring of the monodisperse aerosols. The DMA was by-passed when sampling polydisperse aerosols. Adapted from Trakumas and Salter [34].

The comparison between the mass concentration determined by XRF with number-based data from the SMPS and the DiSCmini required knowledge of the effective densities of the particles. The effective density is defined as the mass of a particle divided by the volume of a spherical particle having the same mobility diameter. It is determined by comparing the mobility diameter, which is independent of particle mass, with the aerodynamic diameter. The effective density of the generated NOAA was estimated using a DMA coupled to an Electrical Low-Pressure Impactor (ELPI)[35]. Polydisperse aerosols of the particles of interest were first classified by a DMA (TSI long model 3081, sheath air 10 L/min) around a set mobility diameter before entering the ELPI for aerodynamic sizing, as shown in Scheme 2. The ELPI was provided with 1 L/min of the size-classified aerosol complemented by 9 L/min of HEPA-filtered air. The mobility diameter was chosen as the modal diameter of the polydisperse aerosol to get sufficiently high aerosol concentrations after size classification.

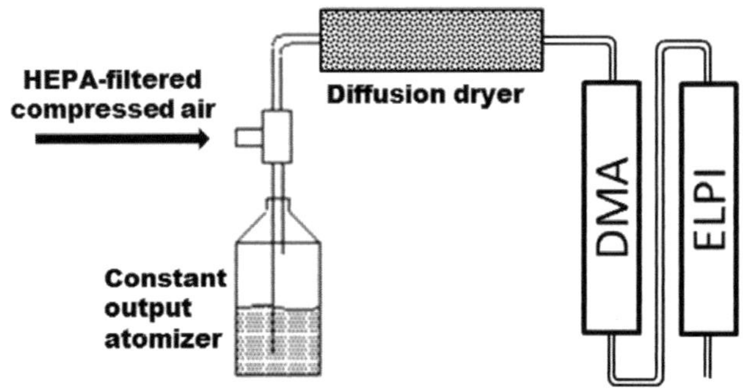

Scheme 2. Tandem DMA-ELPI set-up for the determination of the effective density.

The effective density ρ_{eff} was calculated from the mobility diameter d_m and the aerodynamic diameter d_a as:

$$\rho_{eff}(d_m) = \frac{\rho_0 C_c(d_a) d_a^2}{C_c(d_m) d_m^2}, \quad (2)$$

where $\rho_0 = 1$ g/cm^3,
the Cunningham slip correction

$$C_c(d) = 1 + \frac{\lambda}{d}\left(2.284 + 1.116 \exp\left(-\frac{0.5d}{\lambda}\right)\right) \quad (3)$$

and the mean free path $\lambda = 0.066$ µm [36, 37].

The mean aerodynamic diameter d_a was calculated from the currents and the cut-point diameters of the ELPI impactor stages.

$$d_a = \exp\left(\frac{\sum_i I_i \ln d_i}{\sum_i I_i}\right) \quad (4)$$

In order to compare the quantification by mass of the sampler with number-based metrics provided by the SMPS and the DiSCmini, the particles in the aerosol were assumed to be spherical. The shape of the sampled NOAA was assessed by Scanning Electron Microscopy [5] using a LEO 1530 instrument (LEO Electron Microscopy Ltd, Cambridge, England, acceleration voltage 10 kV, in-lens detector). The polycarbonate filters were coated with platinum to mitigate charging effects.

For the SMPS, the total mass per unit volume of air (g/cm^3) was calculated by summation of the number concentrations for each size channel according to:

$$m = \rho_{eff} \sum_{D_p = 14.6\ \text{nm}}^{D_p = 661.2\ \text{nm}} \frac{\pi D_p^3 N_p(D_p)}{6} dD_p \quad (5)$$

where D_p is the mobility diameter [35], $N_p(D_p)$ is the number concentration of particles having the mobility D_p (cm^{-3}) and ρ_{eff} is the effective density (g/cm^3). The expected mass collected by the sampler can be determined from the sampling flowrate and duration.

For the DiSCmini, we have used the built-in mass concentration provided by the software (Datatool v1.217). The calculation is based on the Hatch-Choate equations [37], which are used to convert the count mean diameter to volume mean diameter:

$$m = \rho N \frac{\pi}{6} d^3 e^{4.5(ln\sigma)^2}, \quad (6)$$

where N is the total particle concentration by number and d is the mean particle size determined by the DiSCmini. The calculation is only valid for polydisperse aerosols having a lognormal size distribution. The software uses a geometric standard deviation σ of 1.9 and a density of ρ of 1.2 g/cm^3 by default [38]. Therefore, experiments made with aerosols that differ significantly from the assumptions made should be analyzed carefully since they might lead to erroneous results. We have scaled the obtained value according to the measured effective density ρ_{eff} for comparison with the sampler.

3. Results and Discussion

The effective densities of ZnO and TiO_2 aerosols were determined using the DMA-ELPI tandem setup shown in Scheme 2. For each aerosol, the electrical mobility diameter of the DMA was matched to the largest modal diameter of the aerosol, which depends on the type of powder and the concentration of the dispersion. The resulting values for the effective density are given in Table 1. The effective densities determined for the NOAA of ZnO and TiO_2 corresponded to approximately 40 % of the values for bulk materials. In both cases the packing density of the NOAA was similar, which was confirmed visually on the SEM images in Figure 2. In addition, the shape of the NOAA was nearly spherical. Thus the assumption of spherical particles made to calculate the mass from the SMPS and the DiSCmini data was expected to be reasonable.

Table 1. Effective density calculated for the aerosols generated by atomization. The effective density of bulk ZnO and TiO_2 is 5.61 g/cm³ and 3.9 g/cm³, respectively. The uncertainty on the density was calculated as twice the relative standard deviation of 100 measurements.

Material	Concentration (mg/L)	Selected mobility diameter [9]	Average aerodynamic diameter [9]	Density (g/cm^3)
ZnO	200	100	181	2.2 ± 2 %
TiO_2	100	60	79	1.4 ± 4 %

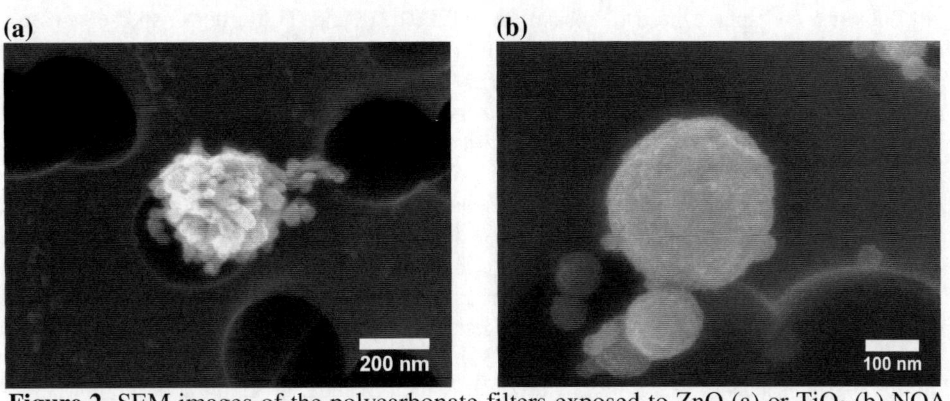

Figure 2. SEM images of the polycarbonate filters exposed to ZnO (a) or TiO_2 (b) NOAA generated by atomization.

The SMPS and the NANOBADGE sampler have been compared on three monodisperse ZnO aerosols which were classified at 75, 100, and 200 nm using a DMA (Scheme 1). The error bars for the SMPS were calculated as twice the average standard deviation of the total number concentration. A value of 20 % was obtained, in good agreement with previous literature [27]. Considering that the error on the density is within 5 %, an accuracy of ± 25 % was used for the SMPS data. A relative error of 10 % on the mass measured by XRF was estimated, due to the calibration procedure and the regulation of the sampling flowrate (more details are given in the supplementary information). As shown in Figure 3, the agreement between the mass calculated from the SMPS and the mass measured by XRF is good. It is worth mentioning that the properties of the test aerosols are ideal for the purpose of this study, since the NOAA are monodisperse, nearly-spherical and in the 20-400 nm size range recommended for the highest accuracy of the DiSCmini.

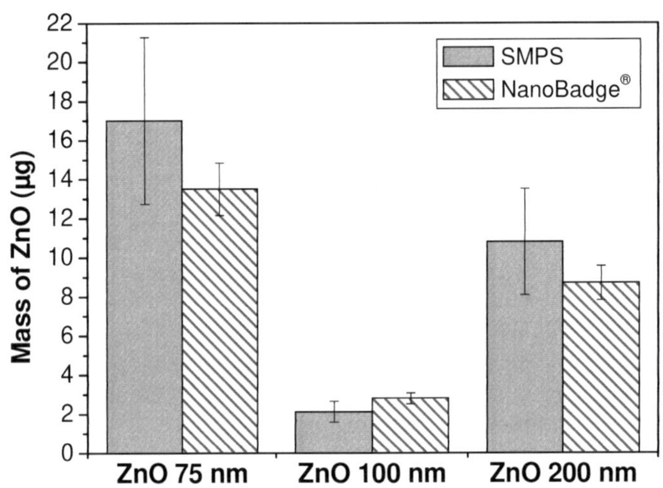

Figure 3. Mass of ZnO calculated from the SMPS data and mass of ZnO measured by XRF analysis of the NANOBADGE filters (calculated from the mass of Zn). An effective density of 2.2 was used.

The DiSCmini portable monitor was compared with the SMPS and the NANOBADGE sampler for test aerosols of ZnO (monodisperse aerosol classified at 200 nm) and TiO_2 (polydisperse aerosol with a geometric mean diameter of 60 nm and standard deviation of 1.8). The results are reported in Figure 4. The accuracy for the SMPS and the NANOBADGE was calculated as described above for ZnO. For the DiSCmini, the accuracy on the mass was estimated as twice the average standard deviation of the number concentration (± 30 %), in good agreement with previously reported values [21, 24]. The uncertainty on the density (± 5 %) was added for a total accuracy of ± 35 %.

All three instruments yielded comparable results on the polydisperse TiO_2 aerosol despite the slight difference regarding the measured geometric standard deviation of the aerosol size distribution (1.8 vs the 1.9 value used by the DiSCmini). However, for the monodisperse ZnO aerosol, the DiSCmini overestimated the mass. This significant deviation originates from assumptions for the aerosol size distribution in the DiSCmini: a lognormal size distribution with a geometric standard deviation of 1.9 is used to calibrate the device response and to convert the data to mass [38, 39]. The monodisperse ZnO aerosol has a narrow size-distribution (σ = 1.3, see Supplementary Material), which results in errors on the particle number concentration and the mean particle size recorded by the DiSCmini. The document "miniDiSC application note 11" provides correction factors for aerosols that do not match the assumed size distribution. The contour plot of the simulated particle number response indicates that the DiSCmini underestimates the number concentration by 15%, leading to an underestimation of the mass by the same amount. In addition the DiSCmini underestimates the particle size by 5%, leading to an underestimation of the sampled mass by 15.8%. The last correction consists in using the measured σ in Equation 6. This calculation shows that the mass was overestimated by a factor of 4.7. Combining these three corrections, a direct use of the DiSCmini data overestimated the mass by a factor of 3.4. Thus the true mass should be 19.7 μg/m^3 which is close to a factor of 2 to the SMPS and NANOBADGE results and not a factor of 7.

The DiSCmini has its highest accuracy for particle sizes between 20 and 400 nm, with an optimum around 100 nm [17, 22, 24]. The properties of the TiO_2 polydisperse aerosol were much more suitable for the DiSCmini, with a mean particle size (60 nm) and geometric standard deviation (1.8) which are close to the assumptions of the device software. Corrections to compensate the slight deviation of σ indicates that the DiSCmini does not affect the particle number response while it underestimates the particle size by 5%. Moreover, the wrong σ used in Equation 6 involves an overestimation of the mass by a factor of 1.35 which consequently implies an overall overestimation of the mass by a factor of 1.157. This overestimation by 13.6% stay in the ± 30 % accuracy margin. Therefore, one can say that in this case all three instruments yielded similar results for the mass.

When the aerosol features deviated from the ideal values (in particular when classified aerosols were studied), the direct use of the DiSCmini data and its conversion tool should be taken with caution unless it can be confirmed by some other method.

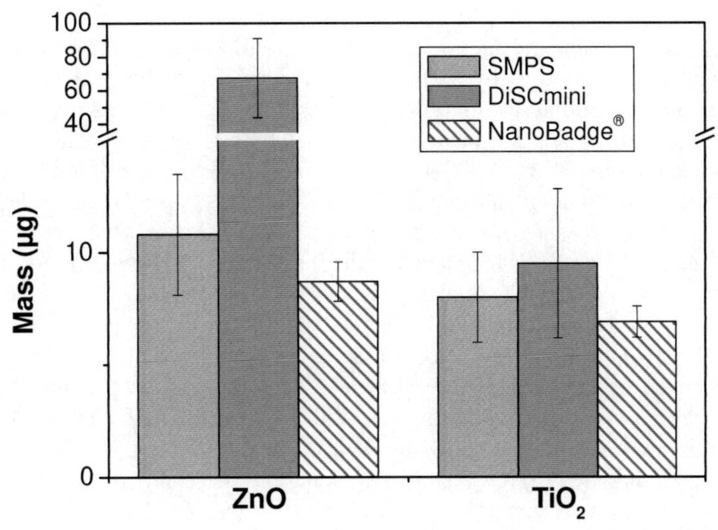

Figure 4. Mass of ZnO and TiO_2 collected on the filters, calculated from the SMPS and the DiSCmini data and measured by XRF for the NANOBADGE sampler. Effective densities of 2.2 and 1.4 were used for ZnO and TiO_2, respectively. *The mass of ZnO calculated from the DiSCmini data was significantly overestimated by using a wrong assumption on the geometric standard deviation of the aerosol. By applying corrections, the mass recalculated from DiSCmini data was 19.7 μg/m^3 and not 67.4 μg/m^3 as shown in the figure. Based on those calculations, the true mass is close to a factor of 2 to the SMPS and NANOBADGE results and not a factor of 7 as it was initially the case.

To further illustrate the validity of the sampler for personal exposure assessments, the LOD for ZnO and TiO_2 were determined. The European Agency for Safety and Health at Work distinguishes long-term and acute exposure, the former being a repeated exposure averaged over working shifts of 8 h and the latter a peak exposure averaged over 15 min [40]. Thus the LOD had been converted to aerosol mass concentrations for a full shift based on the latest recommended exposure levels (REL) of the National Institute for Occupational Safety and Health (NIOSH) [13, 41, 42]. The minimum sampling time required to detect an exposure at or above the REL was also calculated. As shown in Table 2, the limits of detection are much lower than the REL for the two oxides considered in this study. The detection of peak exposure is also possible, since few seconds of sampling at or above the REL are sufficient to exceed the LOD. Since the LOD are several orders of magnitude smaller than the current REL, the NANOBADGE sampler can already accommodate tougher regulation, should the exposure levels be lowered in the future.

Table 2. Comparison between the recommended exposure levels (REL) published by the NIOSH and the limits of detection (LOD) of the NANOBADGE sampler for shift and acute exposure (NIOSH, 2007, 2009).

	REL for ultrafine dust from NIOSH (μg/m^3)	LOD (ng/filter)	LOD (μg/m^3) for 8 h of sampling	Minimum sampling time at the REL
ZnO	5000	30 ± 20 %	0.1 ± 25 %	< 1 min
TiO_2	300	12 ± 25 %	0.04 ± 30%	< 1 min

4. Conclusion

A novel sampler to assess personal exposure to NOAA was developed, based on the collection of particles on filters followed by elemental analysis by XRF. This analysis yields an integrated mass-based exposure value for the chemical elements of interest. In order to assess the validity of the sampler and the associated XRF analysis toolchain, a comparison has been made with stationary and personal instruments, namely a SMPS and a DiSCmini using aerosols in the 20-400 nm size range. The number-based data of the two latter instruments were converted to a mass data using a DMA-ELPI setup. This setup provided an estimation of the effective density of the NOAA present in the test aerosols. In all cases, there was a good agreement between the mass calculated from the SMPS data and the mass measured on the filters. The conversion of the DiSCmini data to mass was reliable on the polydisperse TiO_2 aerosol, for which the particle size distribution had optimal properties for the DiSCmini monitor. In the case of classified aerosols (monodisperse aerosols having a geometric standard deviation far from 1.9) the DiSCmini should not be used, as it is recommended by the manufacturer, unless corrections can be performed using the aerosol specific features. The detection limits of the sampler and the associated XRF analysis toolchain were low enough to reliably detect exposure levels below the REL of the NIOSH and INRS for ultrafine ZnO and TiO_2. We have also shown that the NANOBADGE sampler could be used to detect acute exposure events, which are the most frequent case encountered in an industrial setting. The NANOBADGE sampler is thus well suited for epidemiological studies to establish dose-response relationships with respect to NOAA either for short-term exposure (e.g. 15 min time base measurements) or for covering the full shift exposure.

Future work will involve tests of the sampler on real aerosols in the micro-sized range which are representative of NOAA emitted at workplaces [43] and containing background particles and contaminants. It is expected that the ability to quantify the exposure element-wise will provide a high level of details which cannot be accessed using only number-based aerosol instruments.

5. Supplementary Material

Stability of the NANOBADGE sampling flowrate, uniformity of aerosol concentrations in the tunnel, particle size distributions of the test aerosols, calibration curves of Zn and Ti for quantitative XRF, sample XRF spectra of filters loaded with TiO_2 and ZnO particles. This material is available online free of charge via the journal webpage.

Acknowledgments
We are grateful to Dr. Ana-Maria Todea, Dr. Christof Asbach, Dr. Martin Fierz and Dr. Michel Pourprix for fruitful discussions on aerosol sampling and data analysis. The present work was performed within the frame of the nanoIndEx project, supported by the French National Funding Agency for Research (ANR), the German Federal Ministry of Education and Research (BMBF), the British Technology Strategy Board (TSB) and the Swiss TEMAS AG, under the frame of SIINN, the ERA-NET for a Safe Implementation of Innovative Nanoscience and Nanotechnology. The research leading to these results has also received funding from the European Union's FP7 Grant Agreement n. 310584 (NanoReg project).

References

1. Keller, A A, S McFerran, A Lazareva, and S Suh, Global life cycle releases of engineered nanomaterials. *Journal of Nanoparticle Research*, 2013. **15**(6).
2. Oberdorster, G, E Oberdorster, and J Oberdorster, Nanotoxicology: An emerging discipline evolving from studies of ultrafine particles. *Environmental Health Perspectives*, 2005. **113**(7): p. 823-839.
3. Nel, A, T Xia, L Madler, and N Li, Toxic potential of materials at the nanolevel. *Science*, 2006. **311**(5761): p. 622-627.
4. Limbach, L K, P Wick, P Manser, R N Grass, A Bruinink, and W J Stark, Exposure of engineered nanoparticles to human lung epithelial cells: Influence of chemical composition and catalytic activity on oxidative stress. *Environmental Science & Technology*, 2007. **41**(11): p. 4158-4163.
5. Kreyling, W G, M Semmler-Behnke, and W Moller, Ultrafine particle-lung interactions: Does size matter? *Journal of Aerosol Medicine-Deposition Clearance and Effects in the Lung*, 2006. **19**(1): p. 74-83.
6. Vanwinkle, B A, K L D Bentley, J M Malecki, K K Gunter, I M Evans, A Elder, J N Finkelstein, G Oberdorster, and T E Gunter, Nanoparticle (NP) uptake by type I alveolar epithelial cells and their oxidant stress response. *Nanotoxicology*, 2009. **3**(4): p. 307-318.
7. Seaton, A, L Tran, R Aitken, and K Donaldson, Nanoparticles, human health hazard and regulation. *Journal of the Royal Society Interface*, 2010. **7**: p. S119-S129.
8. Pietroiusti, A and A Magrini, Engineered nanoparticles at the workplace: current knowledge about workers' risk. *Occupational Medicine-Oxford*, 2014. **64**(5): p. 319-330.
9. Ding, Y *et al.*, Airborne engineered nanomaterials in the workplace—a review of release and worker exposure during nanomaterial production and handling processes. *Journal of Hazardous Materials*, 2017(322): p. 17-28.
10. NIOSH, *Approaches to safe nanotechnology: managing the health and safety concerns associated with engineered nanomaterials*, in *Publication No. 2009-125*, N.I.f.O.S.a. Health, Editor. 2009, National Institute for Occupational Safety and Health (NIOSH): Cincinnati: OH.
11. Park, J Y, G Ramachandran, P C Raynor, L E Eberly, and G Olson, Comparing Exposure Zones by Different Exposure Metrics Using Statistical Parameters: Contrast and Precision. *Annals of Occupational Hygiene*, 2010. **54**(7): p. 799-812.
12. Lee, J H, M C Moon, J Y Lee, and I J Yu, Challenges and perspectives of nanoparticle exposure assessment. *Toxicological research*, 2010. **26**(2): p. 95-100.

13. Chazelet, S, J-R Fontaine, S Binet, L Gaté, G Hedelin, S Malard, M Ricaud, D Rousset, and O Witschger, Dioxyde de titane nanométrique: de la nécessité d'une valeur limite d'exposition professionnelle. *Hygiène et sécurité du travail*, 2016. **242**(mars 2016): p. 46-51.
14. Kuhlbusch, T A J, C Asbach, H Fissan, D Gohler, and M Stintz, Nanoparticle exposure at nanotechnology workplaces: A review. *Particle and Fibre Toxicology*, 2011. **8**: p. 18.
15. Zimmermann, E, S Derrough, D Locatelli, C Durand, J L Fromaget, E Lefranc, X Ravanel, and J Garrione, Results of potential exposure assessments during the maintenance and cleanout of deposition equipment. *Journal of Nanoparticle Research*, 2012. **14**(10): p. 17.
16. Cena, L G, T R Anthony, and T M Peters, A Personal Nanoparticle Respiratory Deposition (NRD) Sampler. *Environmental Science & Technology*, 2011. **45**(15): p. 6483-6490.
17. Fierz, M, C Houle, P Steigmeier, and H Burtscher, Design, Calibration, and Field Performance of a Miniature Diffusion Size Classifier. *Aerosol Science and Technology*, 2011. **45**(1): p. 1-10.
18. Tsai, C J, C N Liu, S M Hung, S C Chen, S N Uang, Y S Cheng, and Y Zhou, Novel Active Personal Nanoparticle Sampler for the Exposure Assessment of Nanoparticles in Workplaces. *Environmental Science & Technology*, 2012. **46**(8): p. 4546-4552.
19. R'Mili, B, O L C Le Bihan, C Dutouquet, O Aguerre-Charriol, and E Frejafon, Particle Sampling by TEM Grid Filtration. *Aerosol Science and Technology*, 2013. **47**(7): p. 767-775.
20. Fierz, M, D Meier, P Steigmeier, and H Burtscher, Aerosol Measurement by Induced Currents. *Aerosol Science and Technology*, 2014. **48**(4): p. 350-357.
21. Asbach, C *et al.*, Comparability of Portable Nanoparticle Exposure Monitors. *Annals of Occupational Hygiene*, 2012. **56**(5): p. 606-621.
22. Mills, J B, J H Park, and T M Peters, Comparison of the DiSCmini Aerosol Monitor to a Handheld Condensation Particle Counter and a Scanning Mobility Particle Sizer for Submicrometer Sodium Chloride and Metal Aerosols. *Journal of Occupational and Environmental Hygiene*, 2013. **10**(5): p. 250-258.
23. Bau, S, B Zimmermann, R Payet, and O Witschger, A laboratory study of the performance of the handheld diffusion size classifier (DiSCmini) for various aerosols in the 15-400 nm range. *Environmental Science-Processes & Impacts*, 2015. **17**(2): p. 261-269.
24. Todea, A M, S Beckmann, H Kaminski, and C Asbach, Accuracy of electrical aerosol sensors measuring lung deposited surface area concentrations. *Journal of Aerosol Science*, 2015. **89**: p. 96-109.
25. Standard, *BS EN 1540:2011 Workplace exposure. Terminology*. 2012.
26. Asbach, C, S Clavaguera, and A M Todea, *Measurement Methods for Nanoparticles in Indoor and Outdoor Air*, in *Indoor and Outdoor Nanoparticles: Determinants of Release and Exposure Scenarios*, M. Viana, Editor. 2016, Springer International Publishing: Cham. p. 19-49.
27. Asbach, C *et al.*, Comparison of four mobility particle sizers with different time resolution for stationary exposure measurements. *Journal of Nanoparticle Research*, 2009. **11**(7): p. 1593-1609.
28. Todea, A M *et al.*, Inter-comparison of personal monitors for nanoparticles exposure at workplaces and in the environment. *Submited to Science of the Total Environment*, 2016.
29. Andermann, G and J W Kemp, Scattered X-rays as internal standards in X-ray emission spectroscopy. *Analytical Chemistry*, 1958. **30**(8): p. 1306-1309.
30. Reynolds, R C, Matrix corrections in trace element analysis by X-ray fluorescence - Estimation of mass absorption coefficient by compton scattering. *American Mineralogist*, 1963. **48**(9-1): p. 1133-&.
31. Giauque, R D, R B Garrett, and L Y Goda, Determination of trace-elements in light-element matrices by X-ray fluorescence spectrometry with incoherent scattered radiation as an internal standard. *Analytical Chemistry*, 1979. **51**(4): p. 511-516.

IOP Conf. Series: Journal of Physics: Conf. Series **838** (2017) 012006 doi:10.1088/1742-6596/838/1/012006

32. Motellier, S, K Lhaute, A Guiot, L Golanski, C Geoffroy, and F Tardif. *Direct quantification of airborne nanoparticles composition by TXRF after collection on filters.* in *Journal of Physics: Conference Series.* 2011. IOP Publishing.
33. Rousseau, R M, Detection limit and estimate of uncertainty of analytical XRF results. *The Rigaku Journal*, 2001. **18**(2): p. 33-47.
34. Trakumas, S and E Salter, *Parallel particle impactor - novel size-selective particle sampler for accurate fractioning of inhalable particles*, in *Inhaled Particles X*, L. Kenny, Editor. 2009, Iop Publishing Ltd: Bristol.
35. Kelly, W P and P H McMurry, Measurement of particle density by inertial classification of differential mobility analyzer generated monodispersed aerosols. *Aerosol Science and Technology*, 1992. **17**(3): p. 199-212.
36. Maricq, M M, D H Podsiadlik, and R E Chase, Size distributions of motor vehicle exhaust PM: A comparison between ELPI and SMPS measurements. *Aerosol Science and Technology*, 2000. **33**(3): p. 239-260.
37. Hinds, W C, *Aerosol technology: properties, behavior, and measurement of airborne particles.* 2012: John Wiley & Sons.
38. Fierz, M. *miniDiSC application note #9: mass and surface area calculation.* 2016 [cited 2016 21.01.2016]; Available from: http://fierz.ch/minidisc/pdf/miniDiSCApplicationNote9.pdf
39. Fierz, M. *miniDiSC application note #11: Errors due to the size distribution shape.* 2016 [cited 2016 21.01.2016]; Available from: http://fierz.ch/minidisc/pdf/miniDiSCApplicationNote11.pdf.
40. ECHA, *Guidance on information requirements and chemical safety assessment Chapter R.8: Characterisation of dose [concentration]-response for human health.* 2012, European Chemical Agency. p. 186.
41. NIOSH, *NIOSH Pocket Guide to Chemical Hazards.* 2007.
42. NIOSH, *Current Intelligence Bulletin 63: Occupational Exposure to Titanium Dioxide.* 2011. p. 120.
43. Tsai, C-J, C-Y Huang, S-C Chen, C-E Ho, C-H Huang, C-W Chen, C-P Chang, S-J Tsai, and M J Ellenbecker, Exposure assessment of nano-sized and respirable particles at different workplaces. *Journal of Nanoparticle Research*, 2011. **13**(9): p. 4161-4172.

Towards an optimal adaptation of exposure to NOAA assessment methodology in Multi-Source Industrial Scenarios (MSIS): the challenges and the decision-making process

JM López de Ipiña[1], C Vaquero[1], C Gutierrez-Cañas[2]

[1]TECNALIA – Industry and Transport Division, Parque Tecnológico de Alava, 01510 Miñano/Spain
[2]Universidad del Pais Vasco – Euskal Herriko Unibertsitatea, Faculty of Engineers, Chem. and Env. Eng. Department, Alameda de Urquijo s/n, 48013 Bilbao/Spain

E-mail: jesus.lopezdeipina@tecnalia.com

Abstract. It is expected a progressive increase of the industrial processes that manufacture of intermediate (iNEPs) and end products incorporating ENMs (eNEPs) to bring about improved properties. Therefore, the assessment of occupational exposure to airborne NOAA will migrate, from the simple and well-controlled exposure scenarios in research laboratories and ENMs production plants using innovative production technologies, to much more complex exposure scenarios located around processes of manufacture of eNEPs that, in many cases, will be modified conventional production processes. Here will be discussed some of the typical challenging situations in the process of risk assessment of inhalation exposure to NOAA in Multi-Source Industrial Scenarios (MSIS), from the basis of the lessons learned when confronted to those scenarios in the frame of some European and Spanish research projects.

1. Motivation and context

According to the "Project on Emerging Nanotechnologies" inventory [1] [2], there are currently about 1830 final nano-products on the market. Lux Research Inc. [3] estimated that sales of end products incorporating nanotechnology (end nano-enabled producs, eNEPs) grew by 90% over the period 2012-2014, from $850 billion to $1.6 trillion. In the same period, sales of intermediary products incorporating nanotechnology (intermediate nano-enabled products, iNEPs) also grew by 170%, reaching around $360 billion in 2014. The global value of manufactured nanomaterials (ENMs) between 2012 and 2014 increased 35% from 2012 to reach $ 2.12 billion in 2014. As a result, total nanotechnology sales in 2014 would be close to $2 trillion. Other projections by the same emerging technologies consulting firm [4] estimate that the global market for nanotechnology products will reach $4.4 trillion in 2018. But sales of "*true nanotechnology products*" (the sum of ENMs and iENPs) only accounted for 18% of the global nanotechnology market in 2014. In this sense, it is clear that the value of these products manufactured with breakthrough processes is little relevant compared to consumer finished goods incorporating nanotechnology.

Although market estimates may vary substantially depending of the source consulted, all sources point to a relevant and sustained growth of the nanotechnology market in the coming years, but less optimistic than expected at the beginning of the global recession period. Research and Markets [5], expects the global nanotechnology market to grow at a CAGR of about 17.5% during 2016-2022.

These projections are consistent with the longitudinal evolution of nanotechnology that is reflected in the global increase of patents and scientific publications [6].

With respect to the potentially exposed workforce, some projections estimate between 2 and 6 million in 2015 and 2020 the number of workers involved in some domain of nanotechnology (800,000 and 2 million respectively in the United States) [7]. PEROSH [8] - based on these same estimates - amounts to a workforce in Europe of 400,000 workers in 2015.

All these data suggest in the short-medium term a progressive increase of the industrial processes that use ENMs, but mainly of those processes of manufacture of end products that incorporate ENMs to provide improved properties (eENPs). It is therefore expected that the problem of assessing occupational exposure to airborne NOAA will migrate, from the simple and well-controlled exposure scenarios in research laboratories and ENMs production plants using innovative production technologies, to much more complex exposure scenarios located around processes of manufacture of eENPs that, in many cases, will be modified conventional production processes. A significant increase in exposure scenarios related to the production of iNEPs is also expected, either using new production technologies or modifying existing ones (Table 1). A third relevant group of occupational exposure scenarios to consider in the life cycle of ENMs will be linked to the progressive increase in the professional use of all these products in the different industrial and service sectors.

Manufacturing process	**Nanotechnology product**		
	Engineered nanomaterial (ENM)	Intermediate nano-enabled product (iNEP)	End nano-enabled product (eNEP)
New	+	++	++
Existing		+	+++
Example	*MWCNT*	*MWCNT-based composite*	*Automotive component*

Table 1. The nanotechnology value chain and the potential increase in number of occupational exposure scenarios to airborne NOAA associated with existing and new manufacturing processes [(+) Relevant increase, (++) Very relevant increase, (+++) Highly relevant increase].

In this industrial context, the expected proliferation of Multi-Source Industrial Scenarios (MSIS) [9], characterized by spatially complex distributions of aerosol sources as well as for potential differences in dynamics due to the feasibility of multi-task configuration at a given time, will increase the complexity of the occupational risk assessment to airborne NOAA. In addition, in these real-life complex scenarios uncertainty can significantly increase, mainly due to the lack of an effective distinction of a dynamic background aerosol [9,10,11].

The assessment of risks related to airborne exposure to chemical agents at work by comparison with limit values, is a well-established process regulated by OHS European Directives (e.g. Directives 89/391/EC, 98/24/EC, 2004/37/EC) and national legislations transposing the previous ones. The availability of a suitable instrumentation, well-standardized procedures and regulated limit values, makes risk assessment a standardized routine tool in the practice of industrial hygiene. In addition, this methodology provides security and confidence in compliance with legislation, both for the employer, the worker and its representatives, as well as for the competent public administration responsible for regulation.

For airborne NOAA - chemical agents for all purposes-, the already enforced occupational regulatory framework should be also routinely applied to these particular situations by industrial hygienists. Despite the lack of regulated limit values, the use of recommended and benchmark values [12,13] allows the implementation of this methodology for the risk management of the exposure to airborne NOAA in companies. However, despite the significant research efforts in recent years – measurement strategies and new tiered approach [14,15,16,17,18,19], portable and personal instrumentation [12,20,21,22], proposed NRVs/OELs [12,13,23,24] - uncertainties in the risk assessment to NOAA (inhalation) in complex scenarios still affect significantly the robustness of the results, thus limiting the wide application of existing methods and strategies by industrial hygienists.

Among the set of different metrics for the assessment of exposure to airborne NOAA, particle number concentration (PNC) is a simple and useful parameter for industrial hygienists, because its simplicity of measurement through direct-reading portable and personal instrumentation [12,25], as well as the availability of mature sampling procedures [14,17] and recommended reference values [13] for the comparison with measurements, then allowing a quantitative approach to the hygienic risk assessment. However main limitation of PNC is the lack of selectivity, so depending of the airborne NOAA assessed and the exposure scenario, it will be distinguishable from the background.

The use of PNC in the assessment of low-medium complexity exposure scenarios (e.g. research laboratories, ENMs manufacturing plants) following the well-established tiered approaches has been well demonstrated and documented [25,26,27,28,29,30,31,32,33]. However, the use of this metric in MSIS can lead to erroneous results in the assessment of exposure due to unforeseen influences of simultaneously running processes. In addition, as reviewed in [15] the choice of the instrument, if possible, determines the measurement strategy and the final accuracy of the data.

In this work some examples of typical challenging situations using PNC for risk assessment of occupational exposure to airborne NOAA will be presented, from the basis of the lessons learned when confronted to those complex industrial scenarios, in the frame of some European and Spanish research projects [24,34].

2. Results on PNC temporal records

Here are presented four cases, very common in MSIS, leading to disambiguation problems of the PNC temporal record: the dynamic evolution of the background signal, the spatial differences of this signal, the instrument choice and the correct equivalent dimension.

2.1 Temporal background signal masking the potential contribution of NOAA release from the raw series of PNC.

Figure 1 shows raw PNC signals captured in source (blue) and background (red) by a CPC TSI 3075 and a portable CPC TSI 3007, located at a distance of four meters from each other, inside a maintenance room (6 x 4 x 3 m), during the measurement of occupational exposure to nanoTiO_2 (nTiO_2)in two working tasks: 1) the central drilling of cylindrical nTiO_2 enabled steel tablets (task T3) and its subsequent packaging with aluminum paper in packs of five (task T4) (see section 3 in this paper) [9,34].

It is observed that the background signal is larger than the measured in source; the maximum amplitude of the background signal is around 42,000 particles/cm^3. Both signals have the same shape and evolve in parallel with an offset of about 2,000-5,000 particles/cm^3. Their amplitude is greater before starting the working tasks and none of them seems influenced by potential NOAA releases during tasks development (tasks 3 and 4).

Temporal background signal overrides the potential contribution of NOAA release during mechanical processing. In such a situation it is impossible to evaluate the occupational exposure to nTiO_2 using only PNC.

2.2 Significant disagreement between signals of the background aerosol (PNC) measured in two positions situated very close to each other.

In the reaction stage of a ENMs manufacturing plant by Flame Spray Pyrolysis (reactor closed), significant disagreements were also found between signals of the background aerosol (PNC) measured simultaneously with similar instrumentation (ELPI, ELPI+) in two positions situated close to each other (7 m) [35].

The two raw PNC signals presented different shapes with uncorrelated fluctuations and an offset between the crests of both signals greater than 50,000 particles/cm^3. Before starting the manufacturing process the offset between the background aerosols measured in both positions was about 20,000 particles/cm^3. This situation alerts on the value of the background selected to calculate the net-exposure or emission PNC by subtraction of the background PNC value of the raw measured PNC.

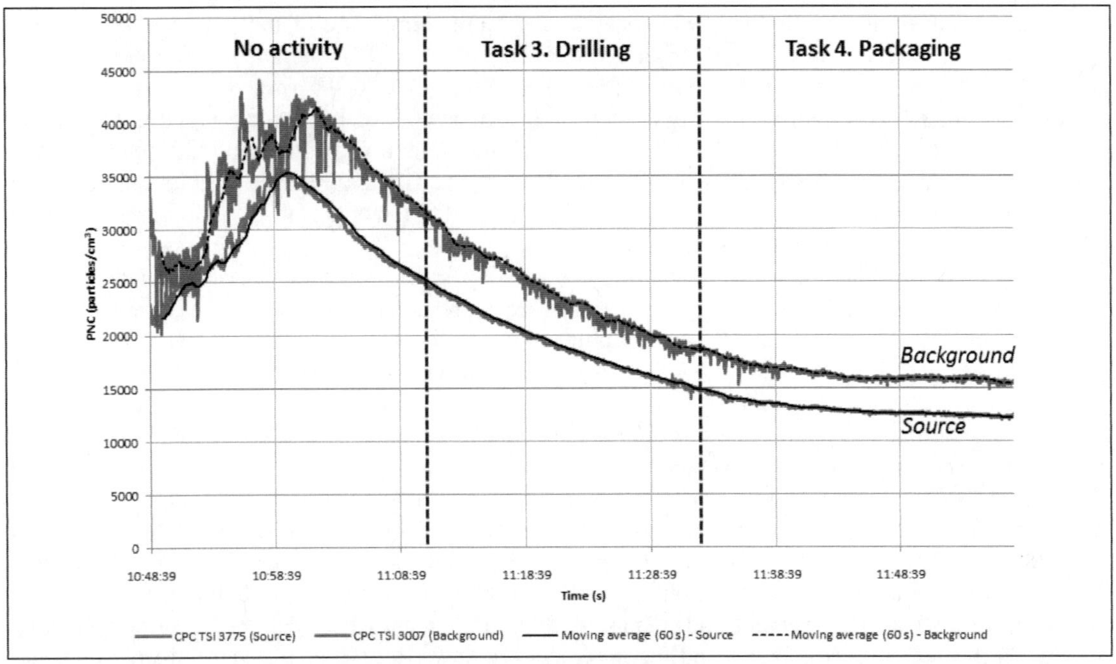

Figure 1. Temporal background signal masking the potential contribution of NOAA release from the raw series of PNC (See the explanation in the text).

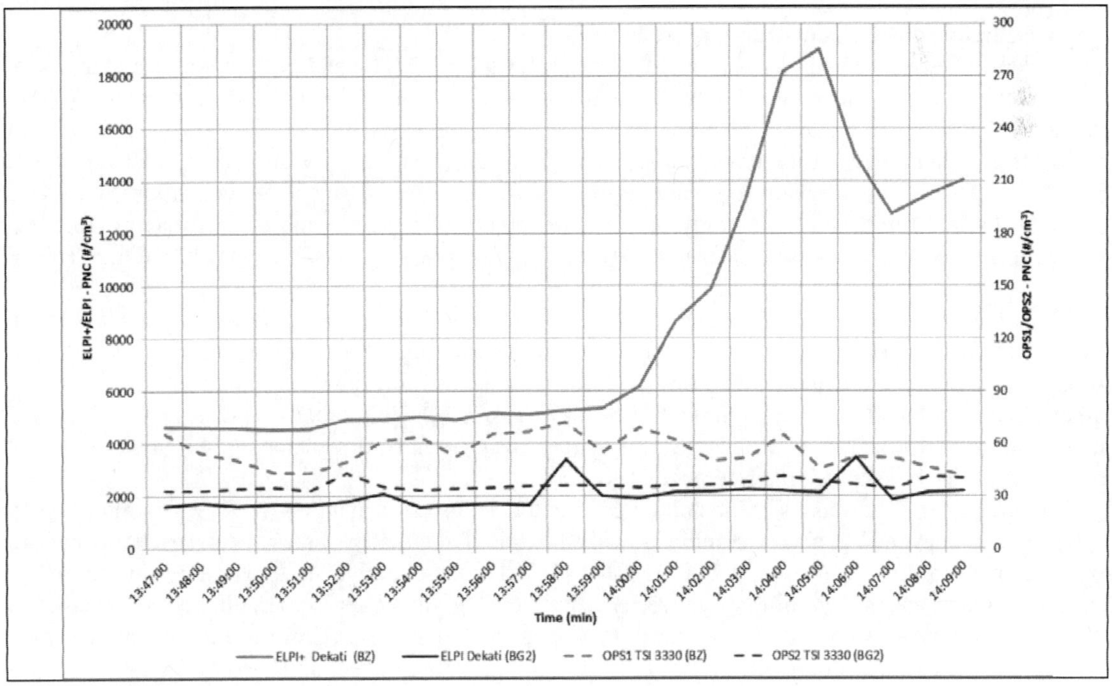

Figure 2. Discrepancies in PNC records from collocated instruments (See the explanation in the text).

2.3 Discrepancies in PNC records from collocated instruments. Instrument choice.

Figure 2, registered during the exposure assessment in a task of packaging nano-SiO_2 cans under an extraction hood, shows raw signals captured by four DRI, deployed in pairs (ELPI + OPS), in two measurement positions: the first one was located at the breathing zone of the worker (BZ) and the second one at a fixed point just 2 m from BZ (BG2). The four DRI operate in the following size ranges: ELPI (7 nm - 10 μm), ELPI+ (6 nm - 10 μm) and OPS1/2 (0.3 - 10 μm).

Analysis of PNC records revealed significant discrepancies between PNC measured at the BZ (in red) and at the fix position close to the BZ (in blue). At the same time, significant discrepancies were also observed among instrumentation measuring PNC at the BZ and sharing a common sampling point. For instance, the event observed by the ELPI+ at the BZ is neither observed by the collocated OPS nor at the close-background position (BG2).

This alerts on the importance of measuring worker exposures in the BZ, in accordance with the well-established industrial hygiene practices, for truly representative exposure measurements. The extrapolation of measured values at a fixed point close to the worker as if they were true values measured in the BZ can lead to significant errors in the assessment of the exposure to NOAA.

2.4 Aerosol characteristics retrieval (PNC) from simultaneously by measured metrics, such as aerodynamic and mobility equivalent diameters.

The analysis of the signals from a case study that deployed a multi-instrumental approach for the measurement of occupational exposure to nanoSiO_2 (ELPI +, CPC TSI 3007, SMPS NanoScan TSI 3910, CPC TSI 3775, NSAM TSI Aerotrak 9000, OPS TSI 3330, among others), showed that although the temporal evolution of the simultaneous signals captured in a fixed position by this instrumentation presented a satisfactory agreement and a similar shape, the signal offset between instruments using different measurement technologies and measurement windows (e.g. ELPI + - aerodynamic diameter and SMPS NanoScan TSI 3910 - mobility diameter) could exceed 50% of the measured value.

This alerts on the selection of the measuring technology and instrument used to provide the PNC exposure value.

3. Discussion: the decision making process

Pitfalls described in previous section can introduce relevant levels of uncertainty in the risk assessment process of exposure to airborne NOAA, by comparing PNC measured values with recommended limit values.

Table 2 summarizes a multi-risk approach (both quantitative - with several metrics - and qualitative) followed in the decision-making process on the risk assessment of occupational exposure to nTiO_2 by inhalation, in a case study of the manufacture of a pilot batch of perforated cylindrical tablets (diameter 90 mm) for the steel industry, by cold compressing of nTiO_2 (AEROXIDE® TiO_2 P 25) [9,34].

Task 1 (Weighting TiO_2) and task 2 (Cold compressing TiO_2) were developed in a first MSIS (production hall) where the process evaluated coexisted with other routine manufacturing processes of products for casting, obtained by mixing and compacting metallic powders, among which alloying tablets and mini tablets for the aluminum industry. Task 3 (Drilling tablets) and task 4 (Packaging tablets) were performed in a second MSIS, a maintenance room of dimensions 6x4x3 m without any additional activity except for the referenced tasks.

The quantitative risk assessment approach directly deployed a quantitative Tier 3 strategy [12]. This quantitative approach was combined with a parallel qualitative assessment based on control banding methodology according to ISO 12901-2 [36]. The quantitative strategy was mainly based in the continuous measurement of PNC, corrected with the background aerosol simultaneously measured. The instrumentation deployed for this purpose was an advanced direct-reading instrumentation rack for aerosol source measurement (ELPI +, CPC TSI 3775, among other) and a portable CPC TSI 3007 for background measurement (Figure 3). In addition, personal samples were collected at the workers breathing zone (BZ) for the off-line analysis of the respirable fraction (SEM, ICP-MS, EDX) (Figure 4).

For the decision-making process and the comparison of exposure measurements with the selected occupational limit values, a NRV of 40,000 particles/cm^3 for PNC (8 hours TWA concentration) [13] and a NIOSH OEL of 0.3 mg/m^3 for $nTiO_2$ mass concentration (TWA concentration for up to 10 hours per day) [23], were selected as reference values.

Given the uncertainty about the risk level of exposure to $nTiO_2$ in the two selected scenarios and to prevent any potential damage to health of monitored workers, during working tasks they were appropriately protected by PPE, such as Tyvek® suit, closed safety glasses, FFP3 filter mask and double nitrile glove.

Figure 3. Rack of DRI located between weighing (right) and cold compressing (left) stations.

Figure 4. Picture taken during a break period in the manufacturing activities, showing a worker carrying two personal samplers (cyclones) for the capture of the aerosol respirable fraction, addressed to off-line analysis (gravimetric, chemical and microscopic).

Quantitative risk approach

Two metrics were used for comparison with limit values: PNC and mass of nTiO2. In the first case the high background camouflaged any potential NOAA emission and the resulting PNC signals were an indistinguishable combination between background and true NOAA releases. Therefore PNC was not adequate to make a decision on the level of occupational exposure risk.

Using the second quantitative approach [23], although the gravimetry was below the limit of detection (not adequate), the chemical analysis of filters by ICP-MS allowed to determine the airborne concentrations and translate them into exposure values to $nTiO_2$ weighted in time, for their subsequent comparison with the reference limit value (0.3 mg/m^3). The calculated values were between 0.01 and 0.07 mg/m^3, representing a SI (Substance Index, EN 689 [37]) in the range <0.1 to 0.2.

These data suggested that the limit value was not exceeded in any of the assumptions evaluated, although according to EN689, additional measurements would be necessary to guarantee the statistical robustness of the results, something in our case was not feasible, since it was a single case study, of limited duration and production, and not of routine operations of the company (exposure scenarios to $nTiO_2$ were operative for 2 hours).

The SEM analysis of filters provided additional evidence of the presence in the breathing zone (BZ) of $nTiO_2$ aggregates in the 1-10 μm range. However, the presence of NOAA in the BZ does not justify itself that the exposure limit value has been exceeded.

Qualitative risk approach

The hazard band (HB) selected in all cases is HB = C (Moderate hazard). The exposure band (EB) assigned in all evaluated tasks is EB = 3. Consequently the control band category (CB) to prevent exposure to NOAA is CB = 3 in all cases. According to ISO 12901-2 the decision for risk management control in a proactive manner would be the implementation of enclosed ventilation in the process (ventilated booth, fume hood). In this case, the qualitative approach overestimated the level of risk.

Risk assessment	Qualitative risk assessment			Quantitative risk assessment							
Working tasks	Control banding (ISO 12901-2)			Comparison with limit values: NRV (2012) and NIOSH (2011)							
				DRI		Off-line analysis					
	HB	EB	CB	PNC	NRV #/cm^3	G (1)	ICP-MS (2)	SEM	Exp.$_{8h}$ mg/m^3	OEL mg/m^3	SI
1.Weighting TiO_2	C	3	3	-	40,000	-	+	+	0.01-0.07	0.3	<0.1-0.2
2.Cold pressing TiO_2	C	3	3	-		-	+	+			
3.Drilling tablets	C	3	3	-		-	+	+			
4. Packaging tablets	C	3	3	-		-	+	+			

(1) NIOSH 0600: All values below the detection limit.
(2) NIOSH 7300 modified (ICP-MS).
Abbreviations: HB=Hazard Band, EB=Exposure Band, CB=Control Band, DRI=Direct Reading Instrumentation, PNC=Particle Number Concentration; NRV=Nano-reference Value (SER 2012), G=Gravimetry, ICP-MS=Inductively Coupled Plasma Mass Spectrometry, SEM=Scanning Electron Microscope, OEL=Occupational Exposure Level (NIOSH 2011), SI=Substance Index, EN 689.

Table 2. Decision-making on the risk assessment of occupational exposure to airborne NOAA ($nTiO_2$), in a case study of the manufacture of a pilot batch of perforated tablets for the steel industry, by cold compressing of $nTiO_2$ [(-) Not suitable method, (+) Suitable method].

4. To summarize

With the progressive introduction of ENMs into the industry, exposure scenarios are evolving from well-controlled R&D laboratories and ENMs production plants to industrial processes that incorporate ENMs for the production of intermediate and final NEPs.

Complex exposure scenarios (MSIS) associated with these processes – which will also coexist with conventional non-nanotechnology processes - will be numerous in near future making the assessment of occupational exposure to NOAA more complex and probably expensive for industrial hygienists.

The significant development achieved during recent years in the field of occupational exposure assessment to NOAA (availability of direct-reading instrumentation (DRI) - portable and personal -, tiered approaches, recommended NRVs/OELs), allows a widespread use of new methods and instruments in industrial scenarios. However the complexity of some workplaces highlighted the next step directions in research:

- The determination of the most reliable strategy for the background assessment
- The identification/development of cost effective and chemical selective methods
- The use of DRI as the core block of the engineering controls

References

[1] PEN (2016) *Project on Emerging Nanotechnologies.* (http://www.nanotechproject.org/inventories/consumer/)

[2] Vance ME, Kuiken T, Vejerano EP, McGinnis SP, Hochella MFJr, Rejeski D and Hull MS (2015) Nanotechnology in the real world: redeveloping the nanomaterial consumer products inventory. *Beilstein J. Nanotechnol.* 2015, 6, 1769–1780.

[3] Lux Research (2013) *Nanotechnology update: corporations up their spending as revenues for nano-enabled products increase* (https://www.nsf.gov/crssprgm/nano/reports/LUX14-0214_Nanotechnology%20StudyMarketResearch%20Final%2017p.pdf)

[4] Lux Research (2015) *Nanotechnology update: U.S. leads in government spending amidst increased spending across Asia.* (http://www.luxresearchinc.com/sites/default/files/AM_Nanotechnology_KTA_12_15.pdf)

[5] Research and Markets (2015) *Global Nanotechnology Market Outlook 2022.* (http://www.researchandmarkets.com/reports/3512791/global-nanotechnology-market-outlook-2022).

[6] Chen H, Roco MC, Son J, Jiang S, Larson CA, and Gao Q (2013) Global nanotechnology development from 1991 to 2012: patents, scientific publications, and effect of NSF funding. *J Nanopart Res* (2013) 15:1951.

[7] Roco MC (2011) The long view of nanotechnology development: the National Nanotechnology Initiative at 10 years. *J Nanopart Res* (2011) 13: 427–445.

[8] PEROSH (2012) *Sustainable workplaces of the future – European Research Challenges for occupational safety and health.* Perosh, Brussels, 35 pp. (http://www.perosh.eu/wp-content/uploads/2013/05/Perosh-Research-Challenges_lowres.pdf)

[9] Lopez de Ipiña JM, Vaquero C, Gutierrez-Cañas C and Pui DHY (2015) Analysis of multivariate stochastic signals sampled by on-line particle analyzers: Application to the quantitative assessment of occupational exposure to NOAA in multisource industrial scenarios (MSIS) *J. Phys. Conf. Ser.* Vol. 617.

[10] Ono-Ogasawara M, Serita F and Takaya M (2009) Distinguishing nanomaterial particles from background airborne particulate matter for quantitative exposure assessment. *J Nanopart Res* 2009, 11, 1651-1659.

[11] Peters TM, Elzey S, Johnson R, Park H, Grassian VH, Maher T, and O'Shaughnessy P (2009) Airborne monitoring to distinguish engineered nanomaterials from incidental particles for environmental health and safety. *J. Occup. Environ. Hyg.* 6 (2) (2009) 73–81.

[12] ISO/TS 12901-1:2012 *Nanotechnologies -- Occupational risk management applied to engineered nanomaterials -- Part 1: Principles and approaches.*

[13] SER (2012) *Provisional nano reference values for engineered nanomaterials.* The Hague: Social Economic Council Netherlands, 15 pp.

[14] Asbach C, T K, Kaminski H, Stahlmecke B, Plitzko S, Götz U, Voetz M, Kiesling H and Dahmann D (2012) *Standard operation procedures for assessing exposure to nanomaterials, following a tiered approach.* Nano GEM, 90 pp. (http://www.nanogem.de/cms/nanogem/upload/Veroeffentlichungen/nanoGEM_SOPs_Tiered_Approach.pdf)

[15] Asbach C (2015) *Exposure Measurement at Workplaces.* In Nanoengineering, Elsevier (2015), 523-555.

[16] ISO/TR 27628:2007 *Workplace atmospheres—Ultrafine nanoparticle and nano-structured aerosols—Inhalation exposure characterization and assessment.*

[17] OECD (2015) Harmonized tiered approach to measure and assess the potential exposure to airborne emissions of engineered nano-objects and their agglomerates and aggregates at workplaces. *Series on the Safety of Manufactured Nanomaterials* No. 55, ENV/JM/MONO(2015)19. (https://nanotechnology.americanchemistry.com/Nanotechnology/Resources/OECD-Harmonized-Tiered-Approach.pdf)

[18] Ramachandran G, Ostraat M, Evans DE, Methner MM, O'Shaughnessy P, D'Arcy J, et al. (2011)

A strategy for assessing workplace exposures to nanomaterials. *Journal of occupational and environmental hygiene* 2011; 8(11):673–685.
[19] Schubauer-Berigan MK, Dahm MM, Schulte PA., Hodson L and Geraci CL (2016) Refinement of the Nanoparticle Emission Assessment Technique into the Nanomaterial Exposure Assessment Technique (NEAT 2.0). *Journal of Occupational and Environmental Hygiene*, Volume 13, 2016 - Issue 9, pp 708-717.
[20] Asbach C, Meyer-Plath A, Clavaguera S, Fierz M, Dahmann D, MacCalman L, Alexander C, Todea AM and Iavicoli I (2016) *Assessment of Personal Exposure to Airborne Nanomaterials*. A Guidance Document. Nanoindex project document, 47 pp.
(http://www.nanoindex.eu/wp-content/uploads/2016/06/Nano_Brosch%C3%BCre.pdf)
[21] Kaminskia H, Kuhlbuscha TAJ, Rathb S, Götzb U, Sprengerb M, Welsb D, Polloczekb J, Bachmannc V, Dziurowitzc N, Kieslingd HJ, Schwiegelshohnd A, Monze C, Dahmanne D and Asbach C (2013) Comparability of mobility particle sizers and diffusion chargers. *J. Aerosol Sci.* Volume 57, March 2013, Pages 156–178.
[22] Price HD, Stahlmecke B, Arthur R, Kaminski H, Lindermann J, Däuber E, Asbach C, Kuhlbusch TAJ, BéruBé KA, Jones T P (2014) Comparison of instruments for particle number size distribution measurements in air quality monitoring, *J. Aerosol Sci.* Vol. 76 (2014), 48-55.
[23] NIOSH (2011) *Occupational Exposure to Titanium Dioxide*. Current Intelligence Bulletin 63, 119 pp.
(https://www.cdc.gov/niosh/docs/2011-160/pdfs/2011-160.pdf)
[24] López de Ipiña JM, Vaquero C, Boutry D, Damlencourt JF, Neofytou P, Pilou M, Jankowska E, Larraza I, Pina R, Fernández S (2015) Strategies, methods and tools for managing nanorisks in construction. *J. Phys. Conf. Ser. Vol. 617.*
[25] Clark K, Van Tongeren M, Christensen FM, Brouwer D, Nowack B, Gottschalk F, Micheletti C, Schmid K, Gerritsen R, Aitken R,Vaquero C, Gkanis V, Housiadas C, López de Ipiña JM and Riediker M (2012) Limitations and information needs for engineered nanomaterial-specific exposure estimation and scenarios: recommendations for improved reporting practices. *J. Nanopart. Res.* 14 (9) (2012) 1–14.
[26] Bekker C, Kuijpers E, Brouwer DH, Vermeulen, R and Fransman, W (2015) Occupational exposure to nano-objects and their agglomerates and aggregates across various life cycle stages; a broad-scale exposure study. *Ann. Occup. Hyg.* (2015), 1-24.
[27] Brouwer D (2010) Exposure to manufactured nanoparticles in different workplaces. *Toxicology* 269 (2–3) (2010) 120–127.
[28] Ding Y, Kuhlbusch TAJ, Van Tongeren M, Sánchez Jiménez A, Tuinman I, Chen R, Larraza Alvarez I, Mikolajczyk U, Nickel C, Meyer J, Kaminski H, Wohlleben W, Stahlmecke B, Clavaguera S and Riediker M (2017) Airborne engineered nanomaterials in the workplace—a review of release and worker exposure during nanomaterial production and handling processes. *Journal of Hazardous Materials* 322 (2017) 17–28.
[29] Kuhlbusch TAJ, Asbach C, Fissan H, Göhler D and Stintz M (2011) Nanoparticle exposure at nanotechnology workplaces: A review. *Part Fibre Toxicol* 2011, 8:22.
[30] Pietroiusti A and Magrini A (2014) Engineered nanoparticles at the workplace: current knowledge about workers' risk. *Occupational Medicine* 2014;64:319–330.
[31] Plitzko S (2009) Workplace exposure to engineered nanoparticles. *Inhal. Toxicol.* 21 (Suppl 1) (2009) 25–29.
[32] Safe Work Australia (2012) *Measurements of particle emissions from nanotechnology processes, with assessment of measuring techniques and workplace controls*. Safe Work Australia, 140 pp.
(http://www.safeworkaustralia.gov.au/sites/SWA/about/Publications/Documents/714/Measurements_Particle_Emissions_Nanotechnology_Processes.pdf)
[33] Tsai CJ, Huang C-Y, Chen S-C, Ho C-E, Huang C-H, Chen C-W, Chang C-P, Tsai S-J and Ellenbecker MJ (2011) Exposure assessment of nano-sized and respirable particles at different workplaces. *J. Nanopart. Res.* 13 (9) (2011) 4161–4172.

[34] Vaquero C, Gutierrez-Cañas C, Galarza N and Lopez de Ipiña JM Exposure assessment to engineered nanoparticles handled in industrial workplaces: The case of alloying nano-TiO_2 in new steel formulations. *J. Aerosol Sci.* Vol. 102 (2016), 1–15.
[35] Scaffold (2015) *Implementation of the RMM in IUC1*. Scaffold FP7 project, deliverable D6.1, 52 pp.
[36] ISO/TS 12901-2:2014 *Nanotechnologies -- Occupational risk management applied to engineered nanomaterials -- Part 2: Use of the control banding approach.*
[37] EN 689:1996 *Workplace atmospheres. Guidance for the assessment of exposure by inhalation to chemical agents for comparison with limit values and measurement strategy.*

Acknowledgments
This paper exploits the results of the research developed by projects SCAFFOLD and EHS Advance. Project SCAFFOLD received funding from the European Union's FP7 research and innovation programme, under grant agreement No. 280535. This paper reflects only the authors' views and the Commission is not responsible for any use that may be made of the information contained therein. Project EHS Advance received funding from the Basque Government research programme (Etortek). The authors would also like to express their grateful to companies BOSTLAN (Bizkaia, Spain) and TECNAN (Navarra, Spain) for the support provided to carry out the exposure measurements in their respective facilities.

Exposure assessment method for products containing nanomaterials using a gas sample introduction system for ICP-MS

Yasuto Matsui [1], Nobuyuki Kato [2], Kohe Nishiguchi [3], Minoru Yoneda [4]

(1) Safety Research Center, Kyoto University, Rakuyu Bekkan, Yoshida, Sakyo, Kyoto, 606-8316 Japan (2) J-SCIENCE LAB co., ltd., 3-1 Kamitoba, Minami, Kyoto, 601-8144 Japan (3) National Institute of Occupational Safety and Health, 6-21-1, Tama, Kawasakai, 214-8585 Japan (4) Graduate School of Engineering, Kyoto University, Katsura, Nishigyo, Kyoto, 615-8540 Japan

Abstract

General aerosol-measuring instruments allow real-time measurements of air particle concentrations. However, these measurements cannot distinguish free particles from target nanomaterials because they do not differentiate nanomaterials. The purpose of this study is investigation of the quantitative nature of atmospheric nanoparticles using GED (Gas Exchange Device)-ICP-MS to detect and measure nanoparticles as an element. The per particle signal intensity increased proportionally to the volume until the particle size reaches 120 nm. For all particle sizes from 20 nm to 160nm, the measured values of FMPS (Fast Mobility Particle Sizer) were consistently higher than those for ICP-MS. The system will be able to adapt to an exposure assessment of CNT (Carbon Nanotube) because carbon-base materials can be identified and quantified as long as an index element can be found.

1. Introduction

Nanomaterials in the air are usually measured using aerosol-measuring instruments or by collecting particulates in filters and impactors. ISO and NIOSH in 2008 and 2009, respectively, published reports on measurement methods of nanomaterials in work environments [1, 2]. They suggest using Condensation Particle Counters (CPC) and Optical Particle Counters (OPC) for the preliminary evaluation of nanomaterials in a work environment. These aerosol-measuring instruments allow real-time measurements of air particle concentrations in a matter of seconds to minutes. However, these measurements cannot distinguish free particles from target nanomaterials because they do not differentiate nanomaterials.

The New Energy and Industrial Technology Development Organization (NEDO) has provided guidance on measurement methods for nanomaterials such as carbon nanotubes (CNT). These methods include instrumental measurements, quantification using filter-collected materials, and observations using an electron microscope. NIOSH recommends the aerosol carbon analysis method (NIOSH 5040) to quantify filter-collected materials [3], and has reported practical measurement examples [4, 5]. Another study quantified nanomaterials using contaminating metal catalysts in CNTs [6]. All of these examples evaluated CNT by weight concentration. On the other hand, other studies have estimated the atmospheric concentrations of CNT by measuring the number of fibers and aggregates obtained via aerosol-measuring instruments or collected on filters [7].

Although exposure assessments of nanomaterials in actual products have been reported, simulated discharge tests using aerosol-measuring instruments show that small amounts of particles are difficult to detect because they are masked by background particles measuring in the range of 10^3–10^4 parts/cm^3. Additionally, nanosized particles other than the target nanomaterial may be present. For example, an increase in particulate concentration has been observed due to the simulated test environment from sources such as active motors in machinery and particles derived from friction of volatile components [8].

For these reasons, a highly sensitive method to differentiate small amounts of target nanomaterials from other nanosized particles is required. Although many previous exposure assessments of nanomaterials have been conducted in work environments, they have focused on measuring one single particle at relatively high concentrations. However, few studies have been published regarding the evaluation of nanomaterials contained in products in the marketplace.

2. Objective

A Gas Exchange Device (GED; GED-01, J-SCIENCE LAB) [9, 10] is developed to quantify atmospheric particles by direct ICP-MS. This device negates the need for pre-processing steps such as sample collection, breakdown, and dilution, enabling continuous and highly sensitive measurements. Moreover, shortening the integration time of ICP-MS allows single particles to be detected because the signal intensity can be used to estimate the particle count and particle size.

Conventional ICP-MS analysis of gas samples involves dissolving atmospheric particulates collected on a filter in acid and using the dissolved solution for measurements. These methods reduce the weight concentrations of samples such as nanomaterials, and consequently require long-term collection to increase the weight. Other challenges include the need for pre-processing steps and contamination risks during pre-processing.

This study investigates the quantitative nature of atmospheric nanoparticles using GED-ICP-MS. To measure nanoparticles introduced to the plasma of ICP-MS, liquidification, vaporization, and ionization must occur at 1-ms intervals. However, ionization cannot be completed if the particle size exceeds the critical particle size, which reduces the analytical precision. Additionally, the rate of ionization depends on particle composition. In particular, metal oxides with high boiling and melting points undergo incomplete ionization. Specifically, we aim to identify the critical particle size by investigating the correlation between particle size and particle counts obtained through ICP-MS of titanium oxide nanoparticles. Additionally, we can increase the signal intensity of GED-ICP-MS by nebulizing a Ti solution with a known concentration. In this way, we determine the amount of elemental metal per element count and the quantitative factor for Ti.

3. Methods

3.1. Generation of titanium oxide nanoparticles

We generated titanium oxide nanoparticles and measured the particle size distribution using a Fast Mobility Particle Sizer (FMPS; FMPS3091, TSI). The sample was a titanium oxide MPT-429 (Ishihara Sangyo Kaisha, Ltd.) diluted 100-fold in a buffer solution. The buffer solution was an ammonium acetate solution (20 mM, pH 8) prepared by dissolving ammonium acetate (0.77 g) in ultrapure water (500 mL) and adding ammonium hydroxide (0.75 mL). Titanium oxide nanoparticles were generated using the Electrospray 3480 (TSI) and applying a voltage of 2.30 kV to the particulate suspension to create small charged droplets.

The electrospray was connected to the FMPS via a conductive tube in order to measure the size distribution of titanium oxide particles. The flow rate of the electrospray was 1.1 L/min, while the sample intake rate of the FMPS was 10.0 L/min; we adjusted the inter-device flow rate by creating branching pathways between the devices and attaching HEPA filters (ADVANTEC) in the tubes. Additionally, we removed the water content released with the particles by passing the released particles through a diffusion drier (TSI).

3.2. Ionization of particles

The signal intensity (counts per second, cps) using an ICP-MS (iCAP-Q, Thermo Fisher Scientific) when particle ionization is complete within the plasma is proportional to the particle volume. We used ICP-MS to confirm the ionization by measuring the signal intensity and particle counts of standard particles.

A 100-fold diluted titanium oxide solution in a buffer was supplied to the electrospray to generate particles. By introducing this output to a Differential Mobility Analyzer (DMA, Model EMC, Tsukasa Sokken Co., Ltd), we changed the applied voltage and classified the particle size in 10-nm intervals from 20 nm to 160 nm. The sheath flow rate was 5.0 L/min controlled by solenoid valve flow controller. After classification by DMA, we simultaneously measured the particle counts by FMPS and the signal intensity of ^{48}Ti by ICP-MS. The flow rates of the electrospray and ICP-MS were 1.1 L/min and 0.23

L/min, respectively, while the aerosol flow rate to the DMA and sample flow rate to the FMPS were 0.25 L/min and 8.0 L/min, respectively. A branch and HEPA filter (ADVANTEC) was introduced before the sample entrance to the FMPS to adjust flow rate (Figure 1).

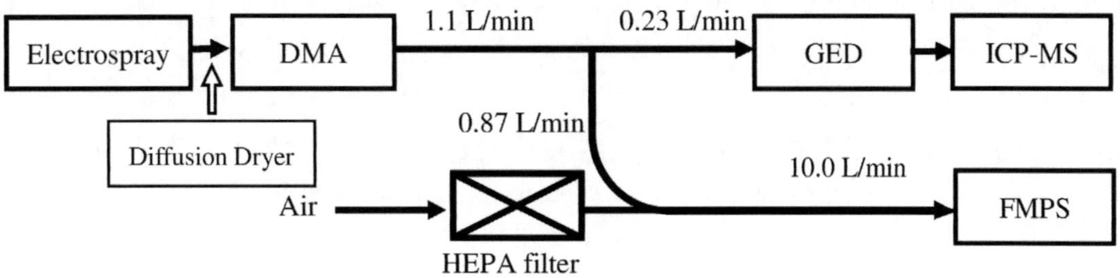

Figure 1. Device setup to determine the ionization of particles.

3.3. Quantification of titanium

To determine the amount of elemental metal per 1 cps signal intensity, we first created a standard particle solution. A Ti standard solution (Kanto Kagaku), which was used as a standard solution for metals during quantitative analysis with ICP-MS, was diluted in 3% nitric acid and an In standard solution (Kanto Kagaku) was added to this dilution to provide an index element. We adjusted the concentration of In and Ti to 250 ppb and 10 ppb, respectively. We also created an In standard solution with a concentration of 1 ppb by diluting the original standard solution in 6% nitric acid.

These solutions were nebulized using a desolventizing sample introduction device (APEX, Elemental Scientific Inc.) for ICP-MS. The solvent in the sample was removed by passing the aerosol through a film. In this way, only the particulate component of the solution was measured. A four-way stopcock was attached to the exit mouth of the APEX with one of the outlets attached between the gas exchange device and the ICP-MS, and the other outlet attached to a syringe filter. Argon gas was introduced to the particle nebulizing device at a flow rate of 300 mL/min (Figure 2).

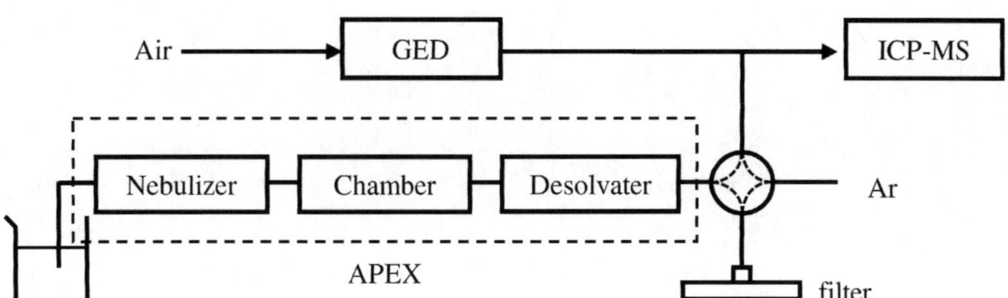

Figure 2. Device setup to quantify elemental titanium.

We initially measured the signal intensity of elemental Ti using the APEX to nebulize the desolventized standard solution and introduced this processed solution to the ICP-MS. Next, we redirected the APEX output through the stopcock to the syringe filter (DISMIC®-25CS080AS, ADVANTEC) for five minutes to collect particulate matter. Afterwards the filter was attached to a syringe dripping 10 mL of 6% nitric acid at a rate of 1 drop per second. The extracted liquid was nebulized again. Then we determined the per second metal emission rate by measuring the signal intensity of In by ICP-MS. Next we nebulized the In standard solution diluted to 1 ppb and measured the signal intensity of In by ICP-MS.

4. Results

Figure 3 shows the particle size distributions before and after drying when the particles are introduced directly to FMPS. A reduction in the particle count concentration occurs for particles between 30 and 60 nm after passing through a diffusion drier. Furthermore, the 60-second average for the total particle count concentration decreases 18.5% from 65,500 cm^{-3} prior to drying to 53,400 cm^{-3} after drying.

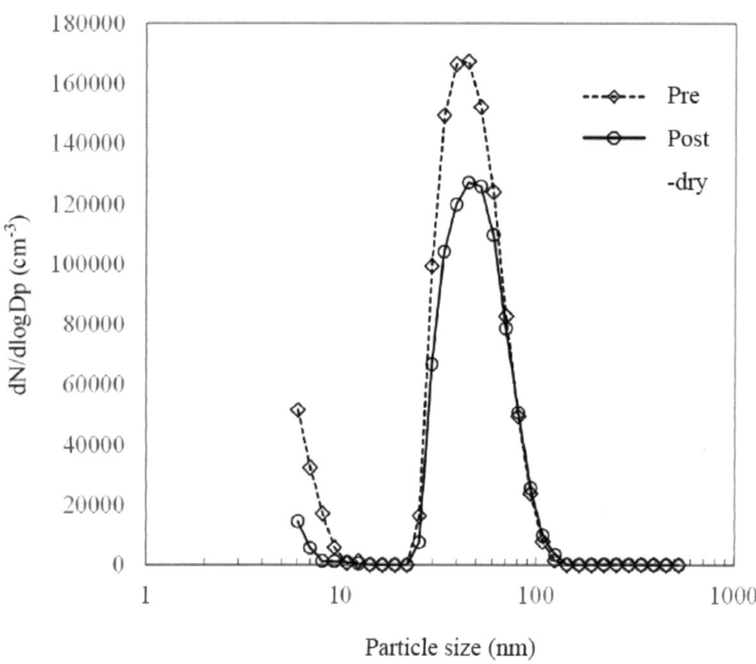

Figure 3. Particle size distribution before and after drying.

IOP Conf. Series: Journal of Physics: Conf. Series **838** (2017) 012008 doi:10.1088/1742-6596/838/1/012008

Particle concentrations for each size classification determined by DMA were measured using FMPS as well as the signal intensity by ICP-MS (Table 1). The blank reading taken before particle generation was subtracted from each measurement to determine the particle concentration and the signal intensity.

Table 1. Particle count and Ti signal intensity.

Particle size (nm)	Number of particles (particles)	Signal intensity (cps)
20	73	0
30	58	0
40	94	0
50	189	243
60	317	2122
70	461	5871
80	437	9336
90	392	12148
100	256	12506
110	172	9828
120	92	6359
130	75	3916
140	50	2591
150	40	1656
160	26	1419

We assumed that the particles are spherical, and then determined the volume per particle as well as the total volume for all particles of a given size (Table 2).

Table 2. Number of particles per size classification and the total volume.

Particle size (nm)	Volume ($\times 10^4$ nm^3)	Number of particles (particles)	Total volume ($\times 10^6$ nm^3)
20	0.419	73	0.306
30	1.41	58	0.814
40	3.35	94	3.15
50	6.54	189	12.4
60	11.3	317	35.9
70	18.0	461	82.7
80	26.8	437	117
90	38.2	392	150
100	52.4	256	134
110	69.7	172	120
120	90.5	92	83.3
130	115	75	86.0
140	144	50	71.7
150	177	40	70.0
160	214	26	56.1

Figure 4 shows the total volume and the signal intensities per particle size classification. A clear peak

IOP Conf. Series: Journal of Physics: Conf. Series **838** (2017) 012008 doi:10.1088/1742-6596/838/1/012008

is observed for both the total volume and the signal intensity at particle sizes of 90–100 nm. Moreover, the patterns for the total volume and the signal intensity begin to separate at particle sizes greater than 120 nm.

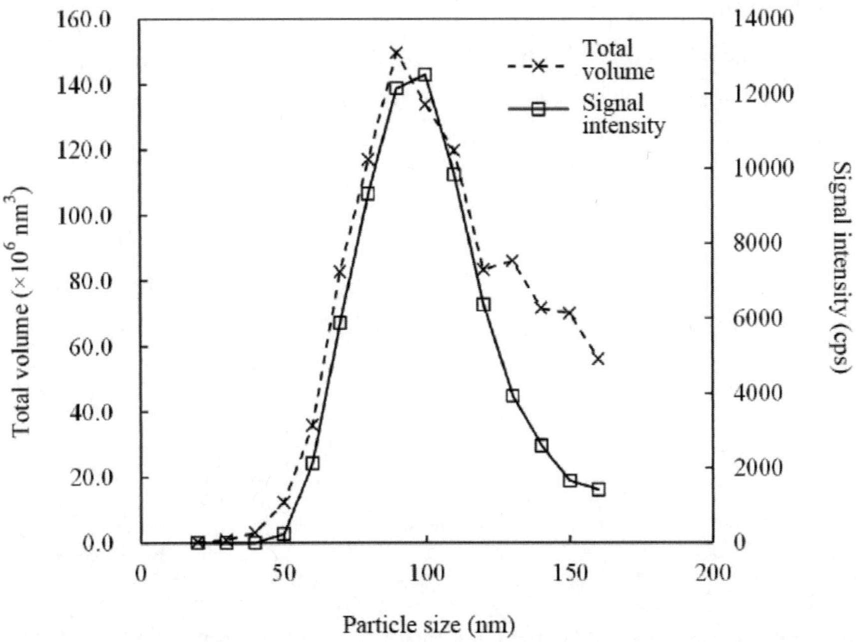

Figure 4. Relationship between the ICP-MS signal intensity and the total volume of particles.

The ICP-MS-measured signal intensity of a Ti standard solution adjusted to 10 ppb is 17,963 cps. Using this intensity, we calculated the amount of Ti per count by determining the amount of Ti nebulized per second. The signal intensities of In in the standard and extracted solutions are 44,718 cps and 55,309 cps, respectively. Furthermore, the extracted volume is 10 mL and the concentration of the standard solution is 1.0 ng/mL. Using these values, we calculated the amount of In (ng) in the extracted solution using Equation 1-1.

$$\begin{aligned}&\text{Extracted Solution In Amount (ng)}\\ &\quad= \text{Concentration Standard Solution (ng/mL)}\\ &\quad\times \frac{\text{Extracted Solution In signal intensity (cps)}}{\text{Standard Solution In signal intensity (cps)}}\\ &\quad\times \text{Extracted Solution volume (mL)}\\ &\quad= 12.4\ \text{ng}\end{aligned} \qquad \text{(Equation 1-1)}$$

This is equivalent to the amount of In nebulized in five minutes. We then determined the amount of

Ti in the extracted solution as well as the rate per second of nebulization. The concentration of In in the standard particle solution is 250 ng/mL, whereas the concentration of Ti is 10 ng/mL. Using these values, the rate per second of Ti nebulization is calculated using Equation 1-2.

$$\begin{aligned}&\text{Per second Ti nebulization (ng)}\\&\quad= \text{Extracted Solution In Amount (ng)}\\&\quad\times \frac{\text{Extracted Solution Ti concentration (ng/mL)}}{\text{Extracted Solution In concentration (ng/mL)}} \div 300\ \text{sec}\\&\quad= 0.00165\ \text{(ng/sec)}\end{aligned}$$

(Equation 1-2)

Since this value corresponds to the signal intensities shown in Table 1, the amount of Ti per count can be determined using Equation 1-3.

$$\text{Ti amount per count (ag/count)} = \frac{\text{Rate of Ti nebulization per second (ng/sec)}}{\text{Standard Solution signal intensity (cps)}} \times 10^9$$

(Equation 1-3)

Given the equation above, we calculated the conversion factor for Ti amounts per count as 91.8.

5. Discussion

First, we considered the ionization of titanium oxide particles. Figure 5 plots the per particle relationship between the volume and the signal intensity to confirm a proportional relationship between the particle volume and the signal intensity.

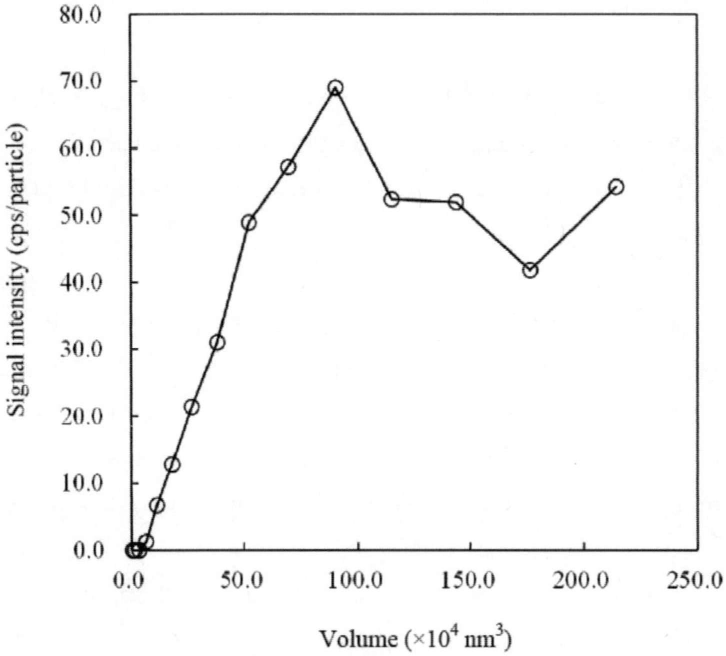

Figure 5. Signal intensity per particle volume.

The per particle signal intensity increases proportionally to the volume until the particle size reaches 120 nm (volume: 90.5×10^4 nm^3). To confirm the correlation, we extracted the particle sizes between 20 nm and 120 nm from Fig. 5 and determined the relationship between the particle volume and the signal intensity (Figure 6).

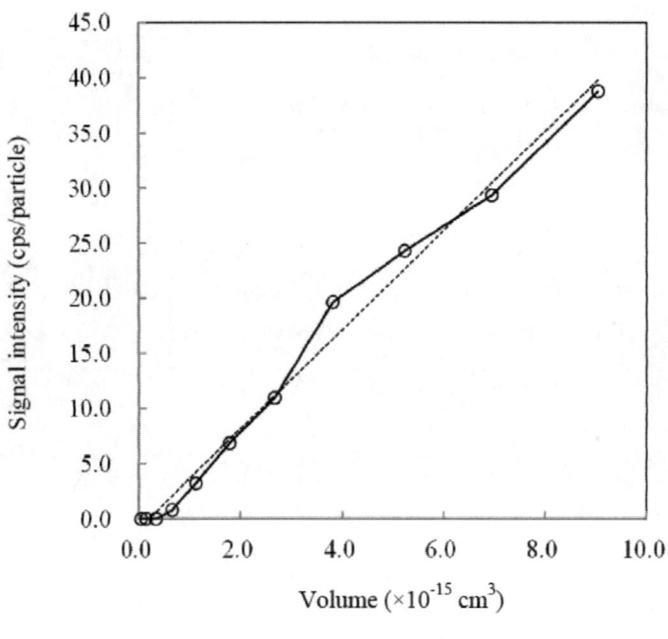

Figure 6. Relationship between the volume and the signal intensity for particle sizes 20–120 nm.

A plot of the linear approximation for the points indicates that the correlation coefficient is R = 0.994. The critical particle size is 120 nm, and we believe that particles below this size are completely ionized in the plasma. Consequently, GED-ICP-MS effectively measures nanoparticles as defined by ISO (Dp < 100 nm).

Next, we considered the validity of the conversion factor for the Ti amounts per count as determined by ICP-MS (91.8). Because gases discharged at 1.1 L/min from the DMA are branched to the ICP-MS at 0.23 L/min and FMPS at 0.87 L/min (Figure 1), we converted the values measured by ICP-MS and FMPS to the particle amounts, and determined the particle amounts by classification by DMA after adjusting for the flow rate. Similar values confirm the validity of the conversion factor. First, we determined the particle amount from the measured values of ICP-MS. Using the signal intensities in Table 1 and Equation 1-4, we calculated the per second amount of Ti introduced to ICP-MS. As a measure of the detection sensitivity for ICP-MS, we used a value of 0.1% for the ion transmittance (detected ions/introduced ions).

$$\text{Amount of introduced Ti (g/sec)} = \text{Ti signal intensity (cps)} \times 91.8\ \text{(ag/count)} \times 10^{-18} \div \text{Ion transmittance}$$

(Equation 1-4)

Next, we determined the per second particle mass introduced to FMPS from the particle counts shown in Table 2. For this calculation, we used 10,000 cm^3/min as the flow rate for FMPS and 4.23 g/cm^3 as the density of titanium oxide.

$$\text{Particle amount (g/sec)} = \text{Total volume } (nm^3/cm^3) \times \text{Flow rate } (cm^3/\text{sec}) \times 10^{-21} \times \text{Density of titanium oxide } (\text{g/cm}^3)$$

(Equation 1-5)

Finally, we adjusted the Ti amounts for ICP-MS calculated using Equation 1-4 and the particle amounts for FMPS calculated using Equation 1-5 by the flow rate. We assumed that the distribution of particles classified by DMA for each device is proportional to its flow rate. Figure 7 and Table 3 show the adjusted values after classification.

Table 3. Particle amounts introduced into each device per particle size classification.

Particle size (nm)	FMPS ($\times 10^{-10}$ g/sec)	ICP ($\times 10^{-10}$ g/sec)
20	0.164	0
30	0.348	0
40	1.35	0
50	5.29	1.07
60	15.3	9.32
70	35.4	25.8
80	50.1	41.0
90	64.0	53.3
100	57.3	54.9
110	51.2	43.2
120	35.7	27.9
130	36.8	17.2
140	30.7	11.4
150	30.0	7.27
160	24.0	6.23

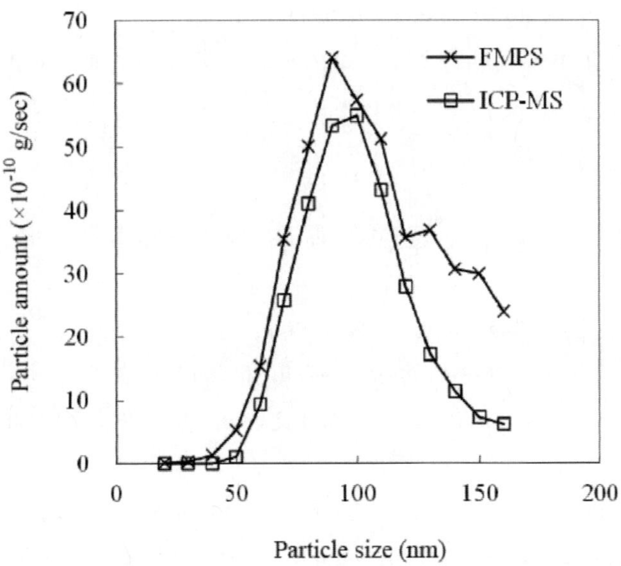

Figure 7. Calculated particle amounts from the measured values per size classification.

For all particle sizes, the measured values of FMPS are consistently higher than those for ICP-MS, with a tolerance of approximately 20.1% for the peak particle size; however, both sets of values roughly align. One reason for the higher FMPS values may be that measurements are made without distinguishing the particles, meaning that the values may contain counts for contaminants such as silica in the dispersant for titanium oxide or other unrelated particles. Regardless, our results show that GED-ICP-MS can be used to quantify nanomaterials continuously with a high sensitivity. Carbon-based materials such as CNT can be identified and quantified as long as an index element can be found in the CNT.

The selection of 1) intentionally included nanomaterials in a given product, 2) composite nanomaterials containing intentionally included nanomaterials and other materials, and 3) spontaneously generated nanomaterials remain challenging in per aerosol assessments of products containing nanomaterials. Previously proposed assessment methods, including particle concentrations and weight concentrations, are inadequate to distinguish the different types of nanomaterials. Although electron microscopy can differentiate among these types of nanomaterials, quantification requires an enormous amount of effort. In our proposed method, the status of nanomaterials described in 1) or 2) can be monitored, enabling safe aerosol assessments, even though it cannot yet distinguish among these types. By applying our setup to a simulated emission test, we expect that our method will be used for aerosol assessments of actual products containing nanomaterials.

6. Acknowledgements

This work was supported by a Ministry of Economy, Trade and Industry (METI) Grant from 2013 to 2014: Exposure Assessment of Market Products Including Nanomaterial in General Use and Disposal Process.

7. References

1) ISO. Nanotechnologies – Health and safety practices in occupational settings relevant to nanotechnologies. (2008) ISO TR 12885.
2) NIOSH. Approaches to safe nanotechnology – Managing the health and safety concerns associated with engineered nanomaterials, DHHS (2009) NIOSH Publication No.2009-125.
3) NIOSH. Method 5040 Issue 3, Diesel particulate matter as Elemental Carbon (2003) NIOSH Manual of Analytical Methods (NMAM; 4th ed.).
4) Birch ME, Ku BK, Evans DE and Ruda-Eberenz TA 2011 Exposure and emissions monitoring during carbon nanofiber production-Part I: elemental carbon and iron-soot aerosols. *Ann. Occup. Hyg.* **55** 1016–1136.
5) Dahm MM, Evans DE, Schubauer-Berigan MK, Birch ME and Fernback JE 2012 Occupational exposure assessment in carbon nanotube and nanofiber primary and secondary manufacturers. *Ann. Occup. Hyg.* **56** 542–556.
6) Reed RB, Goodwin DG, Marsh KL, Capracotta SS, Higgins CP, Fairbrother DH, Ranville JF (2013). Detection of single walled carbon nanotubes by monitoring embedded metals. Environmental Science: Processes & Impacts **15** 204–213.
7) Ogura I, Sakurai H, Mizuno K, Gamo M (2011). Release potential of single-wall carbon nanotubes

produced by super-growth method during manufacturing and handling. *Journal of Nanoparticle Research* **13**:1265–1280.

8) Ogura I, Kotake M, Shigeta M, Uejima M, Saito K, Hashimoto N and Kishimoto A Potential release of carbon nanotubes from their composites during grinding. *J. Phys. Conf. Series 2013*,429, doi:10.1088/1742-6596/429/1/012049.

9) Nishiguchi K, Utania K, Fujimori E (2008). Real-time multielement monitoring of airborne particulate matter using. *J. Anal. At. Spectrom.* **23**, 1125–1129.

10) http://www.sumitomo-chem.co.jp/rd/report/theses/docs/20080103_s52.pdf

IOP Conf. Series: Journal of Physics: Conf. Series **838** (2017) 012009 doi:10.1088/1742-6596/838/1/012009

Dynamics of nanomaterials released from polymer composites in the pelletizing process

Nobuyuki Kato[1], Yasuto Matsui [2] and Minoru Yoneda[1]
(1) Graduate School of Engineering, Kyoto University, Katsura, Nishikyoku, Kyoto, 615-8540, Japan;
(2) Agency for Health, Safety and Environment, Kyoto University, Yoshidahonmachi, Sakyoku, Kyoto, 606-8501, Japan
E-mail: katoh@risk.env.kyoto-u.ac.jp

Abstract. Measures against exposure to carbon nanotubes (CNT) are necessary, especially in workplaces that handle nanomaterials, because adverse health effects are a concern. This study focuses on the dynamics of CNT released from CNT/polymer composites during the pelletizing process at a pilot factory. It is difficult to identify CNT and the base resin. By characterizing the possibility of separating CNT from the composite with a kinetic weighting coefficient, estimation can be carried out using a Computational Fluid Dynamics (CFD) simulation. The mass concentration of black carbon and the particle number concentration by diameter were measured using two different measurement apparatuses. The simulation results were then compared to the measured data. The model was verified by the correlation between the simulation and measured results. The model provided a strong correlation, indicating that the dynamics of CNT and the base resin released from the polymer composite can be simulated. It is expected that the model using the CFD simulation can be applied to the occupational health field.

1. Introduction

The applications of nanomaterials in various fields have vastly increased in the past decade. On the one hand, the small size of nanomaterials which at least in one dimension are smaller than 100 nm allows for new functions. On the other hand, many reports have indicated negative impacts on experimental animals and cells [1–4]. For example, composite materials with carbon nanotubes (CNT) are being produced because adding CNT to a base material to a concentration of several percentages realizes new functions such as conductivity. However, the health effects of CNT remain a concern. Therefore, measures against exposure are necessary, especially in workplaces that handle nanomaterials.

Numerous exposure assessments in workplaces handling nanomaterials have been reported. A survey has been conducted in a CNT pilot factory. However, identifying and assessing the exposure of scattered CNT in pilot factories that deal with CNT/polymer composite resins are difficult by thermal carbon analysis using NIOSH 5040 [5] due to the fact that carbon is already in the air and CNT is composed of carbon resin materials.

Elemental analysis using metal impurities in CNT as a marker is being attempted to identify CNT [6–9]. At CNT/polymer composite pilot factories, CNT goes through the following process. First, CNT materials are repacked into smaller bags (repacking), which are fed into an extruder (feeding). Then the resin and CNT are extruded (extruding). After the composite resin is cooled, it is cut into pellets (pelletizing). During pelletizing, the cutting process has been reported to produce a momentary high concentration, but the specific release has yet to be resolved.

This study focuses on the pelletizing process. By characterizing the possibility of separating CNT from the base resin with a kinetic weighting coefficient, estimations can be conducted using Computational Fluid Dynamics (CFD) simulations.

Measurements were conducted using a light scattering dust counter (LSDC) for area sampling of dust, black carbon monitoring (BCM) to measure the black carbon concentration, and an optical particle sizer (OPS) to measure the suspended particulate matter concentrations (size range: 0.3–10 μm). The rate of CNT released from the composite resin was calculated using the advection-diffusion model, revealing the dynamics of CNT.

2. Methodology

2.1. Sample design

Figure 1 depicts the factory layout. During working hours, all doors are closed. To measure the airborne particulate matter concentrations, LSDCs (Model LD-3, SHIBATA) were placed at six different sampling points (No.1–6) for 25 minutes. Measurements were taken every second. The mass concentration conversion coefficient (K factor) was set to 0.001 mg/m^3/cpm. K factor was based on the reference value. Measurements taken at points No. 2, 4, and 5 were during the pelletizing process.

In addition, the particle concentrations during each process was measured close to the processes (sampling points No. 7–9 in Figure 1) using an OPS (Model 3330, TSI Inc, size range 300 nm–10 μm). The OPS obtained the particle concentrations at a rate of 10 sec/scan. The equipment except in the fume hood was set at 1.1 m above the ground, which is the height of the breathing area.

The mass concentrations of airborne CNT was measured using BCM (Model AE51, AethLabs) at sampling points No. 10–12 in Figure 1. The running condition of BCM was at a flow rate of 0.1 L/min and a time interval of 60 seconds. To compare the base line of the ambient concentration before and after pelletizing, this study focused on the pelletizing process.

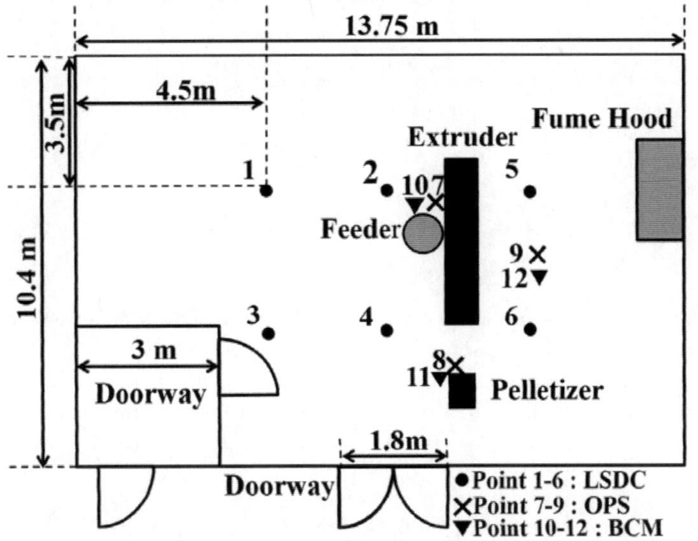

Figure 1. Outline of the factory showing approximate positions of the processes and mobile measurement locations (1–12).

2.2. CFD simulation calculation conditions

A computation mesh was created using a mesh generator. The temperature of the extruder was set rather high (280°C) for calculation purposes. Figure 2 shows the calculation area.

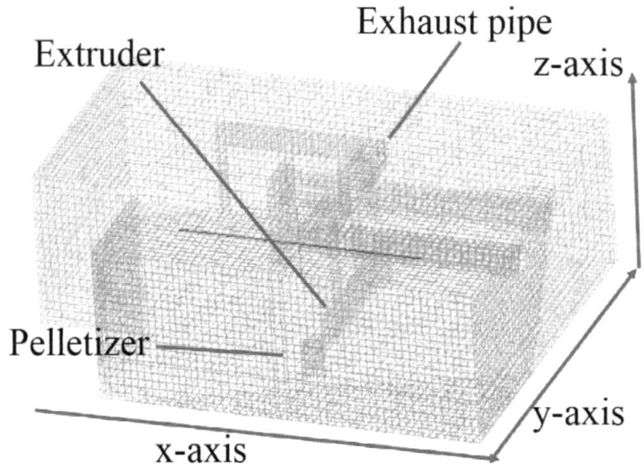

Figure 2. Surface mesh of the extruder and the pelletizer in the manufacturing factory.

The mesh size was selected to reduce the calculation costs. However, dynamics analysis was simulated using 20-cm intervals (69×52×42 [cell]). In this study, the simulation was conducted using the free model (OpenFOAM [10]). As for the flow field, the differential equation was discretized and then the 3D field was decided by steady-state analysis. The fluid was assumed to be a Newtonian. The equations of OpenFOAM, continuity equation, Navier-Stokes equation, and energy transfer equation (Equation 1–3) were used. The concentration field was analyzed by time integrating the scalar transfer equation that does not consider interactions. In addition to the aforementioned equations, the ideal gas equation was also used.

$$\nabla \cdot (\rho u) = 0 \tag{1}$$

$$\frac{\partial \rho u}{\partial t} + \nabla \cdot (\rho u u) = -\nabla p - \nabla \cdot \left[\mu\left\{\nabla u + (\nabla u)^T\right\}\right] - \nabla\left(\frac{2}{3}\mu \nabla \cdot u\right) \tag{2}$$

$$\frac{\partial \rho E}{\partial t} + \nabla \cdot (\rho E u) = -\nabla \cdot (\rho u) + \nabla \cdot (k \nabla T) \tag{3}$$

E is the total energy per mass, k is the thermal conductivity, and T is the absolute temperature. Here, the gravity term is ignored. The total energy E is the sum of the internal energy and the kinetic energy. The gravity buoyancy can be considered under the assumption that a hydrostatic equilibrium is possible using the Boussinesq approximation. The gravity term is calculated using Equation 4.

$$\rho g = \left\{1 - \beta(T - T_0)\right\}\rho_0 g \tag{4}$$

ρ_0 : basic density, T_0 : basic temperature, β: coefficient of cubic expansion.

For the turbulent model in this survey, Reynolds-averaged Navier-Stoles (RANS) was used. The RANS model comes in various forms. To confirm the mesh quality in obtaining y+ from the wall, the most stable potential fluid was achieved around 100–150. Therefore, by defining the Reynolds stress using the eddy viscosity, the standard *k-ε* model modeled by the wall-law logarithmic space was selected. Shown below is the equation.

$$\frac{\partial \bar{u}}{\partial t} + \bar{u} \cdot \nabla \bar{u} = -\nabla \bar{p} - \nabla \cdot (v_{eff} \nabla \bar{u}) \tag{5}$$

where v_{eff} is the coefficient of the eddy viscosity, whereby v_t is added to v, which is the constant of the turbulent term. $\bar{u}$ in the equation is the mean velocity of the flow (Equation 5).

Constants k and ε obtained from the standard k-ε model were derived from the Equations 6 and 7. K is the turbulent energy and ε is the energy disposal rate. Constants k and ε were obtained from the standard k-ε model. For the standard k-ε model, the eddy viscosity coefficient was calculated by obtaining k and ε when solving the two transfer equation modeled by two variables (k and ε).

$$\frac{\partial k}{\partial t} + \frac{\partial \overline{u_i} k}{\partial x_i} = S - \frac{\partial}{\partial x_i}(v + \frac{v_t}{\sigma_k}\frac{\partial \varepsilon}{\partial x_i}) - \varepsilon \tag{6}$$

$$\frac{\partial \varepsilon}{\partial t} + \frac{\partial \overline{u_i} \varepsilon}{\partial x_i} = C_{\varepsilon 1}\frac{\varepsilon}{k}S + \frac{\partial}{\partial x_i}(v + \frac{v_t}{\sigma_k}\frac{\partial \varepsilon}{\partial x_i}) - C_{\varepsilon 2}\rho\frac{\varepsilon^2}{k} \tag{7}$$

Each coefficient is defined by Launder-Spalding (1974). $C_\mu = 0.09$, $C_{\varepsilon 1}=1.44$, $C_{\varepsilon 2}=1.92$, $\sigma_k=1.0$, $\sigma_\varepsilon=1.0$.

The initial values (k and ε) were derived from Equations 8 and 9.

$$k = \frac{3}{2}(UI)^2 \tag{8}$$

$$\varepsilon = \frac{C_\mu^{0.75} k^{1.5}}{L_m} \tag{9}$$

I is the turbulent magnitude. In this survey, the fluctuation value of 0.08, which was measured with an anemometer, was used as the UI value. The default value C_μ of the standard k-ε model (0.09) was used. L_m is the mixing-length. For the representative length L, which is the length that receives the greatest impact, 7% was selected. In this survey, the width of the extruder, which is the turbulent source, was used as the representative length.

Using these values, the turbulent viscosity coefficient μ_t and turbulent heat diffusion coefficient α_t were determined using Equations 10 and 11. In the case of air, the experimental value was around 0.9 at the wall boundary. In this model, the calculation was carried out with P_{rt}=0.85.

$$\mu_t = \rho C_\mu \frac{k^2}{\varepsilon} \tag{10}$$

$$\alpha_t = \frac{\mu_t}{p_{rt}} \tag{11}$$

The scalar transfer equation is as shown below (Equation 12). S is the source term.

$$\frac{\partial \phi}{\partial t} + u \cdot \nabla \phi = \nabla \cdot (D \nabla \phi) + S \tag{12}$$

In addition, the equilibrium velocity of the normal force and gravity was obtained as the terminal velocity to model the characteristic of the particulate matter (Equation 13).

$$v_t = \frac{\rho_P \, d_{ve}^{\;2} \, g \, C_C(d_{ve})}{18 \mu \chi} \tag{13}$$

Here, d_{ve} is the equivalent volume diameter, g is the gravitational acceleration, μ is the viscosity, and χ is the kinetic weighting coefficient. Spherical shape is 1. C_C is the Cunningham weighting factor, and is shown as the Knudsen number (Kn) function (Equation 14). Kn (= $2\lambda / d$) can be represented as the rate between the molecular mean free path λ and the diameter d. For λ, the value adjusted to the temperature of the experiment condition was used. This velocity was added to u in Equation 12 in the virtual down direction.

$$C_C = 1 + 1.257\,Kn + 0.400\,Kn \exp\left(\frac{-1.10}{Kn}\right) \tag{14}$$

The diffusion coefficient D is represented by the Stokes-Einstein equation (Equation 15)

$$D = \frac{k\,T\,C_C(d_{ve})}{3\pi\,\mu\,\chi d_{ve}} \tag{15}$$

Here, k is the Boltzmann constant and T is the absolute temperature. This D was used as the constant D in Equation 12. For stable state conditions, the SIMPLE method was used. The provisional velocity was calculated by the semi-discrete scheme of the motion equation. After this, the pressure equation was solved from the provisional velocity. Subsequently, the velocity was updated using that pressure. This process was repeated until residual value was minimized.

Table 1 shows the calculation conditions. All factory doors were closed. The internal current was simulated from the thermal convection produced by the Extruder.

As previously mentioned, the pelletizing process was first conducted with a resin to investigate the concentration distribution. By focusing on this process and the steady flow, the concentration of CNT and resin released from the pelletizer were simulated. Because the operation time of the resin was 30 minutes, the simulation of the concentration transfer was carried out for 30 minutes. For the real-time measurements, the measurement time of LSDC was considered and the gradient of the increasing function over 20 minutes from the start to before the concentration rapidly increases was compared to the simulation value. In addition, the correlation coefficient between measurement value and simulation value was obtained.

With the CNT aspect ratio used for this process and the bulk density, two kinds of simulations were conducted. For an aspect ratio of 50, $\chi = 2.0$. Using the aerodynamics diameter calculated based on the catalog data of CNT used in the study, $\chi = 3.5$. These two values were used. For the resin value, 1.04 g/m^3 was used according to the physical properties, and a bulk density of 0.12 g/m^3 was used for CNT.

Considering the shape, the terminal velocity was calculated. For OPS, to match the units of the calculation, the number concentration was transferred to the mass concentration as 1 mg/cm^3. Table 1 shows the initial calculation conditions.

Finally, the CNT containing rate was assumed to be 5.0%. Then, the simulation result of the CNT model released from the pelletizer was compared to measured value from BCM. For each material (CNT of shape $\chi = 2.0$, 3.5 and resin), different simulation scenarios were prepared using nine types of diameters (50 nm, 100 nm, 300 nm, 500 nm, 1 μm, 2 μm, 4 μm, 7 μm, 10 μm). The simulation scenarios were selected based on the measurement apparatus and analytical purpose.

IOP Conf. Series: Journal of Physics: Conf. Series **838** (2017) 012009 doi:10.1088/1742-6596/838/1/012009

Table 1. Calculation conditions.

	Parameter
Calculation area	69×52×42 [grid]
Discrete method	Finite volume method
Initial air condition (*U*, *p*)	$U = 0$ [m/s](air velocity) $p = 0.1$[Mpa]
Initial temperature	Air = 293.15[K], Extruder = 553.15 [K]
Initial condition (*k*,*ε*)	$k = 0.0096$, $\varepsilon = 0.0003$
Turbulent kinematic velocity	$\nu_t = 0.0268$ [m²/s]
Boundary condition	*U*wall = 0 [m/s] Pressure gradient = 0
Prandtl number	$Pr_t = 0.85$

3. Results

Figure 3 depicts the calculation result of the steady flow with the simple method, and shows the magnitude of the potential flow in the center and its vector flow.

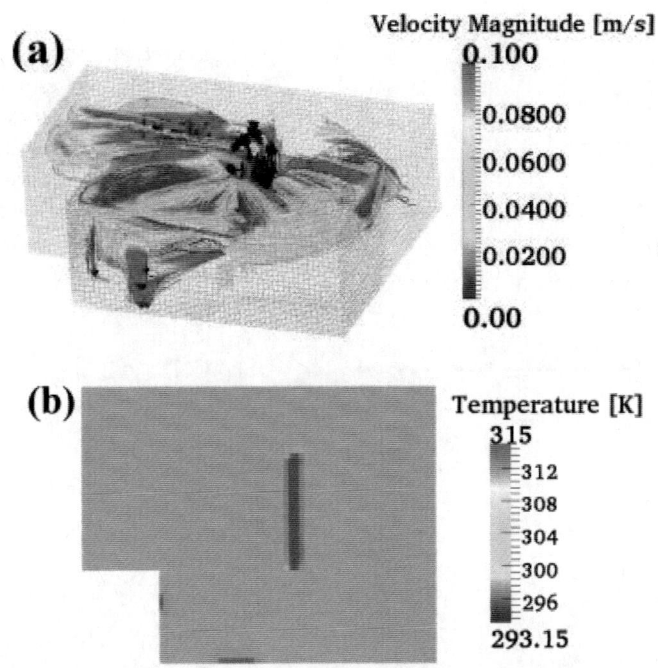

Figure 3. Velocity magnitude destribution of the air stream line and the temperature desutribution at the breathing area height. (a) Air stream line velocity and vector. (b) Temperature cross section of the breathing line.

Due to thermal convection, the flow moves upwards and around the room in general. In addition, by measuring the temperature distribution at points 7–9 at the start of the pelletizing process, the mean temperature from a thermometer at a breathing height is 34.8°C, which approximately matches the calculated temperatures of 35.8°C, 34.3°C, and 35.4°C (Figure 4b). From these results, the mass advection-diffusion simulation from the pelletizer was simulated by assuming that the simulation of the potential flow is reasonable.

IOP Conf. Series: Journal of Physics: Conf. Series **838** (2017) 012009 doi:10.1088/1742-6596/838/1/012009

Next, to grasp the trend of the mass diffusion model, the simulated calculations at Points 1–9 in Figure 1 were conducted. CNT (χ = 2.0, 3.5) and resin were generated every second from the pelletizer. To determine the distribution difference between each particulate matter, the distribution was calculated every 30 minutes. Figures 4 and 5 show the simulation results. The calculated concentration was normalized in order to obtain the tendency of the difference of diameters without depending on generating quantity of particulate matters.

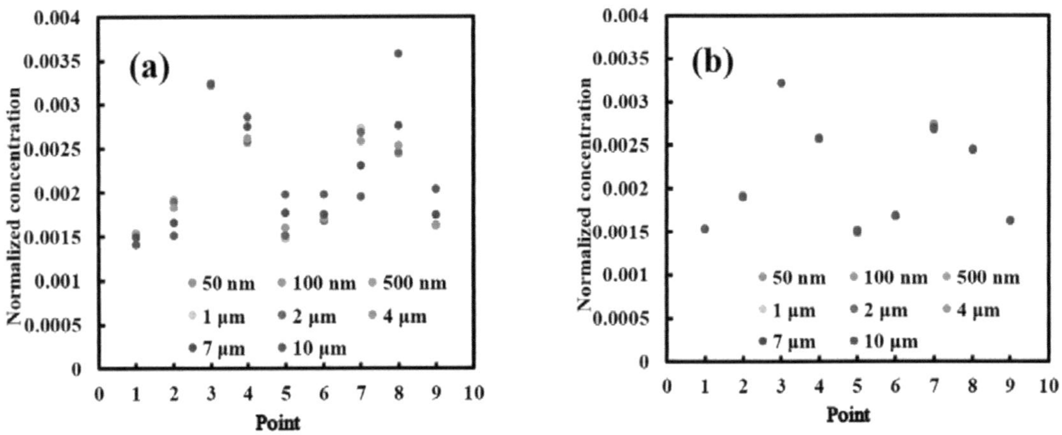

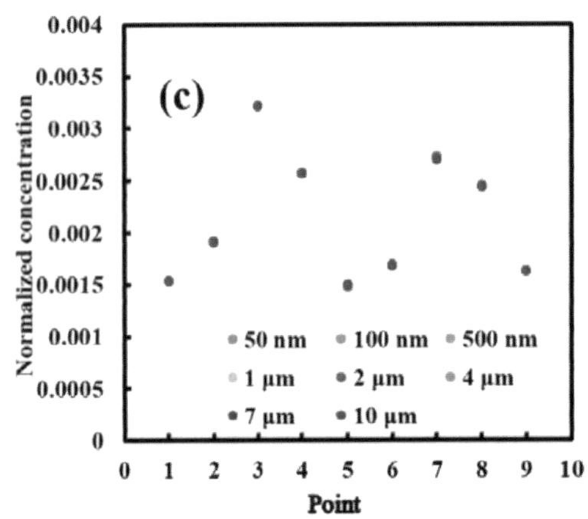

Figure 4. Distributions of the normalized concentration measured using the simulation model at the measurement points. The concentration is normalized using source concentration with the source being 1 scalar value by each particulate matter (a-c) which is generated every second from the pelletizer cells.

(a) Resin,
(b) CNT (χ=2.0),
(c) CNT (χ=3.5).

Figure 5. Resin normalized concentration distributions after 30 min. (a) Resin diameter is 500 nm and (b) Resin diameter is 10 μm.

Regarding the Resin particles, some integrations of measurement points are easier for heavier particles, while other integrations are easier for lighter particles. However, in terms of CNT, the dispersion is independent of the diameter and the kinetic weighting coefficient.

Next, to assess the simulation model, the correlation coefficient of the time series data was obtained by time series data using LSDC and simulations at the measurement points. Figure 6 and Table 2 show the simulation results.

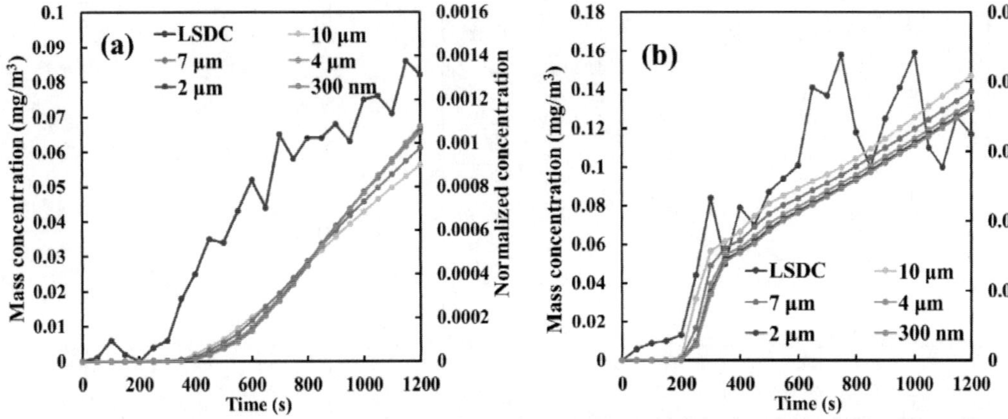

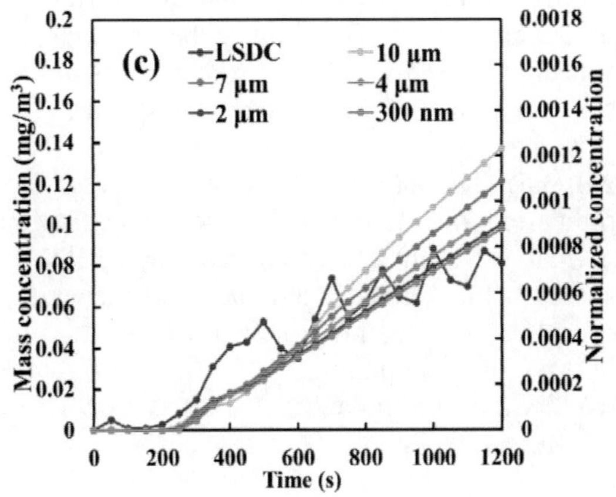

Figure 6. Time series of the mass concentrations measured using LSDC and the normalized concentrations by diameter calculated using the simulation model at measurement points. The baseline is adjusted to zero. The right side of the y-axis is the concentration normalized using source concentration with the source being 1 scalar value by diameter which is generated every second from the pelletizer cells.

(a) Measurement point 2,
(b) Measurement point 4,
(c) Measurement point 5.

As a result, at each measurement point, the time of increase in concentrations measured using LSDC, corresponds reasonably well with that of increase in concentrations calculated using the model. Moreover, gradients of the concentration curve measured at the point 4 and 5 correspond well with the concentrations calculated using the model.

Table 2. Correlation coefficients of the LSDC concentration data and the simulation concentrations data.

Measurement point	Resin diameter	Correlation coefficient
2	300 nm	0.88
4	300 nm	0.88
5	300 nm	0.93

LSDC was calibrated using stearic acid particles (diameter range: 300 nm). Table 2 shows the results calculated by 300 nm. As a result, the correlation coefficients indicate that all the points show a strong correlation. In addition, at Point 5, which is relatively far, a particle of size 300 µm has the strongest correlation because of the light weight.

Next, the correlation coefficients of the time series data were obtained by time series data using OPS and a simulation at points of measurement. Figure 7 and Table 3 show the simulation results.

The particle number concentrations measured using OPS were converted to the mass concentrations by the assumed density (1 g/cm^3). The particles assumed to be spherical for the mass determination.

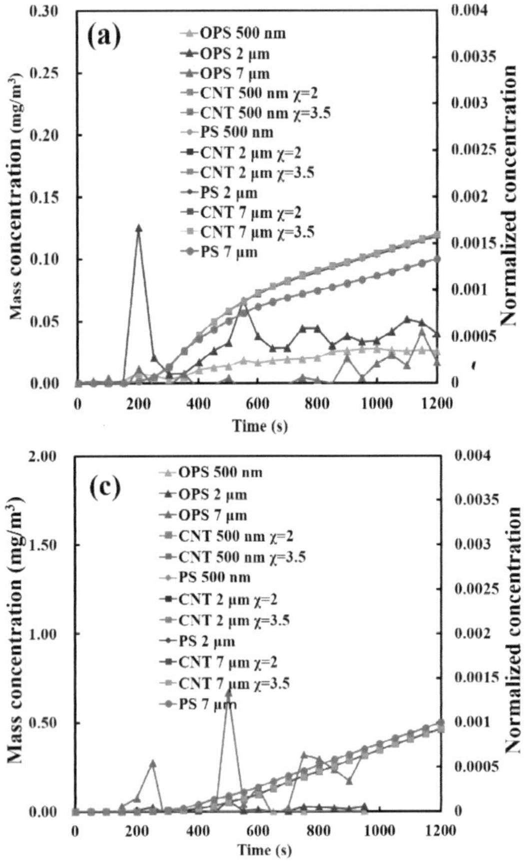

Figure 7. Time series of mass concentrations measured using OPS and the normalized concentrations by diameter calculated using the simulation model at measurement points. The simulation data has Resin particle and CNT (χ = 2.0, 3.5). The baseline is adjusted to zero. The right side of the y-axis is the concentration normalized using source concentration with the source being 1 scalar value which is generated every second from the pelletizer cells.
(a) Measurement point 7,
(b) Measurement point 8,
(c) Measurement point 9.

Table 3. Correlation coefficients of the OPS concentration data and the simulation concentration data.

Measurement point	Diameter	Material and shape	Correlation coefficient
7	500 nm	Resin	0.98
7	500 nm	CNT χ=2	0.98
7	500 nm	CNT χ=3.5	0.98
7	2 μm	Resin	0.98
7	2 μm	CNT χ=2	0.98
7	2 μm	CNT χ=3.5	0.98
7	7 μm	Resin	0.97
7	7 μm	CNT χ=2	0.98
7	7 μm	CNT χ=3.5	0.98
8	500 nm	Resin	0.85
8	500 nm	CNT χ=2	0.85
8	500 nm	CNT χ=3.5	0.85
8	2 μm	Resin	0.86
8	2 μm	CNT χ=2	0.85
8	2 μm	CNT χ=3.5	0.85
8	7 μm	Resin	0.91

8	7 μm	CNT χ=2	0.85
8	7 μm	CNT χ=3.5	0.85
9	500 nm	Resin	-0.28
9	500 nm	CNT χ=2	-0.28
9	500 nm	CNT χ=3.5	-0.28
9	2 μm	Resin	0.47
9	2 μm	CNT χ=2	0.47
9	2 μm	CNT χ=3.5	0.47
9	7 μm	Resin	0.48
9	7 μm	CNT χ=2	0.47
9	7 μm	CNT χ=3.5	0.47

Point 7 around the feeder shows a stronger correlation with CNT. Point 8 around the pelletizer shows a stronger correlation with resin, confirming that particles with a spherical shape and higher density are detected. At Point 9, 500-nm particles show a negative correlation, suggesting that the influence of particulate matter rather than the materials released from the pellet during the pelleting process is detected. Spherically shaped particles with higher densities and diameters greater than 2 μm show a stronger correlation at Point 9.

Finally, the correlation coefficient of the time series data was obtained by time series data using the BCM and CNT simulation data converted into 5 wt. % at measurement points. Figure 8 and Table 4 show the simulation results.

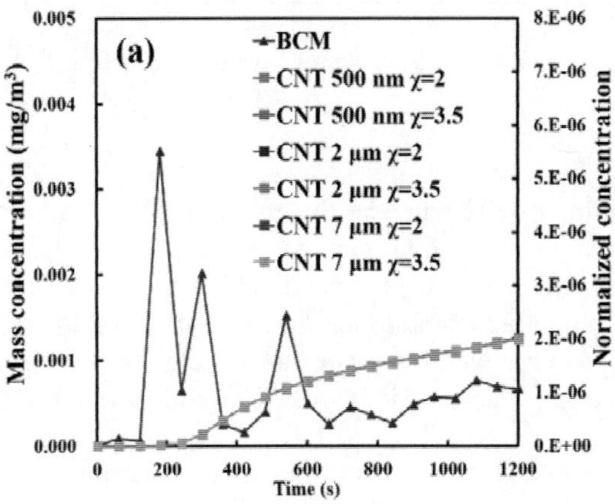

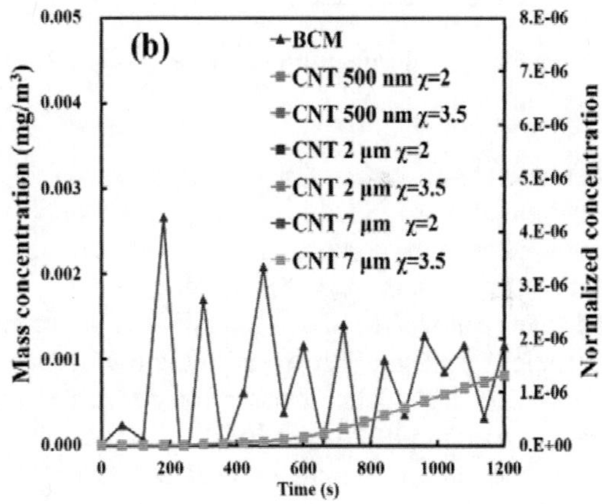

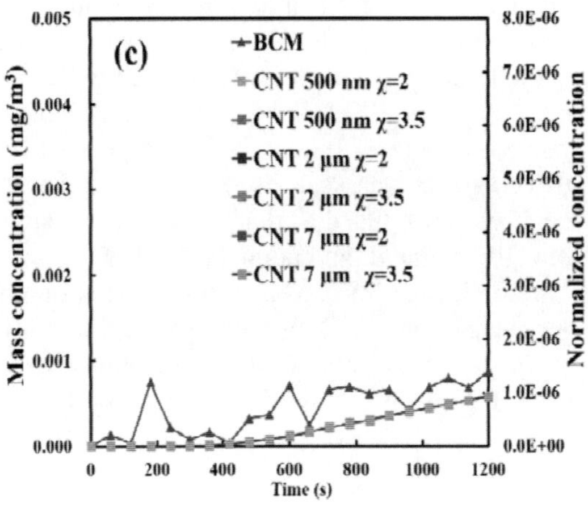

Figure 8. Time series of mass concentrations measured using BCM and the normalized concentrations by diameter calculated using the simulation model at measurement points. The baseline is adjusted to zero. The right side of the y-axis is the concentration normalized using source concentration with the source being 0.05 scalar value which is generated every second from the pelletizer cells.

(a) Measurement point 10,
(b) Measurement point 11,
(c) Measurement point 12.

Table 4. Correlation coefficients of the BCM concentration data and the simulation concentration data.

Measurement point	Diameter	Shape	Correlation coefficient
10	500 nm	CNT $\chi = 2$	-0.197
10	500 nm	CNT $\chi = 3.5$	-0.197
10	2 μm	CNT $\chi = 2$	-0.197
10	2 μm	CNT $\chi = 3.5$	-0.197
10	7 μm	CNT $\chi = 2$	-0.197
10	7 μm	CNT $\chi = 3.5$	-0.197
11	500 nm	CNT $\chi = 2$	0.056
11	500 nm	CNT $\chi = 3.5$	0.056
11	2 μm	CNT $\chi = 2$	0.056
11	2 μm	CNT $\chi = 3.5$	0.056
11	7 μm	CNT $\chi = 2$	0.056
11	7 μm	CNT $\chi = 3.5$	0.056
12	500 nm	CNT $\chi = 2$	0.736
12	500 nm	CNT $\chi = 3.5$	0.736
12	2 μm	CNT $\chi = 2$	0.736
12	2 μm	CNT $\chi = 3.5$	0.736
12	7 μm	CNT $\chi = 2$	0.737
12	7 μm	CNT $\chi = 3.5$	0.737

Point 10 and 11are not correlated with CNT. However, Point 12 shows a stronger correlation with CNT.

4. Conclusion

By modelling CNT and resin with a kinetic weighting coefficient, the dynamics of CNT and resin released from the composite during the pelletizing process was verified. The higher the kinetic weighting coefficient was, the weaker the dependence on the diameter. Since the terminal velocity was slower, the particles easily floated. Comparative simulations of the diameters and materials revealed that the resin particles dispersed in relation to the diameter and the kinetic weighting coefficient. However, no such dispersion dependence existed for CNT. The times series concentrations of LSDC and the simulation data as well as OPS data were strongly correlated.

The measurement points closer to the pelletizer showed a strong correlation with larger particles, whereas smaller particles showed the strongest correlation with measurement points at a relatively far distance. In contrast, BCM showed a negative correlation around the feeder and no correlation around the pelletizer. Since a local ventilation was used during the feeding process, it was difficult to measure the higher CNT concentration continuously.

However, when extruding and pelletizing started, a sharp peak in the OPS data with a diameter below a few micrometers temporarily appeared. Since the BC concentration scale is smaller than the OPS concentration scale, BCM was considered to be more sensitive to residual dust than OPS. It was assumed that the simulation result and BCM data were not correlated at point 10 and point 11[11].

Point 12, which was relatively far from the extruder and pelletizer, showed a strong correlation with the simulation results. In particular, there was a stronger correlation with larger particles as well as the OPS data. Moreover, the maximum correlation was detected in the resin particle data. Therefore, there is a low probability that free and isolated CNT were released from CNT/polymer composites during the pelletizing process.

Next, the reproducibility of mass advection-diffusion modeling in the workplace was verified. The measured concentrations showed an increasing function, but the gradient of the curve at 800 seconds decreased. Our simulation model has yet to reproduce this trend. Since the peak of concentrations measured using OPS with a different diameter was approximately consistent in Figure 7, the aggregation in air was not estimated in this study. It was assumed that by extruding continuously, the real temperature field was separated from the temperature field obtained using steady-state analysis. Drying the surrounding air showed the terminal velocity of airborne particulate matter was slower than in the moist environment, allowing the particles to float easily. Other possible assumptions include the deposition of particulate matter on the factory surfaces (exhaust pipe, wall etc.) and instability of the pelletizing amount. To reproduce such conditions, it is necessary to model the interaction of evaporation, which requires a precise mesh of the factory surroundings.

However, this study revealed that the simple model is sufficient to determine the correlation. In conclusion, this model can be used to simulate dispersion of particulate matter in workplaces.

Acknowledgements

This study is based on the results obtained from a project commissioned by the New Energy and Industrial Technology Development Organization (NEDO).

References

[1] Brown DM, Wilson MR, MacNee W et al. 2001 *Size-dependent proinflammatory effects of ultrafine polystyrene particles. A role for surface area and oxidative stress in the enhanced activity of ultrafines. Toxicol.Appl.Pharmacol* **175**: 191-199.

[2] Lam CW, James J J, McCluskey R et al. 2004 *Pulmonary toxicity of single-wall CNT in Mice 7 and 90 days after intratracheal instillation. Toxicol Sci* **77**: 126-134.

[3] Takagi A, Hirose A, Nishimura T et al. 2008 *Induction of mesothelioma in p53 ± mouse by intraperitoneal application of multi-wall carbon nanotube. J Toxicol Sci* **33**: 105-116.

[4] Poland CA, Duffin R, Kinloch I et al. 2008 *Carbon nanotubes introduced into the abdominal cavity of mice show asbestos-like pathogenicity in a pilot study. Nat Nanotechnol* **3**: 328-333.

[5] NIOSH: current intelligence bulletin 65 2013 Occupational exposure to carbon nanotubes and nanofibers. Cincinati, Ohio: US Department of Health and Human Services Centers for Disease Control and Prevention, National Institute for Occupational Safety and Health; Publication No.2013-145.

[6] Birch ME, Ku BK, Evans DE, Ruda-Eberenz TA . 2011 *Exposure and emissions monitoring during carbon nanofiber production-Part 1: elemental carbon and iron-soot aerocols. The Annals of Occupational Hygiene* **55**: 1016-1136.

[7] Maynard AD, Baron PA, Foley M, Shvedova AA, Kisin ER, Castranova V. 2004 *Exposure to carbon nanotube material: aerosol release during the handling of unrefined single-walled carbon nanotube material. Journal of Tocicology and Environmental Health* Part A **67**: 87-107.

[8] Rasmussen PE, Jayawardene I, Gardner HD, Chenier M, Levesque C, Niu J. 2013 *Metal impurities provide useful tracers for identifying exposures to airborne single-wall carbon nanotube paritcles released from work-related processes. Journal of Physics: Conference Series* **429**: 012007.

[9] Reed RB, Goodwin DG, Marsh KL, Capracotta SS, Higgins CP, Fairbrother DH, Ranville JF. 2013 *Detection of single walled carbon nanotubes by monitoring embeded metals. Environmental Science: Processes & Impacts* **15**: 204-213.

[10] OpenFoam http://www.openfoam.com/

[11] Kuhlbusch TA, Fissan H. 2006 *Particle characteristics in the reactor and pelletizing areas of carbon black production. J Occup Environ Hyg* **3**: 558-567.

Propagation modelling based on airborne particle release data from nanostructured materials for exposure estimation and prediction

Daniel Göhler[1], Ralf Gritzki[2], Michael Stintz[1], Markus Rösler[2], Clemens Felsmann[2]

(1) Research Group Mechanical Process Engineering, Institute of Process Engineering and Environmental Technology, Technische Universität Dresden, D-01062 Dresden, Germany (2) Chair of Building Energy Systems and Heat Supply, Institute of Power Engineering, Technische Universität Dresden, D-01062 Dresden, Germany

E-mail: daniel.goehler@tu-dresden.de

Abstract. The gap between release and exposure is limiting the current risk assessment of nanostructured materials. Both, release and exposure were connected to each other by transport and transformation processes and require therefore the description/specification of complex exposure scenarios. Within this study, propagation modelling based on experimentally determined airborne particle release data was used for exposure estimation and prediction in a defined model room. Therefore, 9 different exposure scenarios based on 3 release scenarios and 3 ventilation scenarios were analysed. Results for near field considerations have shown that the level of inhalation exposure is fundamentally defined by the present exposure scenario, that personal heat can cause particle availability in the breathing zone and that highest exposure levels arise immediate during material processing.

1. Introduction

Risk assessment on nanostructured materials requires data on both exposure and substance-specific hazard [1]. Robust data for exposure estimation can be determined systematically by release analyses in laboratory [e.g. 2, 3], which are currently became standardised [4]. The state of dispersion (i.e. concentration, size) after release (condition at the particle source) can change significantly until exposure (condition at the entrance to the subject of protection) due to transport (e.g. dilution) and transformation (e.g. agglomeration) processes. Thus, one has to distinguish strictly between release and exposure data, because only in exceptional cases (i.e. accidents, misuse) release quantities can become exposure ones. Accordingly, exposure characterisation suffers usually on poorly reproducible, often cross-contaminated measurement data and requires in addition numerous contextual information (e.g. on environment, ventilation conditions, particle source characteristics, sampling strategy) for data interpretation [5]. In this context, release characterisation aimed to provide information solely on the particle source, i.e. process-specific and preferably quantitative release data as the origin of exposure.

During the last years we have performed several experimental studies on the characterization of airborne particle release from nanostructured materials along their life-cycle, i.e. for the handling of powders [6], due to the spray application of liquid coatings [7, 8] and during weak and moderate mechanical stress on solid composite materials (i.e. coatings, plastics) by abrasion [9] and sanding

[10-12]. Additionally, the impact of artificial aging and weathering on the release from solid composite materials due to wind erosion, dynamic friction and sanding was investigated [11, 12]. In order to provide quantitative data for further use, we renounced the use of measured particle concentration values and evaluated instead fractional numbers of released particles as described in [11] and which were related for the purpose of data-transfer (i.e. for scale-up/scale-down use) to process-specific values, like the stressed sample mass in the case of powder handling or the stressed surface area due to sanding [cf. 13, 14]. Note, that release rates (i.e. release quantities related to time), which are often requested in this context [e.g. 3, 5], are seen as derived semi-quantitative data of quantitative release data and thus are only valid for one case considerations.

Despite numerous release and exposure studies dealing with nanostructured materials [e.g. 2, 3, 15-17], there is still a gap concerning the use of release data for exposure estimation and thus for risk assessment [3]. To start with filling the gap, propagation modelling based on experimental determined release data was performed within the actual study for different exposure scenarios.

2. Details on propagation modelling

2.1. Computational details

2.1.1. Calculation modules. Propagation modelling was performed by coupling different calculation modules [e.g. 18] within a "parallel virtual machine" (PVM). The simulation of the indoor air flow is done by the computational fluid dynamic module ParallelNS [e.g. 19], which is based on a GALERKIN-Least-Squares finite element approach (GLS-FEM). This module uses the unsteady REYNOLDS-averaged NAVIER-STOKES equations (URANS) combined with various turbulence models, specific iterative wall functions and boundary layer approaches. To consider both fluid dynamics and thermodynamics within a geometric envelop, an enhanced version of a commercial thermal building simulation module called TRNSYS-TUD was coupled with ParallelNS.

2.1.2. Grid and time resolution. For the purpose of modelling, the grid of the chosen model room described in chapter 2.2 (cf. surfaces in Figure 1 b) was endued with 800000 tetrahedron elements, whereof 350000 ones accrued around the person. Modelling was performed to observe the air quality over a time frame of 120 min. The first 30 min were solely used to receive stationary air-flow conditions before the actual release scenarios (see chapter 2.4) were launched.

2.1.3. Simplifications on release processes and airborne particles. The main appointed simplifications regarding release were that the release processes do not interact with their environment (i.e. particles are supplied interference-free without directional momentum in a source volume) and that the released airborne particles behave gas-like. The former one allows the embedding of any release quantity after completed propagation modelling. The latter one means that particle-particle interactions (e.g. agglomeration), particle-wall interactions (e.g. loss of airborne particles by deposition) and some other particle interactions (e.g. loss of airborne particles by settling) were at first disregarded. Interestingly, a subsequent verification of the simplifications regarding the behaviour of airborne particles after modelling, which will not be discussed in this paper, has shown that the particle interactions (e.g. agglomeration) cause a comparable minor impact on the level of exposure.

2.2. Model room specifications

2.2.1. Geometric specifications. Figure 1 (a) shows the chosen model room (5 m x 6 m x 3 m), which contains of a triple-part window (5 m x 1.5 m), a door, a heated floor, a workbench (1 m x 4 m x 0.95 m) and a person (1.8 m). Dark blue and red highlighted areas illustrate the positions for air inlet as well as for air outlet, which were used for the different ventilation scenarios described in chapter 2.3. Green highlighted areas show the particle source regions for the release scenarios (cf. chapter 2.4), which were positioned in front of the person. Figure 2 (b) shows also the sensor location (red circle) in the breathing zone (1.6 m) of the person.

2.2.2. Thermal specifications. Except for the north wall with the window and the floor, all other walls as well as the ceiling are surrounded by other rooms. All walls fulfil the requirements of the German Energy Saving Ordinance (EnEV 2009). The outdoor temperature was set to $\vartheta_{outdoor}$ = 5°C that corresponds to the mean European outdoor temperature in winter time. The underfloor heating provides a mean room temperature of ϑ_{room} = 20°C, that lead to varying floor temperatures in the dependence of the ventilation scenario (cf. chapter 2.3). Excepting head (ϑ_{head} = 35 °C) and hands (ϑ_{hands} = 30 °C), the thermal-active person has a surface temperature of $\vartheta_{clothes}$ = 26°C as typical for a clothed one at moderate activity. All data were chosen according to DIN EN ISO 7730:2006 [20].

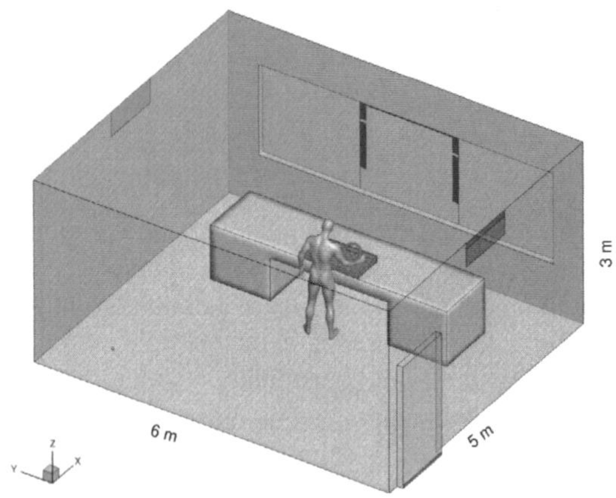

(a)

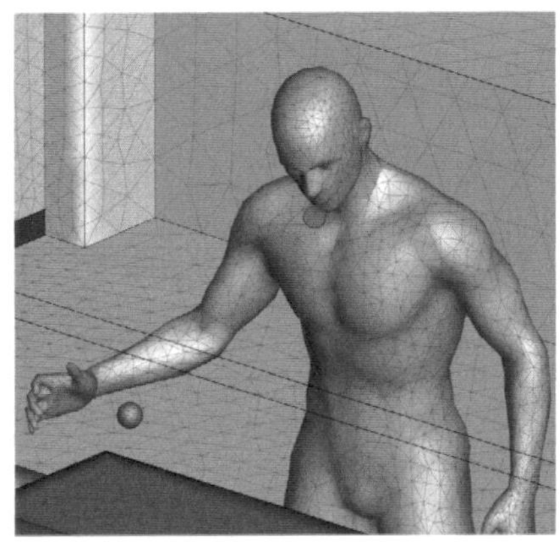

(b)

Figure 1. Model room envelope and interior (a) as well as positions of release sources and breathing zone sensor (b); blue areas = air inlets, red areas = air outlets, green regions = particle release sources.

2.3. Ventilation scenarios

To consider both consumer protection and occupational safety, three different ventilation scenarios were studied, i.e. natural ventilation by door-slit infiltration, natural ventilation by a pivot-hung window and by an improved technical ventilation system. The main specifications of the ventilation scenarios are summarised within Table 1.

Table 1. Main specifications of the artificially designed ventilation scenarios.

ventilation scenario [-]	**ID** [-]	**air exchange rate** [h^{-1}]	**air inlet temperature** [°C]	**comments on air inlet and outlet** [-]
natural ventilation by door-slit infiltration	NVD	0.5	20	air inlet at door slit on east wall, air-outlet at ceiling on west wall
natural ventilation by pivot-hung window	NVW	1.5	5	air inlet and outlet at window on north wall
improved technical ventilation system	TVS	8.0	20	air inlet and outlet at ceiling of east and west wall, flow orientation on workbench

2.4. Release scenarios

Three different near field release scenarios were chosen in this study, i.e. dry wiping of a coated work piece (1 m x 0.5 m = 0.5 m²) for 10 s, belt sanding of a coated work piece (1 m x 0.5 m = 0.5 m²) for 60 s and spray can application of a liquid coating for 60 s. The main specifications of the release scenarios are summarised in Table 2.

Table 2. Main specifications of the release scenarios.

release scenario	ID	ref.	original sample	area-spec. particle release number < 10 µm	particle release rate < 10µm	processing time	processed area	source volume
[-]	[-]	[-]	[-]	[#/m²]	[#/s]	[s]	[m²]	[L]
wiping	WIP	[9]	UV-ZnO[a]	5.0E+05	n.r.	10	0.5	50
sanding	SAN	[12]	UV*-ZnO[b]	1.0E+11	n.r.	60	0.5	50
spraying	SPR	[7]	PU-ZnO[c]	n.r.[d]	7.5E+09	60	n.r.	8

[a] UV-curable parquet coating doped with ZnO nanoparticle additive (UV-ZnO)
[b] 2000 h artificially aged UV-curable parquet coating doped with ZnO nanoparticle additive (UV*-ZnO)
[c] two-pack polyurethane coating doped with a ZnO nanoparticle additive (PU-ZnO)
[d] not relevant

For the purpose of traceability, the measured number-weighted particle size distributions (q_0) of the chosen original samples were fitted at first to lognormal number-weighted density distributions ($q_{0,LNVT}$). This led in the case of sanding to a number-weighted median diameter of $x_{50,0,SAN}$ = 240 nm and a geometric standard deviation of GSD_{SAN} = 1.4, for spraying to $x_{50,0,SPR}$ = 120 nm and GSD_{SPR} = 2.0. Furthermore, the $q_{0,LNVT}$ were converted according to ISO 9276-5:2005 [21] into volume/mass-weighted ones ($q_{3,LNVT}$) to provide volume/mass-weighted data. Note that the conversion of particle size distributions (q_r) between the different types of quantities can lead to huge errors, especially by operating lognormal fits [22]. However, $q_{0,LNVT}$ and $q_{3,LNVT}$ were combined with a gender and activity averaged modified ICRP 66 model for particle deposition in the human respiratory tract of adults as provided in [23] to determine the fractions of total deposition ($\eta_{dep,tot,r}$) as well as regional deposition for the head airways ($\eta_{dep,HA,r}$), the tracheobronchial region ($\eta_{dep,TB,r}$) and the alveolar region ($\eta_{dep,TB,r}$) based on the fraction of inhaled particles ($\eta_{inh,r}$). This results in the following set of empirical equations, wherein the parenthesized terms are the deposition efficiencies provided in [23].

$$
\begin{aligned}
\eta_{inh,r} &= \int_{1\,\mathrm{nm}}^{100\,\mu\mathrm{m}} \left[\left(1-0.5\cdot\left(1-\frac{1}{1+0.00076\cdot x^{2.8}}\right)\right)\right]\cdot q_r(x)\mathrm{d}x \\
\eta_{dep,tot,r} &= \int_{1\,\mathrm{nm}}^{100\,\mu\mathrm{m}} \left[\left(1-0.5\cdot\left(1-\frac{1}{1+0.00076\cdot x^{2.8}}\right)\right)\cdot\left(0.0587+\frac{0.911}{1+\exp(4.77+1.485\cdot\ln(x))}+\frac{0.943}{1+\exp(0.508-2.58\cdot\ln(x))}\right)\right]\cdot q_r(x)\mathrm{d}x \\
\eta_{dep,HA,r} &= \int_{1\,\mathrm{nm}}^{100\,\mu\mathrm{m}} \left[\left(1-0.5\cdot\left(1-\frac{1}{1+0.00076\cdot x^{2.8}}\right)\right)\cdot\left(\frac{1}{1+\exp(6.84+1.183\cdot\ln(x))}+\frac{1}{1+\exp(0.924-1.885\cdot\ln(x))}\right)\right]\cdot q_r(x)\mathrm{d}x \\
\eta_{dep,TB,r} &= \int_{1\,\mathrm{nm}}^{100\,\mu\mathrm{m}} \left[\left(\frac{0.00352}{x}\right)\cdot\left(\exp\left(-0.234\cdot(\ln(x)+3.40)^2\right)+63.9\cdot\exp\left(-0.819\cdot(\ln(x)-1.61)^2\right)\right)\right]\cdot q_r(x)\mathrm{d}x \\
\eta_{dep,AL,r} &= \int_{1\,\mathrm{nm}}^{100\,\mu\mathrm{m}} \left[\left(\frac{0.0155}{x}\right)\cdot\left(\exp\left(-0.416\cdot(\ln(x)+284)^2\right)+19.11\cdot\exp\left(-0.482\cdot(\ln(x)-1.362)^2\right)\right)\right]\cdot q_r(x)\mathrm{d}x
\end{aligned}
\tag{1}
$$

A summary of the determined number- and mass-weighted fractions for total and regional deposition in the human respiratory tract is provided in Table 3. Note that in [9] no particle size distribution could be determined based on the low numbers of released particles.

Table 3. Number- and mass-weighted fractions for regional deposition in the human respiratory tract based on the gender and activity averaged modified ICRP 66 model for adults [23].

				parameter of $q_{0,LNVT}$		number-weighted fractions for total and regional deposition				mass-weighted fractions for total and regional deposition			
release scenario	ID	ref.	original sample	$x_{50,0}$	GSD	$\eta_{dep,tot,0}$	$\eta_{dep,HA,0}$	$\eta_{dep,TB,0}$	$\eta_{dep,AL,0}$	$\eta_{dep,tot,3}$	$\eta_{dep,HA,3}$	$\eta_{dep,TB,3}$	$\eta_{dep,AL,3}$
[-]	[-]	[-]	[-]	[nm]	[-]	[no.%]	[no.%]	[no.%]	[no.%]	[wt.%]	[wt.%]	[wt.%]	[wt.%]
sanding	SAN	[12]	UV*-ZnO	240	1.4	11.30	3.80	0.80	6.70	14.10	6.50	0.70	6.90
spraying	SPR	[7]	PU-ZnO	120	2.0	21.52	3.37	3.04	15.12	26.21	15.67	1.56	8.98

IOP Conf. Series: Journal of Physics: Conf. Series **838** (2017) 012010 doi:10.1088/1742-6596/838/1/012010

3. Results

3.1. Ventilation scenarios

To assess the quality of air exchange in steady state flow conditions, the ventilation scenarios were compared using the local air exchange index ε, which is defined as the ratio of the mean air age at the exit $\tau_{mean,out}$ (i.e. reciprocal of air exchange rate n_{AER}) to the local air age τ_{local} in the room [e.g. 18, 24]:

$$\varepsilon = \frac{\tau_{mean,out}}{\tau_{local}} = \frac{1}{n_{AER} \cdot \tau_{local}} \tag{2}$$

Local air exchange indices of $\varepsilon < 1$ indicate regions of poor air exchange, while for $\varepsilon > 1$ regions with effective air exchange can be identified. Figure 2 provides beside the local air exchange index, the velocity field in the room as well as the mean velocity in the sectional plane through the working area for the three simulated ventilation scenarios in stationary state.

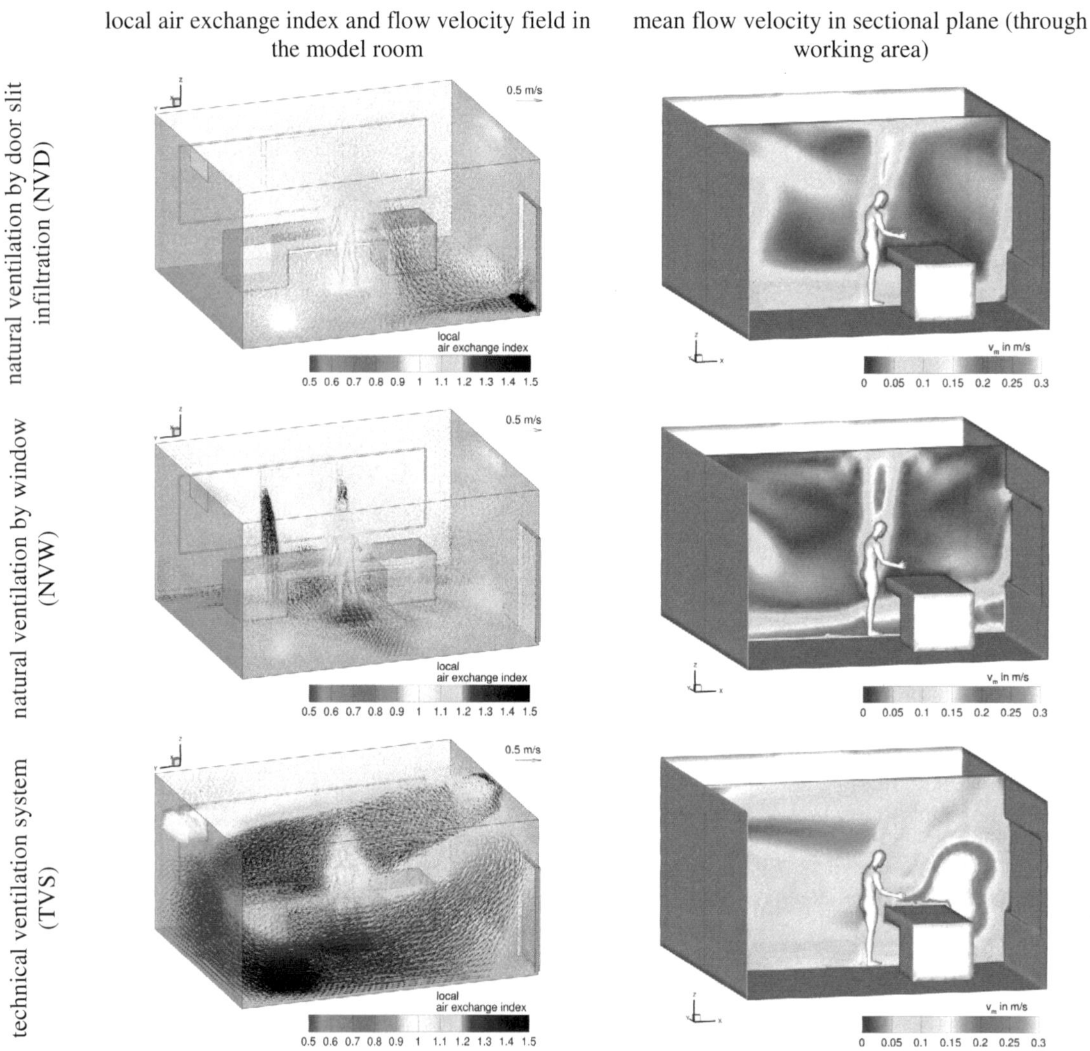

Figure 2. Quality of air exchange and mean velocity in the sectional plane at stationary state for the different ventilation scenarios.

Figure 2 reveals that in the case of NVD an effective air exchange can only be found on the floor behind the person, while in case of NVW regions of effective air exchange are limited to the window and the floor. The best spatial distributed regions of effective air exchange arose for TVS. Regardless the kind of ventilation, a local air exchange index of about 1 can be observed at the breathing zone, i.e. the local air age correlates with the air exchange rate. As it can be deduced from the mean velocity illustration, both NVD and NVW suffer from a poor flow through the working area in contrast to TVS. Furthermore, comparable high flow velocities on the person caused by personal heat can be observed.

3.2. Propagation scenarios and exposure scenarios

When reaching steady state ventilation conditions after 30 min, the release scenarios (cf. chapter 2.4) were launched for ≤ 1 min. A visualization of the particle cloud propagation at 100 s after release initiation for sanding and spraying based on the three ventilation scenarios is provided in Figure 3.

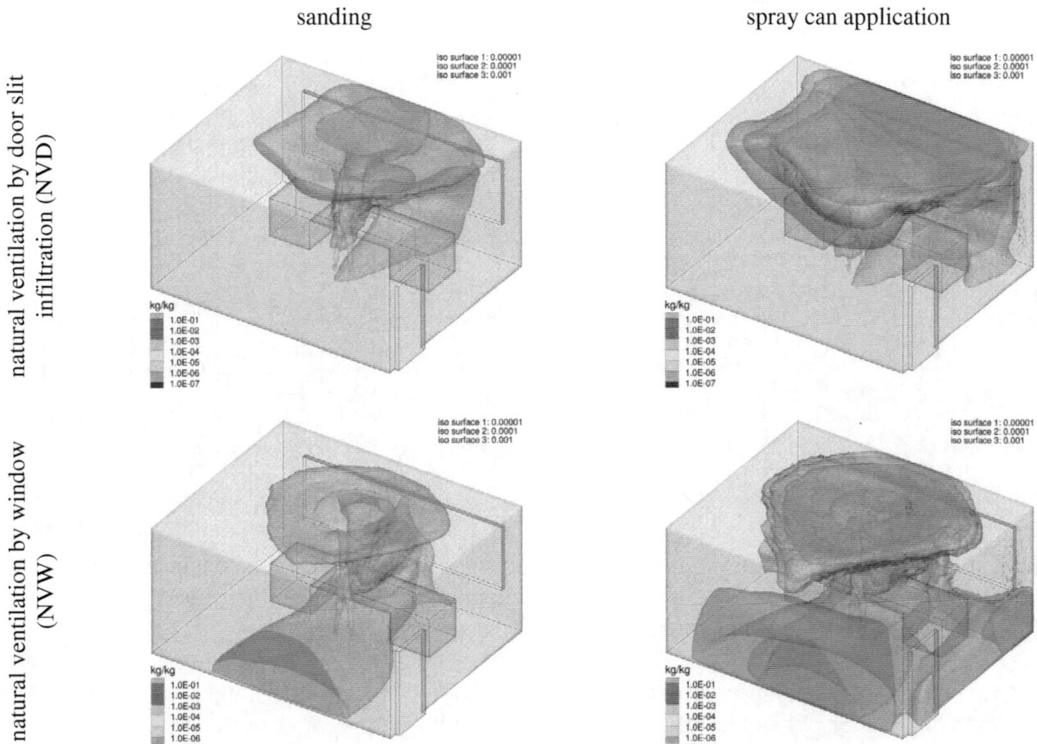

Figure 3. Visualisation of the particle cloud propagation at 100 s after release initiation based on the illustration of three different iso-surfaces for the relative gas exchange.

As it can be deduced from Figure 3 for all propagation scenarios, the flow velocities of around 0.2 m/s on the person caused by personal heat (cf. Figure 2) lead to an up-flow and thus to an aerosol propagation in direction to the ceiling via the breathing zone. In contrast to NVD and NVW, the TVS led to a faster dilution of the released airborne particles that is indicated by considerable higher spatial aerosol propagation.

Based on the recorded data of the relative gas exchanges at the sensor and the experimentally determined release data, the temporal courses of the particle number concentration in the breathing zone (i.e. exposure) was determined as exemplarily shown in Figure 4 (a) for spraying. The temporal courses of the particle number concentration for wiping and sanding show a similar shape, but differ significantly in the concentration levels.

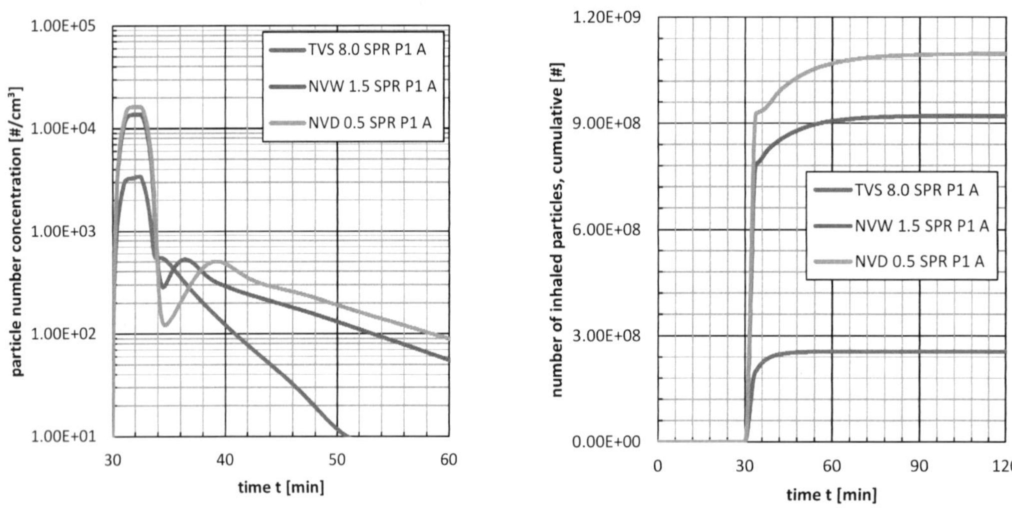

(a) (b)

Figure 4. Particle number concentration in the breathing zone over the first 30 min after launch of release (a) and cumulative number of inhaled particles over the whole modelling time (b) for the three ventilation scenarios on the example of spray can application.

Figure 4 (a) reveals that the highest exposure levels arise immediate during or rather short after release with a slight time delay. After the release, the particle number concentration in the breathing zone decreases at first rapidly. Afterward the concentration increases slightly for a short period of time before a gradually decrease over time can be observed. Both, the magnitude of the rapid decrease in the first phase and the strength of the slope in the third phase depend clearly on the ventilation scenario.

Based on a gender and activity averaged breathing rate of 383 cm^3/s, cumulative numbers of inhaled particles were determined. The values of fractions for total and regional particle deposition in the human respiratory tract (cf. Table 3) and the assumption of unit density (i.e. 1000 kg/m^3) were used for the calculation of cumulative numbers and masses of inhaled and deposited particles. These data are shown in Figure 5, while the corresponding values are provided in Table 4.

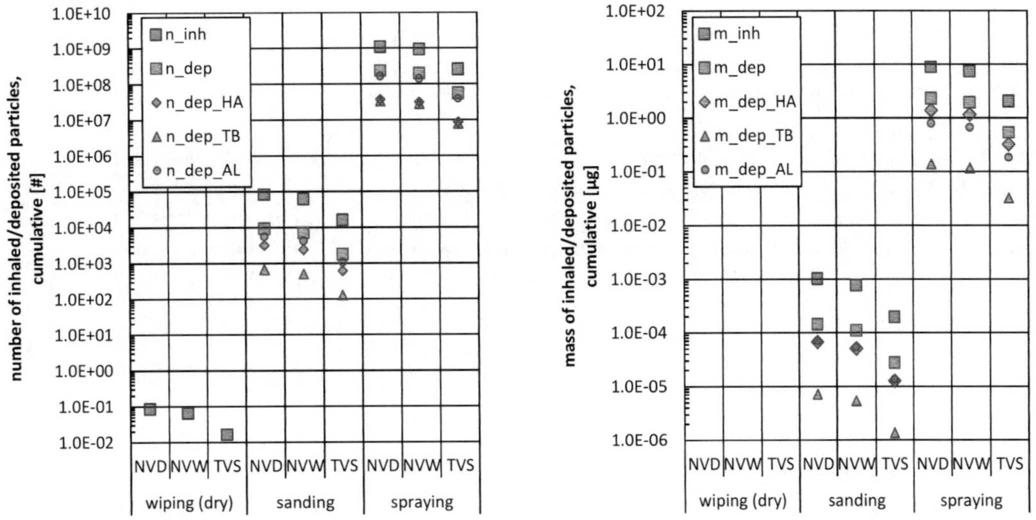

(a) (b)

Figure 5. Cumulative numbers of inhaled/deposited particles (a) and cumulative masses of inhaled/deposited particles (b) of the exposure scenarios.

Figure 5 and Table 4 reveal that the exposure levels due to wiping are negligible (i.e. $n_{inh,WIP} < 1$ #), while the spray can application led to the highest exposure levels (i.e. $n_{inh,SPR} = 2.6 \cdot 10^8$ # - $1.2 \cdot 10^9$ #; $m_{inh,SPR} = 2.0$ µg – 8.8 µg), especially due to poor ventilation by door slit infiltration. The exposure levels due to sanding (i.e. $n_{inh,SAN} = 1.6 \cdot 10^4$ # - $8.3 \cdot 10^4$ #; $m_{inh,SAN} = 0.2$ ng – 1.2 ng) were found to be around 4 magnitudes of orders lower than the ones for spraying, but 6 magnitudes of orders higher in comparison to wiping.

Table 4. Calculated scenario-specific cumulative numbers/masses of inhaled and regional deposited particles in the human airways based on averaged values for gender (adult breathing rate of 383 cm^3/s), activity (i.e. moderate activity) and fractional deposition (modified ICRP 66 deposition model according to [23]).

release scenario	ventilation scenario	n_{inh} [#]	n_{dep} [#]	$n_{dep,HA}$ [#]	$n_{dep,TB}$ [#]	$n_{dep,AL}$ [#]	m_{inh} [µg]	$m_{dep,tot}$ [µg]	$m_{dep,HA}$ [µg]	$m_{dep,TB}$ [µg]	$m_{dep,AL}$ [µg]
wiping	NVD	8.5E-2	n/a	n/a	n/a	n/a	n/a	n/a	n/a	n/a	n/a
	NVW	6.5E-2	n/a	n/a	n/a	n/a	n/a	n/a	n/a	n/a	n/a
	TVS	1.6E-2	n/a	n/a	n/a	n/a	n/a	n/a	n/a	n/a	n/a
sanding	NVD	8.3E+4	9.4E+3	3.2E+3	6.6E+2	5.6E+3	1.0E-3	1.4E-4	6.6E-5	7.1E-6	7.0E-5
	NVW	6.3E+4	7.1E+3	2.4E+3	5.0E+2	4.2E+3	7.7E-4	1.1E-4	5.0E-5	5.4E-6	5.3E-5
	TVS	1.6E+4	1.8E+3	6.0E+2	1.3E+2	1.1E+3	1.9E-4	2.7E-5	1.3E-5	1.4E-6	1.3E-5
spraying	NVD	1.1E+9	2.4E+8	3.7E+7	3.3E+7	1.7E+8	8.8E+0	2.3E+0	1.4E+0	1.4E-1	7.9E-1
	NVW	9.2E+8	2.0E+8	3.1E+7	2.8E+7	1.4E+8	7.4E+0	1.9E+0	1.2E+0	1.2E-1	6.6E-1
	TVS	2.6E+8	5.5E+7	8.6E+6	7.7E+6	3.9E+07	2.0E+0	5.4E-1	3.2E-1	3.2E-2	1.8E-1

In order to quantify the magnitude of dilution for each exposure scenario, ratios between the numbers of released particles to the numbers of inhaled particles (designated as RIR = release-inhalation-ratio) were determined (cf. Table 5). Across all the analysed exposure scenarios, the RIR varied between $4 \cdot 10^2$ up to $1.5 \cdot 10^7$, because of the different local positions of the release sources in the flow field. Considering separately the change of RIR within the exposure scenarios for each release scenario, the dilution efficiency due to the different ventilation scenarios can be examined. Thus, the NVW lead to a

1.2 – 1.3 times higher dilution or respectively lower exposure level than the NVD, while the dilution efficiency increases by a factor of 4.3 - 5.2 due to switching from NVD to TVS. Data on the RIR change between the ventilation scenarios show furthermore, that with decreasing distance to the source the dilution efficiency decreases.

Table 5. Magnitude of dilution from release to exposure based on the ratio between released and inhaled particles (RIR) for each exposure scenario and comparison between the different ventilation scenarios.

release scenario	distance from sensor to source	release duration	ratio between released and inhaled particles (RIR)			change of RIR between the ventilation scenarios		
			NVD	NVW	TVS	NVW/NVD	TVS/NVW	TVS/NVD
[-]	[cm]	[s]	[-]	[-]	[-]	[-]	[-]	[-]
wiping	65.0	10	2.9E+6	3.8E+6	1.5E+7	1.3	4.0	5.2
sanding	65.0	60	6.0E+5	8.0E+5	3.2E+6	1.3	4.0	5.2
spraying	52.5	60	4.1E+2	4.9E+2	1.8E+3	1.2	3.6	4.3

4. Summary and conclusion

Propagation modelling based on experimentally determined airborne particle release data was performed for exposure estimation and prediction. Overall 9 different exposure scenarios based on 3 near-field (< 1 m distance from source to breathing zone) release scenarios (i.e. wiping, sanding, spraying) and 3 ventilation scenarios (i.e. technical ventilation, natural ventilation by door-slit infiltration and by natural pivot-hung-window ventilation) were analysed within a 90 m^3 model room. For the purpose of exposure estimation, the air condition in the breathing zone was observed for a time frame of 120 min, whereof the first 30 min served solely for reaching almost steady state ventilation conditions before the actual release scenarios were launched for ≤ 1 min. On the basis of modelled data for the relative gas exchange, particle number and mass concentration over time as well as cumulative numbers and masses of inhaled and deposited particles in the human respiratory tract were determined.

Despite a lot of assumptions and specifications for propagation modelling, the chosen exposure scenarios show impressive that the level of exposure is affected fundamentally by the release scenario, the ventilation scenario and the positioning of the subject of protection (i.e. consumer/worker). The latter one comprises beside the distance from source to breathing zone also the localisation within the flow field. Within this study, the ratio between the number of released particles to the number of inhaled particles, which can be seen as a kind of dilution factor, varied between $4 \cdot 10^2$ up to $1.5 \cdot 10^7$ for the analysed exposure scenarios. The highest levels of exposure revealed for the modelled spray scenario, where around one billion particles with a mass of ca. 9 µg were inhaled during and after 60 s spray can application in a room at poor ventilation conditions. Additionally, it could be observed for all exposure scenarios that convective flow structures due to person-caused heat can lead to particle availability in the breathing zone and that highest exposure levels arise immediate during material processing.

Accordingly, it could be illustrated how measured particle release data have to be added into well-considered surrounding scenarios to get consumer-relevant or occupational exposure data. This underlines the requirement on a sufficient set of contextual information [e.g. 3, 5], which are necessary for the sampling strategy regarding exposure measurement, the interpretation of measured exposure data or for the prediction/assessment of exposure based on experimentally determined release data.

Acknowledgement
This work was supported by the German Paint Industry Association (VdL e.V., Frankfurt/Main, Germany, representing about 180 German companies). The authors wish to thank all involved members of the VdL e.V. for the suggestions and the discussions during the project.

IOP Conf. Series: Journal of Physics: Conf. Series **838** (2017) 012010 doi:10.1088/1742-6596/838/1/012010

References

[1] National Research Concil Committee on the Institutional Means for Assessment of Risks to Public Health 1983 *Risk Assessment in the Federal Government: managing the process*, Washington, National Academic Press

[2] Kuhlbusch T A J, Asbach C, Fissan H, Göhler D and Stintz M 2011 *Part. Fibre Toxicol.* **8** 22

[3] Ding Y, Kuhlbusch TAJ, Tongeren MV, Jiménez AS, Tuinman I, Chen R, Alvarez IL, Mikolajczyk U, Nickel C, Meyer J, Kaminski H, Wohlleben W, Stahlmecke B, Clavaguera S, Riediker 2016 *J. Hazard. Mater.* **322** 17-28

[4] Stintz M and Göhler D 2015 *Procedia Eng.* **102** 233-239

[5] Clark K, van Tongeren M, Christensen FM, Brouwer D, Nowack B, Gottschalk F, Micheletti C, Schmid K, Gerritsen R, Aitken R, Vaquero C, Gkanis V, Housiadas C, de Ipina JML and Riediker M 2012 *J. Nanopart. Res.* **14** 970

[6] Göhler D and Stintz M 2015 *J. Phys. Conf. Ser.* **617** 012029

[7] Göhler D and Stintz M 2014 *J. Nanopart. Res.* **16** 2520

[8] Göhler D, Stintz M, Rommert A and Eichstädt D 2015 *Farbe und Lack* **121** 142-149

[9] Vorbau M, Hillemann L and Stintz M 2009 *J. Aerosol Sci.* **40** 209-17

[10] Göhler D, Stintz M, Hillemann L and Vorbau M 2010 *Ann. Occup. Hyg.* **54** 615-24

[11] Göhler D, Nogowski A, Fiala P and Stintz M 2013 *J. Phys. Conf. Ser.* **429** 012045

[12] Göhler D, Stintz M and Rommert A 2016 *Farbe und Lack* **122** 52-60

[13] Schneider T and Jensen KA 2008 *Ann. Occup. Hyg.* **52** 23-34

[14] Wohlleben W and Neubauer N 2016 *NanoImpact* **1** 39-45

[15] Losert S, von Goetz N, Bekker C, Fransman W, Wijnhoven SWP, Delmaar C, Hungerbuhler K and Ulrich A 2014 *Environ. Sci. Technol.* **48** 5366-5378

[16] Froggett SJ, Clancy SF, Boverhof DR and Canady RA 2014 *Part. Fibre Toxicol.* **11** 17

[17] Harper S, Wohlleben W, Doa M, Nowack B, Clancy S, Canady R and Maynard A 2015 *J. Phys.: Conf. Ser.* **617** 01226

[18] Gritzki R, Richter W and Rösler M 2003 *Int. J. of Vent.* **1** 33-39

[19] Knopp T, Lube G, Gritzki R and Rösler M 2005 *Comput. Method. Appl. M.* **194** 3797-3816

[20] DIN EN ISO 7730:2006. *Ergonomics of the thermal environment - Analytical determination and interpretation of thermal comfort using calculation of the PMV and PPD indices and local thermal comfort criteria*

[21] ISO 9276-5:2005. *Representation of results of particle size analysis – Part 5: Methods of calculation relating to particle size analyses using logarithmic normal probability distribution*

[22] Babick F and Ullmann C 2016 *Powder Technol.* **301** 503-510

[23] Hinds WC 1999 *Aerosol Technology - Properties, Behavior, and Measurement of airborne particles*, New York, John Wiley & Sons, 2, 242-245

[24] Wildeboer J and Müller D *2006 KI Luft- und Kältetechnik* **42** 439-443

Exposure assessment method for products containing carbon nanotubes inside a test chamber with a Taber abrasion machine

Yasuto Matsui[1], Taiki Nagaya[2], Kato Nobuyuki[2], Tomonori Ishibashi[2], Minoru Yoneda[2]

[1]Agency for Health, Safety and Environment, Kyoto University, Japan
[2]Graduate School of Engineering, Kyoto University, Japan

Abstract. Polymer/carbon nanotube (CNT) composites exhibit distinguished properties, but more quantitative risk assessments on CNTs are necessary as research and development advances. One method to assess the exact risk is to evaluate the characteristics of nanoparticles generated from CNT composites during sanding or Taber abrasion tests. Some researchers have applied loads to CNT composites using Taber machines and analysed the particles using aerosol-measuring instruments and electron microscopes. However, employing aerosol-measuring instruments is challenging due to the small amount of generated particles. Additionally, the presence of abundant background nanoparticles in testing environments creates issues in quantitative measurements. Our research strives to develop an examination method to measure even very small amounts of nanoparticles generated by Taber abrasion. In this study, a Taber abrasion machine is miniaturised so that it fits inside a small chamber. A high-efficiency particulate air filter is attached to the chamber to eliminate background nanoparticles. Then CNT composites are abraded with the miniaturised Taber abrasion machine inside the chamber and the generated particles are analysed.

1. Introduction

Due to their outstanding properties, products containing CNTs are actively being researched and developed. Although many studies have focused on composite materials of multiwall carbon nanotubes (MWCNTs) and epoxy resins, more quantitative risk assessments on CNTs, including hazard and exposure assessments, are required. Some researchers have reported hazard assessments of the MWCNT itself. Takagi et al. reported that MWCNT (MITSUI MWCNT-7, Lot NO. 060125-01k) induced mesothelioma upon injecting a 1 ml solution containing of 1.0×10^9 of the MWCNT particles intraperitoneally to p53 heterozygous male mice (age of 9 to 11 weeks) [1]. Other works have measured the released particles from CNT products by sanding, Taber abrasing, shredding, and grinding. Using aerosol measuring instruments and electron microscopes to analyse nanoparticles generated from epoxy/MWCNTs via a hand-held sander, Gupta et al. detected CNT protrusions from the particles [2].

However, a standardised exposure assessment method for products containing CNTs has yet to be established. In general, a Taber abrasion test is a common wear assessment, which uses a Taber

IOP Conf. Series: Journal of Physics: Conf. Series **838** (2017) 012011 doi:10.1088/1742-6596/838/1/012011

abrasion machine to perform accelerated wear testing. It is the global standard for evaluating abrasion resistance. A Taber abrasion machine can evaluate weight loss, the Taber wear index, wear cycles per mil, the depth of wear, the residual breaking force, average breaking strength, and percentage loss in breaking strength. Although the Taber test has been applied to evaluate exposure during abrasion of CNT products, it is designed to assess the abrasion resistance and not the generated particles from the products.

This study verifies whether inhalation dust is generated from products containing carbon nanotubes (CNTs) by accelerated tests using a Taber abrasion machine inside a chamber. Since our research strives to analyse particles generated by wear in this test, a load larger than that used in a general Taber test was prepared.

2. Methods

The experiment consisted of a Taber abrasion machine inside a chamber. Eight samples were used to measure and analyse the particles released by sanding. A Taber abrasion machine, which can evaluate the wear resistance of plane materials by abrasion rings, has been used to examine plastic, rubber, construction materials, smooth-planed board, leather, paper, cloth, and coatings. A general Taber abrasion machine is based on JIS K6264-2, but some aspects of our machine differ. Similar to a general Taber abrasion machine, our machine used the same abrasion rings: CS10 for soft rubber samples and H10 for hard rubber sample. However, the sample size (a square with sides of 50 mm and 1 mm thick) differed from that based on JIS K6264-2 (a circle with 120-mm diameter and from 1 to 5 mm thick). Although the rotation speed of our machine was the same as a general machine (72 r/min), a heavier load (19.6 N) than the general one (2.45 N, 4.9 N or 9.8 N) was applied because we wanted to verify if the samples released nanoparticles. In addition, we examined the released particles using a general Taber abrasion machine.

To control the background nanoparticles in the testing environment, we created a stainless steel chamber with an attached high efficiency particulate air (HEPA) filter to eliminate background nanoparticles. The chamber measured 30 cm deep, 30 cm wide, and 30 cm high. The Taber abrasion machine was placed in the chamber. This setup generated nanoparticle samples, which were barely deposited inside the chamber, allowing the released particles to be precisely measured. Furthermore, the drive part of the Taber abrasion machine was outside the chamber in order to control the particles generated from the drive part. Table 1 lists the samples used in the experiment.

Table 1. Samples used in the experiment.

No.	Corporation	Matrix	Nanomaterial	%
1	A	Polyamide12	CNT_3	4
2		Polycarbonate		3
3		Polybutylene terephthalate		4

2.1. Generation and sampling of nanoparticles

The Taber abrasion machine inside the chamber was operated at a rotation speed of 72 r/min until the total number of rotations reached 1,000 revolutions. Table 2 shows the instruments used to measure the released particles. Figure 1 depicts the system used to analyse the released particles.

Table 2. Instruments used in the experiments.

Item	Instrument	Application
Number of particles and size distribution	FMPS, SMPS, OPS	5.6–560 nm, 10–420 nm, 0.3–10 μm
Morphology observations	SEM	Particles on the filter, abrasion powder, matrix

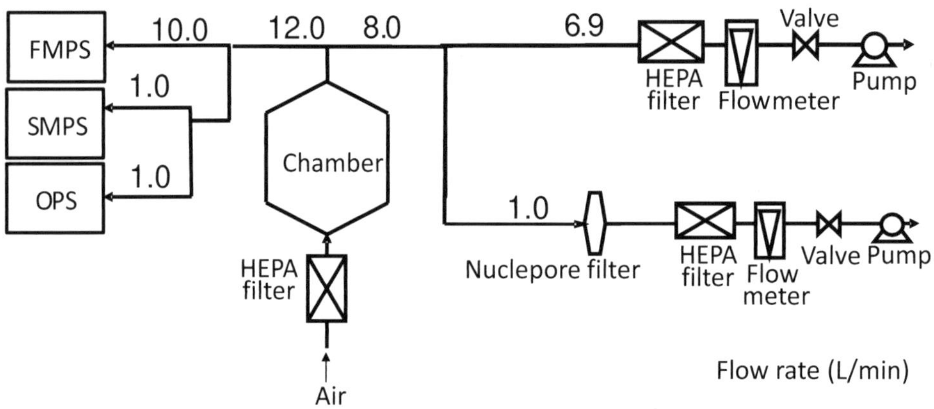

Figure 1[A1]. Sampling system.

Three aerosol measurement instruments were used. The fast mobility particle sizer (FMPS, model 3091, Tokyo Dylec Co.), which can classify 32 total channels at a flow rate of 10 L/min, measured particles between 5.6–560 nm. The scanning mobility particle sizer (SMPS, model 3910, Tokyo Dylec Co.), which can classify 13 channels at a flow rate of 1.0 L/min, had a range of 10–420 nm. The optical particle sizer (OPS, model 3330, Tokyo Dylec Co.), which can classify 16 channels at a flow rate of 1.0 L/min, measured the sub-micron and micron-sized particles with a size distribution between 0.3–10 μm. We also conducted morphological observations of the particles sampled on a Nuclepore filter (PC MB 25 mm 0.2 μm, GE Healthcare Japan Co.), the abrasion powder, and the surface of the samples using a Scanning Electron Microscope (SEM). For the sample on the filter, we used a pump with a flow rate of 1.0 L/min.

2.2. Analysis of the particles

Prior to the actual experiment, particles were eliminated in the chamber using clean air. Then the Taber abrasion machine was used to release particles from the samples. We measured the number of particles released and the size distribution while sanding as well as conducted morphological observations.

2.2.1. Measurements of the number of released particles and their size distribution

Using the aerosol-measuring instruments, we measured the number of particles in the chamber prior to running the machine. The particles were assessed for five minutes to determine the particle background (BG). In addition, we measured the particles in the chamber while the machine was running without sanding a sample for five minutes. Finally, we measured the samples while sanding for about 14 minutes.

2.2.2. Morphology observations

IOP Conf. Series: Journal of Physics: Conf. Series **838** (2017) 012011 doi:10.1088/1742-6596/838/1/012011

We collected the released airborne particles on the filters while sanding the samples. After the experiment, we collected the abrasion powder using carbon tape. SEM was used to observe the filters, powder, and sample surface after abrasion.

3. Results

3.1. Results of the aerosol measuring instruments

Figure 2 shows the number of particles present in the laboratory as well as in the chamber where the particles in the air are controlled. Using FMPS, we measured the total number of particles for both conditions (ambient air and the chamber) for five minutes. The vertical axis represents the average value of the total number of particles per second. The error bars denote the standard deviation.

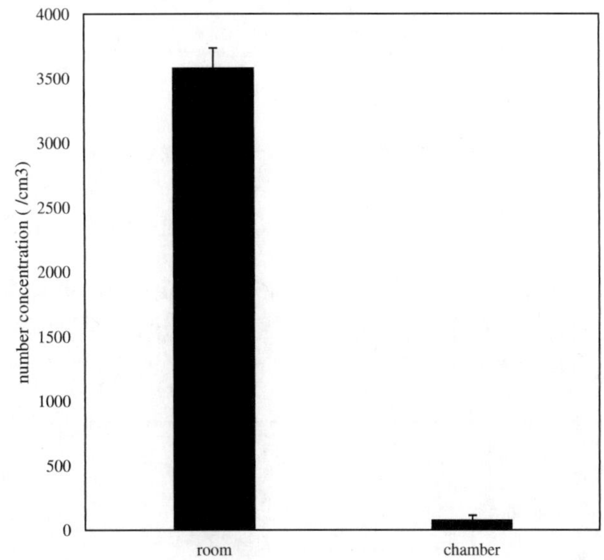

Figure 2. Total number of particles in the laboratory (room) and the chamber.

The chamber can control the number of particles present. Figures 3 and 4 show the measurement results for the number of particles and the size distribution during abrasion using different aerosol measuring instruments. The vertical axis represents the average number of particles (dN/dlogDp) while the horizontal axis represents the diameter (nm). The error[A2] bars represent the standard while sanding for about 14 minutes.

IOP Conf. Series: Journal of Physics: Conf. Series **838** (2017) 012011 doi:10.1088/1742-6596/838/1/012011

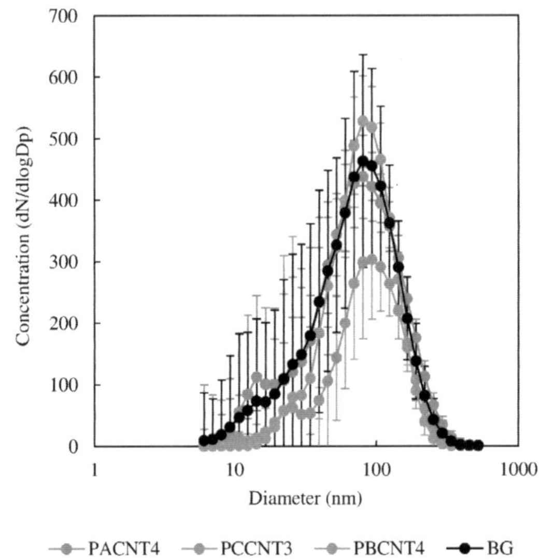

Figure 3. Size distribution of BG and released particles during abrasion (FMPS).

The number of particles is represented by the average size distribution measured by FMPS. In all samples, the concentration near 100 nm is high, but the concentration during abrasion is similar to the detection limit.

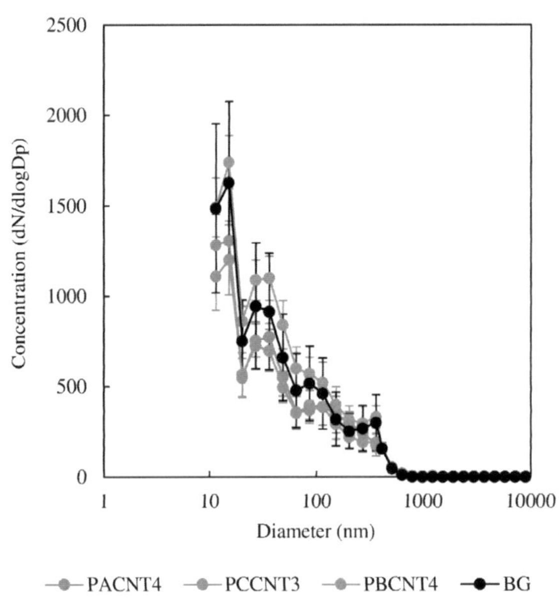

Figure 4. Size distribution of BG and released particles during abrasion (SMPS and OPS).

The number of released particles was also determined by SMPS and OPS. In all samples, the number of particles near 10 nm is very high, and the numbers of particles are high near 40, 100, and 400 nm. However, both the size distribution and concentrations are similar to the BG concentration. Bello et al. reported that drilling composite materials generates smoke with a particle diameter around 10 nm [3].

IOP Conf. Series: Journal of Physics: Conf. Series **838** (2017) 012011 doi:10.1088/1742-6596/838/1/012011

Thus, the high particle concentration near 10 nm may be due to thermal decomposition during abrasion. Because the BG concentration at 10 nm is higher than that during abrasion, the origin of the high concentration near 10 nm is unclear.

There are possible three reasons why BG is high. 1) Particles are generated from the driving part of the Taber abrasion machine. 2) The HEPA filter used to introduce air into the chamber does not sufficiently control the particles. 3) Ambient air outside of the chamber enters the chamber due to a negative chamber pressure. Regarding 1), we measured the concentration near the drive section, but the number of particles is very low. Regarding 2), the concentration measured when the HEPA filter was connected to FMPS indicates that the filter sufficiently controls the particles. Regarding 3), in this study, the total flow rate evacuated from the chamber to the instruments is extremely high at 20 L / min, causing a negative pressure inside the chamber. This pressure difference may result in contamination from the general environment. In the future, it is necessary to repeat the experiment while measuring the pressure difference inside and outside of the chamber with the pressure gauge. If the chamber has a negative pressure, the pressure difference should be reduced by introducing clean air using a pump.

In Figs. 3 and 4, there is an inversion of the curves, which may be due to the measurement principles of FMPS, SMPS, and OPS. Each device charges the particles and then classifies the particles using a high voltage rod. However, the particle detection method varies between FMPS and SMPS･OPS. In the former, the current value of the charged particles is measured and converted into the concentration. Consequently, a high temporal resolution is possible. Thus, FMPS is suitable for measurements requiring high time decomposition. In the latter, particles are detected using a condensation particle counter (CPC), Thus, SMPS･OPS can accurately measure the particle size distribution. Therefore, the inversion of the curve is due to these differences.

3.2. Result of morphological observations

We attempted to confirm the CNT release by observing the shape of the collected particles. Figure 5 shows the SEM results of the collected particles on the Nuclepore filters.

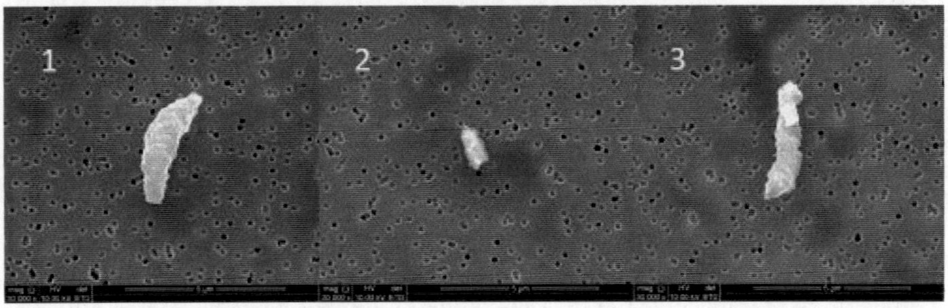

Figure 5. Morphological observations of the released particles collected on the filters.

None of the samples have confirmed protrusions. Although all the samples have high aspect ratios, they are much larger in size than the CNT raw materials. Scattering with CNT alone cannot be confirmed, but it is likely that CNTs are packed in the collected particles. The size of the released particles in No. 2 is smaller than those of No.1 and 3. We can discuss the shape and size of the released particles, but it is difficult to discuss the CNT release as a function of the matrix using only morphological observations. To discuss the scattering of CNTs from the composites, we have to identify whether CNTs are in the released particles simultaneously with the morphological observation.

Figure 6 shows the SEM results of the abrasion powders, while Fig. 7 shows the sample surfaces after abrasion. The abrasion powder and protrusions are confirmed from the surface of the sample, but CNTs cannot be identified.

Figure 6. Each abrasion powder on carbon tape.

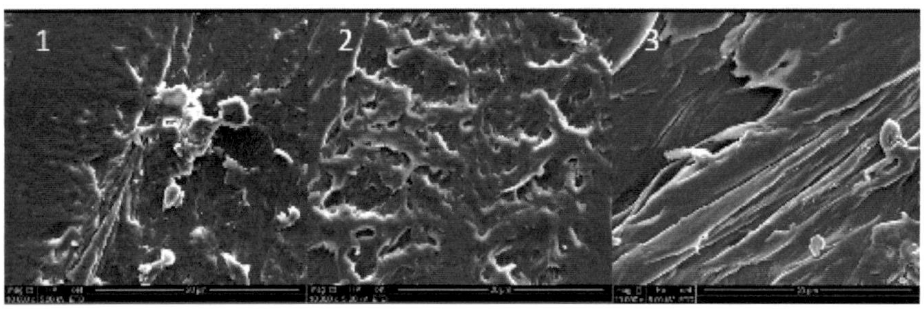

Figure 7. Surface condition after abrasion.

4. Conclusion

The chamber can control the BG concentration to some extent compared to ambient air. We were unable to measure the change in the concentration in the chamber before and after abrasion due to particle contamination. If the chamber setup is improved, it may be possible to confirm particle generation under the load of the Taber test. Thus, we should measure the pressure differential inside and outside the chamber.

The SEM images confirm scattering of the abrasion powder and particles, but not whether CNTs are packed in the abrasion powder or released particles. Elemental analysis of the released particles is necessary to confirm the presence of CNTs. Depending on the production method, a metal catalyst is terminated in the CNTs, and it may be possible to confirm the presence of CNTs by following that metal element. A method combining Inductively Coupled Plasma Mass Spectrometry (ICP-MS) and a Gas Exchange Device (GED) and a method using FE-SEM are options to trace the metal element. Using a GED, the released particles can be introduced directly into an ICP-MS, which identifies and determines the quantities of metals in the released particles. By examining the proportion of catalytic metal in the CNT, we can convert the trace metal amount into the amount of CNT.

We also considered collecting the generated particles by a filter and measuring the weight with an electronic balance. Because the collected particles include the particles derived from the sample and those from the abrasive ring, it is possible that the CNT weight is not accurately measured. Ultimately, the analysis should separate and analyse the released particles because particles are also generated from the abrasion ring of the Taber abrasion instruments. Consequently, we want to further control the number of particles in the chamber and conduct simultaneous analysis of GED-ICP-MS, aerosol measurement equipment, and SEM observations by filter collection in the future in order to verify and quantify the released particles of CNT decisively.

IOP Conf. Series: Journal of Physics: Conf. Series **838** (2017) 012011 doi:10.1088/1742-6596/838/1/012011

References

[1] Takagi A, Hirose A, Nishimura T, Fukumori N, Ogata A, Ohashi N, Kitajima S, Kanno J, "Induction of mesothelioma in p53+/- mouse by intraperitoneal application of multi-wall carbon nanotube" J Toxicol Sci. 2008 Feb; 33 (1): 105-16

[2] Gupta A, Gaspar D J, Yost M G, Gross G M, Rempes P E, Clark M L and Martin J C "Evaluating the potential for release of carbon nanotubes and subsequent occupational exposure during processing of a nanocomposite Nanotechnology occupational and environmental health and safety" Cincinnati, OH, USA, 4-7 December 2006: 110–111

[3] Bello D, Wardie B L, Yamamoto N, deVilloria R G, Garcia E J, Hart A J, Ahn K, Ellenbecker M J and Hallock M "Exposure to nanoscale particles and fibres during machining of hybrid advanced composites containing carbon nanotubes" J. Nanopart. 2009 Res. 11: 231–49

Acknowledgments

This work is based on the results obtained from a project commissioned by New Energy and Industrial Technology Development Organisation (NEDO) Grant.

IOP Conf. Series: Journal of Physics: Conf. Series **838** (2017) 012012 doi:10.1088/1742-6596/838/1/012012

Thermal Stability and Material Balance of Nanomaterials in Waste Incineration

H.-R. Paur, W. Baumann, M. Hauser, I. Lang, N. Teuscher, H. Seifert, D. Stapf

Karlsruhe Institute of Technology, Institute for Technical Chemistry
D-76344 Eggenstein-Leopoldshafen, Hermann-von-Helmholtzplatz 1, Germany
wener.baumann@kit.edu

Abstract. Nanostructured materials are widely used to improve the properties of consumer products such as tires, cosmetics, light weight equipment etc. Due to their complex composition these products are hardly recycled and thermal treatment is preferred. In this study we investigated the thermal stability and material balance of nanostructured metal oxides in flames and in an industrial waste incinerator. We studied the size distribution of nanostructured metal oxides (CeO_2, TiO_2, SiO_2) in a flame reactor and in a heated reaction tube. In the premixed ethylene/air flame, nano-structured CeO_2 partly evaporates forming a new particle mode. This is probably due to chemical reactions in the flame. In addition sintering of agglomerates takes place in the flame. In the electrically heated reaction tube however only sintering of the agglomerated nanomaterials is observed. Ceria has a low background in waste incinerators and is therefore a suitable tracer for investigating the fate of nanostructured materials. Low concentrations of Ceria were introduced by a two-phase nozzle into the post-combustion zone of a waste incinerator. By the incineration of coal dust in a burning chamber the Ceria nanoparticles are mainly found in the size range of the fly ash (1 – 10 µm) because of agglomeration. With gas as a fuel less agglomeration was observed and the Ceria nanoparticles were in the particle size range below 1 µm.

1. Introduction

The amount of consumer products containing engineered nanomaterials is constantly growing. As these products reach their end of life they often end up in the waste incineration. Till now only few data are published concerning a possible release of nanoparticles (NP) during combustion into the environment (Walser et al. 2012; Nowack et al. 2013; Andersen et al. 2014). In the Institute for Technical Chemistry at the Karlsruhe Institute of Technology (KIT) this topic is analysed in fundamental investigations of NP behaviour in lab-scale flames (Lang et al. 2015; Teuscher et al. 2016), in technical investigations at a 3 MW combustion chamber at the KIT (Scherrmann et al. 2016) and in large-scale investigations at an industrial hazardous waste incineration plant in the chemical industry. A new project 'ProCycle' concerning the behaviour and possible release of NPs by incineration of polymer nanocomposites started in 2015. Since 2005, landfilling of municipal waste is no longer allowed in Germany and the only possible disposal routes are therefore recycling or thermal waste treatment (see Figure 1). With every recycle step the product quality is decreasing and a thermal process is always the last possible treatment.

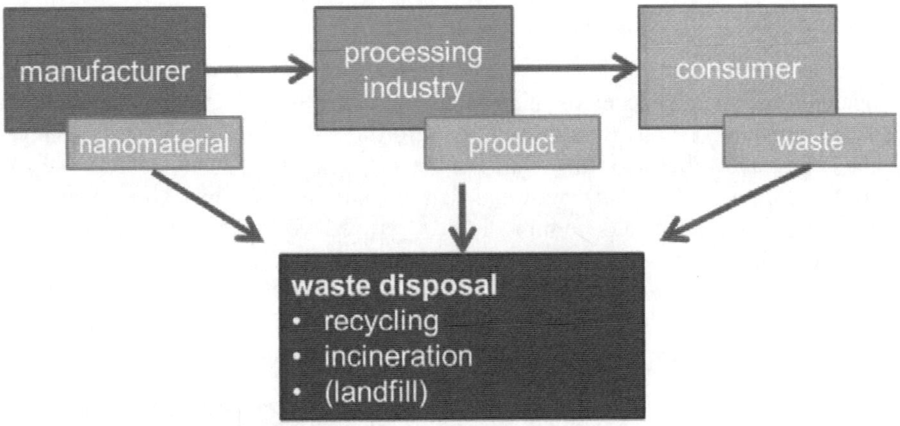

Figure 1: Life cycle of products including nanoparticles.

Figure 2 illustrates the flow sheet of a municipal waste incineration plant with possible paths of nanoparticle release. These can be the exhaust gas, the waste water or the solid residues. First investigations were concentrated on the release via the exhaust path and afterwards also other paths were studied including a mass balance of the incineration plant.

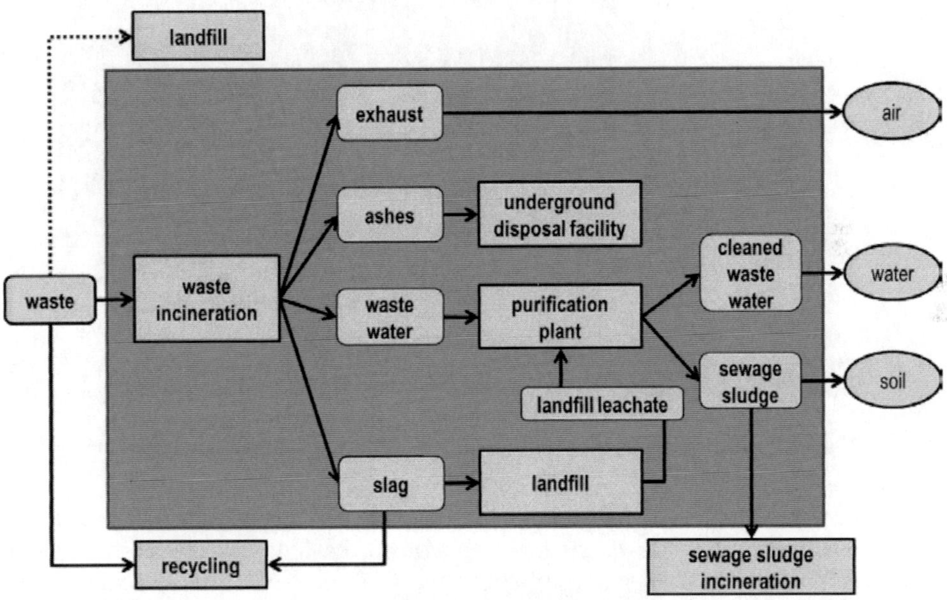

Figure 2: Flow sheet of a municipal waste incineration plant with possible paths of release of nanoparticles.

2. Lab-scale investigations on the thermal behaviour of nanoparticles

The fundamental investigations on the thermal behaviour of nanoparticles were carried out at a lab-scale burner and at a tube furnace. The used temperature in both systems is similar and the comparison of the results allows a differentiation between flame chemistry and temperature effect. The used Ceria suspension (Cerium(IV) oxide) is commercially available at Alfa Aesar (NanoTek CE 6082) and is specified as 18 m.-% Ceria in H_2O. The producer declares the primary particle size to be approximately 30 nm. The melting temperature of the bulk material is around 2000 °C and the density is 7.3 g/cm^3.

IOP Conf. Series: Journal of Physics: Conf. Series **838** (2017) 012012 doi:10.1088/1742-6596/838/1/012012

2.1. Experiments at the lab-scale burner

For this study a so called McKenna burner was used, which provides a laminar premixed flame. The burner was operated with an Ethylene/Air-mixture and the flame is stabilized on a porous bronze plate. The burner is considered to be one-dimensional, therefore the flame properties only change with one coordinate, the height above burner (HaB). The laminar flame speed v_L is supposed to be smaller than the velocity of the cold gases v_{cold}, otherwise the flame would lift off. Through a centered tube in the porous plate it was possible to add cerium dioxide sprayed via an atomizer (see Figure 3). The temperature of the Ethylene/Air-flame was varied by changing either the C/O ratio or the cold gas velocity. The temperature profiles were recorded over the cross section of the burner plate in different HaB. The burner was operated only with non-sooting conditions. The size distribution of the aerosol after passing the flame was measured with a scanning mobility particle sizer (SMPS).

CeO_2 suspension added via centered tube

Figure 3: McKenna burner with centered tube (Holthuis & Associates).

The flame parameters cold gas velocity and stoichiometry can change the flame temperature and their influence on the morphology, size distribution, sintering behaviour, or new particle formation is investigated in different HaB. The size distributions for 4 different cold gas velocities are shown in Figure 4 for CeO_2 nanoparticle injection. The red squares belong to the lowest cold gas velocity and therefore the lowest temperature and refer to the right axis. With increasing cold gas velocity and flame temperature a new particle mode around 10 nm starts to form.
This experiment was executed with a few metal oxides (CeO_2, TiO_2, SiO_2) and the same effect was observed. Only the starting temperature is material dependent and this effect was interpreted as an evaporation and nucleation of the nano-material, which was also found by other authors (Goertz et al. 2011).

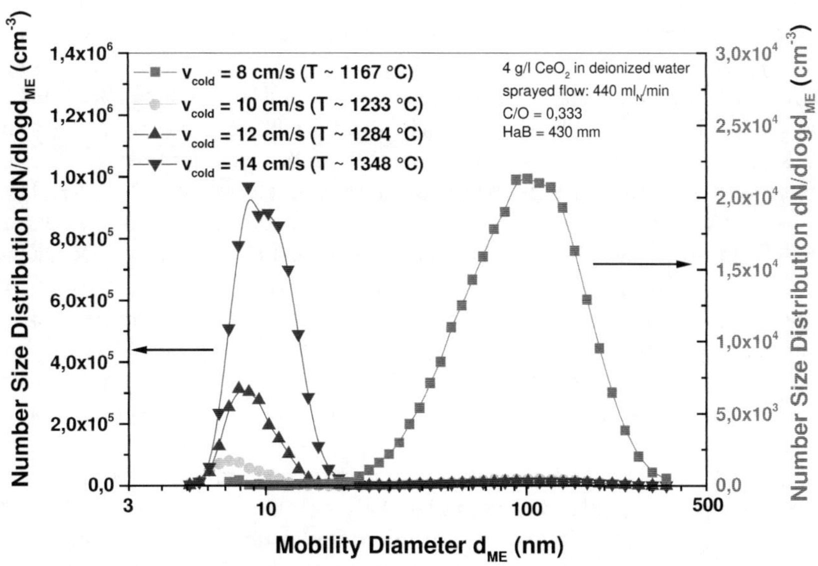

Figure 4: SMPS measurement of a Ceria suspension added to the McKenna burner via center tube. The red curve belongs to the right axis and the other curves to the left axis.

2.2. Experiments at the tube furnace

In addition to the McKenna burner a tube furnace (Gero HTRH 100-600/18 with temperature controller Eurotherm Type 2408) was used. With this furnace temperatures up to 1800 °C were adjustable. The tube material was Al_2O_3 with a heated length of 600 mm and an inner diameter of 30 mm. Either Air or Nitrogen can be used as carrier gas for the sprayed nanoparticle suspension. Additionally the humidity of the carrier gas could be varied via a bubbler system to adjust a similar water-content compared to the flue gas of the flame.
Investigations in the tube furnace with similar temperature don't show a new particle formation. The effect of particle formation in the flame does not only depend on temperature, but depends on gaseous components in the flame, like H_2O, aswell. The most promising component is water and it was possible to add water via a bubbler system to the carrier gas (N_2 or Air) so that the same humidity is adjustable in the tube furnace as in the flue gas of the burner. With this change in the experimental set up a new particle formation was observed. Literature research showed that it is possible for metal oxides to form hydroxides or oxyhydroxides in the presence of water vapour (Golden and Opila 2016).

3. Investigation on the release of nanoparticles during waste incineration

3.1. KIT combustion plant BRENDA

The pilot scale experiments are performed at the KIT combustion plant BRENDA which has a thermal output of 3 MW. BRENDA was formerly used as a hazardous waste incineration plant and is now operated for co-combustion of coal dust and different biomasses in the combustion chamber. BRENDA is a rotary kiln facility and has a combustion chamber equipped with a boiler for heat recovery and a flue gas cleaning system which complies with the German regulations of emissions (17. BImSchV). The pilot plant BRENDA (see Figure 5) provides the opportunity to work on different topics. A combined burner for gases and liquids and an atomizer lance for liquids are installed. Depending on the temperature in the rotary kiln (up to 1300 °C), the combustion residues may be either solid or molten.

The combustion gases from the rotary kiln are fed into the post combustion chamber which is equipped with two combined burners for gases and liquids and an air nozzle for secondary combustion air. The gas residence time in the post combustion chamber is around 9 s. The burners are staggered anti-parallel to each other and this configuration ensures high turbulence and improved mixing of the combustion gases. The post-combustion chamber is designed for gas temperatures up to 1300 °C. The hot flue gas leaves the post-combustion chamber under standard operation conditions with a temperature of 1000 °C and enters the boiler with a residence time of 22 – 25 s. The boiler generates saturated steam of 40 bar and 250 °C. BRENDA is operated with a quasi-dry, waste-water-free flue gas cleaning system, containing spray dryer, fabric filter, scrubbers and an SCR reactor.
The tracer (CeO_2) was dispensed into the combustion chamber via a two-phase nozzle and the tracer concentration along the furnace, boiler and flue gas cleaning system was analysed. The particle size distribution of the fly ash was measured with different methods like Low Pressure Impactor, Electrical Low Pressure Impactor (ELPI+), SMPS and with Single Particle Light Scattering Spectroscopy downstream of the boiler. To determine the concentration of the tracer material along the exhaust path, filter and impactor samples were analysed via inductively coupled plasma – mass spectrometry (ICP-MS). The mass concentration of the fly ash was varied by combustion of either natural gas or hard coal. The influence of the number concentration of fly ash particles on the size distribution of the nano sized tracer material was determined by analysing different impactor plates via ICP-MS. Ceria which has a low background in waste incinerators is a suitable tracer for investigating the fate of nanostructured materials. Low concentrations of Ceria were introduced by a two-phase nozzle into the post-combustion zone of a waste incinerator.

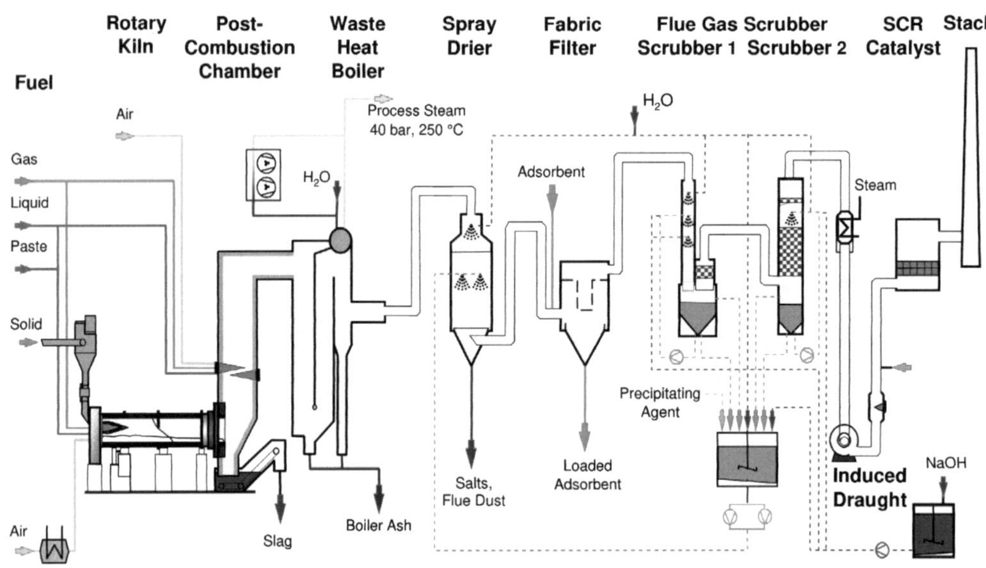

Figure 5: Schematic diagram of the rotary kiln incineration pilot plant BRENDA.

We found that nanostructured Ceria agglomerates with the fly ash, formed by combustion of coal. With gas as a fuel less agglomeration was observed and the Ceria NP were in the particle size range of 0.1 – 1 µm (see Figure 6). The removal efficiency of the flue gas cleaning system for the injected nano sized tracer is in the range of 99.99 %. Additionally the fly ash concentration was measured with gravimetric filter samples and the removal efficiency in the gas cleaning system was 99.99 % which is the exact same range as for the tracer.

IOP Conf. Series: Journal of Physics: Conf. Series **838** (2017) 012012 doi:10.1088/1742-6596/838/1/012012

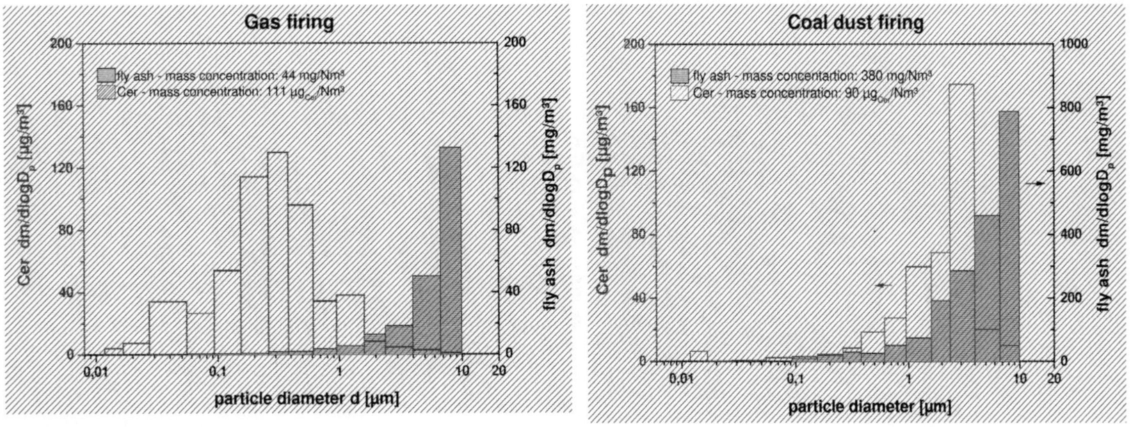

Figure 6: Fly ash and Ceria mass concentration in the off gas of an incinerator for gas and coal dust firing (sampling downstream of the boiler).

3.2. Industrial hazardous waste incineration plant

Large-scale investigations were carried out at an industrial hazardous waste incineration plant with a rotary kiln and combustion chamber. This system can be operated with solid, liquid and gaseous waste. The flue gas temperature at the outlet of the combustion chamber is above 1000 °C and the energy of the hot flue gas is used to generate steam in the boiler. The cerium dioxide suspension was dispensed into the combustion chamber via a two-phase nozzle and the tracer concentration along the plant was measured at different points. The first measurement point in the flue gas was in the boiler and the second at the boiler outlet. The next behind the wet-wall electrostatic precipitator and the last measurement point was at the stack as tracer balance limit in the flue gas. In the boiler the flue gas temperature decreases from 1000 °C to the range of 350 °C. The particle concentration in the raw gas at the boiler outlet is very high and depends on the incinerated waste mixture. The hazardous waste incineration plant has a wet flue gas cleaning system and the first device is a quench, where the flue gas is cooled down to the water dew-point by injection of water. Subsequently two rotary scrubbers are installed, one acid and one alkaline scrubber. Downstream of the rotary scrubbers a wet-wall electrostatic precipitator is operated followed by a heating of the flue gas before it enters the selective catalytic reduction (SCR) to reduce the nitrogen oxide emission (see Figure 7).
At the industrial hazardous waste incineration plant the tracer material (CeO_2) was injected as a suspension into the post-combustion chamber. The size distribution of the fly ash was measured by ELPI+ at the boiler outlet. To determine the concentration of the tracer along the exhaust path, filters were sampled inside the boiler, behind the boiler, behind the wet-wall electrostatic precipitator and at the exhaust chimney and all samples were analysed via ICP-MS. Additionally all relevant material flows were sampled and also analysed via ICP-MS. The mass balance of the injected tracer at the industrial waste incineration plant shows that over 80 % could be retrieved and the largest amount with roughly 70 % is found in the acid scrubber effluent (see Figure 8).

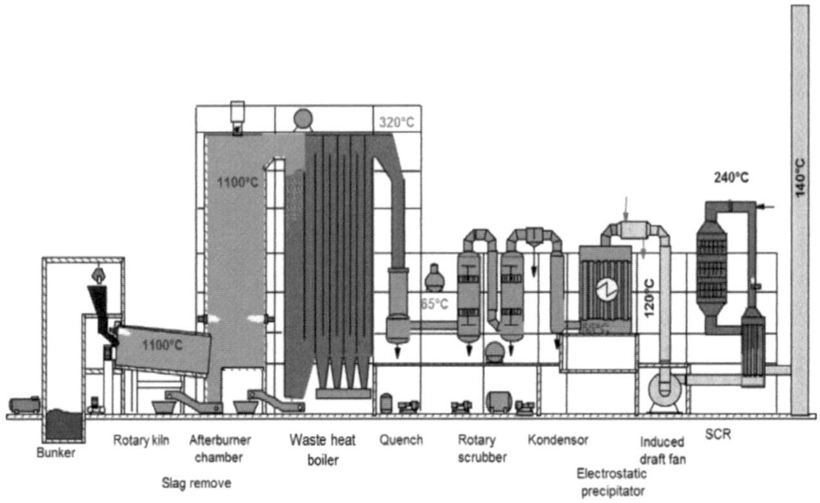

Figure 7: Flow sheet of the industrial hazardous waste incineration plant.

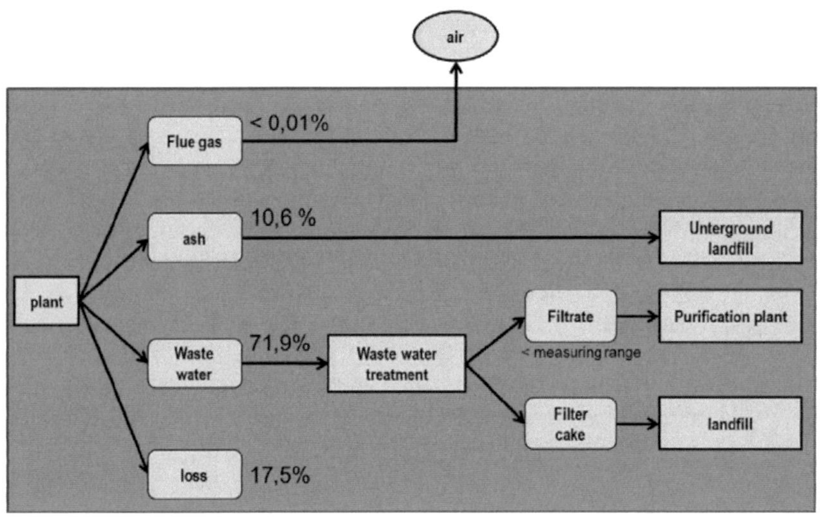

Figure 8: Mass balance of the tracer material at the industrial hazardous waste incineration plant.

4. Summary

Nowadays engineered nanoparticles are used in many products and at the end of their life cycle they often end up in the waste incineration. Therefore deeper knowledge about the behaviour of nanoparticles in thermal treatment is necessary. Basic investigations in a flame show a sintering of metal oxides. With increasing flame temperature the formation of a new particle mode around 10 nm was observed due to a partial evaporation of the material. This formation is not only temperature driven but also dependent on the gaseous components in the flame. Tracer investigations in a semi-technical combustion chamber show that the nano sized CeO_2 tracer interacts with the fly ash particles in the flue gas and the tracer was found in size fraction of the fly ash by ICP-MS analysis of impactor samples. These investigations were repeated on a large scale industrial hazardous waste incineration plant with the same result and measurements at the chimney show a tracer amount below 0.01 % of the injected tracer. Therefore the removal efficiency of the flue gas cleaning system is in the range of 99.99 %. By probing the different material flows of the incineration plant the mass recovery rate of the tracer material was found to be higher than 80 %.

References

[1] Andersen, L. et al. (2014): Nanomaterials in waste. Issues and new knowledge. Environmental Project No. 1608. The Danish Environmental Protection Agency. Denmark.

[2] Goertz, V. et al. (2011): The Effect of Water Vapor on the Particle Structure and Size of Silica Nanoparticles During Sintering. In: Aerosol Science and Technology, 45 (11), pp. 1287–1293.

[3] Golden, R. et al. (2016): A method for assessing the volatility of oxides in high-temperature high-velocity water vapor. In: Journal of the European Ceramic Society 36 (5), pp. 1135–1147.

[4] Lang, I.-M. et al. (2015): Untersuchungen zur Freisetzung von synthetischen Nanopartikeln bei der Abfallverbrennung. In: K. J. Thomé-Kozmiensky und Michael Beckmann (Hg.): Energie aus Abfall, Band 12. neue Ausg. Nietwerder: TK-Vlg, pp. 347–370.

[5] Nowack, B. et al. (2013): Potential release scenarios for carbon nanotubes used in composites. In: Environment international, 59, pp. 1–11.

[6] Scherrmann, A. et al. (2016): Mitverbrennung von Biomasse in einer Kraftwerksstaubfeuerung im Pilotmaßstab. 48. Kraftwerkstechnisches Kolloquium. Dresden, Germany, October 2016.

[7] Teuscher, N. et al. (2016): The influence of temperature and humidity on the thermal stability of nanoparticles. European Aerosol Conference. Tours, France, September 2016.

[8] Walser, T. et al. (2012): Persistence of engineered nanoparticles in a municipal solid-waste incineration plant. In: Nature Nanotech, 7, pp. 520-524 .

IOP Conf. Series: Journal of Physics: Conf. Series **838** (2017) 012013 doi:10.1088/1742-6596/838/1/012013

Release of particles by abrasion of CNT composites using a belt sander

Yasuto Matsui[1], Tomonori Ishibashi[2], Kato Nobuyuki, Taiki Nagaya[2], Minoru Yoneda[2]

[1]Agency for Health, Safety and Environment, Kyoto University, Japan
[2]Graduate School of Engineering, Kyoto University, Japan

Abstract. There have been many reports on the effect of exposure to nanomaterials such as titanium dioxide, silver, and carbon nanotube (CNT) on human health. Several experiments have examined the abrasion of CNT composites, in which CNT nanoparticles are embedded within a resin or rubber matrix, yielding varying results. Separate study of free CNTs and CNT nanoparticles in relation to health is important due to the different physicochemical characteristics of the two types of material. This study investigated the abrasion of CNT composites using a belt sander inside an enclosed chamber, with variation in the applied load and belt sander speed. At lower speeds, the population of particles with diameters of ~100 nm was observed to increase (cf. mode values of ~10 nm), and we found a relationship between the amount of the raising dust and the abrasion conditions. From these results, we propose a robust and widely applicable method to create particles of nanomaterial-containing composite materials of various types in order to conduct accelerated exposure assessment studies.

1. Introduction

Several experiments examining the wear of carbon nanotube (CNT) composites have been reported [1, 2]. For example, Wohlleben *et al.* performed a wear test using a Taber Abraser [1], finding that neither free nor overhanging CNT particles were generated from a CNT-containing composite material; meanwhile, Golanski *et al.* conducted Taber tests using similar apparatus [2] — in this report, we found that although CNTs were not discharged from the polymer, there was evidence of them collecting on the surface during the test.

There are several other types of equipment available for wear testing besides Taber Abrasers: one example is a DIN Abrasion Tester, in which a grinding cloth is wound around a revolving drum; samples are then pressed against the cloth to induce wear with a speed of 3.1 m/s [3]. In this study, we investigated the wear of CNT composites using a modified version of the DIN abrasion test involving a belt sander. While Taber and DIN abrasion testing is generally designed to quantify the wear depth of the materials, these methods are not usually used to analyse the dust generated by wear. First, the loading energy is usually too weak to detect nanomaterials in a resin or rubber. Second, the shaving speeds of these testing methods is quite slow, and thus little dust is generated (ultimately, this class of equipment was never designed with the intent of generating large amounts of dust).

To address the limitations of these testing methods, we used a commercial belt sander that allows the user to change the load and speed. This type of equipment is widely used for shaping processes in chemical factories, and thus the resulting base material wear will be somewhat similar to what would occur during industrial processing.

2. Methods

2.1. Test pieces

The samples used were composed of polycarbonate composite materials prepared without (PC0) and with 3% CNT (PC3). The CNTs distributed in the polycarbonate were multi-walled (CNano Flotube9000). The samples were manufactured by general distribution and kneading methods in a resin factory. All pieces were made from a commercial polycarbonate master batch and requested to Marubeni Information Systems to shape plate piece (10 cm^2).

2.2. Exposure Assessment Chamber

The belt sander was located in a chamber used for exposure assessment experiments that possessed a depth of 1.08 m, width of 0.8 m, and height of 1.64 m (Figure 1) [4]. The air in the chamber was controlled and isolated from the outer environment by two different types of HEPA filters, and an electrically conductive film was used to seal the inner wall to prevent absorption of abrasion dust. A clean airflow was directed from the upper filtration box to filters on both sides of the bottom of the chamber. All manual operations inside the chamber were conducted using either the upper or lower pairs of rubber gloves shown in Figure 1, and a power adapter was located inside the box. Prior to abrasion testing, we ensured that the ambient particle concentration in the chamber was under 100 particles/cm^3 by ventilating the chamber; once an acceptable background particulate level was reached, the ventilation system was switched off and the abrasion test initiated. Different particle detectors, FMPS (Fast Mobility Particle Sizer, Model 3091, TSI), SMPS (Scanning Mobility Particle Sizer, Model 3910, TSI), and OPC (Optical Particle Sizer, OPS 3330, TSI) were connected through holes in the ports in the chamber wall.

Figure 1. Photograph of the exposure assessment chamber containing the belt sander.

2.3. Belt sander abrasion testing

PC0 and PC3 samples were abraded using the belt sander; the sander was equipped with a 90 mm × 915 mm belt with a P80 grit size (aluminium oxide grains, average abrasive grain diameter of 177 μm), and the standard operating speed used in the experiments was 9.1 m/s (Figure 2). Each sample was attached to the bottom of a standard 500 g weight using double-sided tape. The weight was held on a plate seat by a fixing handle specifically designed for this purpose by our group, as shown in the schematic diagrams in Figure 3 (top diagram, overhead view; bottom diagram, horizontal view). The design of the fixing stage allows for changes in the grid size, belt rotation velocity, and loading weight, depending on the type and dimensions of the material being tested.

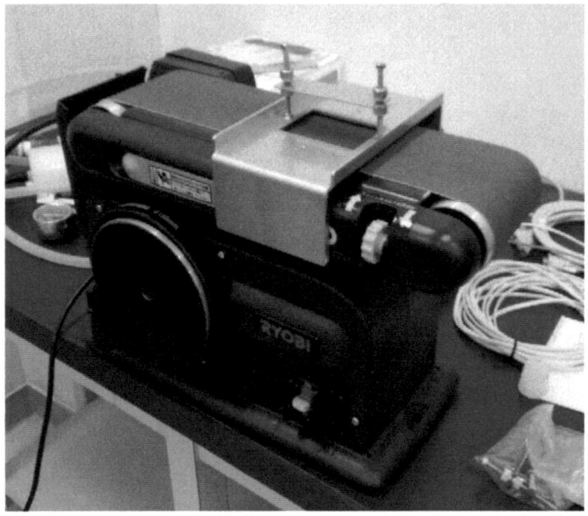

Figure 2. Custom sample fixing stage installed on the belt sander.

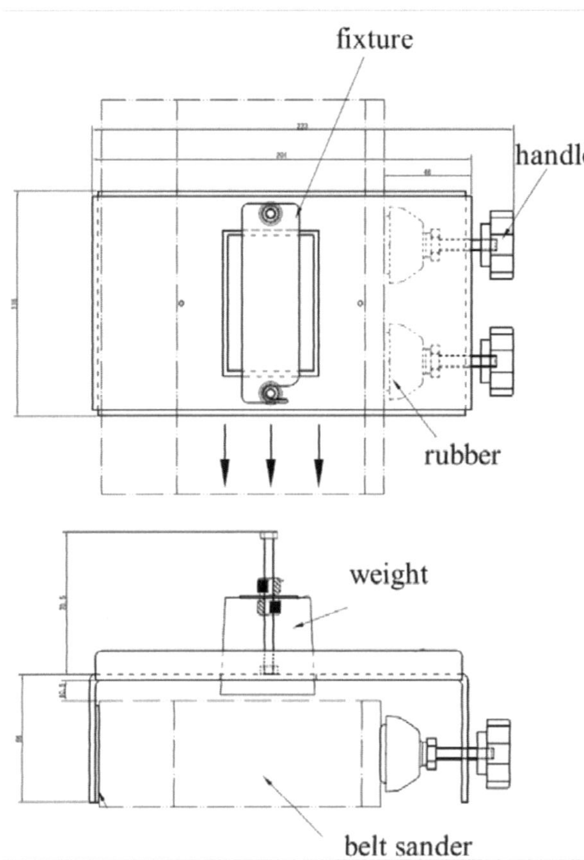

Figure 3. Schematic diagram of the sample fixing stage designed for use with a belt sander.

Each sample was subjected to wear for 10 s, after which the particulate distribution within the enclosed chamber atmosphere was measured using the FMPS. Fine particles were collected from the chamber by a glass fibre filter using a mini pump (MP-Σ300, Shibata Kagaku) with a flow rate of 0.3 L/min over a period of 13 min. Additionally, abraded powder stuck to the belter sander was collected using carbon tape in order to analyse the particle shapes, and to distinguish elemental CNT from embedded CNT with polycarbonate. Scanning electron microscopy (SEM) was used to observe the surfaces of the samples after testing, the fine particles from the filter, and the abraded powder on the carbon tape. Different abrasion conditions were used for each experiment, as shown in Table 1. In the second experiment, the weight used to apply the load to the sander was increased from 500 g to 1000 g, and in the third experiment, the belt speed was decreased from 9.1 m/s to 7.6 m/s (500 g load).

Table 1. Experimental conditions.

	load (g)	grit size	speed (m/s)
Experiment 1	500	P80	9.1
Experiment 2	1000	P80	9.1
Experiment 3	500	P80	7.6

3. Results

3.1. Chamber Atmosphere

Prior to measurement, the background air level in the chamber was measured. It was confirmed that an impurely particle in the chamber decreases by ventilating through the HEPA filter unit located at the top of the exposure chamber (Figure 1). The data in Figure 4 indicate the changes in the particle concentration in the chamber after the filter, showing that the concentration decreased dramatically over a period of 1,000 s, from 500 particles/cm^3 to just several tens of particles/cm^3. Prior to each abrasion test, we confirmed that the background level had been reduced to under 100 particles/cm^3 by ventilation.

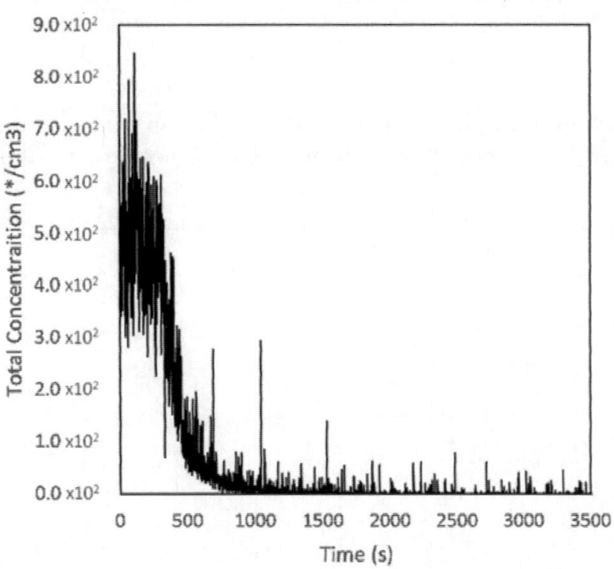

Figure 4. Changes in chamber particle concentration on ventilation.

3.2 Abrasion by belt sander

During the sanding experiments, the change in the total particle concentration was measured over time. Figure 5 shows the results of Experiments 1–3 (Table 1). In the period from 0 s to 60 s, the belt sander was switched on, but not placed in contact with the sample; after the initial 60 s period, abrasion was initiated. In addition to measuring the total particle concentration, the particle grade scale (grain size) was also measured during sanding, as shown in Figure 6. The results of Experiment 2 show that the concentration of sanded particles produced under these conditions was three times higher compared to that of Experiments 1 and 3. From these observations, we conclude that the load weight had a greater effect on particle generation than belt speed.

Figure 6 shows the average particle size distributions for Experiments 1–3 during the period from 60 s to 120 s. While the pattern of particle size distribution was found to be similar for each experimental dataset, the results from Experiment 2 (especially PC3) show a secondary concentration peak at around 30 nm diameter. In the case of Experiment 3, a slight uptick in the distribution was seen at around 100 nm diameter. In general, the composite samples that did not contain CNTs (PC0) were found to produce higher particulate concentrations soon after contact was made between the sample and the sander.

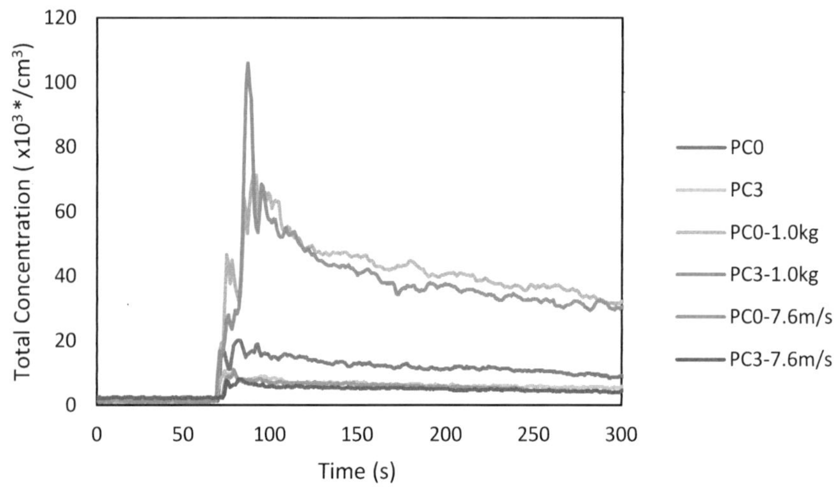

Figure 5. Change in total particle concentration during the abrasion test. PC0 and PC3 = Experiment 1; 1.0kg suffix = Experiment 2; 7.6m/s suffix = Experiment 3.

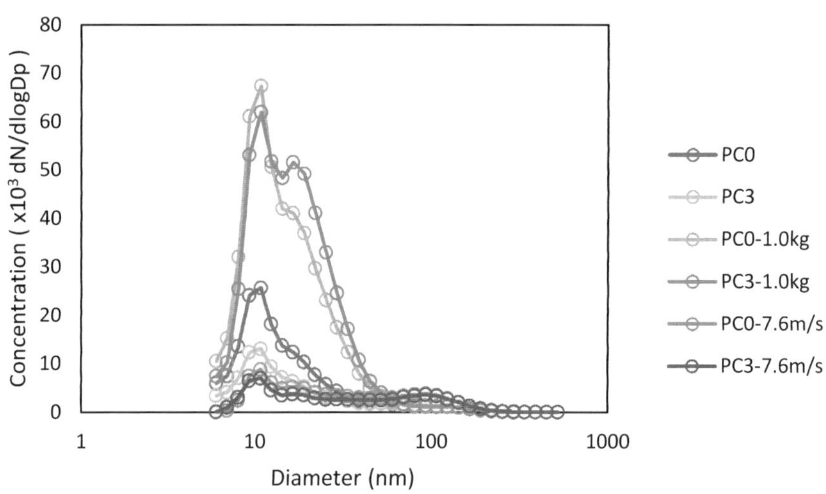

IOP Conf. Series: Journal of Physics: Conf. Series **838** (2017) 012013 doi:10.1088/1742-6596/838/1/012013

Figure 6. Size distribution of sanded particles. PC0 and PC3 = Experiment 1; 1.0kg suffix = Experiment 2; 7.6m/s suffix = Experiment 3.

SEM was used to analyse the abraded particles, and confirmed that the composite produced dust during the experimental. The difference in each experiment was not seen by particles. There were a lot of particles with size bigger than 10 μm. Elemental CNTs, CNTs overhanging from polymer, and embedded CNTs were not identified among the polycarbonate dust particles (Figure 7), and while several white dots were observed on the surface of the particles (as seen in the Figure 7, lower right), it was not confirmed that CNTs were present.

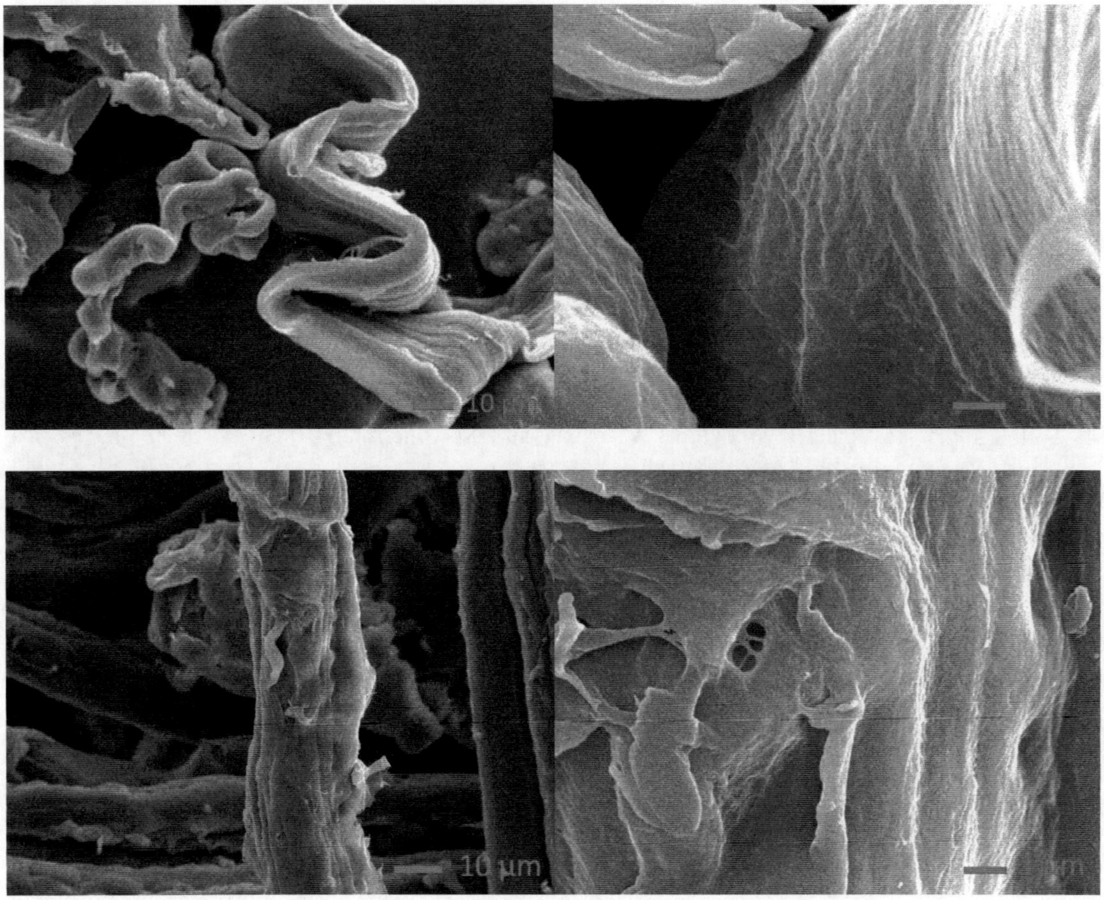

Figure 7. Scanning electron micrographs of the abraded composite powder.

4. Discussion

The results confirm that abrasion by the belt sander led to an increase in particulate concentration, with particles generated from both PC0 and PC3 composites increasing dramatically soon after contact between the belt and the sample. An increase in load led to an increase in the rate of particle generation, while the rate decreased with a decreased belt speed. A relation was found to exist between the amount of generated dust and the abrasion conditions. It was noted that a combination of increase weight loading and presence of CNTs (PC3) led to an increase in the population of particles with ~30 nm diameter; in contrast, high belt velocity increased the population of particles with ~100 nm particle diameter from both species of composite, PC0 and PC3. These results indicate that more high velocity effect a breaking of composites. To assess the exposure dose, we should find suitable load weight and velocity by using belt sander machine.

The presence of dust particles was confirmed by SEM observation. In future studies, after optimising the sanding parameters, we will measure the particle concentration and diameter distribution, followed by SEM analysis to distinguish between elemental, embedded, and overhanging CNTs.

5. Conclusions

When the belt speed was slowed, we observed that the population of particles with diameters of ~100 nm increased, the origins of which require further investigation. We will make a similar experiment by the other loads and the other speed from now on. A standard exposure assessment method for composite materials — including those containing nanomaterials — should be established using Taber Abraser, DIN Abrasion, belt sanders, or some other method. From this study, we also conclude that control of the background particle level is crucial for the chamber method.

From these experiments, it is clear that there are a number of challenges associated with CNT detection. There is some CNT including metal. We are trying the way to determine the quantity of the metal content which is included by wear in the trash which occurred and determine the quantity of CNT. This technique will support a SEM method to distinguish CNT variations after wearing.

References

[1] Wohlleben W, Meier MW, Vogel S, Landsiedel R, Cox G, Hirth S and Tomovie Z 2013 *Elastic CNT-polyurethane nanocomposite* (Nanoscale)
[2] Golanski L, Guiot A, Pras M and Tardif F 2012 *Release-ability of nano fillers from different nanomaterials* (J Nanopart Res)
[3] Japanese Industrial Standards K-6264-2
[4] N Kato, Y koyama, H yokoyama, Y matsui and M yoneda 2015 *Development of exposure assessment method with the chamber* (IOP science)

Acknowledgements

This research was supported by a grant from the Japanese Ministry of Economy, Trade and Industry from FY2012 to FY2014.

IOP Conf. Series: Journal of Physics: Conf. Series **838** (2017) 012014 doi:10.1088/1742-6596/838/1/012014

Quantitative measurement of carbon nanotubes released from their composites by thermal carbon analysis

I Ogura[1,2], M Kotake[2], S Ata[1,2] and K Honda[1,2]

[1] National Institute of Advanced Industrial Science and Technology (AIST), 16-1 Onogawa, Tsukuba, Ibaraki 305-8569, Japan
[2] Technology Research Association for Single Wall Carbon Nanotubes (TASC), 1-1-1 Higashi, Tsukuba, Ibaraki, 305-8565, Japan

E-mail: i-ogura@aist.go.jp

Abstract. The release of free carbon nanotubes (CNTs) and CNTs partly embedded in matrix debris into the air may occur during mechanical and abrasion processes involving CNT composites. Since the harmful effects of CNT-matrix mixtures have not yet been fully evaluated, it is considered that any exposure to CNTs, including CNT-matrix mixtures, should be measured and controlled. Thermal carbon analysis, such as Method 5040 of the National Institute for Occupational Safety and Health, is one of the most reliable quantitative methods for measuring CNTs in the air. However, when CNTs are released together with polymer matrices, this technique may be inapplicable. In this study, we evaluated the potential for using thermal carbon analysis to determine CNTs in the presence of polymer matrices. Our results showed that thermal carbon analysis was potentially capable of determining CNTs in distinction from polyamide 12, polybutylene terephthalate, polypropylene, and polyoxymethylene. However, it was difficult to determine CNTs in the presence of polyethylene terephthalate, polycarbonate, polyetheretherketone, or polyamide 6.

1. Introduction

Carbon nanotubes (CNTs) have unique properties, due to which their use as a filler material in composites is considered promising. However, the release of free CNTs and CNTs partly embedded in matrix debris into the air may occur during mechanical and abrasion processes involving CNT composites. Some studies have indicated that CNT-matrix mixtures are less harmful than free CNTs [1–4]. However, the harmful effects of CNT-matrix mixtures have not yet been fully evaluated. Under the present circumstances, it is considered that any exposure to CNTs, including CNT-matrix mixtures, should be measured and controlled.

Thermal carbon analysis, such as Method 5040 of the National Institute for Occupational Safety and Health (NIOSH) [5], is often used as a quantitative measurement of CNTs in the air [6–8]. This is a method to quantify organic carbon (OC) and elemental carbon (EC). It is one of the most reliable quantitative methods for measuring CNTs in the air. However, when CNTs are released together with polymer matrices, this technique may be inapplicable. In this study, we evaluated the potential for using thermal carbon analysis to determine CNTs in the presence of polymer matrices.

2. Methods

The samples tested in this study are summarized in table 1. Masses of OC and EC in each sample, placed in a boat made of Au foil or a quartz fiber filter, were measured using a thermal-carbon analyzer (CAA-202M-D, Sunset Laboratory Inc., USA). The results were compared with the carbon masses of the samples, calculated by their gravimetric masses—measured using an ultra-microbalance (SE2-F ultra-microbalance, Sartorius AG, Germany)—and carbon content. The temperature-step program was based on either the NIOSH method or the Interagency Monitoring of Protected Visual Environments (IMPROVE) method, with a slight modification (tables 2 and 3). The optical pyrolysis correction for pyrolytically generated carbon soot from OC during the analysis was not used.

Table 1. List of test samples.

CNTs	CNano Flotube9000 (average tube diameter: 10–15 nm, average tube length: 10 µm, carbon purity: 95–97.5%)
	Nanocyl NC7000 (average tube diameter: 9.5 nm, average tube length: 1.5 µm, carbon purity: 90%)
	AIST/TASC super-growth (SG) CNTs (average tube diameter: 3 nm, average tube length: >100 µm, carbon purity: 99.9%)
Polymers	Polyamide 12 (PA12)
	Polyamide 6 (PA6)
	Polybutylene terephthalate (PBT)
	Polypropylene (PP)
	Polyoxymethylene (POM)
	Polyethylene terephthalate (PET)
	Polycarbonate (PC)
	Polyetheretherketone (PEEK)
CNT composites	PA12-Flotube9000 (4%)
	PA12-Flotube9000 (10%)
	PA6-Flotube9000 (2%)
	PBT-Flotube9000 (4%)
	PBT-Flotube9000 (6%)
	PP-Flotube9000 (1%)
	PP-Flotube9000 (2%)
	POM-Flotube9000 (1%)
	PP-NC7000 (1%)
	PP-NC7000 (2%)
	POM-NC7000 (1%)
	PA6-SGCNT (1%)
	PA6-SGCNT (2%)
	PP-SGCNT (1%)
	PP-SGCNT (2%)
	POM-SGCNT (1%)

Table 2. NIOSH 5040 method-based protocol.

	Duration (s)	Temperature (°C)	Gas
OC1	80	310	He
OC2	80	475	He
OC3	80	615	He
OC4	300 (240*)	870	He
OC5	45	550	He
EC1	45	550	2%O_2/He
EC2	45	625	2%O_2/He
EC3	45	700	2%O_2/He
EC4	45	775	2%O_2/He
EC5	45 (300*)	850 (870*)	2%O_2/He
EC6	110 (120*)	870 (900*)	2%O_2/He

* When SGCNT was analyzed.

Table 3. IMPROVE method-based protocol

	Duration (s)	Temperature (°C)	Gas
OC1	300	250	He
OC2	600	450	He
OC3	600	500	He
OC4	300	550	He
EC1	300	550	2%O_2/He
EC2	120	600	2%O_2/He
EC3	120	650	2%O_2/He
EC4	240	700	2%O_2/He
EC5	180 (600*)	870	2%O_2/He

* When SGCNT was analyzed.

3. Results and discussion

First, the CNT powder samples were analyzed (figure 1). The amounts of total carbon (TC = OC + EC) detected were comparable to the carbon masses of the CNT powder, calculated by their gravimetric measured masses and carbon purity. The AIST/TASC super-growth (SG) CNT powder was detected mostly in the EC fraction in both cases of the NIOSH- and IMPROVE-based temperature-step protocols. In contrast, for Flotube9000 and NC7000, several tens of percent of the CNT powder was detected in the OC fraction when the NIOSH-based temperature-step protocol was used. When the IMPROVE-based temperature-step protocol was used, all the three types of CNTs were detected mostly in the EC2–5 fractions.

Subsequently, the grinding debris of the polymer samples (containing no CNTs) was analyzed using the IMPROVE-based protocol (figure 2). The amounts of TC detected were in good agreement with the carbon masses of the polymers, calculated by their gravimetric measured masses and carbon content (the percentage of carbon atoms). Polyamide 12 (PA12), polyamide 6 (PA6), polybutylene terephthalate (PBT), polypropylene (PP), and polyoxymethylene (POM) were detected mostly in the OC fraction and slightly in the EC1 fraction, indicating that they were potentially distinguishable from the CNTs. On the other hand, polyethylene terephthalate (PET), polycarbonate (PC), and polyetheretherketone (PEEK) were detected partly (or mostly) in the EC2–5 fractions, indicating that they were hardly distinguishable from CNTs. In a similar way, when the NIOSH-based protocol was used, PET, PC, and PEEK were detected partly in the EC fraction (data not shown), indicating that they were hardly distinguishable from CNTs.

Finally, the grinding debris of the CNT composite samples was analyzed using the IMPROVE-based protocol. Although the amount of the grinding debris used for each analysis was only several hundred micrograms, judging from visual observation, the grinding debris seemed to have almost the same CNT content as the original CNT composite. For each sample, the analysis was repeated at least five times to increase the representativeness of the sample. The representative thermal-carbon analyzer thermograms are shown in figure 3. The CNTs in the composites tended to burn at a lower temperature than the CNT powder. The amounts of EC excluding EC1 were comparable to the carbon masses of the CNT composites, calculated by their gravimetric measured masses and CNT content, except for the PA6 composites (table 4). Because the ratio of the amount detected in the EC1 fraction to the TC was higher for the PA6 composites than for the pure PA6, the CNTs in the PA6 composites were probably detected partly in the EC1 fraction.

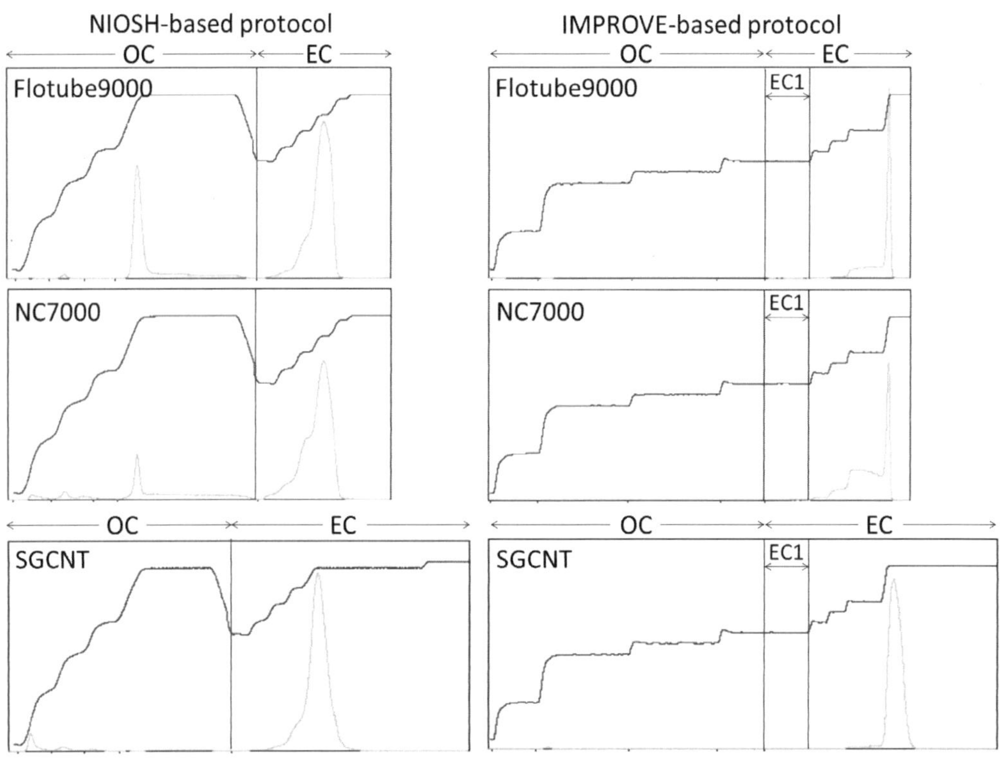

Figure 1. Thermal-carbon analyzer thermograms of CNT powder. Blue and green lines represent the temperature and detector response (arbitrary unit), respectively.

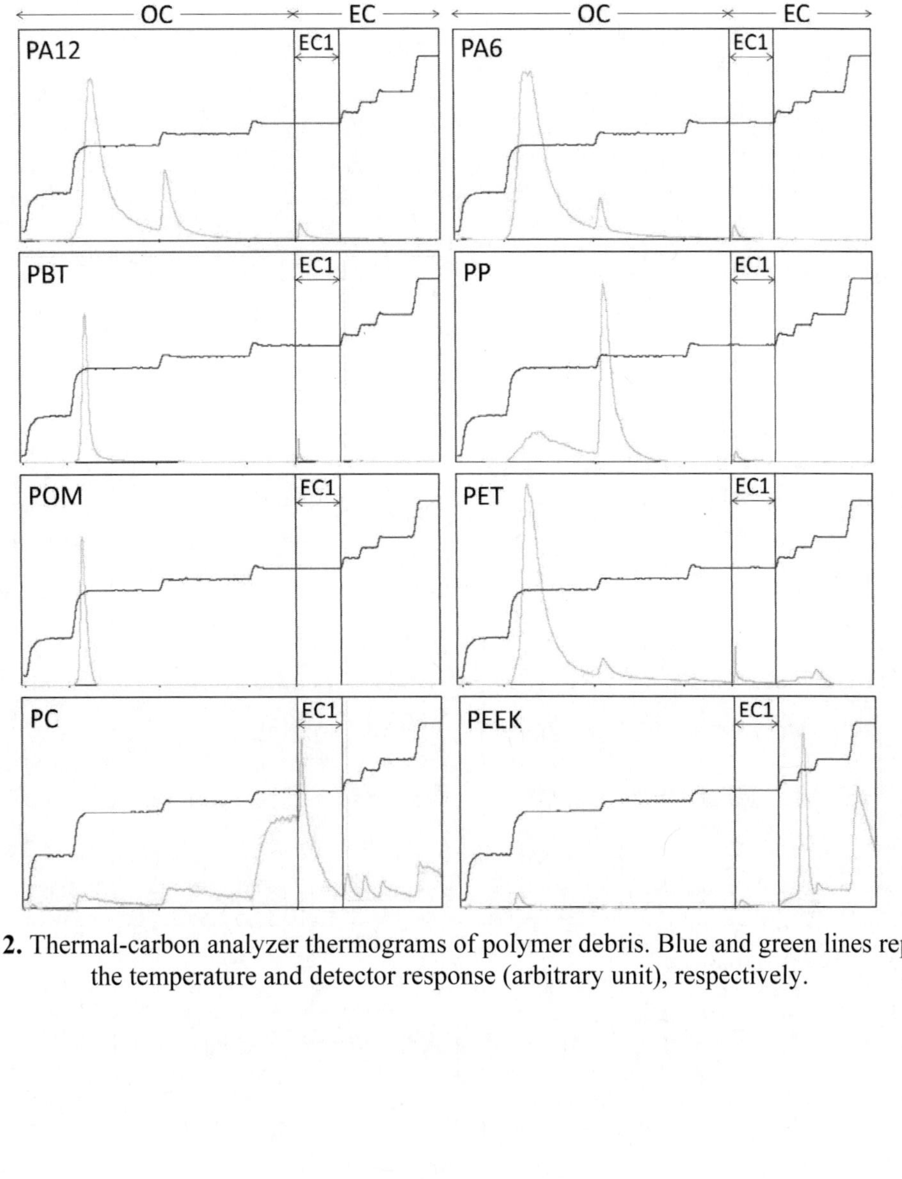

Figure 2. Thermal-carbon analyzer thermograms of polymer debris. Blue and green lines represent the temperature and detector response (arbitrary unit), respectively.

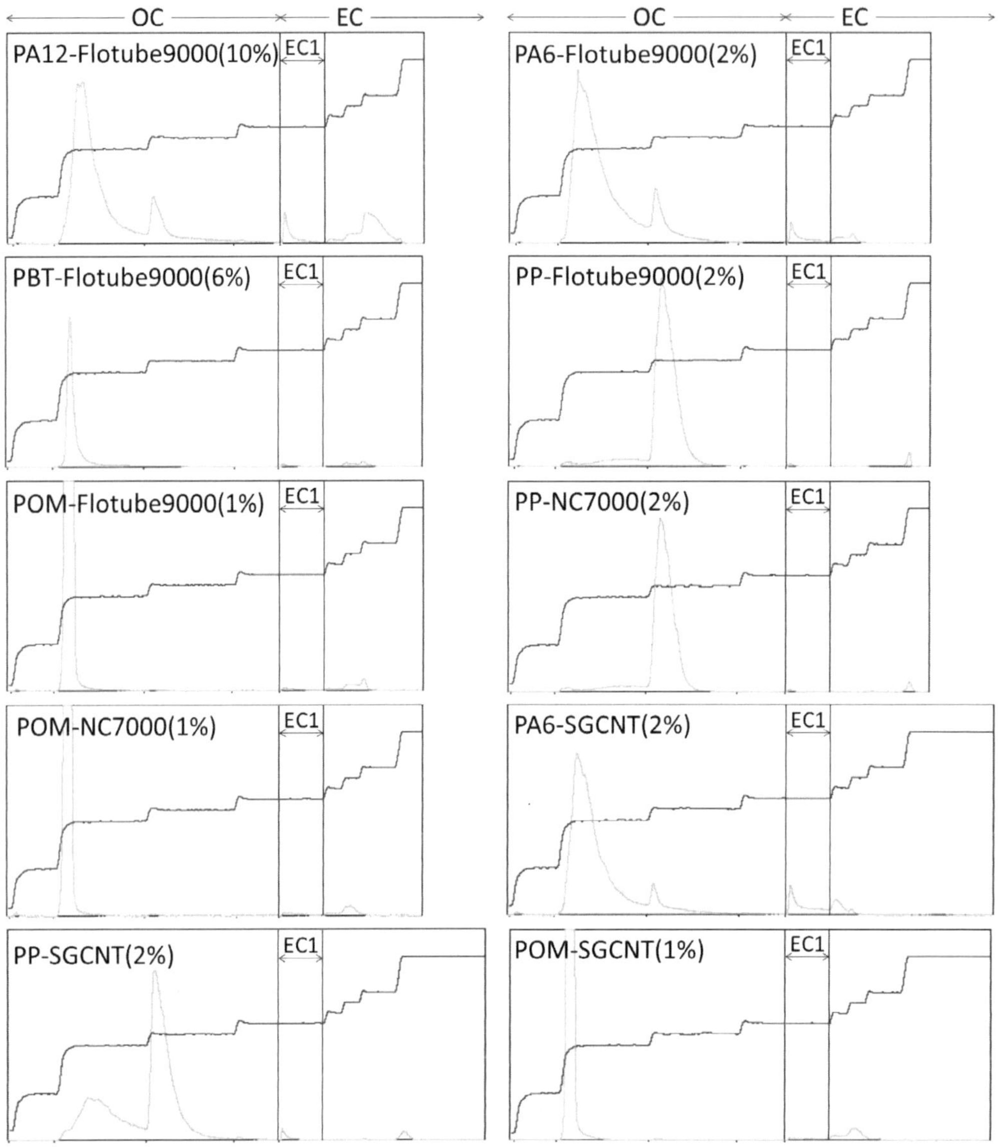

Figure 3. Thermal-carbon analyzer thermograms of CNT composites. Blue and green lines represent the temperature and detector response (arbitrary unit), respectively.

Table 4. Ratio of the amount of EC (excluding EC1) to the carbon mass of the sample (calculated by its gravimetric measured mass and CNT content).

Sample	Ratio
PA12-Flotube9000 (4%)	1.01±0.02
PA12-Flotube9000 (10%)	0.96±0.01
PA6-Flotube9000 (2%)	0.69±0.09
PBT-Flotube9000 (4%)	1.01±0.04
PBT-Flotube9000 (6%)	0.93±0.05
PP-Flotube9000 (1%)	1.09±0.05
PP-Flotube9000 (2%)	1.05±0.04
POM-Flotube9000 (1%)	0.98±0.04
PP-NC7000 (1%)	1.21±0.18
PP-NC7000 (2%)	1.05±0.08
POM-NC7000 (1%)	0.93±0.06
PA6-SGCNT (1%)	0.32±0.15
PA6-SGCNT (2%)	0.86±0.67
PP-SGCNT (1%)	1.09±0.06
PP-SGCNT (2%)	0.86±0.06
POM-SGCNT (1%)	1.10±0.06

$n = 5–7$

4. Conclusions

Our results showed that thermal carbon analysis was potentially capable of determining CNTs in distinction from PA12, PBT, PP, and POM when the CNT content was a few percent or higher. However, it was difficult to determine CNTs in the presence of PET, PC, PEEK, or PA6.

Acknowledgment
This work is based on results obtained from a project commissioned by the New Energy and Industrial Technology Development Organization (NEDO).

References

[1] Wohlleben W, Brill S, Meier M W, Mertler M, Cox G, Hirth S, von Vacano B, Strauss V, Treumann S, Wiench K, Ma-Hock L and Landsiedel R 2011 On the lifecycle of nanocomposites: comparing released fragments and their in-vivo hazards from three release mechanisms and four nanocomposites *Small* 7 2384–2395
[2] Wohlleben W, Meier MW, Vogel S, Landsiedel R, Cox G, Hirth S, Tomović Ž 2013 Elastic CNT-polyurethane nanocomposite: synthesis, performance and assessment of fragments released during use *Nanoscale* **5** 369–380
[3] Schlagenhauf L, Buerki-Thurnherr T, Kuo YY, Wichser A, Nüesch F, Wick P, Wang J 2015 Carbon Nanotubes Released from an Epoxy-Based Nanocomposite: Quantification and Particle Toxicity *Environ Sci. Technol.* **49** 10616–10623
[4] Schlagenhauf L, Kianfar B, Buerki-Thurnherr T, Kuo YY, Wichser A, Nüesch F, Wick P, Wang J 2015 Weathering of a carbon nanotube/epoxy nanocomposite under UV light and in water bath: impact on abraded particles *Nanoscale* **7** 18524–18536
[5] NIOSH (National Institute for Occupational Safety and Health) 2003 Method 5040 Issue 3, Diesel particulate matter (as Elemental Carbon). NIOSH Manual of Analytical Methods (NMAM; 4th ed.)
[6] NIOSH (National Institute for Occupational Safety and Health) 2013 NIOSH Current

intelligence bulletin 65: Occupational exposure to carbon nanotubes and nanofibers. http://www.cdc.gov/niosh/docs/2013-145/

[7] Ono-Ogasawara M, Takaya M, Kubota H, Shinohara Y, Koda S, Akiba E, Tsuruoka S and Myojo T 2013 Approach to the exposure assessment of MWCNT by considering size distribution and oxidation temperature of elemental carbon *J. Phys.: Conf. Ser.* **429** 012004

[8] Ogura (2013). Guide to measuring airborne carbon nanotubes in workplaces. https://en.aist-riss.jp/assessment/2571/

Airborne particles released by crushing CNT composites

I Ogura[1,2], C Okayama[3], M Kotake[2], S Ata[1,2], Y Matsui[4] and K Gotoh[3]

[1] National Institute of Advanced Industrial Science and Technology (AIST), 16-1 Onogawa, Tsukuba, Ibaraki 305-8569, Japan
[2] Technology Research Association for Single Wall Carbon Nanotubes (TASC), 1-1-1 Higashi, Tsukuba, Ibaraki, 305-8565, Japan
[3] Okayama University, 3-1-1 Tsushima-naka, Okayama, 700-8530, Japan
[4] Kyoto University, Yoshida-honmachi, Sakyo-ku, Kyoto, 606-8501, Japan

E-mail: i-ogura@aist.go.jp

Abstract. We investigated airborne particles released as a result of crushing carbon nanotube (CNT) composites using a laboratory scale crusher with rotor blades. For each crushing test, five pellets (approximately 0.1 g) of a polymer (polystyrene, polyamide, or polycarbonate) containing multiwall CNTs (Nanocyl NC7000 or CNano Flotube9000) or no CNTs were placed in the container of the crusher. The airborne particles released by the crushing of the samples were measured. The real-time aerosol measurements showed increases in the concentration of nanometer- and micrometer-sized particles, regardless of the sample type, even when CNT-free polymers were crushed. The masses of the airborne particles collected on filters were below the detection limit, which indicated that the mass ratios of the airborne particles to the crushed pellets were lower than 0.02%. In the electron microscopic analysis, particles with protruding CNTs were observed. However, free-standing CNTs were not found, except for a poorly dispersed CNT-polystyrene composite. This study demonstrated that the crushing test using a laboratory scale crusher is capable of evaluating the potential release of CNTs as a result of crushing CNT composites. The advantage of this method is that only a small amount of sample (several pieces of pellets) is required.

1. Introduction

Carbon nanotubes (CNTs) exhibit unique properties, due to which they are promising candidates as filler materials in composites. Although CNT composites are expected to be used in a wide range of industrial applications, the potential impact of CNTs on human health remains a concern. In order to control the exposure to CNTs, collecting information on the potential release of CNTs throughout their life cycle is important.

In our previous study [1,2], we investigated particle release during the grinding of CNT composites. Many studies have investigated particle release by mechanical processes, such as grinding drilling, cutting, and sanding [3–12]. However, limited study has been carried out regarding particle release by the crushing/shredding of CNT composites [13], which are related to waste treatment processes.

In this study, we investigated airborne particles released as a result of crushing CNT composites. Since commercially available shredding machines for waste treatment are too large to be used for a simulation test, in addition to requiring a large amount of test samples, we developed a crushing test using a laboratory scale crusher.

2. Methods

2.1. Test samples

The samples tested in this study are summarized in table 1. Polystyrene (PS) and polycarbonate (PC) composites containing multiwall CNTs (Nanocyl NC7000) were provided by a company where CNT composites are manufactured. The CNT-PS composite sample was manufactured in a worse condition intentionally, so that the CNTs would disperse poorly in PS (as a peculiar case). PC and polyamide 12 (PA12) composites containing another type of multiwall CNTs (CNano Flotube9000) were purchased from Marubeni Information Systems Co., Ltd., which is a sales agency for CNano Technology, Ltd. Nanocyl NC7000 has an average tube diameter of 9.5 nm, an average tube length of 1.5 μm, and a carbon purity of 90%, while CNano Flotube9000 has an average tube diameter of 10–15 nm, an average tube length of 10 μm, and a carbon purity of 95–97.5%, as provided by the manufacturers.

Table 1. List of test samples.

Sample name	Matrix	CNT type	CNT content
PS	PS	-	0%
Nanocyl-PS	PS	Nanocyl NC7000	A few% (poorly dispersed)
PC	PC	-	0%
Nanocyl-PC	PC	Nanocyl NC7000	A few%
CNano-PC	PC	CNano Flotube9000	3%
PA	PA12	-	0%
CNano(4%)-PA	PA12	CNano Flotube9000	4%
CNano(10%)-PA	PA12	CNano Flotube9000	10%

2.2. Test methods

Figure 1 shows the schematic diagram of the experimental setup. The crushing test was conducted using a laboratory scale crusher with rotor blades (Wonder Crusher WC-3, Osaka Chemical Co., Ltd., Japan). The container (108 mmΦ × 51 mm) of the crusher was modified by providing air intake and exhaust ports for sampling the airborne particles. The crushing power was set to level 3 or level 7 (of 10), at which the samples were roughly and finely crushed, respectively. In these cases, the rotation speed in a steady state was 7,900 and 18,750 r/min, respectively. Considering that the maximum running time for the maximum crushing power (level 10 of 10) recommended by the manufacturer was 60 seconds, the crushing time was set to 120 and 57 s, respectively; so that the total number of rotations reached 14,000 in both conditions. For each crushing test, five pellets (approximately 0.1 g in total; the typical pellet size was 3–4 mm) of the test sample were placed in the container (see figure 2).

Prior to the test, purified air was passed through the container to eliminate background particles. The airborne particles released in the container by the crushing process were then measured using real-time aerosol measuring instruments: a scanning mobility particle sizer (SMPS, model 3936L75, TSI Inc., USA), an optical particle sizer (OPS, model 3330, TSI Inc., USA), and a condensation particle counter (CPC, model 3776, TSI Inc., USA).

In addition, the airborne particles released in the container were collected on (1) polytetrafluoroethylene (PTFE) membrane filters (37 mm diameter, 2 μm pore size, P/N R2PJ037, Pall Corporation, USA) set in the OPS for gravimetric analysis, (2) Nuclepore membrane filters (25 mm diameter, 0.2 μm pore size, GE Healthcare UK) with a stainless holder (1209, Pall Corporation, USA) [14] for scanning electron microscope (SEM) observation, and (3) holey carbon film-coated transmission electron microscope (TEM) grids (Quantifoil R1.2/1.3 on 200 mesh Cu, Agar Scientific, UK) with a TEM grid holder (Mini Particle Sampler, MPS, Ecomesure, France) [15,16] for TEM observation.

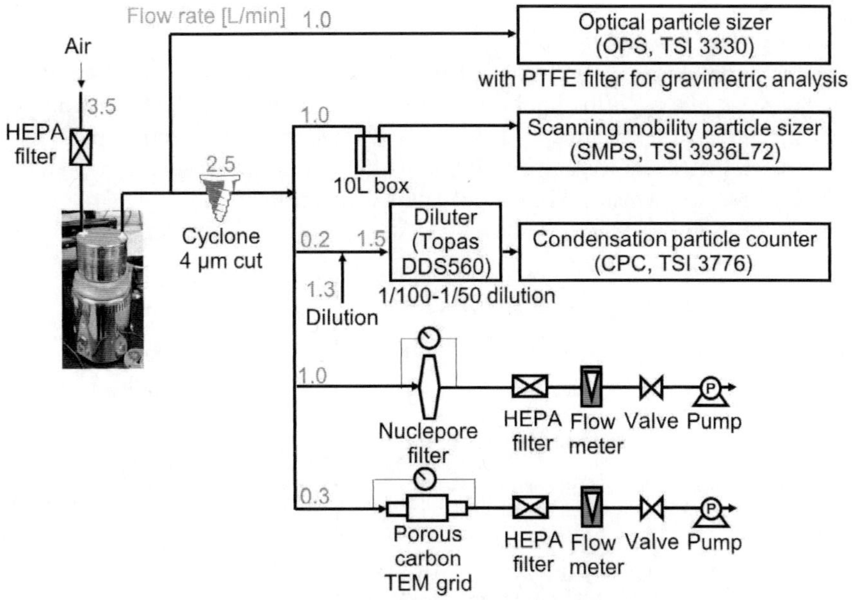

Figure 1. Schematic diagram of the experimental setup.

Figure 2. Photographs of pellets placed in the container of the crusher. Left: before crushing; right: after crushing.

2.3. Gravimetric analysis

The masses of the airborne particles collected on the PTFE filters were determined by weighing the filters before and after sampling, using an ultra-microbalance (SE2-F ultra-microbalance, Sartorius AG, Germany). The limit of detection (LOD, 4.6 μg/filter) was defined as three times the standard deviation of the weight differences of the unused filters prepared and stored along with the filters used (n=10).

2.4. Electron microscope observation

The airborne particles collected on Nuclepore filters were observed using a field-emission SEM (Quanta 250 FEG, FEI Company, USA), under an accelerating voltage of 10 kV. Prior to the SEM observation, the Nuclepore filters were coated with osmium to prevent image charging. The aerosol particles collected on TEM grids were observed using a TEM (JEM-2100F, JEOL Ltd., Japan).

3. Results and discussion

3.1. Real-time aerosol measurements

Figure 3 shows the particle size distributions of the cumulative number of released airborne particles normalized by the mass of the crushed pellets. Compared with the particle concentrations during idle running (i.e., the running of a machine with no pellet), increases in the concentration of nanometer- and micrometer-sized particles were recorded, regardless of the sample type. However, similar increases were also observed when CNT-free polymers were crushed.

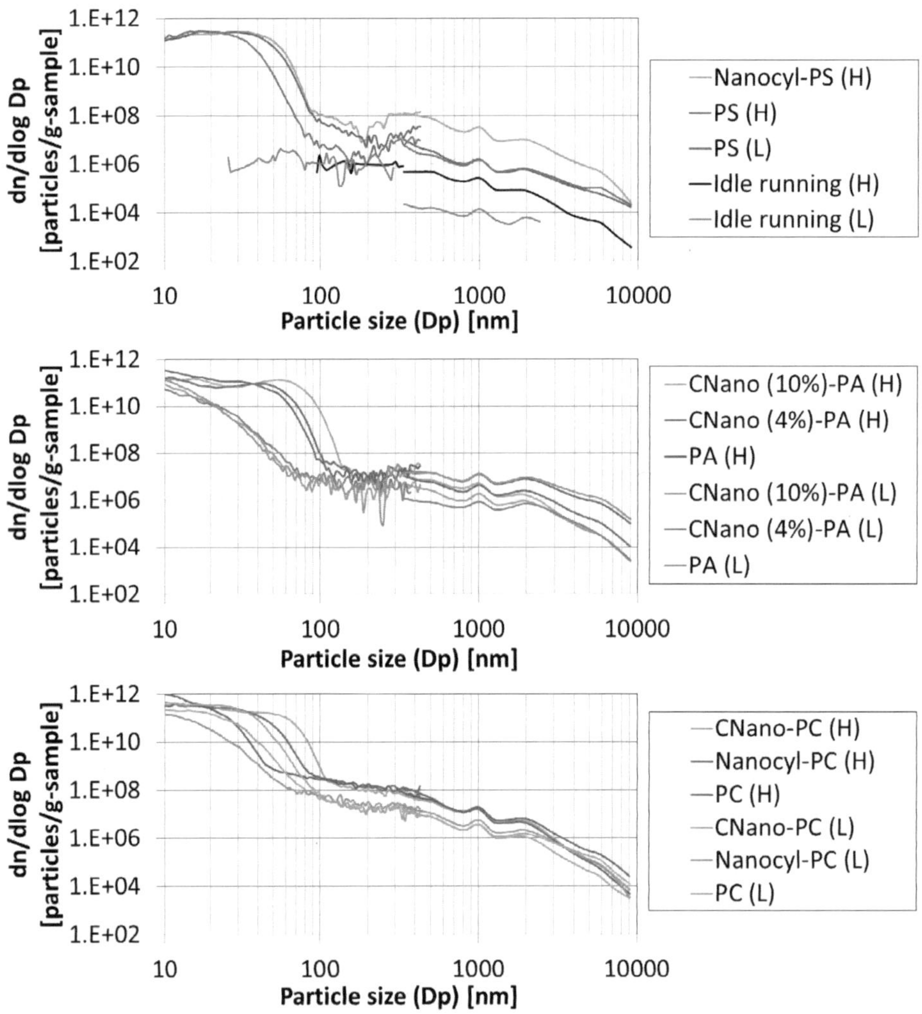

Figure 3. Particle size distributions of the cumulative number of released airborne particles normalized by the mass of crushed pellets. L and H represent low and high rotation speeds (7,900 and 18,750 r/min), respectively.

3.2. Mass ratio of airborne particles to crushed pellets

For all the samples, the masses of airborne particles collected on PTFE filters were below the LOD, which indicated that the mass ratios of the airborne particles to the crushed pellets were lower than 0.02%.

3.3. Electron microscopic observation

Figures 4 and 5 show respectively SEM and TEM images of airborne particles released as a result of crushing CNT composites at a rotation speed of 18,750 r/min. Particles with protruding CNTs were observed for all the CNT composites. On the other hand, free-standing CNTs (isolated or agglomerated CNTs) were found only for the Nanocyl-PS composite. As stated in the 'Methods' section, the dispersion of CNTs in the Nanocyl-PS composite was poor. Figure 6 shows SEM images of the fracture surface of the Nanocyl-PS composite strand (from which pellets were obtained) after tensile breaking. As shown in the areas enclosed by the circles in the image on the left side in figure 6, the CNTs were not uniformly dispersed, partly existing in the form of agglomerates. Thus, the release of free-standing CNTs could be related to the poor dispersion of CNTs in the polymer. Similar results were found in our previous study, in which agglomerated CNTs were found when poorly dispersed CNT-PS composites were ground [2].

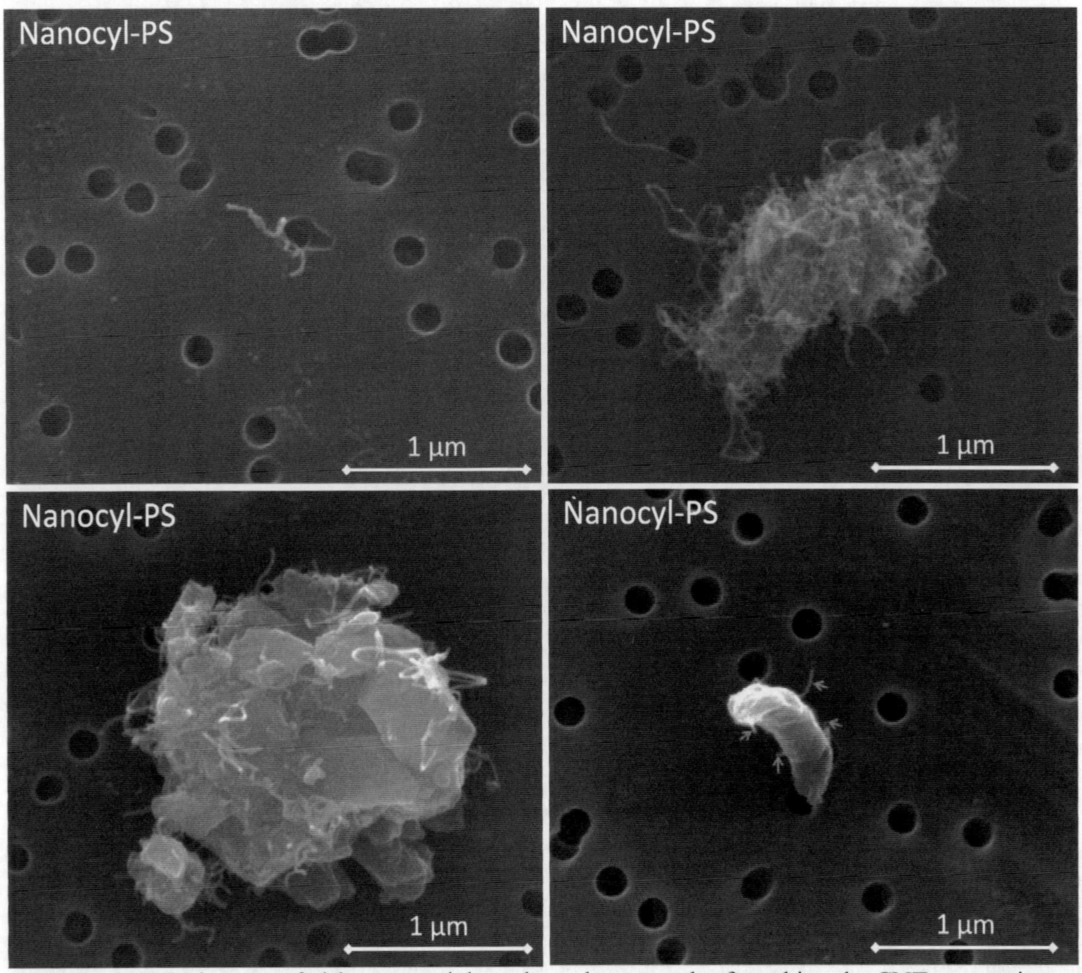

Figure 4. SEM images of airborne particles released as a result of crushing the CNT composites. Arrows indicate protruding CNTs.

IOP Conf. Series: Journal of Physics: Conf. Series **838** (2017) 012015 doi:10.1088/1742-6596/838/1/012015

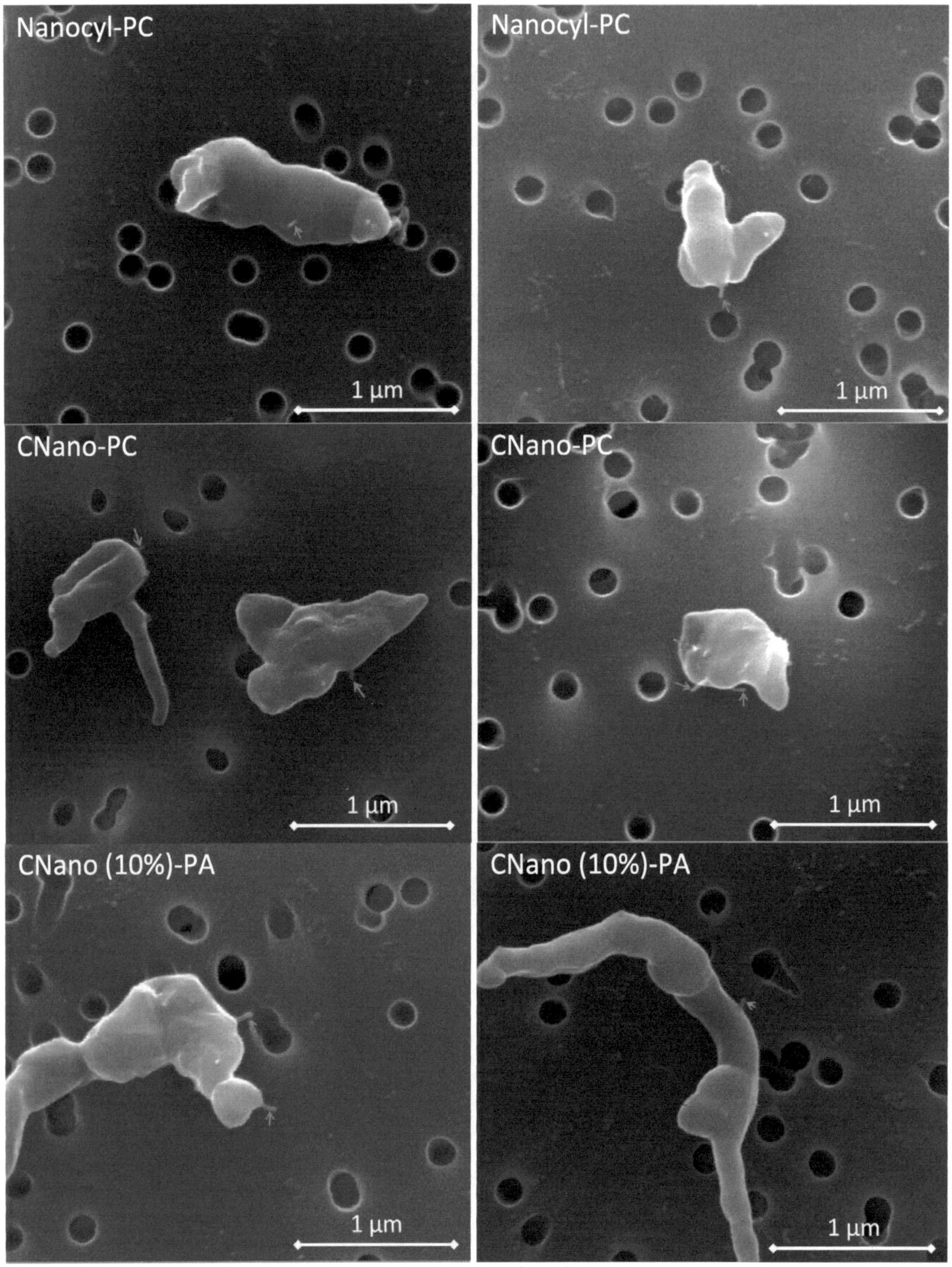

Figure 4. (continued)

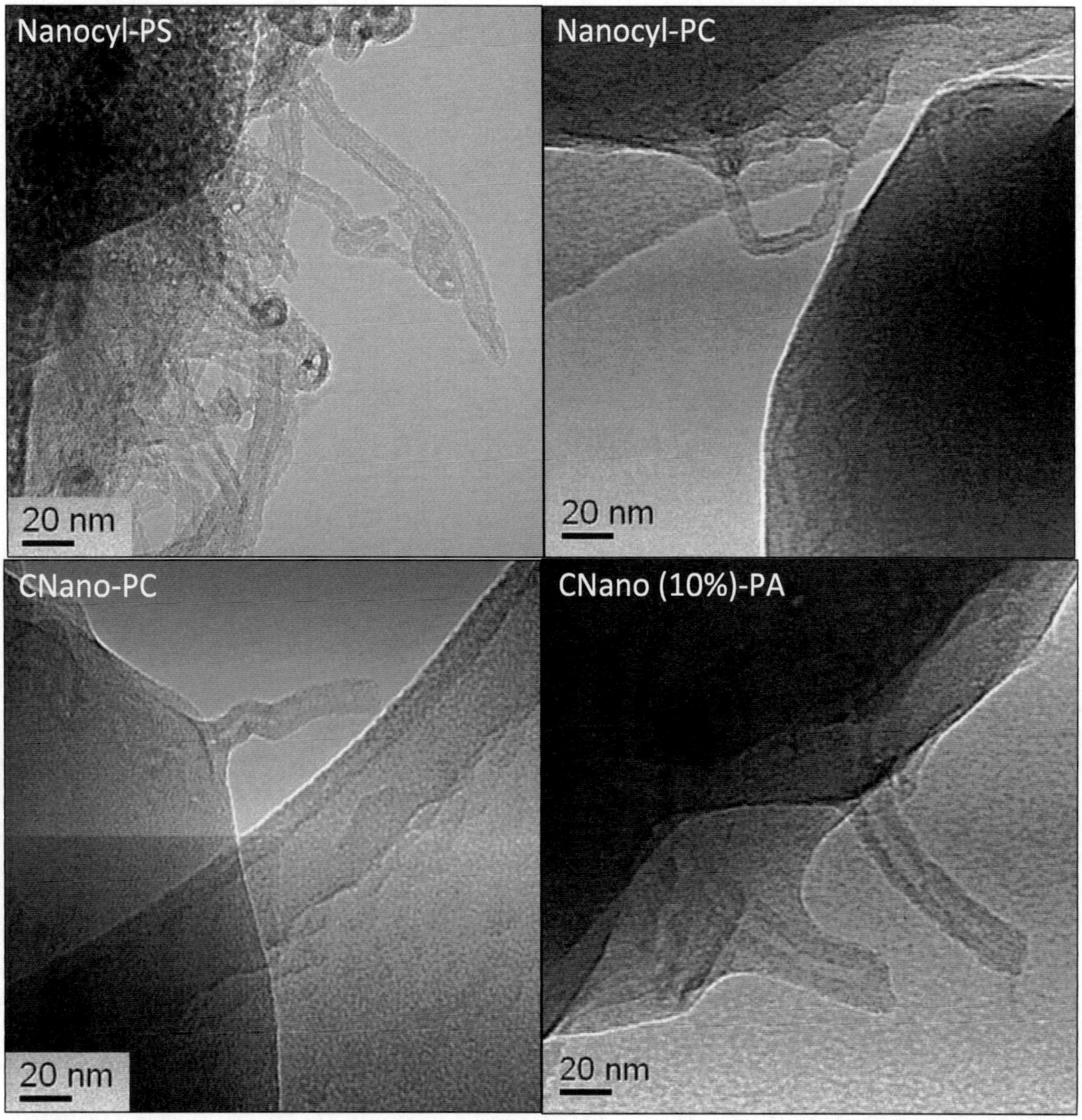

Figure 5. TEM images of airborne particles released as a result of crushing the CNT composites.

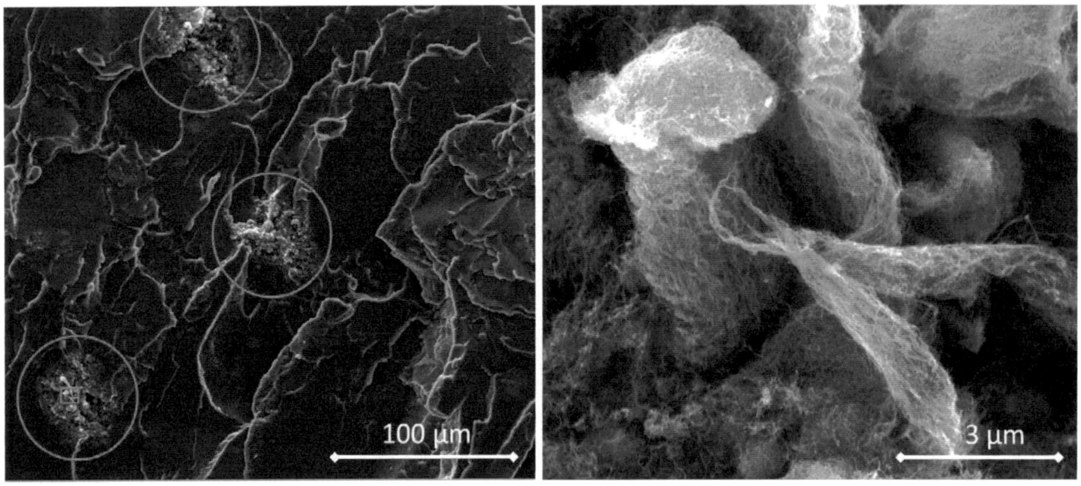

Figure 6. SEM images of the fracture surface of Nanocyl-PS composite strand after tensile breaking. The image on the right is the enlarged image of the area enclosed by the box in the image on the left. CNTs are not uniformly dispersed in PS and partly exist in the form of agglomerates in the areas enclosed by circles in the image on the left.

The percentages of the particle types detected by the electron microscopic observation are summarized in figure 7 for all particles (the graph on the left), as well as particles with a projected area-equivalent diameter less than 300 nm (the graph on the right). The percentages of the particles with protruding CNTs (categorized as Type C in figure 7) were several to several tens of percent and decreased as the particle size decreased. Comparing the CNano (4%)-PA composite with the CNano (10%)-PA composite, the percentages of the particles with protruding CNTs were higher for the CNano (10%)-PA composite than for the CNano (4%)-PA composite.

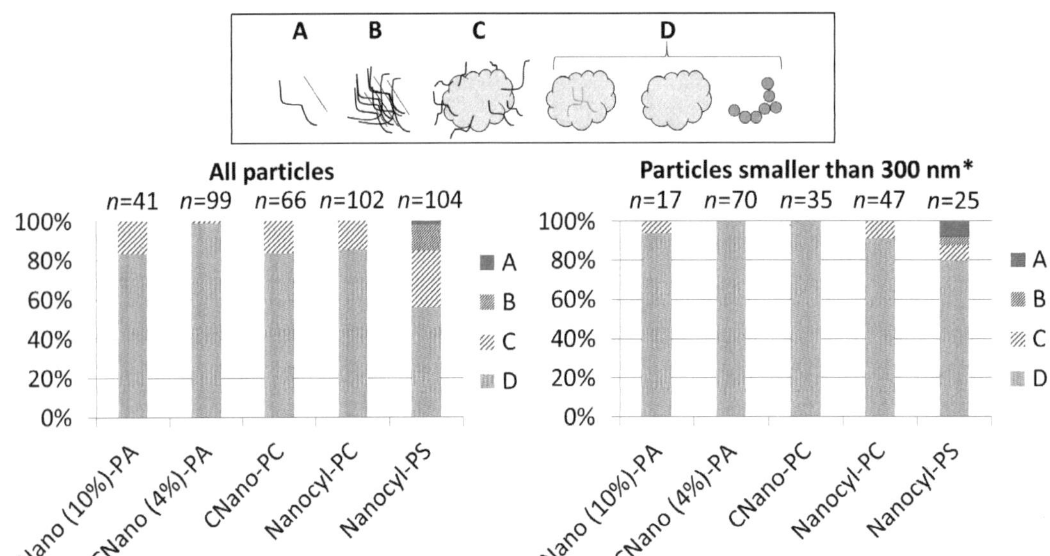

Figure 7. Types of airborne particles detected by electron microscopic observation. A, B, C, and D represent isolated CNTs, agglomerated CNTs, particles with protruding CNTs, and others, respectively.
*Projected area-equivalent diameter.

4. Conclusions

We investigated airborne particles released as a result of crushing CNT composites, using a laboratory scale crusher with rotor blades. The real-time aerosol measurements showed increases in the concentration of nanometer- and micrometer-sized particles, regardless of the sample type, even when CNT-free polymers were crushed. The mass ratios of the airborne particles to the crushed pellets were lower than 0.02%. In the electron microscopic analysis, particles with protruding CNTs were observed for all the tested CNT composites. On the other hand, free-standing CNTs were found only for a poorly-dispersed CNT composite. This study demonstrated that the crushing test using the laboratory scale crusher is capable of evaluating the potential release of CNTs as a result of the crushing of CNT composites. The advantage of this method is that only a small amount of sample (several pieces of pellets) is required.

Acknowledgment

This work is based on results obtained from a project commissioned by the New Energy and Industrial Technology Development Organization (NEDO).

References

[1] Ogura I, Kotake M, Shigeta M, Uejima M, Saito K, Hashimoto N and Kishimoto A 2013 Potential release of carbon nanotubes from their composites during grinding *J. Phys.: Conf. Ser.* **429** 012049

[2] Ogura I, Shigeta M, Kotake M, Uejima M, Honda K 2015 Particle release from single-wall and multiwall carbon nanotubes in polystyrene-based composites during grinding *J. Phys.: Conf. Ser.* **617** 012028

[3] Bello D, Wardle BL, Zhang J, Yamamoto N, Santeufemio C, Hallock M, Virji MA 2010 Characterization of exposures to nanoscale particles and fibers during solid core drilling of hybrid carbon nanotube advanced composites *Int. J. Occup. Environ. Health* **16** 434–450

[4] Golanski L, Guiot A, Pras M, Tardif F 2012 Release-ability of nano fillers from different nanomaterials (toward the acceptability of nanoproduct) *J. Nanopart. Res.* **14** 962

[5] Huang G, Park JH, Cena LG, Shelton BL, Peters TM 2012 Evaluation of airborne particle emissions from commercial products containing carbon nanotubes *J. Nanopart. Res.* **14** 1231.

[6] Wohlleben W, Meier MW, Vogel S, Landsiedel R, Cox G, Hirth S, Tomović Ž 2013 Elastic CNT-polyurethane nanocomposite: synthesis, performance and assessment of fragments released during use *Nanoscale* **5** 369–380

[7] Hirth S, Cena L, Cox G, Tomović Z, Peters T, Wohlleben W 2013 Scenarios and methods that induce protruding or released CNTs after degradation of nanocomposite materials *J. Nanopart. Res.* **15** 1504

[8] Kingston C, Zepp R, Andrady A, Boverhof D, Fehir R, Hawkins D, Roberts J, Sayre P, Shelton B, Sultan Y, Vejins V, Wohlleben W 2014 Release characteristics of selected carbon nanotube polymer composites *Carbon* **68** 33–57

[9] Schlagenhauf L, Buerki-Thurnherr T, Kuo YY, Wichser A, Nüesch F, Wick P, Wang J 2015 Carbon nanotubes released from an epoxy-based nanocomposite: quantification and particle toxicity *Environ Sci. Technol.* **49** 10616–10623

[10] Harper S, Wohlleben W, Doa M, Nowack B,Clancy S, Canady R, Maynard A 2015 Measuring nanomaterial release from carbon nanotube composites: review of the state of the science *J. Phys.: Conf. Ser.* **617** 012026

[11] Boonruksa P, Bello D, Zhang J, Isaacs JA, Mead JL, Woskie SR 2016 Characterization of potential exposures to nanoparticles and fibers during manufacturing and recycling of carbon nanotube reinforced polypropylene composites *Ann. Occup. Hyg.* **60** 40–55

[12] Rhiem S, Barthel AK, Meyer-Plath A, Hennig MP, Wachtendorf V, Sturm H, Schäffer A, Maes HM 2016 Release of ^{14}C-labelled carbon nanotubes from polycarbonate composites *Environ. Pollut.* **215** 356–65

[13] Stahlmecke B, Asbach C, Todea A, Kaminski H, Kuhlbusch TAJ 2014 Investigations on CNT release from composite materials during end of life *Handbook of Nanosafety: Measurement, Exposure and Toxicology* ed Vogel U, Savolainen K, Wu Q, van Tongeren M, Brouwer D, Berge M,. (London, Waltham, San Diego: Elsevier) pp 242–255

[14] Ogura I, Kotake M, Sakurai H, and Honda K 2016 Surface-collection efficiency of Nuclepore filters for nanoparticles *Aerosol Sci. Technol.* **50** 846–856

[15] R'mili B, Le-Bihan OLC, Dutouquet C, Aguerre-Charriol O and Frejafon E 2013 Particle sampling by TEM grid filtration *Aerosol Sci. Technol.* **47** 767–775

[16] Ogura I, Hashimoto N, Kotake M, Sakurai H, Kishimoto A and Honda K 2014 Aerosol particle collection efficiency of holey carbon film-coated TEM grids *Aerosol Sci. Technol.* **48** 758–767

IOP Conf. Series: Journal of Physics: Conf. Series **838** (2017) 012016 doi:10.1088/1742-6596/838/1/012016

Safer by design strategies

M. Cobaleda-Siles,[1] A.P. Guillamon,[1] C. Delpivo,[2] S. Vázquez-Campos,[2] V. F. Puntes[1,3,4,5]

[1]Institut de Recerca Vall d'Hebron (VHIR). Pg. Vall d'Hebron, 119-129 08035 Barcelona (Spain)
[2]LEITAT Technological Center. C/ Innovació 2, 08225 Terrassa, Barcelona, Spain
[3]Institució Catalana de Recerca i Estudis Avançats (ICREA). Pg. Lluís Companys, 23, 08010 Barcelona (Spain)
[4]Institut Català de Nanociencia i Nanotecnologia (ICN2), Campus de la UAB, Av. Serragalliners, s/n, 08193 Barcelona (Spain)
[5]Author to whom any correspondance should be addressed

E-mail: victor.puntes@icn.cat

Abstract. Throughout the EU funded FP7 project GUIDENano, we are trying to control and monitor the evolution of nano-enable products during their lifecycle. Small alterations of the nanoparticle (NP) state may have critical consequences on the NP behaviour and performance. For this reason it is important to highlight the importance of an extensive and proper characterization to define the NP physico-chemical characteristics under several environmental conditions. Furthermore, this characterization is necessary to ensure that obtained results are reproducible and allow understanding the behaviour of the NP on biological systems. In this paper different strategies reported in the literature regarding the safety-by-design concept are summarized. Several strategies from the synthetic point of view that help us to modulate the main factors which determine the safety of nanomaterials are proposed.

1. Introduction.

The growing interest in the field of nanotechnology leads to an increase in the number of nanoparticles (NPs) producers and users, making indispensable the promotion of safe strategies in the development and management of nanomaterials (NMs).[1–3] In fact, the rise of the production of NPs is estimated on thousands of tons by 2020,[4] accompanied by the increase of congresses, publications and research projects related to nanosafety in the last 10 years,[5] and consequently reflected in the increasing number of nano-enabled products that reach the market. However, some concerns have been arisen over current uncertainties on the safety of such products and they need to be carefully addressed to avoid human and environmental damage which could hinder the development of the nanotechnology field. Consequently, the exposure to nano-sized materials (NM) and their potential to cause harm to humans and other biological organisms have focused the attention of several recent studies.[6–9] In these studies, it is agreed that there is a need to understand the NP behavior during its full lifecycle in order to design materials with a high yield and quality at the same time along with well controlled properties. In fact,

IOP Conf. Series: Journal of Physics: Conf. Series **838** (2017) 012016 doi:10.1088/1742-6596/838/1/012016

if there are not more nano-enabled products in the market is partly due to the risks associated with nano enabled-products by consumers, regulators and insurance communities. It is in this context, that safety by design strategies, are translated into *nanosafety by design* strategies, which needs of a precise knowledge of NP behavior in order to properly design safe products.

The need to develop commercial products that are at the same time useful and safe, starts at the very beginning of their conception. The objective is establishing safer by design selection rules and synthetic approaches that can be used for the reduction of nanotechnology associated risks. Taking into consideration all stages of the life cycle of these products, these rules should protect the safety of workers, users and consumers and product end of life. In detail, safer-by-design approaches aim at the reduction of NMs hazard and exposure, the reduction of NMs migration and release, and the controlled degradation of these once they are released from their matrices. Traditionally, several strategies have been employed towards these aims:

1. **Reducing toxicity** of the employed materials. For instance, **i)** avoiding the use of intrinsically toxic elements or substances where possible, e.g. Cd,[10] **ii)** modifications in the size and shape to reduce the toxicity of the NMs in a specific biological system;[11] **iii)** increase in hydrophilicity to decrease the potential to cross biological membranes; or **iv)** changing the oxidation state to mitigate NP reactivity. Regarding size, an important point is reported in the case of non biodegradable inorganic fibers longer than 10 μm where macrophages cannot engulf them, setting then a defense action well known as frustrated phagocytosis.[12] In these cases, the NPs cannot be completely phagocytosed inducing thus chronic inflammation and consequent carcinogenicity (*asbestos-like* effect).[13]
2. **Reducing release** of NMs from the matrix during their life cycle. Normally, NPs are components embedded in a solid or liquid matrix forming part of different products. It is possible to reduce the release of NM from the matrix by controlling van der Waals, ionic, coordination and covalent bondings between NP and matrix using ligands and compatibilizers.[14] In the case of inorganic NPs, not only the release of the NP should be controlled but also the release of its constituent ions, since toxicity of NPs have been often attributed to their leached ions rather than the NPs themselves.[15,16] In this case, encapsulating the NMs makes possible to preserve the properties of the core material along with protecting it from dissolution, as in the case of ZnO NPs coated with a *nanothin* amorphous silica.[17]
3. **Reducing the persistence** of NMs. If NPs are used massively in consumer products they will inevitably end up in the environment. Thus developing strategies which control the end of life of the NPs are needed. As NPs have a high surface energy, they are prone to both aggregation and dissolution.[18] Thus, NPs can be irreversibly aggregated (and easily sintered) until they reach bulk sizes of well-known materials that we know how to deal with safely, or otherwise, NPs can be corroded into ionic species where our knowledge to safely deal with them is broad.

In this work, the aggregation of TiO_2 NPs and corrosion of silver NPs (AgNPs) was monitored in order to establish procedures where the NP is not only defined by its composition but its physical (colloidal) and chemical stability, so they can be properly and safely handled.

2. Experimental procedures: synthesis of nanoparticles

The nanoparticles used in the studies were synthesized as follows.

2.1. TiO_2 nanoparticles

TiO_2 nanoparticles were prepared following a procedure previously described on the literature by precipitation of Titanium tetrachloride ($TiCl_4$) in aqueous medium.[19] The stock solution of Ti^{4+} (0.7 mol/L) was prepared by dissolving the $TiCl_4$ precursor in HCl (3 mol/L) solution. For the production of TiO_2, an aqueous Ti^{4+} stock solution (20 mL) was diluted in Milli-Q water (270 mL) at room temperature. The pH of the mixture was fixed at 5 by the addition of NaOH (3 mol/L). Suspensions were aged at 60 °C for 24 hours and the solid was collected by centrifugation. Samples were further purified by three centrifuge cycles with water of the same pH as used for the synthesis and finally re-suspended in an aqueous solution of TMAOH.

2.2. Silver nanoparticles (AgNPs)

The typical procedure for the synthesis of AgNPs have been previously described on the literature.[20] Briefly, 100 mL volume of aqueous solution containing sodium citrate (SC) (5 mM) and tannic acid (TA) was prepared and heated with a heating mantle in a three-neck round bottomed flask for 15 min under vigorous stirring. A condenser was used to prevent the evaporation of the solvent. After boiling had commenced, 1 mL of $AgNO_3$ (25 mM) was injected into this solution. Then, the reaction was cooled until the temperature of the solution reached 90 °C. Then, 100 μL of SC (25 mM), 250 μL of TA (2.5 mM), and 250 μL of $AgNO_3$ were sequentially injected (time delay 1 min). By repeating this process and adjusting the amount of Ag precursor injected, different AgNP generations of progressively larger sizes were grown. Aliquots were purified by centrifugation at 18000 g for 20 min in order to remove the excess of TA and further redispersed in Milli-Q-water or SC 2.2 mM before sample characterization.

3. Results. NPs Evolution.

A comprehensive knowledge of the NPs evolution during their intended use but also storage and disposal, and knowledge on their most common transformations, opens the possibility of making safer NMs.[21] The importance of a proper and extensive characterization to understand and control possible variations on the starting material is crucial in the development of improved NMs.[22] The use of reliable equipments and commonly present at laboratories such as UV-Vis Spectrophotometer, can provide useful information about physico-chemical properties of NMs. Furthermore, it allows us to characterize variations on NMs properties such as sizes, morphologies or the effectiveness of binding a coating layer to the NP surface quite univocally. Other techniques are needed to fully characterize NMs, as electron microscopy and X ray diffraction, and others spectroscopic techniques are also fast and sensitive to the NMs state, as Dynamic Light Scattering or Nanoparticle Tracking Analysis, however, the former are complex and the latter are simple but difficult to interpret, making UV-Vis a good tool for the rapid assessment of NM sample transformation.

It is well-known that NMs possess unique and different physico-chemical properties compared with their corresponding bulk materials.[3,15] These properties lead to biological effects which could be completely different to their reference materials **(Fig. 1)**.[23] In fact, initially, the European Commission Scientific Committee on Emerging and Newly Identified Health Risks (SCENIHR) has concluded that NMs might have different toxicological properties compared to bulk substance and thus the risks of the NMs should be assessed case by case.[15,24] However this would be extremely expensive and cumbersome, so general it is expected that as more NMs are analyzed and studied, behavioral trends of different families of NMs will arise allowing for class grouping and simplify tracking and monitoring of NMs. This has been done previously with pharmaceuticals, nutriceuticals and chemicals and there is no reason why this could not happen with NMs. Thus, classification of NPs in families depending on their biological effects, and identification of the specific parameters that controls NP toxicity, will be a more realistic approach to NP safety and it is being already addressed by the scientific community.[25]

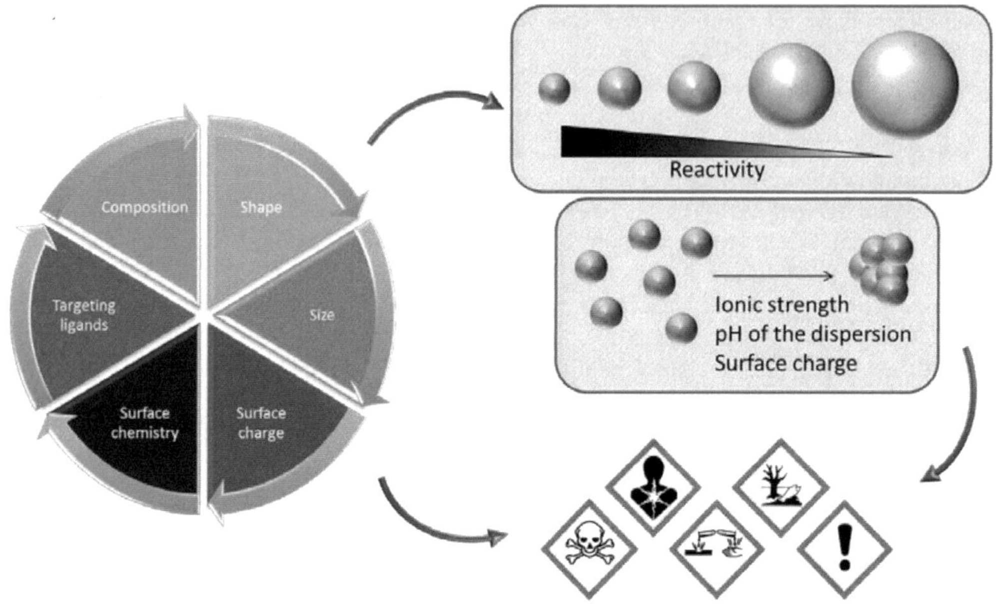

Figure 1. Illustration of physico-chemical characteristics of nanoparticles associated with their possible toxic properties.

Two of the most common transformations that NPs can undergo when dispersed in biological media, are aggregation and chemical degradation (oxidation, corrosion, dissolution). Controlled colloidal stability (aggregation, sedimentation and stickiness) or controlled chemical reactivity and integrity of the material are needed to study and to monitor the evolution of the NPs during their lifecycle. Note that the loss of colloidal stability, aggregation and sedimentation as much as the corrosion and dissolution may influence biological assays and knowledge of these processes are therefore necessary to make a correct interpretation of them.[26]

3.1. Aggregation

Aggregation of NPs is a very common phenomenon in biological media. Parameters such as concentration of NPs and chemical nature and the ionic strength of the dispersing medium can cause the cross-linking and aggregation of colloidal NPs.[27] As NPs tend to agglomerate/aggregate after relatively short incubation times in different buffers and biological media, coating them with stabilizers, such as the well known biocompatible FDA approved PVP or PEG, to provide colloidal stability in physiological media, is a very common strategy.[28] Despite that, the stability of NPs is not yet well understood and controlled and it is therefore important to study the aggregation behaviour of NPs to predict their fate in different aqueous media and environments in order to analyze their biological effects.

In the experiments performed with TiO_2 NPs, the state of aggregation can be fine-tuned in the presence of different TMAOH concentrations during the synthesis process. The addition of TMAOH can modulate the state of aggregation by stabilization of TiO_2 at different degrees of aggregation.[29]

The stability of the NPs was studied over a period of 48 h in the absence of light. This is because illumination of TiO_2 NPs may promote catalytic TMAOH degradation and consequently favour aggregation. Aggregation has a direct effect in buoyancy and therefore sedimentation profiles can be employed as a measure of the degree of aggregation. Thus, analysis of samples colloidal stability was obtained by performing a straightforward sedimentation assay by UV-Vis spectroscopy. Note that the sample was left unperturbed between analysis inside the UV-Vis Spectrophotometer. **Fig. 2** shows how the sedimentation of large aggregates is faster compared with the other two samples. In the specific case of small aggregates the sedimentation is negligible. It is interesting that from 20 h onwards the sedimentation is almost imperceptible even in the case of large aggregates, since the effective concentration of NPs has decreased indicating how aggregation is directly related to concentration (NPs have to meet before they aggregate).

The sedimentation study for AgNPs was performed following the same methodology for a period of 48h **(Fig. 2)**. A similar behavior in AgNPs can be observed compared to the previous ones, since the sedimentation process of small AgNPs is almost imperceptible during the complete assay **(inset figures)**. However, in the case of AgNPs the sedimentation process is more evident compared with TiO_2 NPs, because the higher density of silver. It is important to highlight that in both cases, the sedimentation process is reversible, after a hand shake the NPs restore their initial spectra, which indicates that NPs are not irreversibly aggregated.

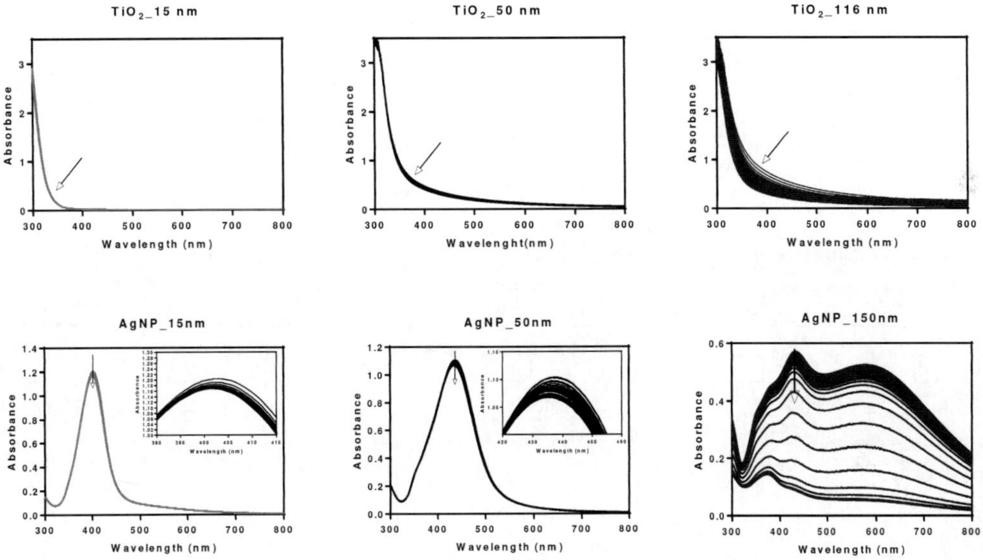

Figure 2. UV-Vis spectra profiles of sedimentation process of TiO_2 nanoparticles and AgNP during 48 h recorded at time intervals of 1h (from t_0 to t_{48} from top to bottom).

3.2. Dissolution

It is important to analyze the processes and mechanisms of dissolution in order to define strategies to control it, because it is well known that the release of cadmium from quantum dots or silver ions from AgNP contribute to the toxicity of these materials.[15,30] Not only because the released atoms, but also the redox process itself alters the metabolic status of cells.[31] Corrosion critically depends on the presence of dissolved oxygen and ions (nucleophiles) in solution. In this experimental case the behaviour of three different sizes of AgNPs when exposed to different environmental conditions were analyzed as it is shown in **Fig. 3**.

AgNP were exposed for 48 h to: **(1)** nitrogen atmosphere (AgNP as synthesized but preserved under inert atmosphere used as a control of non-oxidation), **(2)** Dulbecco´s Modifed Eagle Medium (DMEM) and **(3)** hydrogen peroxide (H_2O_2) at 1 mM concentration. During the process the samples were studied by UV-Vis spectroscopy as it is shown in **Fig. 3**. To ensure that the results observed are only due to the dissolution process the samples were always shaken before each UV-Vis acquisition.

As it is shown in the different graphs **(Fig. 3)** it can be concluded that the AgNP reactivity depends strongly on size and environment. The smaller the size, the more reactive the nanoparticle. We can observe this effect more notably in the presence of H_2O_2, where the oxidation occurs much faster in the AgNP_15 nm than in the other two samples. In the presence of DMEM it can be noted that oxidation and aggregation occurs simultaneously, due to the high ionic strength of this media. It is important to highlight that both processes can occur simultaneously and that one can promote or inhibit the other, since corrosion increases the ion concentration in solution which may lead to aggregation,[32] which in turn decreases dissolution by decreasing surface to volume ratio. In order to decouple the aggregation and dissolution process in our assay and observe a proper corrosion study, the samples were stabilized with PVP (polyvinylpirrolidone) at 0.1 mM, obtaining the spectra profiles shown in **Fig. 3**.

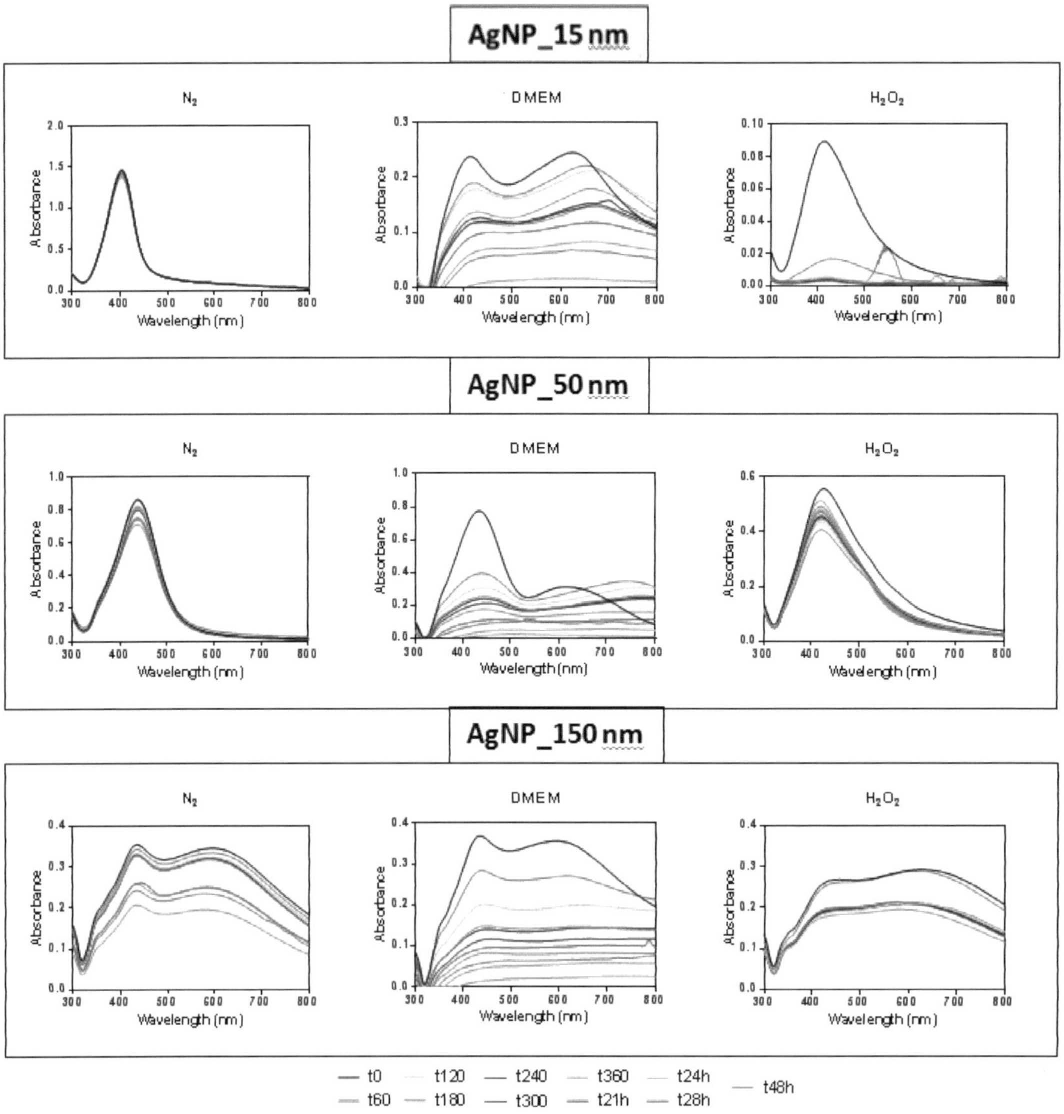

Figure 3. UV-Vis spectra profiles of corrosion process of AgNP 15, 50 and 150 nm under inert atmosphere, DMEM and H_2O_2 during 48 h.

4. Conclusions.

Regarding the **experimental data**, it has been shown that aggregation and corrosion process of TiO_2 and Ag NPs with different diameters can be presented by monitoring sedimentation and dissolution studies using an UV-Vis spectrometer. We would like to highlight that among all approaches available to study aggregation and sedimentation processes, in this paper it was presented an affordable, easy and accurate method that provide useful information to study both processes thanks to the optical properties of colloids.

It is also demonstrated that the reactivity depends on the size of the NPs. The smaller is the size, the more reactive is the nanoparticle. Furthermore, it can be concluded that as NP size increases the NP becomes more stable chemically but is more difficult to maintain it in suspension. Besides, it was observed that aggregation/sedimentation and corrosion are interlinked. Understanding the aggregation process of the NPs or controlling the state of aggregation due to the presence of surfactants, it is possible to understand the corrosion process of the material.

The behavior of NPs in biological media or its release to the environment is complex and depends on several factors in which aggregation and corrosion are included. For this reason, it is important to carry out a proper and extensive characterization of NPs to obtain an optimal and safe design. Three parameters have to be taken into account in the characterization of NPs: **i)** controlling and monitoring our NPs "from cradle to grave", changes in their size, shape or oxidation states during lifecycle; **ii)** controlling the release of the NPs from the solid or the ions from the NPs; **iii)** controlling the persistence of NPs, including possible transformations and corrosion.

Finally, it is important to highlight that the concept of safety by design starts in the inception point and it has to be considered during all lifecycle of the NPs and bearing in mind that sometimes is not possible to make something completely safe but safer is a more realistic goal.

References

1. Klaine, S. J. *et al.* Nanomaterials in the environment: behavior, fate, bioavailability, and effects. *Environ. Toxicol. Chem.* **27,** 1825–1851 (2008).
2. Vance, M. E. *et al.* Nanotechnology in the real world: Redeveloping the nanomaterial consumer products inventory. *Beilstein J. Nanotechnol.* **6,** 1769–80 (2015).
3. Stark, W. J., Stoessel, P. R., Wohlleben, W. & Hafner, A. Industrial applications of nanoparticles. *Chem. Soc. Rev.* **98,** 2035–2044 (2015).
4. Lewinski, N., Colvin, V. & Drezek, R. Cytotoxicity of nanoparticles. *Small* **4,** 26–49 (2008).
5. SafeNano. http://www.safenano.org.
6. Ai, J. *et al.* Nanotoxicology and nanoparticle safety in biomedical designs. *Int. J. Nanomedicine* **6,** 1117–27 (2011).
7. Elsaesser, A. & Howard, C. V. Toxicology of nanoparticles. *Adv. Drug Deliv. Rev.* **64,** 129–137 (2012).
8. J Nanopart Res. *et al.* Perspectives on the design of safer nanomaterials and manufacturing processes. *September* **17,** 366 (2015).
9. Bakand, S. & Hayes, A. Toxicological Considerations, Toxicity Assessment and Risk Management of Inhaled Nanoparticles. *Int. J. Mol. Sci.* **17,** 929 (2016).
10. Godt, J. *et al.* The toxicity of cadmium and resulting hazards for human health. *J. Occup. Med.* **6,** 1–6 (2006).
11. Oberdörster, G. Safety assessment for nanotechnology and nanomedicine: concepts of nanotoxicology. *J. Intern. Med.* **267,** 89–105 (2010).

12. Donaldson, K., Murphy, F. a, Duffin, R. & Poland, C. a. Asbestos, carbon nanotubes and the pleural mesothelium: a review of the hypothesis regarding the role of long fibre retention in the parietal pleura, inflammation and mesothelioma. *Part. Fibre Toxicol.* **7,** 5 (2010).

13. Napolitano A., Jube S., Gaudino G., Pass Harvey I., C. M. and Y. H. *Cancer and Inflammation Mechanisms: Chemical, Biological, and Clinical Aspects. Chapter 16. Asbestos-Induced Chronic Inflammation and Cancer.* (2014).

14. Hsu, P. *et al.* Personal Thermal Management by Metallic Nanowire-Coated Textile. *Nano Lett.* **15,** 365 (2015).

15. Beer, C., Foldbjerg, R., Hayashi, Y., Sutherland, D. S. & Autrup, H. Toxicity of silver nanoparticles—Nanoparticle or silver ion? *Toxicol. Lett.* **208,** 286–292 (2012).

16. Kirchner, C. *et al.* Cytotoxicity of Colloidal CdSe and CdSe / ZnS Nanoparticles. *Nano Lett.* **5,** 331–338 (2005).

17. Georgios A. Sotiriou, Christa Watson, Kimberly M. Murdaugh, Thomas H. Darrah, Georgios Pyrgiotakis, Alison Elder, J. D. B. and P. D. Engineering safer-by-design silica-coated ZnO nanorods with reduced DNA damage potential. *Environ. Sci. Nano,* **1,** 144–153 (2014).

18. Casals, E., Gonzalez, E. & Puntes, V. F. Reactivity of inorganic nanoparticles in biological environments: insights into nanotoxicity mechanisms. *J. Phys. D. Appl. Phys.* **45,** 443001 (2012).

19. Pottier, A. *et al.* Size tailoring of TiO2 anatase nanoparticles in aqueous medium and synthesis of nanocomposites. Characterization by Raman spectroscopy. *J. Mater. Chem.* **13,** 877–882 (2003).

20. Bastús, N. G., Merkoçi, F., Piella, J. & Puntes, V. Synthesis of Highly Monodisperse Citrate-Stabilized Silver Nanoparticles of up to 200 nm: Kinetic Control and Catalytic Properties. *Chem. Mater.* **26,** 2836–2846 (2014).

21. Mitrano, D. M., Motellier, S., Clavaguera, S. & Nowack, B. Review of nanomaterial aging and transformations through the life cycle of nano-enhanced products. *Environ. Int.* **77,** 132–147 (2015).

22. Hirn, S. *et al.* Particle size-dependent and surface charge-dependent biodistribution of gold nanoparticles after intravenous administration. *Eur. J. Pharm. Biopharm.* **77,** 407–416 (2011).

23. Rivera Gil, P., Oberdörster, G., Elder, A., Puntes, V. & Parak, W. J. Correlating physico-chemical with toxicological properties of nanoparticles: The present and the future. *ACS Nano* **4,** 5227–5231 (2010).

24. Hansen, S. F. & Baun, A. When enough is enough. *Nat. Nanotechnol.* **7,** 409–411 (2012).

25. Fadeel, B. Nanosafety: Towards safer design of nanomedicines. *J. Intern. Med.* **274,** 578–580 (2013).

26. Krug, H. F. Nanosafety Research-Are We on the Right Track? *Angew. Chemie Int. Ed.* **53,** 12304–12319 (2014).

27. Bian, S. W., Mudunkotuwa, I. A., Rupasinghe, T. & Grassian, V. H. Aggregation and dissolution of 4 nm ZnO nanoparticles in aqueous environments: Influence of pH, ionic strength, size, and adsorption of humic acid. *Langmuir* **27,** 6059–6068 (2011).

28. Li, X. & Lenhart, J. J. Aggregation and dissolution of silver nanoparticles in natural surface water. *Environ. Sci. Technol.* **46,** 5378–5386 (2012).

29. Chemseddine, A. & Moritz, T. Nanostructuring titania: Control over nanocrystal structure, size, shape, and organization. *Eur. J. Inorg. Chem.* 235–245 (1999). doi:Doi 10.1002/(Sici)1099-0682(19990202)1999:2<235::Aid-Ejic235>3.0.Co;2-N

30. Xiu, Z., Zhang, Q., Puppala, H. L., Colvin, V. L. & Alvarez, P. J. J. Negligible Particle-Speci fi c Antibacterial Activity of Silver Nanoparticles. *Nano Lett.* **12,** 4271–4275 (2012).

31. Bastus, N. G., Casals, E., Vazquez-Campos, S. & Puntes, V. Reactivity of engineered inorganic nanoparticles and carbon nanostructures in biological media. *Nanotoxicology* **2,** 99–112 (2008).

32. Wang, D., Tejerina, B., Lagzi, I., Kowalczyk, B. & Grzybowski, B. A. Bridging interactions and selective nanoparticle aggregation mediated by monovalent cations. *ACS Nano* **5,** 530–536 (2011).

Banding approach for engineered nanomaterial risk assessment and control

Nausikaä Van Hoornick[1], Dimiter Prodanov[1], Alain Pardon[1]

[1]Imec, Kapeldreef 75, 3001 Heverlee, Belgium
Email: nausikaa.vanhoornick@imec.be

Abstract. Since there is a lack of sufficient toxicological information about engineered nanomaterials, and since within the semiconductor research and manufacturing use is already made of these materials, an alternative for the classical quantitative risk assessment was sought. Within the research facilities of imec use is made of a banding technique to determine the risks associated with the nanomaterial research. The method and the measures that are taken are discussed in the paper. The method has been benchmarked with other available techniques.

1. Introduction

In contrast to the quantitative risk assessment approach based on sufficient toxicological data, which is used into risk assessments of bulk materials, engineered nanomaterials (ENM) represent a challenge because their effects on the human health and environment are poorly understood. Despite the fact that the effects of these materials are scarcely known, novel ENMs are designed, synthesized and used on a regular basis.

One way to deal with this would be to look at the properties of the bulk materials and base the risk assessment of the ENMs on this. However it is known that the properties and thus the effects of the ENM's are quite different than the properties of their parent bulk materials.

Obtaining all necessary data to do a quantitative risk assessment requires considerable resources. Moreover, health effects of nanomaterials are still generally uncertain because of the lack of standardization in testing protocols and in most of the cases there are no established reference values. Since even a best case scenario introduces a considerable time lag between the actual work and the availability of the input information for a quantitative risk assessment other techniques of risk assessment should be used.

2. Risk and Control Banding approaches

In order to manage the environment, health and safety (EHS) risks in circumstances where information is missing **various risk and control banding tools** have been developed by a variety of academic and industrial groups [1]. Control banding was originally developed by the pharmaceutical industry as a way to safely work with new chemicals that had little or no toxicity information. These new chemicals were classified into "bands" based on other more-studied materials' toxicity and anticipated safe work practices, taking into consideration exposure assessments. Each band was then aligned with a control scheme. At present, the concept is used in the ISO Standard ISO/TS 12901-2:2014 Nanotechnologies - Occupational risk management applied to engineered nanomaterials [3]. ISO/TS 12901-2:2014 is focused on intentionally produced nano-objects such as nanoparticles, nanosheets, nanofibres, nanotubes, nanowires, as well as of aggregates and agglomerates of the same. Currently, the risk and control banding approach is applied for emerging risks such as nano-objects (e.g. nanoparticles), by the development of various control banding approach tools, for example ISO/TS 12901-2, ANSES, Stoffenmanager Nano, NanoSafer etc.

The EHS team of *Imec* – a large research center for semiconductor nanotechnology – uses a banding approach to deal with risk assessments of ENMs. The risk control approach is based on the one proposed originally in Stoffenmanager[4], but has been geared for application within the semiconductor research and development. Our approach parallels the accepted practices in life sciences (e.g. biosafety) where the intrinsic hazard of the biological entity is difficult to quantify precisely.

Within the risk control approach we look at the hazard potential of the ENM and to the exposure potential during specific activities. Both aspects are divided into three groups: low, moderate and high. By classifying the ENM and the exposure to the ENM the appropriate risk control bands can be defined, as is shown in figure 1. Based on these risk control bands measures are proposed in order to work safely during the activity at hand.

		Exposure potential		
Hazard potential	*RISK*	High	Moderate	Low
	High	Likely	Likely	Likely
	Moderate	Likely	Likely	Potential
	Low	Likely	Potential	Unlikely

Figure 1: Risk banding approach based on hazard potential of the ENM and exposure potential of the activity as adapted from Stoffenmanager.

2.1 Hazard potential

A number of different characteristics of the ENM are taken into account in order to determine the hazard potential of the material. The most important properties taken into consideration for this classification are:

- size and shape
- water solubility
- dispersion capacity
- toxicity of parent material
- toxicity of the chemical functionalization.

The hazard is considered high when the particle size is less than 10 nm, the particles have high aspect ratio and the ENM is dispersed and hydrophobic.

The intrinsic hazard is considered moderate if the particle size is in the range of 10 nm - 100 nm, the ENM aggregates or agglomerates or it has irregular shape.

The intrinsic hazard is considered low if the parent material is water soluble or dispersed in liquid, if it is embedded in solid matrix or has a low dispersion capacity.

2.2 Exposure potential

The probability of exposure (e.g. the potential of a process step to generate airborne ENM) is banded in three categories. Banding is based on the assumption that the greatest health hazard comes from airborne particles. The most important properties taken into consideration for this classification are:

- quantity of ENM
- proximity of the activity (near breathing zone or not)
- duration of the activity
- kind of activity (grinding, milling, polishing, ...)

A high exposure potential is introduced by manipulations that mechanically disrupt a surface functionalized with ENM, manipulations done within the breathing zone (i.e. less than half a meter away), involving large amount (area factor) of ENM, steps producing aerosols or mechanical cleaning of process chambers and HVAC systems.

The moderate exposure potential is used when manipulations are done in the near field (in between half and one meter from the breathing zone), when the duration of the activity is taking less than 4 hours a day, 3 days a week or when cleaning of process chambers is done with wet tissues.

A low exposure potential is met when the manipulations are taking place more than 1 meter from the breathing zone and when less than 1 g ENM is used.

2.3 Risk control band

Finally, the activities involving ENM are grouped in three risk bands of unlikely, potential and likely risk (see Figure 1). The classification is realized by a combination of structured questionnaires, sample characterization and workplace assessments where the presence of risky operations are identified.

3. Safety measures

Different safety measures are prescribed based on the assigned risk control band. The measures are about:

- work practices in storage and shipment
- proper labelling and signposting
- primary containment
- use of appropriate personal protective equipment (PPE)

Table 1 gives an overview of the prescribed measures depending on the risk control band.

Table 1: Safety measures that are taken when a certain risk control band is defined.

Risk control band	**Measures**
Unlikely	• Signposting on receptacles • Signposting on doors and labs • Storage of materials in closed receptacles • Use of appropriate PPE (lab coats, chemical resistant gloves, safety goggles) • All personnel involved with working with nanomaterials should be properly trained • Practices of good housekeeping
Potential	• All of the above + • Workplace assessment to be conducted • Air sampling to identify possible routes of exposure
Likely	• All of the above + • Working in closed environments (i.e. containment of manipulations) • Working with breathing apparatus

When the risk control band is defined as being likely, this means there either is a high or potential hazard associated with the ENM or there is a high possibility of exposure to the product. This means protecting yourself against the material is of the utmost importance.

Looking into working in closed environments means as such that it is recommended to use for example glove boxes to do the activity. If this is not possible – and only then – ventilated and closed environments, like biosafety cabinets are allowed. If this is not possible either, use can be made of semi-closed environments with ventilation, like fume hoods and wet benches. If course in the latter cases additionally use has to be made of the appropriate personal protective equipment.

In some cases – i.e. when cleaning process chambers – use has to be made of self-contained breathing apparatus to do the job. In case of doing experiments in labs use is made of FPP3 masks.

4. Discussion and Conclusions

Several types of control banding tools are in use at present. In order to determine the hazard and exposure bands, two major systems are used:

i. A decision tree or a binary system that utilizes decision criteria to allocate specific nanomaterial or workplace condition into certain hazard categories or exposure categories.
ii. A scoring system that evaluates the toxicity data or workplace conditions. The hazard band and the exposure band are then determined according to the obtained score.

Afterwards, the risk band and corresponding control band can be determined by a matrix.

Typical representatives of decision tree and scoring system are the ISO approach and the CB NanoTool [4] respectively. However, the used type of decision systems varies. Some are a mixture of a binary decision system and a scoring system such as Stoffenmanager. Some are solely decision trees, such as the NanoValid approach [6].. The ways of dealing with unknown toxicity data are also different. For instance, the ISO approach allocates an unknown toxicity parameter to the most hazardous category while CB NanoTool presumes an unknown hazard property between medium and high.

As Table 2 indicates our approach compares with several other approaches. The imec approach uses a binary system to determine the hazard and exposure bands. Imec decided to work with 3 distinct risk bands making it clear and easy for researcher to determine the measures that need to be taken. The method has proven to be very effective within the imec organization to reduce the risks associated with the research on nanomaterials.

Acknowledgements
The authors wish to thank Ms Yange Liu who helped with the literature search and comparison. Publication has been supported in part by NanoStreeM (Horizon 2020 grant agreement No 688194).

Table 2: Comparison of different banding approaches.

Name	Hazard band and exposure band determin ation	Number of hazard bands	Number of exposure bands	Risk band/ control band determina-tion	Number of risk/ control bands	Method for dealing with unknown information	Ref
imec approach	Binary system	3	3	Matrix	3	-	-
ISO approach	Decision tree	5	4	Matrix	5	Allocating an unknown toxicity parameter to the most hazardous category	[3]
CB NanoTool	Scoring system	4	4	Matrix	4	Assigning score to an unknown hazard property between medium and high. e.g. score for carcinogenity- Yes: 6 pts, No: 0 pts, Unknown: 4.5 pts.	[5]
Stoffen-manager Nano	Binary system	5	4	Matrix	3	- Setting default choice if user selected 'unknown'. - Importing data from OECD's testing program for available nanomaterials.	[4][7]
NanoValid approach	Decision tree				5 categories	Classifying nanomaterials into categories without knowing detailed properties.	[6]

References

[1] Bhushan, D. (2014). Handbook of Nanomaterials Properties. Berlin: Springer Berlin Heidelberg.
[2] Brouwer D., Control Banding Approaches for Nanomaterials, Ann. Occup. Hyg. 2012, 56 (5), p506– 514.
[3] ISO Standard ISO/TS 12901-2:2014 ; http://www.iso.org/iso/catalogue_detail.htm?csnumber=53375
[4] Cosanta, B. *Stoffenmanager Nano module 1.0.* Retrieved from stoffenmanager 6®:

https://nano.stoffenmanager.nl/

[5] Zalk, D. M., & Paik, S. Y., Summary instructions for CB NanoTool 2.0. retrieved from http://controlbanding.net/Services.html

[6] NanoValid. (2015). *Safe handling of nanomaterials and other advanced materials at workplace.* Dortmund: Federal Institute for Occupational Safety and Health

[7] Duuren-Stuurman, B. V., Vink, S. R., Verbist, K. J., Heussen, H. G., Brouwer, D. H., Kroese, D. E., Van Niftrik, M.F., Tielemans, E. and Fransman, W. (2012). Stoffenmanager Nano Version 1.0: A Web-Based Tool for Risk Prioritization of Airborne Manufactured Nano Objects. *The annuals of occupational hygiene*, July 2012, 56(5), p525-541. doi:10.1093/annhyg/mer113

Implementation of a safe-by-design approach in the development of new open pilot lines for the manufacture of carbon nanotube-based nano-enabled products

Jesús M López de Ipiña[1], Angel Hernan[2], Xabier Cenigaonaindia[2], Mario Insunza[2], Sonia Florez[3], Richard Seddon[3], Antonios Vavouliotis[4], Vasilios Kostopoulos[5], Paulina Latko[6], Paweł Durałek[7], Nadir Kchit[8]

[1]TECNALIA – Parque Tecnológico de Alava, 01510 Miñano/Spain
[2]Sisteplant SL - Parque Científico y Tecnológico de Bizkaia, 48160 Derio/Spain
[3]TECNALIA - Parque Científico y Tecnológico de Gipuzkoa, 20009 Donostia-San Sebastián/Spain
[4]Adamant Composites Ltd. - Agias Lavras & Stadiou Str., 26504 Platani-Patras/Greece
[5]University of Patras - Mechanical Engineering and Aeronautics Department, 26500 Patras/Greece
[6]Technology Partners Foundation - Ul. Pawinskiego 5a, 02507 Warsaw/Poland
[7]TMBK Partners - Ul. Czarnieckiego 44, 05126 Nieporęt/Poland
[8]NANOCYL SA - Rue de l'Essor 4, 5060 Sambreville/Belgium

E-mail: jesus.lopezdeipina@tecnalia.com

Abstract. The project PLATFORM (H2020, GA 646307) aims to develop three new pilot lines (PPLs) for the manufacture of carbon nanotube-based nano-enabled products (buckypapers, treated prepregs, doped veils), for the European aeronautics and automotive industries (a Technology Readiness Level 6 - TRL6 - is expected at the end of the project). The Machinery Directive 2006/42/EC (MD) - transposed into the respective national legislations - is the European regulatory framework for the design and construction of new machinery, as the future PPLs. PPLs are not required to comply with the provisions of the MD until they are put into service - expected in 2020, after project completion - but then, the MD will be fully applicable. In this regulatory context, the project PLATFORM is aligning the design of the PPLs according to the MD requirements, in order to facilitate the CE marking in 2020 (TRL9) and avoid potential economic costs associated with future re-adaptations or modifications needed to ensure compliance with the MD. This paper discusses the methodological approach followed by the project PLATFORM to integrate all the nanosafety aspects in the design of the PPLs, in order to achieve safe designs in conformity with the relevant Essential Health and Safety Requirements (EHSRs) of the MD. Since machinery must be designed and constructed taking into account the results of the risk assessment (RA), this paper describes the systematic and iterative approach for RA and risk reduction followed to eliminate hazards as far practicable and to adequately reduce risks by the implementation of protective measures. This process has been guided by the harmonized standards EN ISO 12100 and EN ISO 14123, taking the relevant phases of life cycle, expected uses and operation modes of the PPLs into account. A specific tool to guide the safe design of the PPLs and facilitate the RA process has also been produced by the project (PLATFORM – SbD toolkit).

1. Motivation and context

Pilot lines are the physical infrastructure and equipment needed to produce small series of pre-commercial products. They can drastically contribute to bridge the gap between nanotechnology research and markets, facilitate large scale market introduction of innovative, safe and sustainable nano-enabled products (NEPs) as well as provide pilot infrastructures and advisory services for small and medium-sized enterprises (SMEs) and start-ups, in order to maximize the impact of upscaling activities, boosting investment and facilitate market access. Thus they are essential tools for the strategic deployment of Key Enabling Technologies (KETs) [1].

The project PLATFORM (H2020, GA 646307) aims to develop three new open pilot lines (PPLs) for the industrial production of carbon nanotube-based nano-enabled products (CNT-based NEPs), such as buckypapers, CNT-treated prepregs and CNT-doped non-woven veils, for composite applications in sectors such as aerospace and automotive [2] (Figure 1).

Buckypapers are self-supporting sheets comprised of entangled carbon nanotubes. The PPL for buckypapers will have the capability to manufacture more than 40 m^2/week of rolls of 300 mm width to any sub-multiples (150 mm, 75 mm width), with controllable thickness in the range of 30-80 μm (as customer demand) and a length of more than 100 m. The encapsulation of the buckypaper can be also configurable (as customer demand) from protective paper to resin or thermoplastic films in one or both sides of the product.

Prepreg are reinforcement carbon fiber materials which have been pre-impregnated with a resin system. CNT-doped prepregs are conventional carbon fiber prepregs doped with CNTs on it, in a way that are homogeneously distributed over the whole surface and deeply integrated on it. The current production capacity of this second PPL is established in 120,000 m^2/ year. The format of the rolls can be the same as conventional prepreg (in width and length) or sub-multiples of these. The protective films can be the same that uses the conventional prepreg of any other different (as customer demand).

CNT-doped veils are non-woven thermoplastic textile materials containing CNTs, which can be easily used as interlayers in Carbon Fiber Reinforced Polymer (CFRP) composites. At this moment the CNT-doped veil pilot line can produce veils with 3.5 wt. % CNTs based on different thermoplastic polymers. The minimum areal weight is 15g/m^2 with 0.5 m width and 1.9 m length. The current productivity of pilot line is 9,000 m^2 veils/year.

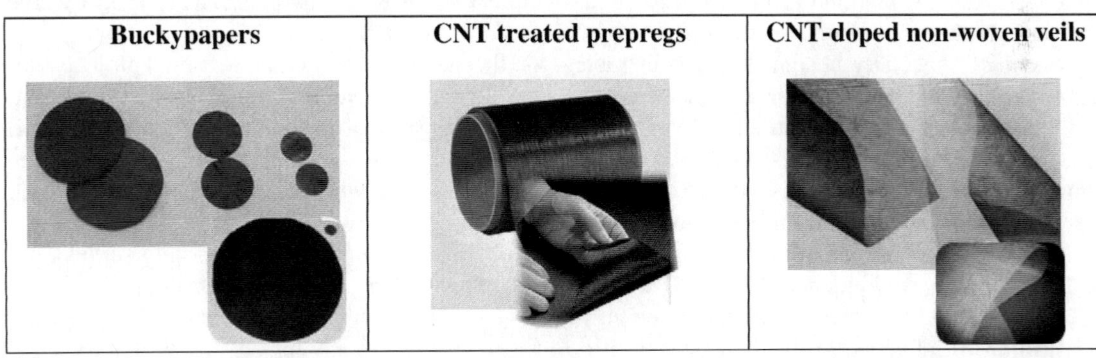

Figure 1. Three of the most promising intermediate CNT-based NEPs to be manufactured by the PLATFORM Pilot Lines (PPLs).

All these CNT-based NEPs can be used for the manufacture of laminate composites materials with improved properties (e.g. light-weight, strength, antistatic, electromagnetic shielding), for the production of structural components and devices in the European aeronautics, automotive, military, medical and electronic industries [2].

The input materials of the PPLs will be CNT-based formulated products supplied by NANOCYL SA such as waterborne dispersions or thermoplastic masterbatches [3]. Output products will be CNT-based NEPs, such as buckypapers, CNT treated prepregs and CNT doped non-woven veils. PPLs can include wet or dry process steps as well as thermal stages (e.g. extrusion, drying). Taking their relevant phases of life cycle, expected uses and operation modes into account, the main potential environmental impacts will be emissions to atmosphere, wastewaters and wastes containing CNT, and the main potential OHS risks will be associated with the exposure of workers to CNT and products containing CNT by inhalation and dermal contact [4,5,6,7,8,9,10]. Safety risks related to fire will be also considered.

Contact of workers with CNTs and CNT-based NEPs can occur during the normal operation of the PPLs, in the stages of preparation of the raw materials (waterborne dispersions, sonication, mixing, filling of tanks, filling of hoppers), extrusion, cutting, packaging, transport and storage of raw materials and final products; also in tasks of adjustment, maintenance or cleaning - particularly of pollution control equipment - as well as during the cleaning of accidental spills and the management of wastes containing CNTs generated by the PPLs

PLATFORM will end in 2018 with the goal of demonstrate the overall feasibility and competitiveness of the new products and production technologies (Technology Readiness Level 6, TRL6). However the project's impact goes beyond, and PLATFORM is also developing a business platform to commercialize NEPs manufactured by the PPLs in the European market from 2020 onwards (TRL9) [2].

These future commercialization needs as well as the intended use of PPLs by third parties - particularly SMEs - force the design and construction of the PPLs in conformity with legal requirements, before putting them into service and made available to workers in 2020 (after project completion).

The Machinery Directive 2006/42/EC (MD) - transposed into the respective national legislations of the Member States - is the European regulatory framework for the design and construction of new machinery, as the future PPLs, guaranteeing a high level of protection of the health and safety of persons and environment [11].

The CE marking on a machine (PPL) is a conformity mark by which the manufacturer indicates that the product is in conformity with the applicable requirements set out in Community harmonization legislation providing for its affixing (in our case, basically the requirements of the MD) [12]. At the same time, the CE marking affixed by each PPL-manufacturer is the visible consequence of a whole process comprising conformity assessment (Figure 3).

Machinery specially designed and constructed for the needs of particular research projects, for temporary use in laboratories, is excluded from the scope of the MD. But this exclusion only applies to equipment designed and constructed for temporary research use, that is to say, equipment that will cease to be used when the research projects for which it was designed and constructed have been completed [13]. However this is not the case of the PPLs that will be fully operative in 2020 manufacturing NEPs for the European market.

PPLs are not required to comply with the provisions of the MD until they are put into service, but in 2020, all the requirements of the MD that apply to PPLs will be mandatory for them. Evidently PPLs will need to be tested before they are put into service (e.g. during assembly, installation, testing, verification or adjustment). In these stages, it will be mandatory to implement the necessary OHS preventive and protective measures to protect the safety and health of workers and other exposed persons when performing such tasks [13].

In this industrial and regulatory context, the project PLATFORM is aligning the PPLs design to comply with all the provisions of the MD, in order to facilitate the CE marking (TRL9) in 2020 and avoid potential economic costs associated with future re-adaptations or modifications needed to ensure compliance with MD when PPLs are put into service.

Thus the project's main challenge is how to integrate all the nanosafety aspects into the well-systematized process established by the MD to meet the Essential Health and Safety Requirements (EHSRs) in the design of new machinery, as the PPLs.

2. Discussion: the methodological framework for the safe design and construction of the PPLs

2.1 The PLATFORM approach for Safe-by-Design (SbD)

Prevention through Design (PtD), Safe-by-Design (SbD) or Safety Integration (SI) are similar concepts that refer to design out hazards or minimize risks early in the design process [11,13,14,15,16,17,18]. From the perspective of the design and construction of new machinery (PPLs), the SbD concept is established by the legislation itself. Thus the MD defines a set of fundamental principles for intrinsically safe design of machines – called Principles of Safety Integration - sometimes referred to as SbD [11]. These principles are encapsulated in the Annex I of the MD (Section 1.1.2) and have been summarized in Table 1.

Table 1. Principles of Safety Integration established by the MD and applying the SbD of the PPLs (MD-Annex 1, section 1.1.2) [11].

	Principles of Safety Integration
1	Machinery must be designed and constructed so that it is fitted for its function, and can be operated, adjusted and maintained *without putting persons at risk* when these operations are carried out under the conditions foreseen (*intended use*) but also taking into account any *reasonably foreseeable misuse* thereof
2	Machinery must be designed and constructed in such a way as to *prevent abnormal use* if such use would engender a risk.
3	The aim of measures taken must be to *eliminate any risk* throughout the foreseeable lifetime of the machinery including the phases of transport, assembly, dismantling, disabling and scrapping.
4	In selecting the most appropriate methods, the following principles must be applied, in the order given, often referred to as the *3-step method*: 1. Eliminate or reduce risks as far as possible (*inherently safe machinery design and construction*) 2. Take the necessary *protective measures* in relation to risks that cannot be eliminated, 3. *Inform users of the residual risks* due to any shortcomings of the protective measures adopted, indicate whether any particular training is required and specify any need to provide personal protective equipment.
5	Machinery must be designed and constructed to take account of the constraints to which the operator is subject as a result of the necessary or foreseeable use of *personal protective equipment.*
6	Machinery must be supplied with all the *special equipment and accessories* essential to enable it to be adjusted, maintained and used safely.

The table 1 shows the framework deployed by the project PLATFORM for SbD and summarizes the main pillars of this approach. This vision is aligned with the model proposed by the project NanoReg2 which defines three basic pillars underpinning SbD: 1) Safe design, 2) Safe production and 3) Safe use [14]. However, the project PLATFORM only considers two pillars - safe design and safe production -, since the professional use of the PPLs by workers has been integrated into the safe production pillar.

Table 2. Main pillars of the PLATFORM - SbD framework

	Pillar		**Description**
1	Product safety		Focused on achieving the best *SbD formulations for NEPs* manufactured by the PPLs (buckypapers, prepregs and doped veils).
2	Safe production	2.1 Safety of machinery	Addressed to ensure that the new *PPLs (machinery) put into service in the EU are safe*. This section includes issues related to machinery certification according to EU harmonized legislation and CE marking.
		2.2 Occupational Health and Safety (OHS)	Focused on the *safety and health at work of workers* potentially exposed to CNT during the professional use of PPLs.
		2.3 Environmental safety	Addressing the *protection of the environment* against potential hazardous emissions to atmosphere, wastewaters and wastes containing CNTs.

In this context, this paper focuses on the safety pillar of machinery (item 2.2 in the PLATFORM-SbD model), thus in the design, construction and putting into service of safe PPLs for the manufacture of CNT-based NEPs [19].

2.2 Main steps in the process of PPLs certification in conformity with the requirements of the Machinery Directive

The MD (2006/42/EC) sets out the mandatory EHSRs for the PPLs while detailed technical specifications for fulfilling these EHSRs are given in European harmonized standards (usually voluntary).

With regard to the design and construction of the PPLs, there are no specific EHSRs for nanosafety risks, but some of the EHSR listed in the Annex I of the MD are directly connected with nanosafety, such as e.g. [11,13]: 1) Materials and products used or created by the PPLs, 2) Risks of fire and explosion posed by the PPLs itself or by gases, liquids, dust, vapours or other substances produced or used by these machines, 3) Hazardous materials and substances emitted by the PPLs (airborne and non-airborne emissions) (Table 3). In addition, they are of particular importance those EHSRs applying the design of parts of the PPLs control systems related to safety functions, that in case of failure can lead to hazardous situations (e.g. accidental airborne releases or spills of wastewaters or wastes containing CNT) [20,21].

Table 3. EHSRs established by the MD for the emissions of hazardous materials and substances and applying the SbD of the PPLs (MD-Annex 1, section 1.5.13) [11].

	EHSRs - Emissions of hazardous materials and substances
1	Machinery must be designed and constructed in such a way that *risks of inhalation, ingestion, contact with the skin, eyes and mucous membranes and penetration through the skin* of hazardous materials and substances which it produces can be avoided.
2	Where a hazard cannot be eliminated, the machinery must be so equipped that hazardous materials and substances can be *contained, evacuated, precipitated by water spraying, filtered or treated by another equally effective method.*
3	Where the process is not totally enclosed during normal operation of the machinery, the devices for containment and/or evacuation must be situated in such a way as to have the *maximum effect.*

The machinery manufactured in conformity with a harmonized standard (HS) shall be presumed to comply with the EHSRs covered by such standard. And this presumption of conformity conferred by the application of a HS gives a certain legal security for the manufacturer.

There are no specific HS for NEPs manufacturing machinery (type C standards) or for the nanosafety risks involved in the design of such production machinery. When harmonized standards are not available, the designer can use other reference documents such as international and national standards, draft standards, recommendations, best practices and guidelines from professional organizations [13]. In any case, the designer should use his/her best professional judgment in the application of existing standards (harmonized and non-harmonized) and guidelines, in order to achieve safe designs for the PPLs.

Before putting the three PPLs into service (TRL9), the manufacturer of each PPL will ensure that [11,13] (Figure 2):

- the relevant EHSRs applying to the PPL have been satisfied
- the PPL - Technical File (TF) is available. The TF will contain the documentation to demonstrate the conformity of the PPL with the requirements of the MD (general description of the machine, drawings of the machine, technical reports, risk assessment, list of EHSRs applying to the PPL, description of the protective measures implemented to eliminate identified hazards or to reduce risks, standards and other technical specification used for the design, instructions handbook,

IOP Conf. Series: Journal of Physics: Conf. Series **838** (2017) 012018 doi:10.1088/1742-6596/838/1/012018

declarations of conformity, etc). Part of this documentation will demonstrate the solution of nanosafety design aspects.

- the necessary information, such as instructions, has been provided;
- the appropriate procedure for assessing the PPL conformity has been carried out. None of the three PPLs is listed in the Annex IV of the MD (Particularly dangerous machinery) and thus the project PLATFORM will use the simplest procedure for conformity assessment, based on internal checks, with the additional obligation to include in the respective Technical Files - if necessary - the information required with regard to Partial Completed Machinery (PCM) and second-hand machines incorporated in the PPLs.
- the Declaration of Conformity (DoC) has been drawn up and accompanies the PPL. The DoC is a legal statement by the manufacturer attesting that the PPL complies with all of the relevant provisions of the MD.
- The CE marking has been affixed on the PPL, attesting the conformity of the product (machine) with the applicable requirements of the relevant EU harmonized legislation.

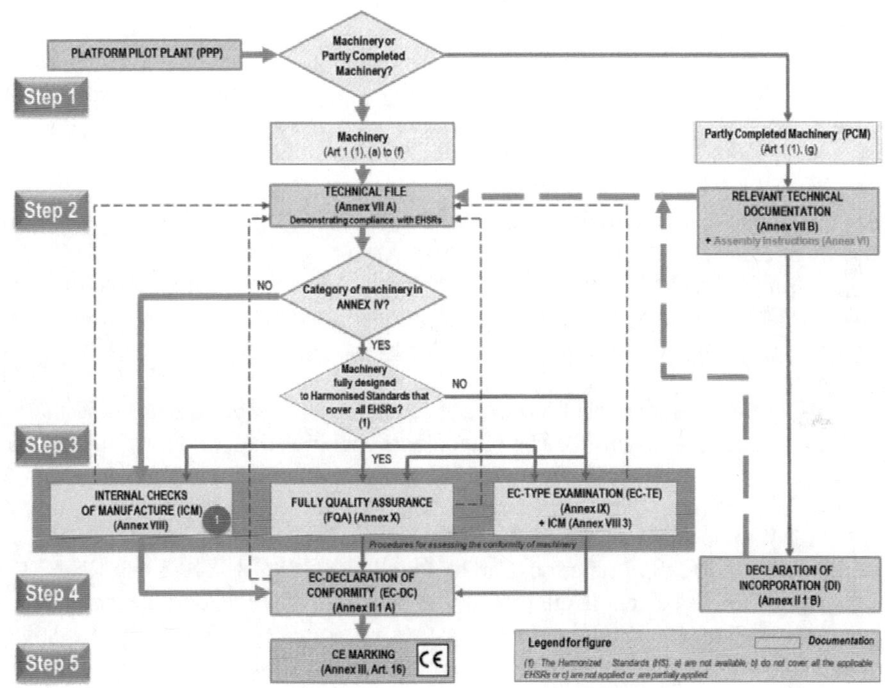

Figure 2. Main steps in the process of PPLs certification in conformity with the requirements of the Machinery Directive (2006/42/EC): Step 1. Definition of the extension of the PPL; Step 2. Development of the PPL-Technical File; Step 3. Carry out the procedure for assessing the conformity; Step 4. Draw-up the Declaration of Conformity and finally; Step 5. Affix the CE marking on the PPL. Thick green arrows show the procedure selected by the project PLATFORM for the conformity assessment of the PPLs (Internal checks of manufacture) ([13] modified).

2.3 The risk assessment process and the PLATFORM-SBD tool

PPLs must be designed and constructed taking into account the results of a risk assessment (RA), in order to identify those EHSRs which apply to the design of each PPL. The purpose of the RA is to identify hazards, and to estimate and evaluate risks so that they can be reduced (nanosafety risks among them) [21,22]. Obviously this process involves design aspects that go beyond specific aspects of nanosafety.

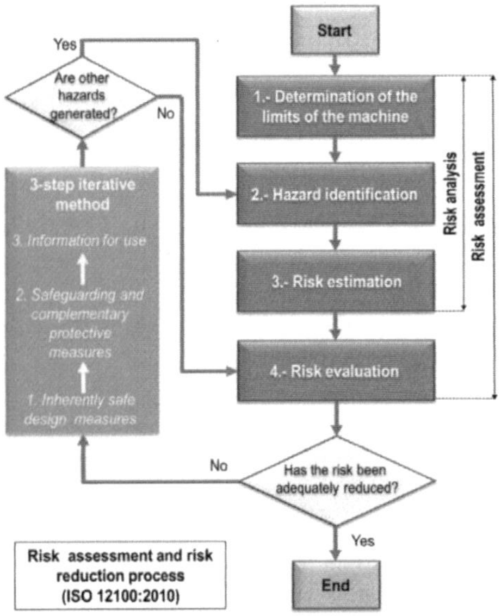

Figure 3. Main steps in the process of PPLs risk assessment and risk reduction, according to EN ISO 12100 [22].

To implement the risk assessment and risk reduction process (Figure 4), the designer will take the following actions, in order given [11,22]:

1. Determine the limits of the PPL (use, space and time limits, other limits), including the intended use and any reasonably foreseeable misuse thereof.
2. Identify the hazards that can be generated by the PPL, the associated hazardous situations and hazardous events.
3. Estimate the risks, for each identified hazard and hazardous situation to determine the highest risk arising from each hazardous situation. The judgements shall be supported by a qualitative or, where appropriate, quantitative estimation of the risk [24]. Some of more extended methods to conduct risk estimation for machinery (risk matrix, risk graph, hybrid tool, control banding) are described in ISO/TR 14121-2 [24] and ISO/TS 12901-2 [25]. Specific tools for risks related to the parts of the control systems related to safety functions can be found in EN ISO 13849-1 and EN 62061 [20,21].
4. Evaluate the risks, and take decisions about the need for risk reduction and the selection of appropriate protective measures. Protective measures are a combination of the measures taken by the designer and the user, but measures incorporated at the design stage are preferable to those implemented by the user and usually prove more effective [8,22,24,26].
5. Eliminate the hazards or reduce the risks associated with these hazards by application of protective measures, in the order of priority (3 steps process) [22,23,24] :
   1) *Inherently safe design measures*, addressed to eliminate hazards or reduce the associated risks by a suitable choice of design features of the machine itself and/or interaction between the exposed persons and the machine [24].
   2) *Safeguarding and/or complementary protective measures.* If hazards cannot be eliminated or risk cannot reduced adequately by design measures, safeguarding (protective/risk reduction measures using guards and protective devices) should be applied that results in restricting exposure to hazards, lowering the probability of the hazardous event, or improving the possibility of avoiding or limiting harm [24]. Particular attention will be given to those protective/risk reduction measures related to control system safety functions [20,21].
   3) *Information about the residual risks* to the user and advice on additional protective measures to be implemented to reduce exposure [24].

Particularly in the case of the protective measures implemented to prevent and control the releases of airborne NOAA (nano-objects, and their aggregates and agglomerates greater than 100 nm) emitted by the PPLs [8,23,26], the strategy for the evaluation and verification of these emissions as well as the performance of the control systems integrated in the PPLs, will be conducted, in general, according to EN 1093-1, taking into account the nature of the pollutant (aerosol containing CNT) and the nature of environmental test conditions (usual working environments). For this purpose a combination of measurement methods will be explored (Direct-reading instruments/DRI, Laser-induced incandescence/LII and tracers) [27,28,29,30,31]. If the deployment of such strategy is technically unfeasible in the locations where the PPLs will be installed, then according to ISO/TR 14121-2, an indirect technique will be used to demonstrate the conformity of the PPLs with the EHSRs, by measuring the occupational exposure of workers operating the PPLs and comparing such measurements with suitable criteria [24,32,33,34,35,36]. In this case, personal respirable exposure concentrations to CNT will be determined as elemental carbon (EC) by NIOSH method 5040 and compared with the NIOSH REL of 1 $\mu g/m^3$ 8-hr TWA [10] (NC7000™ case from NANOCYL recommends an OEL of 2.5 $\mu g/m^3$ [3]).

To guide the safe design of the PPLs and facilitate the RA process, the project PLATFORM developed a simple and friendly Microsoft Excel tool (PLATFORM-SbD tool), that can be applied to the overall risk assessment of the PPLs or only focus on the nanosafety risks [19]. This tool is structured in 11 Excel spreadsheets covering aspects such as: technical file, emissions inventory, limits of the machine, risk assessment and control system, risk estimation tools and EHSRs verification. In the absence of a C-type standard, the methodology deployed by the tool is fundamentally based on the harmonized standards EN ISO 12100, EN ISO 13849-1 and EN ISO 14123-1 (A&B-type standards) and other international standards such as EN 62061, ISO/TR 14121-2 and ISO/TS 12901-2 [20,21,22,23,24,25]. The RA takes all phases of life cycle of PPLs, their operating modes and hazardous areas into account. At the moment, the current version of the PLATFORM-SbD Tool (v1) is only available for project partners. However the final version of the tool will be free available at the end of the project (2018) for any interested party, along with a guideline for safe design of PPLs.

3. Conclusions

The strategy deployed by the project PLATFORM for the safe design, construction and putting into service of three PPLs for the manufacture of CNT-based NEPs, can be summarized as follows:

- **PPLs - Design**: In order to facilitate the CE marking in 2020 (TRL9) and avoid potential economic costs associated with future re-adaptations needed to ensure compliance with the MD, the design of the PPLs has been developed following the principles for safety integration (SbD) established by the MD, incorporating at the same time all the relevant nanosafety issues, to achieve final safe designs in conformity with the relevant EHSRs. RA and the design of preventive and protective measures against hazardous substances emitted by the PPLs have been guided by the harmonized standards EN ISO 12100 and EN ISO 14123, integrating in this process the best available practices to prevent, assess, control and manage CNT emissions and workers exposures. The PLATFORM-SbD tool has been used to facilitate the overall RA process [19]. The TFs produced at this stage will be essential elements to demonstrate the safe design of the PPLs in conformity with the MD.
- **PPLs - Construction**: PPLs will be constructed according to the SbD produced at the design stage and following the specifications contained in their respective TFs. In order to prevent any risk of the new designs (Precautionary principle) [37], the PPLs will be initially built inside dedicated rooms, that will allow testing and verification of the effectiveness of the implemented preventive and protective measures under controlled conditions.
- **PPLs - Testing and verification**: For the operation of the PPLs during the testing & adjustment phase (TRL6), specific measures to protect the health and safety of the potential exposed workers will be taken in accordance with the national legislations and the best available practices. The verification of the safety requirements and the measures for risk reduction of hazardous substances

emitted by the PPLs (CNTs) will be done by several procedures: 1) Visual inspection, 2) Measurement, 3) Functional testing, 4) Design validation or 5) Calculation. Special attention will be given to the verification by measuring, of potential airborne CNT emissions and exposures.

- **PPLs - Putting into service:** In 2020 PPLs manufacturers will play a crucial role in ensuring that their machines are safe. Before putting the PPLs into service (TRL9), all the obligations related to the conformity with the MD will have to be properly fulfilled by the manufacturers. Thus a Declaration of Conformity will be drawn up and the CE marking affixed on PPLs, attesting the responsibility of manufacturers for the conformity of PPLs with the applicable requirements. And only then, the three PPLs will be put into service in the European market.

Acknowledgments
The project PLATFORM has received funding from the European Union's Horizon 2020 research and innovation programme, under grant agreement Nº 646307. This paper reflects only the authors's views and the Commission is not responsible for any use that may be made of the information contained therein.

References

[1] EC (2015) *Pilot production in Key Enabling Technologies. Crossing the Valley of Death and boosting the industrial deployment of Key Enabling Technologies in Europe.* DG Internal Market, Industry, Entrepreneurship and SMEs, 27 pp.
[2] PLATFORM (2016) *H2020 - Project Platform* (http://www.platform-project.eu/)
[3] NANOCYL (2016) *Safety Data Sheets* (http://www.nanocyl.com/product/)
[4] Clark K, Van Tongeren M, Christensen FM, Brouwer D, Nowack B, Gottschalk F, Micheletti C, Schmid K, Gerritsen R, Aitken R,Vaquero C, Gkanis V, Housiadas C, López de Ipiña JM and Riediker M (2012) Limitations and information needs for engineered nanomaterial-specific exposure estimation and scenarios: recommendations for improved reporting practices. *J. Nanopart. Res.* 14 (9) (2012) 1–14.
[5] Ding Y , Kuhlbusch TAJ, Van Tongeren M , Sánchez Jiménez A , Tuinman I , Chen R , Larraza Alvarez I , Mikolajczyk U , Nickel C , Meyer J , Kaminski H , Wohlleben W , Stahlmecke B , Clavaguera S and Riediker M (2016) Airborne engineered nanomaterials in the workplace—a review of release and worker exposure during nanomaterial production and handling processes. *J. Hazard. Mater.* Volume 322, Part A, Pages 17-28.
[6] Guseva Canu I, Bateson TF.,Bouvard V, Debia M, Dion C, Savolainen K and Yu IJ (2016) Human exposure to carbon-based fibrous nanomaterials: A review. *International Journal of Hygiene and Environmental Health,* Volume 219, Issue 2, March 2016, Pages 166-175.
[7] Kingston C, Zepp R, Andrady A, Boverhof D, Fehir R, Hawkins D, Roberts J, Sayre P, Shelton B, Sultan Y, Vejins V and Wohlleben W (2014) Release characteristics of selected carbon nanotube polymer composites. *Carbon,* Volume 68, March 2014, Pages 33-57.
[8] NIOSH (2013) *Occupational Exposure to Carbon Nanotubes and Nanofibers.* Current intelligence bulletin 65. Publication No. 2013–145, 156 pp.
[9] Safe Work Australia (2012) *Measurements of particle emissions from nanotechnology processes, with assessment of measuring techniques and workplace controls.* Safe Work Australia, 140 pp.
[10] Virji MA and Stefaniak AB (2015) *A Review of Engineered Nanomaterial Manufacturing Processes and Associated Exposures.* Reference Module in Materials Science and Materials Engineering, from Comprehensive Materials Processing, Volume 8, 2014, Pages 103-125.
[11] EC (2006) *Directive 2006/42/EC of the European Parliament and of the Council of 17 May 2006 on machinery, and amending Directive 95/16/EC.*
[12] EC (2008) *Regulation (EC) Nº 765/2008 of the European Parliament and of the Council of 9 July 2008 setting out the requirements for accreditation and market surveillance relating to the marketing of products and repealing Regulation (EEC) No 339/93.*
[13] EC (2010) *Guide to application of the Machinery Directive* 2006/42/EC, 2nd Edition, June 2010 .
[14] NANOREG2 (2016) *H2020 - Project NanoReg2* (http://www.nanoreg2.eu/safe-design)

[15] NIOSH (2010) *Prevention through design: plan for the national initiative*. Pub. N° 2011–121,44 pp.
[16] Schulte P, Rinehart R, Okun A, Geraci C, Heidel D (2008) National prevention through design (PtD) initiative. *J Safety Res* 39:115–121.
[17] Srinivasan R and Natarajan S (2012) Developments in inherent safety: A review of the progress during 2001–2011 and opportunities ahead. *Safety and Environmental Protection*, Volume 90, Issue 5, September 2012, Pages 389-403.
[18] Szymberski, R., (1997) Construction Project Safety Planning. *TAPPI Journal*, 80(11), 69-74.
[19] PLATFORM (2016) *Methodological Toolkit for safe-by-design and management of EHS risks. Part 2.1: Safety of machinery*. Project internal report D9.1, 174 pp.
[20] EN ISO 13849-1:2015 *Safety of machinery - Safety-related parts of control systems - Part 1: General principles for design.*
[21] EN 62061:2005 *Safety of machinery - Functional safety of safety-related electrical, electronic and programmable electronic control systems.*
[22] EN ISO 12100: 2010 *Safety of machinery – General principles for design – Risk assessment and risk reduction.*
[23] EN ISO 14123-1:2015 *Safety of machinery. Reduction of risks to health from hazardous substances emitted by machinery. Principles and specifications for machinery manufacturers.*
[24] ISO/TR 14121-2:2012 *Safety of machinery. Risk assessment. Part 2: practical guidance and examples of methods.*
[25] ISO/TS 12901-2:2014 *Nanotechnologies -- Occupational risk management applied to engineered nanomaterials -- Part 2: Use of the control banding approach.*
[26] NIOSH (2013) *Current strategies for engineering controls in department of health and human services nanomaterial production and downstream handling processes.* Pub. N° 2014–102, 79 pp.
[27] EN 1093-1: 2008 *Safety of machinery - Evaluation of the emission of airborne hazardous substances - Part 1: Selection of test methods.*
[28] Hristozov D, Gottardo S, Semenzin E, Oomen A, Bos P, Peijnenburg W, Van Tongeren M, Nowack B, Hunt N, Brunelli A, Scott-Fordsmand J J, Tran L and Marcomini A (2016) Frameworks and tools for risk assessment of manufactured nanomaterials. *Environment International*, Volume 95, October 2016, Pages 36-53.
[29] ISO/TR 27628:2007 *Workplace atmospheres—Ultrafine nanoparticle and nano-structured aerosols—Inhalation exposure characterization and assessment.*
[30] OECD (2016) *Strategy for using metal impurities as carbon nanotube tracers.* Series on the Safety of Manufactured Nanomaterials No. 79, ENV/JM/MONO(2016)62.
[31] Schubauer-Berigan MK, Dahm MM, Schulte PA., Hodson L and Geraci CL (2016) Refinement of the Nanoparticle Emission Assessment Technique into the Nanomaterial Exposure Assessment Technique (NEAT 2.0). *Journal of Occupational and Environmental Hygiene*, Volume 13, 2016 - Issue 9, pp 708-717.
[32] Asbach C, T K, Kaminski H, Stahlmecke B, Plitzko S, Götz U, Voetz M, Kiesling H and Dahmann D (2012) *Standard operation procedures for assessing exposure to nanomaterials, following a tiered approach.* Nano GEM, 90 pp.
[33] EN 689:1995 *Workplace atmospheres. Guidance for the assessment of exposure by inhalation to chemical agents for comparison with limit values and measurement strategy.*
[34] ISO/TS 12901-1:2012 *Nanotechnologies -- Occupational risk management applied to engineered nanomaterials -- Part 1: Principles and approaches.*
[35] OECD (2015) *Harmonized tiered approach to measure and assess the potential exposure to airborne emissions of engineered nano-objects and their agglomerates and aggregates at workplaces.* Series on the Safety of Manufactured Nanomaterials No. 55, ENV/JM/MONO(2015)19.
[36] SER (2012) *Provisional nano reference values for engineered nanomaterials.* The Hague: Social Economic Council Netherlands, 15 pp.
[37] EC (2000) *Communication from the Commission on the precautionary principle* COM (2000) 1 final Brussels, 2.2.2000.

IOP Conf. Series: Journal of Physics: Conf. Series **838** (2017) 012019 doi:10.1088/1742-6596/838/1/012019

Implementation of the NANoREG Safe-by-Design approach for different nanomaterial applications

C Micheletti[1], M Roman[1], E Tedesco[1], I Olivato[1] and F Benetti[1]

[1]ECAMRICERT-ECSIN - Viale Porta Adige 45, 45100 Rovigo (RO), Italy

c.micheletti@ecamricert.com

Abstract. The Safe-by-Design (SbD) concept is already in use in different industrial sectors as an integral part of the innovation process management. However, the adopted approach is often limited to design solutions aiming at hazard reduction. Safety is not always considered during the innovation process, mainly due to the lack of knowledge (e.g. in small and medium companies, SMEs) and the lack of dialogue between actors along the innovation chain. The net result is that safety is considered only at the end of the innovation process at the market authorization phase, with potential loss of time and money. This is especially valid for manufactured nanomaterials (MNM) for which the regulatory context is not completely developed, and the safety knowledge is not readily available. In order to contribute to a sustainable innovation process in the nanotechnology field by maximising both benefits and safety, the NANoREG project developed a Safe Innovation approach, based on two elements: the Safe-by-Design approach which aims at including risk assessment into all innovation stages; and the Regulatory Preparedness, focused on the dialogue with stakeholders along the innovation chain. In this work we present some examples about the implementation in our Laboratory of this approach for different MNM applications, covering different steps of the innovation chain. The case studies include: the feasibility study of a medical device including substances, for topical application; the testing of two potential nanotech solutions for the consolidation of cultural heritage artifacts; the testing of coatings already on the market for other uses, which was tested as food contact materials (FCM) to evaluate the conformity to food applications. These three examples represent a good opportunity to show the importance of NANoREG SbD and Safe Innovation Approach in general, for developing new nanotechnology-based products, also highlighting the crucial role of EU ProSafe project in promoting this concept to industries and interested stakeholders.

1. Introduction

The European Commission, together with National Authorities is working toward a research agenda supporting the regulatory implementation of nanosafety. This effort is included in the European innovation agenda, including the creation of a sustainable job environment, re-industrialization of Europe, the digital single market, and the energy union. All these goals may benefit from the implementation of nanotechnology. Regulatory approach to nanosafety was pursued in a systematic way starting with the FP7 project NANoREG, which involved a huge number of partners from inside and outside Europe. The scope of NANoREG was to develop tools supporting the generation of robust data for regulatory use, and concepts allowing a rational and faster regulatory assessment of nanomaterials.

IOP Conf. Series: Journal of Physics: Conf. Series **838** (2017) 012019 doi:10.1088/1742-6596/838/1/012019

Among the concepts developed in NANoREG, the Safe Innovation Approach (SIA) is a systematic framework supporting pre-regulatory and regulatory safety assessment as well as the dialogue between regulators and decision makers, researchers, and especially industry. The SIA is designed to be used mainly by small and medium companies (SMEs), since they are lacking the means to properly tackle the regulatory and sometimes the technical challenges posed by the implementation of nanotechnology in products and processes. This scenario calls for the intervention of private organizations acting as intermediaries between the European and national authorities and companies. These organizations concentrate the technical and regulatory knowledge to be transferred to companies in a context of Responsible Research and Innovation (RRI). Safe Innovation is a mean to implement the RRI in SMEs, and in this work it was evaluated through three case studies the applicability of SIA by an excellence center, discussing what was possible and where were the hurdles and limitations in relation to: efficacy of innovation, nanosafety assessment, and the dialogue with regulators.

2. Safe Innovation approach

The SIA approach, as developed by RIVM and TEMAS in FP7 NANoREG project, disseminated in the H2020 ProSafe project, and under implementation in H2020 NANOREG[2] project, is described in Figure 1. It includes two parts, a Safe-by-Design (SbD) approach, which aims at reducing uncertainties and risks as early as possible in the innovation chain, and the Regulatory Preparedness concept aiming at building a structured and integrated dialogue between decision makers and industry to allow the risk governance of nanotech-based innovation.

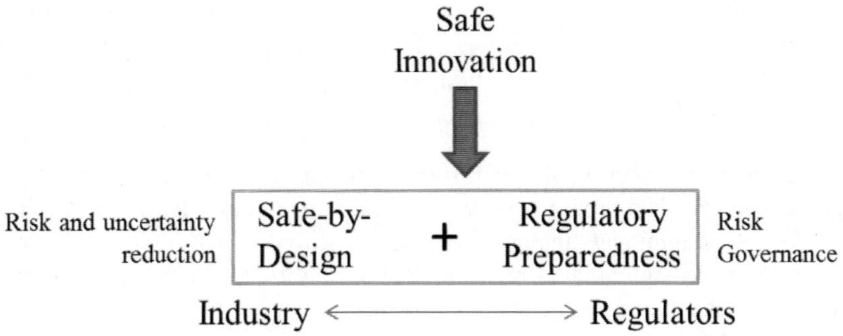

Figure 1: Overview of Safe Innovation Approach elements

SbD is built on the stage-gate approach (Figure 2), which is used by companies to manage innovation. Stage-gate model is structured as a set of steps (stages) where some activity is carried out, from the idea stage to the prototyping and placing to the market. Stages are interlaced by decision points (gates) where the decision makers decide whether the innovation project has to be cancelled, to be modified, or it can go on to the next stage. For each stage, a set of tools can be used to assess safety of the novel product/process for human health, environment, and consumers. The information is organized following the current vertical and horizontal regulatory framework, allowing a pre-regulatory and regulatory dossier compilation, and also avoiding the collection of unnecessary information in relation to the specific stage.

Regulatory Preparedness is based on the definition of "trusted environments", where industry, regulators, and researchers generating the regulatory safety data can exchange views, anticipate information (also sensitive information), and decide upon a course of action to demonstrate safety of the product/process in a robust way. This approach can work both at specific level to deal with specific company needs, and at general level to establish EU-wide policies, leading to improvement of the regulatory safety assessment process and the implementation of a innovation risk governance.

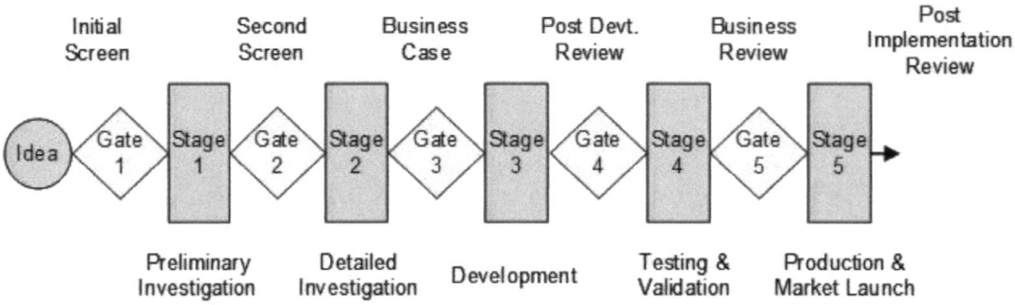

Figure 2. Stage-gate model. Adapted from [1]

3. Case studies

The applicability of the SIA to SMEs by an excellence center was evaluated considering three case studies on the basis of three main criteria: i) efficacy; ii) safety; iii) dialogue with regulators.

Efficacy is essential for the innovation process, and something that companies would like to assess as early as possible in the innovation process. Therefore, while not an inherent safety issue, it is an important component that needs to be considered in the SIA application. Safety is the core element of the SIA, and in this context what is considered is the availability of tools and data to assess exposure and hazard of nanomaterials. Finally, the dialogue with regulators is essential in the nanotech innovation field, since there is lack of experience in the regulatory world, and the legislation is not always clear. While technical guidance is available in some cases, the dialogue with regulators may give indications to companies about necessary information, and allow regulators to use robust information for their decision-making.

3.1. Topical medical device

The first case study is about a medical device (MD) and it is located at the beginning of the innovation chain, since it is a feasibility study. The idea at the basis of the new product is to develop a nano-enabled topical medical device, including botanical substances with both non-pharmacological and pharmacological mode of action, to support treatment of health conditions. In this case study were considered two related aspects: identification of suitable substances and regulatory feasibility. Both aspects are relevant for the development of the final product, due to specific regulatory requirements. Addressing safety data and needs at this stage would save resources for the next stages of the product development, but the regulatory context under development is posing issues of innovation governance.

The selection of suitable substances was based on both efficacy and safety, and also in terms of mode of action (in reference to the regulatory context). At this stage of innovation it was verified that published literature was sufficient to identify the suitable substances among a large panel of possibilities. Mode of action, efficacy, and potential toxicity were all well described, and supported the expert selection of the best candidates with examples of potential nano formulations. For some applications, such as botanicals, existing data are often sufficient to properly define the prototype. This may not be always the case, and computational models may be useful to fill some relevant data gaps. However, these models are completely missing right now for nanomaterials.

Concerning the regulatory feasibility, the regulatory framework of MD based on substances is complicated, since this kind of products fall between Medical Device directive [2] and Medicinal Products directive [3]. In addition, there is a proposal for a European Medical Device Regulation [4] which is posing stricter requirements on safety assessment of substances and of nanomaterials used in MDs. Therefore, a dialogue with national authorities was initiated, leading to a good relationship with national Notified Body which supported the enquiries about the feasibility as MD of the proposed product. However, the regulatory development with the new European Regulation is posing a challenge to this dialogue, since industry would need timely guidance from competent authorities about the implementation of the new legislation requirements, which in absence of the official technical documentation can be difficult to get.

3.2. Cultural heritage restoration

Cultural heritage has a great socio-economic importance, both representing the history and culture of a country, but also representing one of the main drivers of tourism. Therefore, its conservation and restoration is paramount, especially in Italy. Cultural heritage is often exposed to environmental stressors, both outdoors and indoors, and there is the need of more effective and respectful treatment approaches. Nanotechnology, as key enabling technology, is already providing solutions for cleaning and conservation of paintings, frescos, and stone manufactures. Consolidation is a physical stabilization of the work of art aiming at binding constituting materials, and ensuring both any further treatment procedure and its stability over time. This case study is focused on the comparison of efficacy and toxicity of two different SiO_2 nanoparticles used in consolidation: unmodified silica and chemically modified silica. This case study is an example of chemical composition modification of a product to reduce impacts and increase efficacy, with nanotechnology playing a positive role. The modification of the chemical composition is one of the common strategies to address SbD, and here it is demonstrated that there are tools to apply this strategy for nanomaterials in a SIA context.

The substitution of solvent-based consolidants in favor of solvent free nanoparticle-based products is already a trend in the market. However, for the same nanoparticle different formulations are possible, and the nano-specific assessment of toxicity and safety for both professional users and the environment may lead to the selection of the best product. This case study is located at the end of the innovation chain, because these products are already on the market, and it is focusing on the efficacy and safety assessment [5]. Also, some considerations about the regulatory context in relation to the testing requirements are addressed.

The nano-product efficacy was measured taking into account the consolidation properties (Figure 3) and the antibacterial activity, since bacteria can cause the works of art degradation. Antibacterial efficacy of the two formulations was compared against both gram+ and gram- bacteria usually present on paintings and monuments. At professional use concentrations, both types of nanoparticles inhibit bacterial growth, with some difference in terms of bacterial species.

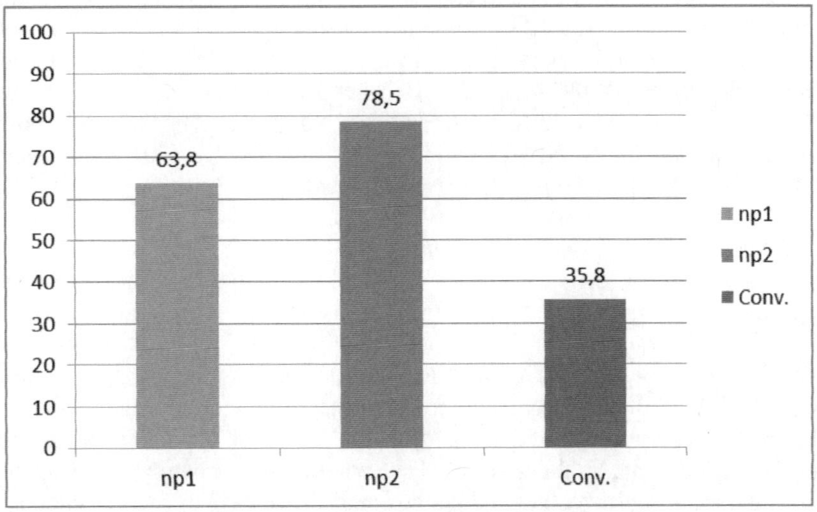

Figure 3. Cohesion factor of different products. np1 and np2 are the two silica nanoparticles, compared to the conventional product (Conv.)

After a proper pchem characterization, measuring shape, particle size distribution, and surface charge, *in vitro* toxicity tests were used to test for potential exposure routes and internal target organs (i.e. lungs, skin, and kidney). The results showed that, while chemically modified SiO_2 did not induce toxicity, unmodified SiO_2 showed toxic effects at realistic concentrations on all *in vitro* models. An example is shown in Figure 4.

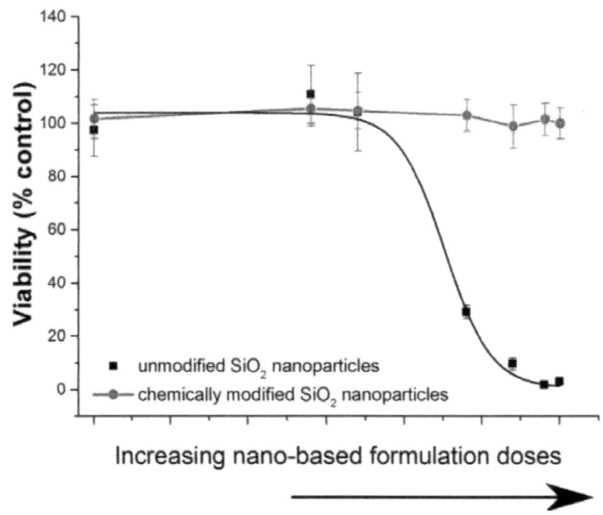

Figure 4. Example of effects on cell viability of the two nano formulations

The results of the study highlighted that it is possible to develop more effective and safer products at the same time, taking into account that small changes in nanoparticle features, like surface chemical modification, deeply affect biological behavior of nanomaterials. *In vitro* tools provide answers that can support the design of safer nanoparticles at early stages, but also to support the marketing strategy of a company, focusing on the best product and advertising the advantages in terms of sustainability.

From a regulatory point of view, this material is covered by REACH. However, nano SiO_2 is only one of the many forms of the substance SiO_2 as registered in REACH, but the dossier cannot realistically cover all the possible existing nanoforms. Therefore, the safety of a nanoform is not necessarily adequately covered by the REACH Chemical Safety Report. This underlines the importance of risk governance also in terms of industry voluntary actions, where the innovation is based also on social responsibility. SIA can support this approach in SMEs, optimizing the use of resources besides or in addition to the regulatory requirements.

3.3. Food contact material

Food industry is one of the sectors in which nanotechnology is increasingly employed. While silica dioxide (SiO_2) is one of the most used additive in food, its use as food contact material (FCM) is not so common.

This case study is dealing with the request to evaluate the compliance of a SiO_2 coating to the Italian Food Contact Material legislation. The product was already on the market for other uses, and therefore in this case we are covering the last part of the innovation chain. To assess the applicability of this coating as FCM entails the need to characterize the coating assuring that it is not nanostructured, and measuring (detecting and quantifying) the release of SiO_2 as chemical, but also as nanomaterial. This approach is a key requisite for improving its safe use and to protect consumer's health [6]. Therefore the case study is focused on the dialogue with regulators and the assessment of potential exposure.

The first step was to contact regulators to collect information on the national provisions about the use of SiO_2 in coatings, and what was the orientation of the Italian Ministry with regard to nanomaterials. The dialogue was fruitful, and all necessary and pertinent clarifications were provided in time. Therefore, in this case, regulatory preparedness dialogue worked very well, allowing the excellence center to provide the services to the client.

Concerning the exposure assessment, in this case study we explored the potential of Single particle Inductively Coupled Plasma-Mass Spectrometry (spICP-MS) for detecting and characterizing SiO_2 NPs in food simulating media. Determination of Si in biological matrices by standard ICP-MS results in a very high background level. On the contrary, the spICP-MS with the adoption of a very short dwell time allowed to significantly increase the signal-to-background ratio (S/B) for SiO_2 nanoparticles. The low limit of detection (LOD) for number concentration (~104 particles per mL are sufficient for robust statistics) allowed high sample dilution, with further reduction of matrix effects. The estimation of nanoparticles' size by spICP-MS is indeed concentration independent. Simultaneous determination of dissolved Si (not possible by AF4-MALS, DLS and TEM) within a single analysis was achieved by implementing a statistical procedure [7] conceived to separate the signals generated by dissolved species and nanoparticles, the latter being detected individually.

Validation and direct comparison to TEM, DLS and AF4-MALLS was performed by analyzing standard suspensions prepared in ethanol 95%. All techniques except TEM effectively detected nanoparticles with consistent size distribution (Figure 5). Beside the unique information of dissolved Si level, spICP-MS provided the most narrow particle size distribution (70-120 nm) with respect to AF4-MALLS (40-200 nm) and DLS (60-400 nm considering the number fraction). Since direct measurement of individual particles mass can be achieved by spICP-MS, this technique is not affected by dominating signals of large particles like in DLS, or by shape effects like can occur in AF4-MALS. Overall, its integration in multi-technique approaches was established as an efficient strategy to achieve the most robust assessment of nanoparticles in complex media within a regulatory framework.

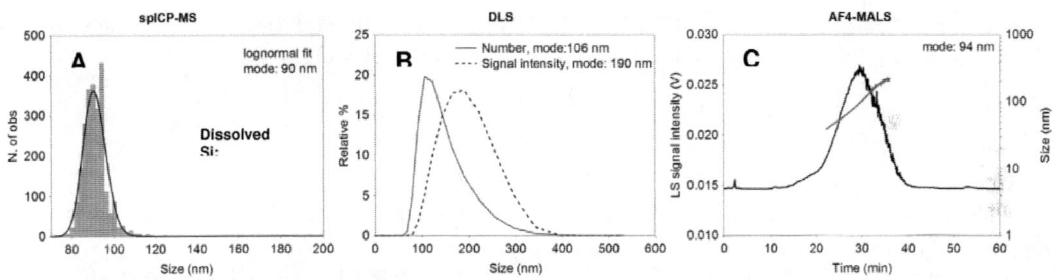

Figure 5. Size distribution of SiO_2 nanoparticles in pre-concentrated food simulating medium (ethanol 95%) after migration test from a Si-coated food-contact inox support, as obtained by spICP-MS (A), DLS (B) and AF4-MALS (C)

Real-case testing of such an approach was carried out to assess the potential migration of SiO2 nanoparticles from a silica-coated stainless steel support into food simulating media. Migrations were prepared in acetic acid 3% w/v, ethanol 10 and 95% v/v according to the UNI EN 1186-3,14 standard procedures. Preliminary determination of global and specific migration revealed significant (still, well below the limits of Reg. CE 10/2011) results only for ethanol 95%, but no significant signals attributable to nanoparticles were detected by TEM, DLS or spICP-MS.

4. Conclusions

The application of Safe-by-Design and Safe Innovation Approach to three case studies involving the development of products for small companies led to the identification of positive aspects and issues still to be resolved. As a positive note, the case studies showed that SbD and SIA can be effectively integrated into the innovation processes of SMEs, and there is no additional burden but instead the rationalization of the whole innovation approach. In fact, SIA can very well work as an innovation safety management system, identifying development phases, responsibilities, regulatory requirements, data needs, and data providers. Robust methods and tools for the SIA application on nanomaterials are available, allowing the measurement of all risk assessment components: from physicochemical characterization (e.g. spICP-MS) to hazard assessment (e.g. *in vitro* tests). In fact, the available knowledge and instruments are sufficient to answer the requests of the companies, from pre-regulatory assessment to the preparation of the dossier for the authorization to place on the market the product. However, acceptability of these methods by regulators is not granted. While standardized approaches and SOPs for nanomaterials are increasingly available, not all endpoints can be measured with e.g. OECD methods, and anyway adaptations could be necessary. While interaction with companies and technical partners from the beginning is useful to develop a sound experimental plan, more structured exchanges with technical authorities to define an agreed experimental approach could be useful to generate regulatory acceptable data. Part of the experimental approach could include in silico approaches, such as QSAR, grouping, and read across, that could reduce the need of testing in the early stages of the innovation process. However, these approaches are still in their infancy with regard to nanomaterials, and their development is expected to be slow despite the global efforts.

Regarding the regulatory preparedness, often competent authorities are not prepared to deal with a proactive approach toward safety, adhering strictly to regulations timing and requirements. In addition, communication is not always fruitful, especially in borderline cases, where different regulations may apply to the same product or regulations are changing. There is the need to define specific moments at different stages of the innovation process when industry can consult regulators, to be able to share information and come to an agreed market authorization process which is satisfactory for all parties. For example, the Food and Drug Administration in the United States strongly advice companies that want to put nano foods on the market to consult the agency before starting the actual development. An interactive system like this would improve the communication between competent authorities, industry, and researchers, streamlining the effectiveness of regulatory interpretations and implementation, and accelerating the transfer of the innovations benefits to the society.

References

[1] Cooper R G 1990 *Business Horizons* **33(3),** 44-54
[2] Council Directive 93/42/EEC of 14 June 1993 concerning medical devices, OJ L 169 of 12 July 1993
[3] European Parliament and The Council, Directive 2001/83/EC of 6 November 2001 on the Community code relating to medicinal products for human use, OJ L 311, 28.11.2001, p. 67
[4] Proposal for a REGULATION OF THE EUROPEAN PARLIAMENT AND OF THE COUNCIL on medical devices, and amending Directive 2001/83/EC, Regulation (EC) No 178/2002 and Regulation (EC) No 1223/2009, /* COM/2012/0542 final - 2012/0266 (COD) */
[5] Tedesco et al. 2015 *Toxicol In Vitro* **29(7),** 1736-44
[6] Huang et al. 2015 *Trends Food Sci Tech* **45**, 187-199
[7] Roman et al. 2016 *Anal Bioanal Chem* **408**, 5109–5124

IOP Conf. Series: Journal of Physics: Conf. Series **838** (2017) 012020 doi:10.1088/1742-6596/838/1/012020

Nanomaterials, and Occupational Health and Safety—A Literature Review About Control Banding and a Semi-Quantitative Method Proposed for Hazard Assessment.

Kaotar Dimou[1], Claude Emond[1,2]

(1) Department of Environmental and Occupational Health, University of Montreal School of Public Health (ESPUM, École de santé publique de l'Université de Montréal), C.P. 6128, Succ. Centreville, Montreal, Quebec, Canada
(2) BioSimulation Consulting Inc., 2915 Ogletown Road, #1182, Newark, DE, 19713, USA
E-mail: claude.emond@biosmc.com

Abstract. In recent decades, the control banding (CB) approach has been recognised as a hazard assessment methodology because of its increased importance in the occupational safety, health and hygiene (OSHH) industry. According to the American Industrial Hygiene Association, this approach originates from the pharmaceutical industry in the United Kingdom. The aim of the CB approach is to protect more than 90% (or approximately 2.7 billion) of the world's workers who do not have access to OSHH professionals and traditional quantitative risk assessment methods. In other words, CB is a qualitative or semi-quantitative tool designed to prevent occupational accidents by controlling worker exposures to potentially hazardous chemicals in the absence of comprehensive toxicological and exposure data. These criteria correspond very precisely to the development and production of engineered nanomaterials (ENMs). Considering the significant lack of scientific knowledge about work-related health risks because of ENMs, CB is, in general, appropriate for these issues. Currently, CB can be adapted to the specificities of ENMs; hundreds of nanotechnology products containing ENMs are already on the market. In this context, this qualitative or semi-quantitative approach appears to be relevant for characterising and quantifying the degree of physico-chemical and biological reactivities of ENMs, leading towards better control of human health effects and the safe handling of ENMs in workplaces. The need to greater understand the CB approach is important to further manage the risks related to handling hazardous substances, such as ENMs, without established occupational exposure limits. In recent years, this topic has garnered much interest, including discussions in many technical papers. Several CB models have been developed, and many countries have created their own nano-specific CB instruments. The aims of this research were to perform a literature review about CBs, to classify the main approaches that were developed worldwide, and then to suggest an original methodology based on the characterisation of the hazard. For this research, our team conducted a systematic literature review over the past 20 years. This approach is important in understanding the conceptual basis for CB and the model's overall effectiveness. These considerations will lead to the proposal of an original hazard assessment method based on physico-chemical and biological characteristics. Such a method should help the entire industry better understand the ability of the CB approach to limit workers' exposure, while identifying the strengths and weaknesses of the approach. Developing this practice method will help to provide relevant recommendations to workers who handle hazardous chemicals such as ENMs and to the general population.

1. Introduction

Since 2000, nanotechnology has grown dramatically in all industrialised countries. Back then and even now, the advent of the infinitely small particle through nanotechnology is considered to be the industrial revolution of the twenty-first century [1]. However, the idea to manipulate matter at the atomic and molecular levels was raised explicitly by American physicist Richard Feynman during his famous lecture from 1959 that was titled "There Is Plenty of Room at the Bottom" [2]. In 1974, Japanese professor Norio Taniguchi coined the term of "nanotechnology" [3]. However, the era of nanotechnology really arrived in 1986, thanks to the development of new sophisticated tools such as the scanning tunnelling microscope and the atomic force microscope [4, 5]. Therefore, remarkable progress is being made in a variety of fields such as energy, environment and health [6].

Nowadays, nanotechnology is a multidisciplinary research field in which an increasing variety of engineered ENMs are synthesised [7]. According to International Organization for Standardization (ISO)/TS 80004-1, one or more external dimensions of ENMs are in the *size* range of *1 nm to 100 nm.* At the nanoscale level, ENMs have unique physical and chemical properties, which make them an attractive candidate for a wide range applications in a variety of revolutionary scientific fields [8]. Most industrialised countries are investing heavily in nanoscience and nanotechnology research and development around the world (in the United States of America, China, Japan, Russia and European countries) [7, 9]. Nevertheless, rapidly developing nanotechnology and the growing utilisation of ENMs in consumer and industrial products have led to a potential increase of human exposure, thus raising concerns of many scientists and governments regarding health risks to workers and the public [10-12]. Investigators supported some initiatives to implement the industry's and researchers' use of ENMs in nanotechnology. A survey was initiated among industries and researchers working in different specialties that were potentially involved in the development, production, distribution and integration of ENMs and use of ENMs-containing products [13, 14]. Indeed, the fear among public health institutions in several countries has increased regarding the quantitative health risk assessment of ENMs [7].

In 1983, the National Research Council first proposed a health risk assessment [15]. The four steps of this approach include source characterisation, hazard characterisation, exposure characterisation and risk characterisation. The principle of the health risk assessment is based on "*the use of the factual base to define the health effects of exposure of individuals or populations to hazardous materials and situations*" [15]. In the case of ENMs specifically, it will be virtually impossible to assess the potential occupational health hazard, given the almost limitless uncertainties of ENMs [16]. In such an uncertain context, in which current data regarding the toxicity of ENMs remain fragmented [17], quantitative health risk assessment (QHRA) is purely hypothetical. Thus, a risk assessment method for controlling ENMs exposures remains mainly qualitative or semi-quantitative in nature [18]. The qualitative risk assessment method, known as "control banding (CB)," is an alternative approach to QHRA that has some drawbacks that limit its use in nanotechnology; therefore, the current work is in this context. It is important to develop an approach of CB based on characterisation of potential hazards posed by ENMs. Characterisation is necessary to identify and quantify the degree of hazardousness of ENMs to establish a better framing of their health implications and to ensure safe handling of these ENMs. Emond et al. [19, 20] designed the first portion of the methodology to integrate QHRA and CB, thereby integrating a semi-quantitative approach for the assessment of ENMs. The second portion of the methodology is underway and is a part of the initiative in this current paper.

The core structure of this paper is organised in a step-wise manner. In Section 2 (Methods), we provide the literature search strategy that includes the study selection and data collection processes. This review provides the information that we observed regarding why these matrices were developed and their limitations, In Section 2, we also discuss the CB strategy and describe the main CB tools that have been developed specifically for ENMs. This synthesis is important in understanding the key concepts for CB and the implications of this approach's overall effectiveness, and then for proposing our approach. In

Section 3 (Result and Discussion), we present the results from the literature search and our proposed assessment tool of ENMs based on physico-chemical and biological characteristics. In Section 3, we also briefly discuss the approach developed as an innovative hazard assessment method, which is different from the previously developed approaches. This paper discusses the ongoing project, addressing where we are so far, and the perspective of this project. Section 4 (Conclusion) of this paper presents an overview of the state of the project and describes the future work needed to complete it.

The aims of this research were to perform a literature review about CBs, to classify the main CB approaches that were developed worldwide and then to suggest an inventive methodology based on the characterisation of the hazard.

2. Methods

2.1. Literature Search Strategy

The study was conducted to cover the period from January 1996 to November 2016. The review was limited to only French and English papers. The following databases were searched: Toxline, PubMed, Web of Science, ScienceDirect and Google Scholar. The research strategy was based on four major concepts: CB, ENMs, physico-chemical characteristics and biological characteristics. Many keywords were used that referred to different complementary concepts that all linked to ENMs. The keywords were as follows: agglomerate and biological characterisation of ENMs, carbon nanotubes, CB tools, engineered ENMs, exposure banding, hazard assessment, nanotechnology, nanotoxicity, physicochemical parameters of ENMs, quantum dots and risk banding.

Furthermore, the governmental reports of recognised organisations were considered. Some of these organisations included the Robert-Sauvé Research Institute for Occupational Health and Safety (l'Institut de recherche Robert-Sauvé en Santé et en Sécurité du travail [IRSST]), the National Agency for Food Safety, Environment and Labor (Agence nationale de sécurité sanitaire de l'alimentation, de l'environnement et du travail [ANSES]) and NIOSH (the National Institute for Occupational Safety and Health). This review also considered conference abstracts (e.g. Nanosafe International Conferences on Safe Production and Use of ENMs) and theses and dissertations. This search also included specialised databases, taking to account primary and secondary, literature documents. The purpose of this work was not to conduct an exhaustive review of all of the studies published to date, but to identify the major documents, thereby obtaining a better understanding of the prioritisation of work regarding CB strategy, CB tools specifically developed for ENMs for ENMs and physicochemical and biological properties of ENMs. The aims of this review were to understand how the CD tools were developed and to identify the parameters in these matrices.

A total of 932 references were identified through bibliographic databases searches. An additional 50 references were obtained through other Internet sources. After removing duplicates, a total of 832 records were available for title review/abstract screening. Out of the 832 references, 91 (11%) records remained for a comprehensive review. After the comprehensive review of the 91 records, only 51 (6%) of these records were included in this literature review.

2.2. New Methodology Approach Based on CB, but for Health Hazard Characterisations of ENMs

The new method developed during the current work and discussed in this paper aims to evaluate the health hazards resulting from the use of ENMs, with the objective to protect workers and the general population from hazardous situations. This method is based on the CB approach that classifies several levels of toxicological hazards from ENMs; however, as discussed in this paper, our first intend in that project did not take exposure into consideration at this step. This approach is based exclusively on relevant physico-chemical and biological characteristics of ENMs. The levels of toxicological hazards used in the current method will combine four or five physico-chemical characteristics and the same number for the toxicological tests. The combination of these hazard bands will lead to the development of a decision matrix tool to establish a control level, and then to make recommendations for the workers who handle hazardous chemicals such as ENMs and for members of the general population who might be exposed in the workplace.

We believe that the current method can be used to classify the ENMs according to their intrinsic reactivities with biological materials to establish a limit for the ENMs that currently do not have any reference values. It is important to mention that in this paper, we will discuss several potentials tests that could be used in this matrix. However, another exhaustive literature review will be necessary to identify the tests commonly used by different laboratories. All of these tests will be challenged by extensive criteria not discussed in this paper. A validation will then be required before a decision is made regarding the prioritisation of tests to be used to characterise the hazards. Thus, the validation component is not included in this current work.

3. Result and Discussion

3.1. Results from the Literature Search

These CB–type tools for ENMs that were developed between 2008 and 2012 have already been described in detail and reported in published studies for a comparative analysis [37, 38, 45, 46, 52]. The authors of these publications reviewed many of these CB tools for ENMs relating to their scope and applicability, hazard and exposure banding parameters and risk classification or control bands. It is evident from the scope of each of the strategies that they were developed for different purposes. For example, the CB Nanotool was created to protect nanotechnology researchers, whereas the Precautionary Matrix was created to identify and prioritise risks, considering the health of workers and consumers, as well as the environment. The CB methodology should not in any way be compared with the other tools because it is not a matrix that allocates hazard and exposure bands; instead, the methodology is only being used to determine whether there is a need for action. These Guidance were developed to help employers and workers with identifying the risks associated with different work situations. NanoSafer and the Stoffenmanager Nano were applied to assess and manage occupational risks during the synthesis and downstream use of ENMs in workplace settings and at laboratories.

3.2. Origins of the CB Strategy

CB, which is used as a risk management strategy, was developed as a simplified approach to conduct a qualitative risk assessment and to take measures to protect workers and the public. In a historical context, CB has grown from a number of qualitative risk assessment strategies that began to appear in the 1970s because of events that involved explosions, radiation, lasers and biological agents [21, 22]. In the early 1990s, the pharmaceutical industry implemented the CB method that was used to manage chemical risks to employees in the workplace [22]. In 1994, the Health and Safety Executive (HSE) in the United Kingdom initially developed the CB method, which was known as the Control of Substances Hazardous to Health Essentials (COSHH Essentials), primarily to benefit small- and medium-sized enterprises (SMEs) that may not have the expertise of a full-time occupational hygienist on staff or as a consultant [23, 24].

This approach was designed to help SME employers to carry out suitable and sufficient risk assessments for all chemicals used in the workplace and to take steps to ensure that exposure is adequately controlled [23, 25]. This regulation was intended to increase the emphasis on the use of risk assessments in the industry to protect human health [26]; however, many challenges were experienced regarding their effective implementation. An unpublished survey of 2000 companies revealed "widespread ignorance of the regulations and their implications"[27]. The main challenge was that the SMEs did not understand

IOP Conf. Series: Journal of Physics: Conf. Series **838** (2017) 012020 doi:10.1088/1742-6596/838/1/012020

the requirements of the COSHH Essentials and "wanted to be told exactly what they need and do not need to do" [21, 28]. However, there may be little practical use for occupational exposure limits (OELs) in small companies, given that SMEs often lack the financial and technical resources to conduct toxicological and epidemiological research to establish the OELs and collect and analyse samples [21]. With the challenge indicated, the SMEs have experienced increased difficulties with assessing and ensuring appropriate protection for their workers [29]; therefore, an alternative, user-friendly method to protect workers from adverse health effects of hazardous chemicals was needed. This concept was the origin of CB.

Strictly speaking, CB represents a qualitative strategy that originates from industrial hygiene and offers solutions and control measures to ensure the safety of workers who use new chemicals, for which reliable toxicological and exposure data are absent [25, 30]. These new chemicals were classified into "bands," mainly defined according to the hazard level of known products similar to those used, taking into consideration exposure assessments. Each band corresponded to a risk control strategy [31]. In one of the least complex forms of CB, a four-level hierarchy of risk management options for controlling exposures to chemicals includes: (1) effective occupational hygiene practices, which may be supplemented by using appropriate personal protective equipment; (2) engineering controls, including local exhaust ventilation; (3) containment and (4) the need to seek advice from specialists [21].

To determine the appropriate control strategy, the exposure potential may be estimated by quantity of use, the volatility of liquids or dustiness of solids and the frequency and duration of exposure, and the potential hazard may be captured in what is known as a risk phrase [R-phrase] that were required by the European Union and used to rank the hazard of a chemical [21, 22]. The use of R-phrases or their equivalents in the Globally Harmonized System for Classification and Labelling of Chemicals in CB is helpful, but it is not intended to replace OELs, exposure assessments, or classic industrial hygiene protocols [21]. The majority of the CB methods focus the resources on preventing workers' exposure to chemicals; however, the CB method was not intended to be a predictive exposure model, but rather a way to sort work processes into bands. Simply stated, the CB method was a way to link hazards to general control plans [21, 32]. The CB method is recognised by many national and international organisations and is being increasingly applied worldwide as a practical approach to meet the needs of SMEs and developing countries for addressing chemical hazards. CB has been internationalised by the International Labour Organization [25, 33]. Of all of the forces that drive the evolution of the CB model, there was an increasing recognition that the traditional process was losing ground. In fact, reliance on this approach has become increasingly difficult because of the growing number of potentially hazardous materials in the workplace that do not have OELs, specifically ENMs that are currently used in occupational settings [22, 34].

3.3. CB Tools for ENMs

Various countries have developed a variety of CB strategies [26]. In recent years, several nano-specific CB methodologies for different types of workplace environments (e.g. small and large industries and laboratories) have been developed and published, and each instrument has specific strengths and limitations [35]. The remainder of Section 3 of this review presents a brief summary and the purpose of each of the six instruments used for ENMs. These six instruments are the CB Nanotool, the Precautionary Matrix, the ANSES CB Tool for ENMs, NanoSafer, The Guidance and Stoffenmanager Nano.

3.3.1. CB Nanotool (United States of America)

In 2008, Paik et al. developed the CB Nanotool as a strategy for performing a risk assessment and protecting nanotechnology researchers at the Lawrence Livermore National Laboratory [36, 37]. The CB Nanotool is intended to be used by both experts and non-experts for providing recommendations for appropriate engineering controls [35, 36]. The CB Nanotool is a simplified approach based on the paradigm established by the COSHH Essentials that uses a four-by-four factor risk matrix (Table 1) to determine the risk level (RL) [38-40]. The specific RL is determined by a "severity" score on one axis and by a "probability" score on the other axis [36]. The severity score is based on a mix of intrinsic properties of ENMs and on the presence or absence of health effects (four bands for the hazard). The probability score is based on interaction of the worker in the workplace and the ENMs (four bands for the exposure) [36]. The lack of data resulted in high scoring. The issue with this scoring, it is difficult to apply a summation of items that were different such as a dermal hazard and solubility or particle shape and carcinogenicity. The maximum probability/severity score is 100 (Table 1). In fact, the combination of severity and probability leads to an overall four possible RL control bands with corresponding specific control strategies that can be classified in RL 1 to RL 4 [36, 41].

Table 1: The CB Nanotool matrix (modified based on research from Paik et al. 2008)

Severity	Probability			
	Extremely unlikely (0–25)	**Less likely (26–50)**	**Likely (51–75)**	**Probable (76–100)**
Very high (76–100)	**RL 3**	**RL 3**	**RL 4**	**RL 4**
High (51–75)	**RL 2**	**RL 2**	**RL 3**	**RL 4**
Medium (26–50)	**RL 1**	**RL 1**	**RL 2**	**RL 3**
Low (0–25)	**RL 1**	**RL 1**	**RL 1**	**RL 2**

Note: RL 1 = general ventilation; RL 2 = fume hoods or local exhaust ventilation; RL 3 = containment; and RL 4 = seek specialist advice.

The severity scores are determined by using the physical-chemical characteristics of both the ENM and the parent material. The probability scores are determined by using the factors that determine the potential exposure of workers to ENMs [36]. The CB Nanotool provides indications regarding how to allocate scores for the severity factors and the probability factors. Then, the maximum scores can be attributed to each factor.

3.3.2. Precautionary Matrix (Switzerland)

The Swiss Federal Office of Public Health and the Federal Office for the Environment developed a Precautionary Matrix [42] in 2008 and revised the tool in 2010 [43]. The Precautionary Matrix was published almost at the same time as the CB Nanotool. The Precautionary Matrix is different from the CB Nanotool because the matrix also considers the workplace, consumers and the environment [43, 44]. This CB approach (i.e. the matrix) aims to guide SMEs to apply a precautionary method to identify possible sources of risk from the production, use and disposal of synthetic ENMs [42]. In contrast to the name, the Precautionary Matrix does not distinguish the bands for the hazard from the bands for exposure; however, the matrix does combine the hazard and exposure potential into a single score to determine whether there is a need for action [35, 45]. The total score is subdivided into two categories: Class A (lower score ≤20) and Class B (higher score ≥21) (see Table 2). The major issue is that the categories do not discriminate between the following: (1) hazard and exposure and (2) the sum of the pictograms based on different information about the amount of ENMs and work or consumer information that is related to the exposure. This approach requires some expertise to help ensure accurate interpretation. The precautionary matrix estimates the risk potentials—through the entire life cycle—for

IOP Conf. Series: Journal of Physics: Conf. Series **838** (2017) 012020 doi:10.1088/1742-6596/838/1/012020

the health of workers and consumers and for the environment. The applicability of this CB approach is difficult, and global.

Table 2: Classification of ENMs based on overall score in the Precautionary Matrix (modified based on research from Höck et al. 2010)

Score	Classification	Importance
≤20	A	The nano-specific need for action can be rated as low, even without further clarification.
≥20	B	The nano-specific action is needed. Existing measures should be reviewed, further clarification undertaken and, if necessary, measures to reduce the risk associated with manufacturing, use and disposal should be implemented.

3.3.3. ANSES CB Tool for ENMs (France)
ANSES developed an operational CB approach for small- to large-sized enterprises during the synthesis and downstream use of ENMs in workplace settings and in laboratories [31, 35, 45]. The CB Tool for ENMs is only available as a paper version, and it is explicitly recommended that the users (academics and industrials) are adequately qualified to conduct work regarding chemical risk prevention [31, 39]. In the approach from ANSES, the control bands (levels) are determined by combining the bands for the hazard and the exposure (emission potential). ANSES uses five hazard bands that range from HB1 (very low [no significant risk to health]) to HB5 (very high [a severe hazard that requires an expert to perform a comprehensive hazard assessment]) [31]. The exposure bands are allocated based on the emission potential and can be grouped according to the following four levels: EP1 (solid), EP2 (liquid), EP3 (powder) and EP4 (aerosol) [31]. These five hazard bands and four exposure bands are directly linked to the five control bands associated with general recommendations that are ranked from lower CL1 (natural or mechanical general ventilation) to higher CL5 (full containment and review by a specialist are required) [31, 45, 46]. The hazard portion of the CB matrix (from ANSES) is based on the toxicity warning label that is associated with clinical observations. The issue regarding these criteria is that they are not specific enough; therefore, the criteria cannot be used to protect sensitive sub-populations. These criteria are based on general observations of large classifications and are not supported by quantitative data, with the exception of a large characteristic from the labelling that is used.

3.3.4. NanoSafer (Denmark)
Developed by the Danish Technological Institute and National Research Center for the Working Environment, NanoSafer is an online CB tool that helps manage the risks resulting from the use of ENMs [37, 47]. Currently, there is only a Danish version of this tool, which is intended for SMEs and laboratories that do not have any or have limited experience in producing or working with ENMs [35, 47]. NanoSafer focuses on the work environment, and application of the tool is limited to assessing the handling of nanopowder products and fugitive and point-source emissions [35, 45]. In NanoSafer, four bands are allocated for the hazard and five bands are for the exposure. In addition, there are five RLs (control bands) [35] that range from RL1 (low hazard and low exposure potential) to RL5 (high hazard and/or moderate to very high exposure potential) [47]. NanoSafer uses its e-learning tool to make recommendations that are suitable control measures for each RL [47]. Because NanoSafer relies on the size of the ENMs, such as nanopowder, and the OEL of the application, it is relatively accepted that the sizes of ENMs are proportional to the toxicity and related to the surface. However, it also true that two ENMs of the same size may have vastly different toxicities. It is important to note that only considering

IOP Conf. Series: Journal of Physics: Conf. Series **838** (2017) 012020 doi:10.1088/1742-6596/838/1/012020

the size of the ENMs will provide an indication, but this criterion will not be strong enough to ensure an accurate and valid assessment. The issue regarding hazards is much more complex than that.

3.3.5. The Guidance (The Netherlands)

The Guidance was developed by the Confederation of Netherlands Industry and Employers (Verbond van Nederlandse Ondernemingen-Nederlands Christelijk Werkgeversverbond [VNO-NCW]) and two Dutch trade associations, the Dutch Federation Vakbeweiging (Federatie Nederlandse Vakbeweiging [FNV]) and the Christian National Trade Union (Christelijk Nationaal Vakverbond [CNV]). The purpose of The Guidance is to guide employers and employees in identifying the risks associated with the use of ENMs and nanoproducts during different work situations [48]. The Guidance has a list of 10 generic activities for making an inventory of ENMs that are produced or used along their life cycle within a company [35, 48]. In The Guidance, the following three hazard categories for ENMs and nanoproducts are identified: 1 (soluble nanoparticles), 2 (synthetic, persistent nanoparticles) and 3 (fibrous, nonsoluble nanoparticles) [48]. The exposure bands are allocated based on the emission potential from the different activities related to the production of polymeric nanocomposites: I (there are no emissions of free nanoparticles due to working in full containment), II (emissions of nanoparticles embedded in a matrix are possible) and III (emissions of free nanoparticles are possible) [48, 49]. Hazard and exposure data are collected and combined in a decision matrix to establish a control level (Table 3).

Table 3: The decision matrix to determine the risk management class and the control level for activities with ENMs and nano-enabled products (adapted from The Guidance, Version 4.2 dated August 2012)

Description of the hazard category for each nano-enabled product / **Probability of exposure to nanoparticles during activities**	***Hazard category 1*** Rigid, biopersistent nanofibers for which asbestos-like effects are *not* excluded	***Hazard categories 2a and 2b*** Biopersistent granular and fibre-form ENMs with excluded asbestos-like effects	***Hazard category 3*** Non-biopersistent granular or (water) soluble ENM
Exposure category I: Emission of primary nanoparticles is possible.	**A**	**A**	**C**
Exposure category II: Emission of nanoparticles embedded in a larger solid (>100 nm) or liquid matrix is possible	**A**	**B**	**C**
Exposure category III: Emission of free nanoparticles is minimised due to working in full containment	**B**	**C**	**C**

A	The hierarchic Occupational Hygienic Strategy will be strictly applied, and all protective measures that are both technically and organisationally feasible will be implemented. The *reasonableness principle* is not used.
B	According to the hierarchic Occupational Hygienic Strategy, the technical and organisational feasible protective measures are evaluated regarding their economic feasibility. Control measures will be based on this evaluation.
C	Sufficient (room) ventilation, if needed, is applied to local exhaust ventilation and/or containment of the emission source and use appropriate personal protective equipment.

3.3.6. Stoffenmanager Nano (The Netherlands)

Developed by the Netherland's Organisation for Applied Scientific Research (Nederlandse Organisatie voor Toegepast Natuurwetenschappelijk Onderzoek [TNO]) and Arbo Unie, Holland, the Dutch Stoffenmanager Nano [50] is a generic tool that has both Dutch and English versions. The

Stoffenmanager Nano was initially developed as a practical approach for scanning electron microscopy (SEMs) and also for laboratory work to support researchers with assessing, prioritising and controlling occupational risks during the synthesis and downstream use of ENMs [46, 50]. This CB approach is intended to be used by non-expert employees at SMEs [45]. The Stoffenmanager Nano combines five hazard bands (i.e. low, average, high, very high or extreme); the highest band is associated with the highest risk priority, irrespective of the exposure band [50]. The potential hazard level is assessed based on how it relates to the ENMs properties (i.e. size, shape and solubility) and the toxicological data available, together with the properties of the parent material [50, 51]. The Stoffenmanager Nano uses five exposure bands that are expressed by using a sophisticated algorithm and includes the most important exposure parameters, taking into account various aspects from the source of the ENM in the breathing zone of the workers [39, 51]. The combination of these hazard and exposure bands leads to three overall risk prioritisation bands that can be classified as 1 (the highest priority) to 3 (the lowest priority) [50].

The results of the literature review provided interesting information about how and where CB originated, how CB has been used in the past, and how CB is organised or structured. From this review, it was observed that each CB matrix described one or a few parameters related to the exposure. These parameters are usually related to a physical property of the ENMs (e.g. length, size, volatility, solubility). These physico-chemical properties integrated limited information that can be used to characterise the nanoparticles to help control risk. Other methods describe the CB approach that is used to determine the severity of a potential effect on the probability which depends on the manipulation of ENMs [36]. In this case, the matrix is a two-pronged account that is based on criteria of physical and health effects information to determine the severity and on the workplace information about exposure to derive the probability. This matrix is still fuzzy and needs an expert judgment to obtain a good interpretation. More effort should be made to improve the hazard characterisation of the nanoparticles. Thus, one way to improve this approach is to characterise the ENMs by using chemical and biological analysis tests, which, when combined, can provide a discriminate level of hazard. This characterisation of the hazard can be associated with a second banding graph, during which the hazard is correlated to the exposure classification, and then the risk assessment is estimated.

3.4. New Methodology Presented After the Literature Review

After reviewing the CB tools that have been developed specifically for ENMs, all of these tools are qualitative in nature and are based on the exposure. More specifically, all of the hazard criteria used in these different CB methods that were evaluated during this literature review missed the point of really characterising the ENMs based on their major characteristics. In fact, the series of parameters presented for each CB developed and reviewed as part of the current work is incomplete and based on classical occupational safety approaches. We believe that the literature review performed as part of our research about the CB approaches has revealed that the best thing that we can do to protect the workers and the general population exposed to ENMs is to use the established conventional occupational health methods. In fact, currently it is impossible to perform quantitative risk assessments because there is a lack of data. With that said, from the perspective of using a better and more relevant approach to conduct risk assessments, what can be done to improve the tool that is currently being used? This is why a new methodology to improve the hazard assessment is presented in this paper. The criteria of the tests selected for the new methodology are based on the availability of the analysis tools, the cost and the level of information that can be obtained from this analysis. The objectives are to maximise the information collected with tests and minimise the cost of the test. Of course, during the selection and prioritisation, more criteria may be added to obtain the best selection as possible, and then to characterise the reactivity of the ENMs.

IOP Conf. Series: Journal of Physics: Conf. Series **838** (2017) 012020 doi:10.1088/1742-6596/838/1/012020

This new approach is called a semi-quantitative hazard assessment method and is based on the CB approach and on the physico-chemical and toxicological tests of the ENMs. This new approach may help to quantify and interpret the levels of hazards from working with ENMs and with maximum confidence and minimum uncertainty regarding future recommendations to the workers who handle hazardous chemicals such as ENMs and to the general population.

The first phase of this work involved the development of an assessment tool of ENMs, based on the National Research Council's (NRC's) approach [20] in the publication titled *"Integrated Approach to Design and Safe Handling of Nanomaterials—A Program Based on a Dialogue Between Industry and Evaluators of Health Risks,"* which was published on the Web site for IRSST. The second phase of this work is underway (part of this initiative of the current paper). The second phase is needed to review the literature about the development of CB across time and to select the appropriate chemical and biological characterisation tests, which will be standardised and then assessment tools will be developed. Some examples of the potential tests that can be used are presented in Table 4.

For the physico-chemical characterisations, we have selected tests that describe the properties we want to measure. For example, there is a linkage between the size of the ENMs and toxicity. In general, the small nanoparticles have a higher toxicity, partly due to both the reactivity and to the proximity of the nucleus. Another example might be the surface chemistry of the nanoparticles. Therefore, two nanoparticles having the same size, but with different surface chemistries, will yield different toxicities. For each test, Table 4 also lists the equipment that can be used to perform analyses of this physical property. It is important to note that the list in Table 4 is not exhaustive because we are in the first step of this process.

Table 4: Physico-chemical characteristics of ENMs and suitable evaluation equipment

Physico-chemical characterisation	Instruments and methods
Size (size distribution, agglomeration)	Transmission electron microscopy (TEM), scanning electron microscopy and Raman scattering
Shape of nanoparticles	TEM and near-field scanning optical microscopy
Chemical composition	Secondary ion mass spectrometry and nuclear magnetic resonance
Surface (surface charge, surface area)	Zeta potential, TEM and attenuated total reflection–Fourier transform infrared
Stability (colloidal stability, solubility)	Zeta potential, dynamic light scattering and circular dichroism
Crystal structure	X-ray diffraction

The method also considers the biological characterisations of ENMs (Table 5). In this table, we tend to analyse the biological reactivity of the ENMs through interaction with the biology. In this characterisation, only the in vitro tests were considered because there is a budget issue regarding application in the industry. For example, the viability assay and the oxidative stress test are two major tests to analyse for toxicity. The right column in Table 5 lists some examples types of tests that are available to measure this toxicity.

IOP Conf. Series: Journal of Physics: Conf. Series **838** (2017) 012020 doi:10.1088/1742-6596/838/1/012020

Table 5: Toxicological tests of nanoparticles (in vitro toxicity)

Toxicological testing of ENMs	Tests available
Viability assay	Neutral red, resazurin test and Trypan blue exclusion assay (for cell counting)
Oxidative stress and inflammation tests	Superoxide dismutase activity, catalase, glutathione peroxidase, glutathione reductase, glutathione transferase, nuclear transcription factor–kappa B and interleukins (IL-1, IL-6, IL-8)
Apoptosis assay	Caspase activity assay and membrane asymmetry
Mitochondrial activity assay	MTT (3-(4,5-dimethylthiazol-2-yl)-2,5-diphenyltetrazolium bromide) or WST1 (water-soluble tetrazolium salts) assay
Metabolic activity test	Sulforhodamine B (SRB) assay
Membrane integrity test	Lactase dehydrogenase (LDH [or LD]) test

To determine which tests will be incorporated in the assessment tool, more work needs to be done during the third phase. In addition, a validation based on different criteria is required and will be explored during the second phase of the project. The results from the second phase will be presented in another paper.

The list of physical and biological tests selected will represent an original hazard assessment approach that can be used to characterise the ENMs. The selection and the analysis will be used to produce a hazard matrix (Table 6). Nevertheless, at this point, it is impossible to determine with some certainty which tests that correspond to the validation component will be used. To determine the list of tests, several criteria will be considered (e.g. widely used, inexpensive or free, easy to conduct and provides the information needed regarding the reactivity).

Table 6: Assessment tool of nanoparticles based on physico-chemical and biological characteristics

Physicochemical / Biological	TEST A	TEST B	TEST C	TEST D	TEST E
TEST A					
TEST B					
TEST C					
TEST D					
TEST E					

Note: The risk factors are as follows: green = lower; yellow = moderate; orange = less high; and red = high.

The objectives of the literature review were to understand the CB tool and to understand the history of its development and how the CB tool has evolved over time. The goal of our new methodology is to provide a tool that uses an original hazard assessment method to compare the physico-chemical and biological characteristics of ENMs. The CB tool will be used by occupational safety workers who are in charge of conducting hazard evaluations for workers and the general population. The literature review has allowed us to understand what tools have been developed for CB to protect workers who have been potentially exposed to ENMs. The major observations were that the CB matrix is based on one physico-chemical characteristic or on a cluster of physical or chemical properties and clinical observations. These parameters were then compared to the exposure scenario. Reviewing these studies, we came at the conclusion that a better approach should be developed to improve the hazard characterisation of the ENMs.

The new approach developed to characterise the ENMs was very important step in this project. In fact, a comprehensive assessment of the physico-chemical and biological characterisations of ENMs is a key element of any ENM toxicity screening strategy. In this study, we were primarily interested in the characterisation of ENMs to analyse this reactivity with biological materials. Indeed, it is necessary to characterise the ENM that is intended for therapeutic use in both its originally manufactured condition and after its introduction into a physiological environment. The proposed method will contribute towards a better knowledge of ENMs to anticipate their potential effects on humans.

The proposed method discussed in this paper will require validation before it can be used as a semi-quantitative CB approach. It is important to remember that our aim in this paper was not to present the final version of proposed semi-quantitative method. We are currently selecting the most relevant physico-chemical and biological parameters of ENMs. The selection process and the findings will be discussed in upcoming paper. This method is not intended to be a substitution of existing methods currently in use to manage the hazards posed by chemicals. This proposed method is different from the approaches that have been developed by other agencies. The other approaches assess and characterise the risk for workers and the general population on a semi-quantitative basis. Nevertheless, it was important in this paper to identify a minimum number of in vitro studies that could generate a maximum amount of information. In vitro studies can generate reproducible results as quickly as possible at an affordable cost for industrialists. The toxicology and physico-chemical tests that will be selected will be the most suitable and commonly used in the nanotechnology field.

4. Conclusion

During the literature review, we evaluated how the CB models were developed and how they work. There were different observations regarding how the models were developed and how the parameters remained in each CB. The major problem with the CB models reviewed in this current paper is the hazard information required is not always available from the Material Safety Data Sheet; therefore, expert judgment is often required. Following the comprehension of CB, including the strengths and weaknesses, a new semi-quantitative methodology of CB is proposed for the Human Health Hazard Assessment based on physico-chemical and biological characteristics of ENMs. This new approach is original because it is dependent upon the structure of ENMs, but it is not based only on the exposure on the first intend. It is important to develop this CB approach based on the characterisation of potential health hazards posed by ENMs to identify and quantify the degree of hazardousness of ENMs to establish a better framing of their health implications and to ensure safe handling of these nanoparticles. In particular, this proposed method is a practical tool that will to help provide relevant recommendations to the general population and particularly to the workers who handle ENMs and hazardous chemicals, thereby bolstering employees' confidence in adopting a sophisticated hazard assessment approach by using objective and validated criteria. At this point, the limitation of this method is that all tests suggested in this paper must be validated before a decision is made regarding the prioritisation of tests to be used for characterising the hazards. This limitation corresponds to Step 3 of our certification process of this methodology, which was not presented during the Nanosafe 2016 meeting in Grenoble, France. Nevertheless, this CB method, based on characterisations of the potential hazards posed by ENMs, is required in order to develop better guidelines regarding health implications that will ensure safe handling by employees in workplace settings, as well for the general population, which could come into contact with ENMs. From this dependent hazardous assessment, it is possible to combine this assessment and the exposure assessment. The scientific knowledge of the risks associated with ENMs

IOP Conf. Series: Journal of Physics: Conf. Series **838** (2017) 012020 doi:10.1088/1742-6596/838/1/012020

is constantly changing, both with regard to physico-chemical characteristics and toxicology. Thus, it is therefore mandatory to update information regularly by integrating these new data into the CB method used in this study.

5. References

[1] Lungu, M., et al., *Nanoparticles' Promises and Risks*. 2015.

[2] Feynman, R.P., *There's plenty of room at the bottom.* Miniaturization"(HD Gilbert, ed.) Reinhold, New York, 1961.

[3] Camille, L.P., *Les nanotechnologies: Concepts et applications*. 2013: Lavoisier.

[4] Drexler, K.E., *Engines of creation 2.0: the coming era of nanotechnology*. 2006.

[5] Lintz, P., *Analyse et enjeux éthiques des nanotechnologies en médecine: temps et discours, approche éthique systémique: double discours, approche psychanalytique: complémentarité des discours entre science et théologie*. 2014, Université Paul Valéry-Montpellier III.

[6] Iavicoli, I., et al., *Opportunities and challenges of nanotechnology in the green economy.* Environmental Health, 2014. **13**(1): p. 1.

[7] Binet, S., et al., *Les nanomatériaux manufacturés à l'horizon 2030*. 2015, INRS.

[8] Luther, W. and A. Zweck, *Safety aspects of engineered nanomaterials*. 2013: CRC Press.

[9] Ezema, I.C., P.O. Ogbobe, and A.D. Omah, *Initiatives and strategies for development of nanotechnology in nations: a lesson for Africa and other least developed countries.* Nanoscale research letters, 2014. **9**(1): p. 1-8.

[10] Vogel, U., et al., *Handbook of nanosafety: Measurement, exposure and toxicology*. 2013: Elsevier.

[11] Fransman, W., et al., *Comparative Human Health Impact Assessment of Engineered Nanomaterials in the Framework of Life Cycle Assessment.* Risk Analysis, 2016.

[12] Nielsen, E., *Human Health and Nanomaterials in Consumer Products.* Literature Review for the Region of Peel's Public Health Unit, 2008.

[13] Endo, C.-A., et al., *Portrait de la nanotechnologie au Québec dans les milieux industriels et de la recherche universitaire et publique. R-854, 1-90. IRSST.* 2015.

[14] Ostiguy, C., et al., *Mapping the use of engineered NM in Quebec's industries and research laboratories. Journal of Physics: Conference Series 429, 1-8.* 2013.

[15] NRC, *Risk assessment in the federal government: Managing the process*. 1983, Washington (DC), National Academy Press.

[16] Anses, *Méthode d'évaluation des niveaux de risques sanitaires et des dangers écotoxicologiques*. 2015.

[17] Emond, C. and L. Multigner. *Human Risk Assessment and Its Application to Nanotechnology: A Challenge for Assessors*. in *Journal of Physics: Conference Series*. 2015. IOP Publishing.

[18] Anses, *Enjeux et mise à jour des connaissances*. 2014.

[19] Emond, C., S. Kouassi, and F. Schuster, *Development of an integrative program of nanosafety: Promote the coordination between industries and risk assessor. Journal of Physics: Conference Series 429, 1-10.* 2013.

[20] Emond, C., S. Kouassi, and F. Schuster, *Approche intégrée pour une conception et une manipulation sécuritaires des nanomatériaux - Un programme basé sur une concertation entre l'industrie et des évaluateurs des risques sanitaires. R-877 , 1-35. 2015. (Report)*. 2015.

[21] NIOSH, *<Qualitative Risk Characterization and Management of Occupational Hazards Control Banding (CB)-2.pdf>*. 2009.

[22] Zalk, D.M. and D.I. Nelson, *History and Evolution of Control Banding: A Review.* Journal of Occupational and Environmental Hygiene, 2008. **5**(5): p. 330-346.

[23] Russell, R., et al., *An introduction to a UK scheme to help small firms control health risks from chemicals.* Annals of Occupational Hygiene, 1998. **42**(6): p. 367-376.

[24] Naumann, B.D. *Control banding in the pharmaceutical industry*. in *Occupational Hygienists Conference Proceedings*. 2005.

[25] Nelson, D.I. and D.M. Zalk, *Control Banding: Background, Critique, and Evolution*, in *Patty's Industrial Hygiene*. 2011, John Wiley & Sons, Inc.

[26] Skordal, K., *Control Banding Tools: Advanced Regulated Evaluation and Authorization of Chemicals Tool and Agreement with Professional Judgment*, M. Yost, J. Camp, and N. Seixas, Editors. 2012, ProQuest Dissertations Publishing.

[27] Seaton, A., *Control of substances hazardous to health.* BMJ: British Medical Journal, 1989. **298**(6677): p. 846.

[28] Topping, M., *Occupational exposure limits for chemicals.* Occupational and environmental medicine, 2001. **58**(2): p. 138-144.

[29] Oldershaw, P., *Control banding workshop, 4–5 November 2002, London.* Annals of Occupational Hygiene, 2003. **47**(7): p. 531-532.

[30] Zalk, D.M. and G.H. Heussen, *Banding the world together; the global growth of control banding and qualitative occupational risk management.* Safety and health at work, 2011. **2**(4): p. 375-379.

[31] Anses, *Development of a specific control banding tool for nanomaterials.* Expert committee (CES) on physical agents. French Agency for Food, Environmental, and Occupational Health and Safety, Maisons-Alfort Cedex, 2010.

[32] Erbis, S., et al., *Review of Research Trends and Methods in Nano Environmental, Health, and Safety Risk Analysis.* Risk Analysis, 2016.

[33] Chalupka, S., *Control Banding for Risk Management of Source Chemical Agents and Other Occupational Hazards.* Workplace Health & Safety, 2010. **58**(9): p. 404.

[34] Garrod, A. and R. Rajan-Sithamparanadarajah, *Developing COSHH Essentials: dermal exposure, personal protective equipment and first aid.* Annals of Occupational Hygiene, 2003. **47**(7): p. 577-588.

[35] Liguori, B., et al., *Control banding tools for occupational exposure assessment of nanomaterials—Ready for use in a regulatory context?* NanoImpact, 2016. **2**: p. 1-17.

[36] Paik, S.Y., D.M. Zalk, and P. Swuste, *Application of a pilot control banding tool for risk level assessment and control of nanoparticle exposures.* The Annals of occupational hygiene, 2008. **52**(6): p. 419.

[37] Sánchez Jiménez, A., et al., *A comparison of control banding tools for nanomaterials.* Journal of Occupational and Environmental Hygiene, 2016. **13**(12): p. 936-949.

[38] Eastlake, A., R. Zumwalde, and C. Geraci, *Can control banding be useful for the safe handling of nanomaterials? A systematic review.* Journal of Nanoparticle Research, 2016. **18**(6): p. 1-24.

[39] EU-OSHA, *"E-fact 72: Tools for the management of nanomaterials in the workplace and prevention measures,18,"* 2013.

[40] Hansen, S.F., A. Baun, and K. Alstrup-Jensen, *NanoRiskCat–a conceptual decision support tool for nanomaterials.* 2011, Danish Ministry of the Environment.

[41] Zalk, D. and S.Y. Paik. *Risk Assessment Using Control Banding*. in *Elsevier Inc.* 2011.

[42] Höck, J., et al., *Guidelines on the precautionary matrix for synthetic nanomaterials.* Federal Office for Public Health and Federal Office for the Environment, Berne, 2008.

[43] Höck, J., et al., *Guidelines on the precautionary matrix for synthetic nanomaterials.* Federal Office for Public Health and Federal Office for the Environment, Bern, 2010.

[44] Riediker, M., et al., *Development of a Control Banding Tool for Nanomaterials.* 2012. **2012**.

[45] Brouwer, D.H., *Control banding approaches for nanomaterials.* Annals of occupational hygiene, 2012. **56**(5): p. 506-514.

[46] Fleury, D., et al., *Nanomaterials Risk Assessment in the Process Industries: Evaluation and Application of Current Control Banding Methods*, in *Lp2013 - 14th Symposium on Loss Prevention and Safety Promotion in the Process Industries, Vols I and Ii*, E. DeRademaeker, et al., Editors. 2013, Aidic Servizi Srl: Milano. p. 949-954.

[47] Jensen, K.A., et al. *NanoSafer vs. 1.1-nanomaterial risk assessment using first order modeling*. in *6th international symposium on nanotechnology, occupational and environmental health, Japan*. 2014.

[48] Cornelissen, R., et al., *Guidance working safely with nanomaterials and products, the guide for employers and employees.* The Netherlands, Amsterdam, 2011.

[49] Silva, F., et al. *Qualitative risk assessment during polymer mortar test specimens preparation-methods comparison*. in *Journal of Physics: Conference Series*. 2015. IOP Publishing.

[50] Van Duuren-Stuurman, B., et al., *Stoffenmanager nano version 1.0: a web-based tool for risk prioritization of airborne manufactured nano objects.* Annals of occupational hygiene, 2012: p. mer113.

[51] Xing, B., N. Senesi, and C.D. Vecitis, *Engineered Nanoparticles and the Environment: Biophysicochemical Processes and Toxicity*. Vol. 4. 2016: John Wiley & Sons.

IOP Conf. Series: Journal of Physics: Conf. Series **838** (2017) 012021 doi:10.1088/1742-6596/838/1/012021

Experimental Evaluation of the Effectiveness Offered by Different Types of Personal Protective Clothing Against Nanoaerosols

M Domat[1], J Pla[1], M C Cadavid-Rodríguez[2], C Fito[1]

[1]ITENE, C/ Albert Einstein, 1, 46980 Paterna, Spain
[2]Honeywell, ZA du Berret, 30200 Bagnols sur Cèze, France

E-mail: maida.domat@itene.com

Abstract. The rapid expansion of nanotechnology is outpacing health and safety recommendations for engineered nanomaterials. Thus, there is a lack of information about the effects that nanomaterials can induce in the human health. Nevertheless, workers in nanotechnology-related industries are potentially at risk of being exposed to nanomaterials. Therefore, there is a need of characterize the behaviour of personal protective equipment against penetration nanoparticles, in order to provide an adequate protection to the workers. In this study, the efficiency of several protective dermal equipment against water-based NaCl aerosol was evaluated. For this purpose, different protective clothing and gloves were selected to carry out the assays, simulating typical use conditions of protective equipment under occupational settings. Results obtained exposed that the level of protection offered by the distinct types of personal protective coveralls depended not only on the fabric, but also on their fitting to the body of the subject. On the other hand, the efficiency of the protective gloves was set in the range from 95% to 99%, depending on the thickness and the type of material.

1. Introduction

Nanotechnology is a term coined by Norio Taniguchi in 1974 [1], referring to the manipulation of matter on an atomic, molecular or supramolecular level. This discipline has experienced a tremendous growth during lasts years, promoting the development of a new generation of products in most major industry sectors [2–4]. Nevertheless, along with their benefits, engineered nanomaterials (ENMs) raise many questions and generate concerns about their potential effects on the human health owing to their physical, chemical and biological properties, which differ significantly from larger sized bulk materials, even when the basic material is the same [3,5,6]. This is as a consequence related to the increase of the surface area per unit of mass when size decreases, and the fact that the quantum effects play a greater role the smaller a particle becomes [1,7].

ENMs are defined as intentionally manufactured products having, at least, one dimension in the range 1-100 nm [8]. Nanoparticles (NPs) are a subset of ENM, described as single particles with a diameter or dimension below 100 nm [9]. However, apart from the intentionally-manufactured NPs, there are different sources of unintentionally-produced NPs which exist in some workplaces. A variety of airborne dust generating mechanisms can be presents in the occupational settings, including accidental spills, combustion, falling powders, agitation or transferring processes. Grinding and milling bulk materials also generates and suspends small particles as airborne dusts [10]. Consequently, workers in nanotechnology-related industries are potentially at risk of being exposed to ENMs [11,12].

Penetration of NPs into the human body can occur through respiration, dermal or oral routes [13]. Inhalation is the main exposure pathway for particulate matter and, consequently, the most critical for NP intake [14]. Nevertheless, the skin is the largest organ of the body, and it is also a potential route of exposure to ENMs. Oral exposure can also occur from incidental ingestion or unintentional hand-to-mouth transfer [15].

Regarding dermal exposure, the potential for NPs to penetrate the skin has been investigated in previous works found in the literature. Results show that the smaller the nanoparticle is, the easier is to penetrate the skin [16]. Furthermore, damaged or injured skin increases the NPs permeation [17]. However, no clear evidence has been found demonstrating that NPs are transported via the circulatory system through skin penetration to other body organs [16,18].

Although data from toxicological studies have suggested that nanoscale materials can pose new and unconventional risks, it should be noted that current available information about toxicity and hazard is limited, making difficult to carry out a quantitative risk assessment [8–10,19]. Therefore, given the limited amount of information about the health risks associated with occupational exposure to engineered nanomaterials, the precautionary principle suggests to take measures to minimize worker exposures [9,20].

The framework for occupational risk management applied to engineered nanomaterials takes into account the traditional industrial hygiene hierarchy to avoid or control the exposure of workers to these materials. Thus, these approaches comprises elimination or substitution, process enclosure, engineering control, procedural control, and personal protective equipment (PPE) [9,12]. Therefore, the use of PPE, such as protective clothing and gloves, supposes the last line of defence in the hierarchy of exposure control after all other available measures have been implemented [21,22]. Hence, the development and implementation of efficient PPE against NPs is needed in order to maintain the exposure as low as possible [6].

There are several types of protective clothing available in the market to protect workers who are exposed to a chemical risk. Protective clothing is classified using a scale ranged from 1 to 6 as a function of the level protection offered. Thus, the type 1 provides the highest level of protection and 6 the lowest. Suits type 1 are designed to be completely isolated from the environment, protecting the worker against chemicals and gases (UNE-EN 943-1,2); type 2 offers protection against chemicals, but not against gases (UNE-EN 943-1); type 3 is impermeable to liquids (UNE-EN 14605); type 4 is impermeable to aerosols (UNE-EN 14605); type 5 resist the penetration of solid particles suspended in the air (UNE-EN 13982-1); finally, type 6 offers a limited protection against liquid products (UNE-EN 13034) [23]. Although there are no standard regulations available describing the type of protective clothing needed to protect the worker against the exposition to NPs, some recommendations have been published which suggest the use of determined protective equipment. On the one hand, if the NM is used in form of dust, the use of disposable suits type 5 is proposed. In contrast, when the NM is used in a colloidal solution, the use of type 4 or 6 is suggested. It has to be said that there exists disposable protective clothing which is certified for several types simultaneously. Furthermore, the use of nonwoven materials is strongly recommended due it seems that offers better protection than woven fabrics [24].

Moreover, different publications related to the efficiency that protective clothing offers against the exposure to NPs have been published. Golansky et al. [13], Park et al. [21], Gao et al. [25], Vinches et al. [20] and Ben Salah et al. [26] concluded that nonwoven clothes offered better protection than woven fabrics, which is in accordance with the recommendations exposed above.

In the case of protective gloves, recommendation suggest the use of chemical or biological protective gloves (UNE-EN 374). This equipment must be made with "impermeable" polymeric materials such as latex, nitrile, polyethylene or butyl. Nevertheless, although the word "impermeable" is used to define these kind of polymeric materials, they offer a limited resistance against the permeation of chemical products. Sooner or later the chemical can be adsorbed in the external surface of the glove and diffuse through it to reach the skin. Thus, in some cases, the use of double-glove is encouraged to increase the protection level offered. Furthermore, if the NM is used along with other chemical product such as an organic solvent, the effectiveness of the glove may be decreased [27].

As in the case of protective clothing, Vinches et al. [20], Gonlasky et al. [13,28] , Dolez et al. [29] and Park et al. [21], studied the protection offered by distinct protective gloves exposed to different types of NPs. The conclusions achieved in these works showed that protective gloves offers, in general, goods levels of protection against the NPs assayed.

The selection of PPE must be based on an adequate risk assessment, giving rise to a clear picture on the level of protection that is needed. However, penetration efficiencies of protective clothing against NPs have not been explored deeply yet and, in some cases, results obtained are conflictive [12,13,20,29]. Therefore, little is known about the effectiveness of protective clothing when is exposed to NPs and, consequently, the available data for the selection of protective dermal equipment is limited [21,29,30].

This paper reports results obtained related to the penetration of NPs through protective clothing and gloves. The assayed protective clothing consisted in four different non-woven tissues, while the materials of the gloves selected to perform the assays were latex, vinyl, nitrile, neoprene, polyvinyl chloride and butyl. Furthermore, different movements were carried out in order to simulate typical use conditions of protective equipment under occupational settings.

2. Materials and methods

Due to the lack of a standard to test the protective equipment against airborne NPs, ISO norms and standards were reviewed and adapted to the conditions necessary to generate and measure NPs.

2.1. The testing chamber

Experiments were carried out in a test room isolated from the environment, specifically designed and built for this purpose. Entrance to the exposure chamber was performed through a Safe Access System (SAS) room, an isolated room from the others and from the outside where the subjects could put on their PPE safely. This room was maintained at overpressure to avoid leakage or contamination from ENMs from the other testing rooms, which were maintained at negative pressure. Thus, the ENMs were always contained inside the testing room, avoiding the release to the outside or contamination of the testing room. Furthermore, the SAS entrance has a control dispositive which does not allow that the door is unlocked when testing rooms are opened, in order to avoid direct contact with the testing chambers to the outside.

The pilot test chamber was designed to simulate working environments. It has a big, rectangular shape with a total volume of 10 m^3 where industrial instrumentation or processes can be replicated. A well-controlled laminar flow and HEPA filtering at both ends of the room remove completely the background, allowing tests under clean environments.

The tests room is prepared for wet or powdered dispersions of ENMs, since it has a flow director to modify ventilation type and a drainage system. It is coupled to the pilot test chamber through a connection panel where tubes can be plugged. Therefore, it can be measured or nebulized from any of the sides without the necessity of being inside the chamber.

The space of the rooms was designed minimizing the elements that could lead to accumulation of contaminants in edges or ground, building it with round corners and antistatic walls. Likewise, the material of the walls and floor is easily cleanable. Clean air was taken in from above the cabinet and passed through several HEPA filters to remove contaminants. Temperature, pressure, humidity, inlet and exhaust flowrates were controlled by an electronic control panel placed at the entrance.

Initially, in order to select the most appropriate concentration of NaCl to generate the aerosol, solutions with different amounts of this compound, ranged from 0.0001% to 0.1%, were prepared and tested. Two main assumptions related to the number and size of the NPs have to be taken into account; firstly, the number of particles generated depend on the concentration of the solution, following a directly proportional relation. Secondly, the greater the concentration, the bigger the particle diameter.

Thus, it was chosen a concentration of 0.05% of NaCl to carry out the experiments. The geometric mean diameter for the particle count distribution was about 40 nm, and the particle concentration was maintained in about 3,5 to 5,0 x 10^4 particles/cm^3. This concentration was maintained with less than

10% of variability between the beginning and the end of the tests (generally below 5%). The distribution of the aerosol obtained for this level of concentration is shown in Figure 3.

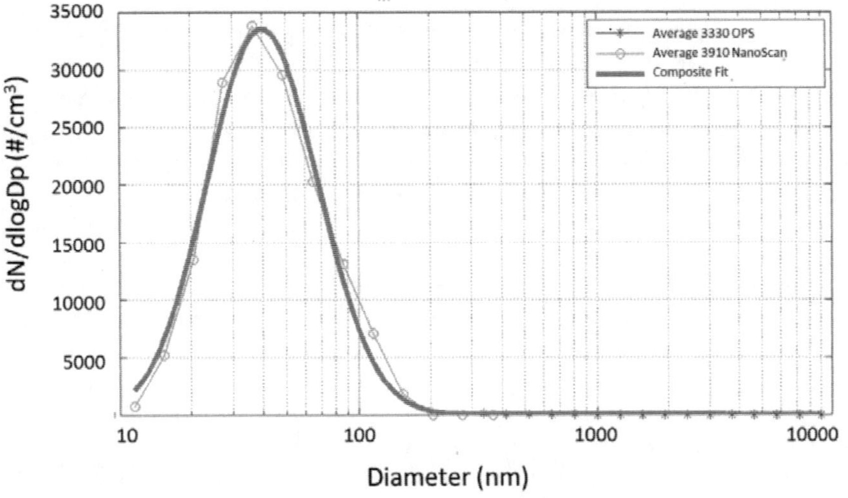

Figure 1: Particle number concentration as a function of size.

2.2. Equipment

Tests were performed generating a NaCl solid aerosol inside a test chamber isolated from the environment [26,31,32]. In order to produce the atmosphere of NPs, 0.5 g of NaCl purchased from Sharlau (Spain) were weighted and dissolved in 1 liter of ultrapure water obtained using a Milli-Q® Integral Water Purification System (Merk, USA). A bottle with the NaCl solution was placed on a magnetic stirrer to avoid deposition and connected to the Constant Output Atomizer (TSI 3076, USA). Flow was adjusted to 4 L/min and the pressure was set in 2.4 bar. The nebulized material was passed through a Diffusion Dryer (TSI 3062, USA), a silica gel desiccant aimed to remove the water droplets and the remaining humidity, in order to avoid agglomeration or aggregation effects. Afterwards, a neutralizer is placed to remove most of the charge of the generated particles and reduce electrostatic effects. Finally, when the concentration of NaCl NPs in the chamber was stabilized, experiments were carried out.

The specific instrumentation used to measure the airborne NPs characteristics were the Optical Particle Sizer (TSI 3330, USA) and Nanoscan SMPS (Scanning Mobility Particle Sizer, TSI 3910, USA), used for size distribution measurements. Two parallel CPCs (Condensation Particle Counter, TSI 3007, USA) were employed to quantify the particle number concentration at different locations, and to monitor the temperature and humidity in the chamber, a Testo 435 Multifunction Measuring Instrument (Testo, Germany) was employed.

2.3. Protective equipment assays

With the purpose of evaluate the effectiveness of the dermal protective equipment against NPs, several types of protective gloves and clothing were selected to carry out the experiments. The selection of the representative PPEs to be studied was carried out distributing questionnaires amongst industrial companies working with ENMs to compile information on the most widely used suits on each case. Samples of the same characteristics as the ones indicated in the answers were selected, trying to cover a wide range of standard levels of protection.

2.3.1. Setup for protective gloves assay. Following the recommendations which suggest the use of gloves against chemical and biological risks to protect workers against NPs, different types of common protective gloves made with impermeable polymeric materials (including reusable and single-use

IOP Conf. Series: Journal of Physics: Conf. Series **838** (2017) 012021 doi:10.1088/1742-6596/838/1/012021

gloves) were selected and tested. Thus, the single-use protective gloves selected were latex (without powder), vinyl (with and without powder) and nitrile (two different thickness). On the other hand, reusable gloves chosen were made of neoprene, polyvinyl chloride or butyl. Before starting the experiment, a previous visual examination was performed with the aim to discard gloves which presented any defect or gash.

Assays were carried out employing a specific device designed for this purpose, consisting in a steel cylinder (35 mm length, 20 mm diameter) opened by the front side. The back and the side were perforated in three points (Ø 5 mm) in order to insert three metal tubes (two in the back and one in the side). Finally, a circular mark (Ø 50 mm) was adapted in the front side.

A circular piece of the glove was fixed in the aforementioned device sealed with a clamp and placed into the chamber described in section 2.1, so that the glove was exposed to the NPs aerosol only by one side. The concentration of NPs in the chamber was measured using a CPC. Another CPC was connected in the back side of the device to measure the concentration of NPs capable of permeate the glove material. At the same time, a sheath flow of clean dry air was supplied inside the glove at the same flow rate as the measuring devices were suctioning, in order to avoid the creation of a depression. Furthermore, a HEPA filter was connected in the third tube to act as a scape/entrance valve: in the case of overpressure, air was allowed to escape from inside the system; on the other hand, if the pressure inside the system was lower than outside, air entered in the system through the HEPA filter. A schematic representation of the montage is illustrated in **Figure 2**.

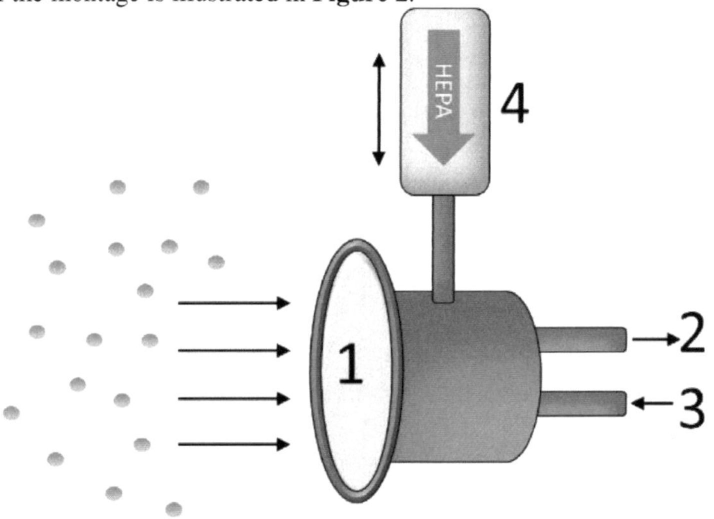

Figure 2: Schematic representation of the device employed to test the penetration of NPs in different gloves. 1: Test samples, 2: Input measurement device, 3: Clean dry air flow, 4: HEPA filter.

2.3.2. Setup for protective suits assay. To study the efficiency of protective clothing against NPs, five types of non-woven protective suits were selected to perform the tests, according to the criteria which encourages the use of this type of fabrics (Table 1): The first suit (PS1), classified as Cat III type 3B, 4B, 5B and 6B, consisted on a material with a high barrier protection against chemicals which offers protection against many concentrated inorganic acids and bases, and against a wide range of organic chemicals. The second one (PS2) was a coverall formed by million fine and continuous fibers made of high density polyethylene, classified as type 5B and 6B. The third coverall (PS3) was made with double-side PVD coated Nylon and classified as type 4 Cat III. Finally, the fourth suit (PS4), made with polypropylene, was classified as type 6, Cat I. The last suit (PS5) was a self-ventilated protective suit for radioactive protection (type 1). All suits were available in sizes ranging from M to XXL in order to cover the different body shapes of the test subjects.

Table 1: Protective suits selected for testing along with their code name and their standard classification.

Code	Material	Cat./Type	Reuse/Disposable
PS1	High barrier protection against chemicals (Tyvek type)	Cat III Types 3B, 4B, 5B and 6B	D*
PS2	Million fine and continuous fibers made of high density polyethylene (Tychem type)	Types 5B and 6B	D*
PS3	Double-side PVD coated Nylon	Cat III Type 4	R
PS4	Polypropylene	Cat I Type 6	D
PS5	Self ventilated radioactive suit	Cat III Type 1	R

* *According to manufacturers' instructions, it could be reused provided that is in good conditions.*

The protocol followed to calculate the inward leakage of NPs into protective clothing was adapted from the international standard UNE-EN ISO 13982-2 2005 "Protective clothing for use against solid particulates – Part 2: Test method of determination of inward leakage of aerosols of fine particles into suits" (ISO 13982-2: 2004). The main variation was the use of NPss to carry out the test, which required different ambient conditions and the use of specific devices to generate and detect them.

Firstly, a visual inspection was carried out to detect any malformation or defect. If so, the garment was discard. At least 10 human subjects with different features and good health were dressed with the suit following the manufacturer's instructions. Besides, in order to protect the subjects during the assays but also simulate the real operative conditions, they were equipped with protective respiratory equipment, gloves and security shoes.

The suit had connexions for silicone tubing in three different points at different heights: chest, waist and knee, and a CPC was connected to the point selected to perform the sampling. The concentration inside the chamber was measured with a second CPC. A pump was connected to supply a flow of dry, clean air inside the suit in the same range as the measuring devices suction rate, with the purpose of equal the drift inside, while, the third point remained closed. The sequence of connexions can be seen in Table 2.

Table 2: Sampling scheme and positions to perform the test of suits and coveralls against NPs.

	Sampling	Sheath Air	Closed
Position 1	Knee	Chest	Waist
Position 2	Waist	Knee	Chest
Position 3	Chest	Waist	Knee

The subject was placed inside the test chamber and asked to perform a series of exercises in the following order: (I) Standing still, (II) walking at 5 km/h, (III) continuous squatting at a frequency of 5 squats/minute. The time for each exercise in each sampling position was 3 minutes, and 1 minute of stabilization was left between exercises. A schematic representation of the assay performed to study the efficiency of protective suits is shown in **Figure 3**.

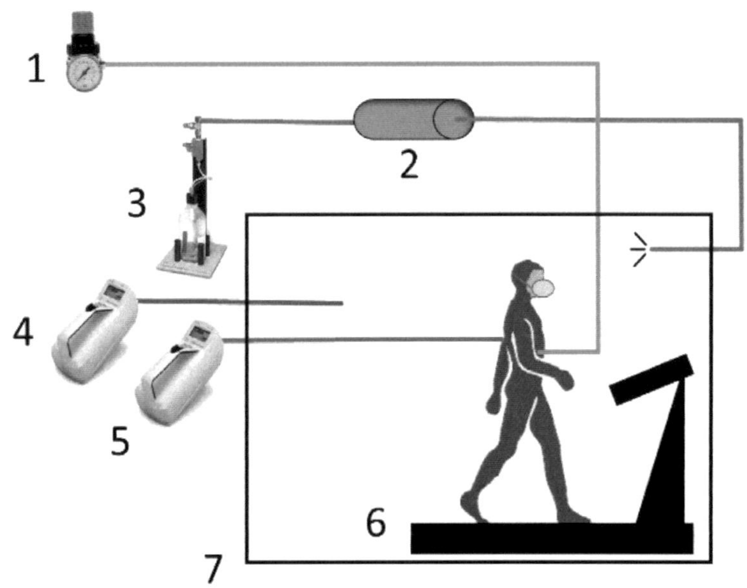

Figure 3: Scheme of the assay carried out to study the efficiency of protective suits. 1: Supply of clean air, 2: Diffusion dryer, 3: Constant output atomizer, 4: CPC measuring outside the suit, 5: CPC measuring inside the suit, 6: Treadmill, 7: Isolated chamber.

Furthermore, in order to evaluate the efficiency offered by the fabric itself, a piece of the protective clothing was placed in the device developed to analyse the leakage in gloves. The protocol followed was the same than the explained in section 2.3.1.

3. Results and discussion

Distinct values of efficiency against NPs were obtained depending on the personal protective equipment tested. Results achieved are described and discussed in the following sections.

3.1. Protective gloves

According to the standards, chemical protective gloves have to be made with impermeable polymeric materials. Although the term "impermeable" is used, impermeable materials offer a limited resistance against chemicals due to, sooner or later, chemicals will permeate them. The standard UNE-EN 374:2004 stablishes some considerations about the characteristics which protective gloves against chemicals and microorganisms have to satisfy. Nevertheless, the classification and protection offered by these gloves, according to this standard, does not take into account the airborne NPs.

The efficiency of several commercial protective gloves made with different materials was evaluated. The percentage of NPs capable to cross the barrier of the glove was determined calculating the percentage of inward leakage (IL%), comparing concentrations at both sides of the glove:

$$IL\% = \frac{C_1}{C_2} \cdot 100$$

Where C_2 represents the concentration of aerosol in the side of the glove exposed to the atmosphere and C_1 is the concentration of NPs found at the other side.

As can be seen in **Table 3**, the efficiency of the gloves increases with the thickness. The surprisingly high penetration of the latex gloves can be due to a deformation or breakage of this specific set of gloves or due to a greater size of the pore of the glove material in comparison with the nanoparticle. In the case of Butyl, results were so unalike each other that were discarded.

IOP Conf. Series: Journal of Physics: Conf. Series **838** (2017) 012021 doi:10.1088/1742-6596/838/1/012021

Table 3: Efficiency of distinct protective gloves against NaCl nanoparticles.

Type	Material	Thickness (mm)	IL (%)
Disposable	Nitrile Thin	0.07-0.09	0.040
	Nitrile Thick	0.11-0.15	0.006
	Vinyl	0.08	0.103
	Vinyl (without powder)	0.08	0.013
	Latex (without powder)	0.15	4.64
Reusable	Neoprene	0.7	1.63
	PVC	1.30	3.17
	Butyl	0.36	-

In order to understand whether previous results depend on the pore size or material of the glove, a close SEM image of the surface of some gloves was taken (***Figure 4***). Performance depends strongly on the material of the glove. Although, generally, there are no pores in their surface, some small defects or gaps can be enough to offer a way to penetrate inside the glove. Nevertheless, materials such as neoprene or butyl seem to be composed with cracks big enough in comparison with NM size. Thus, worst performance is expected with them.

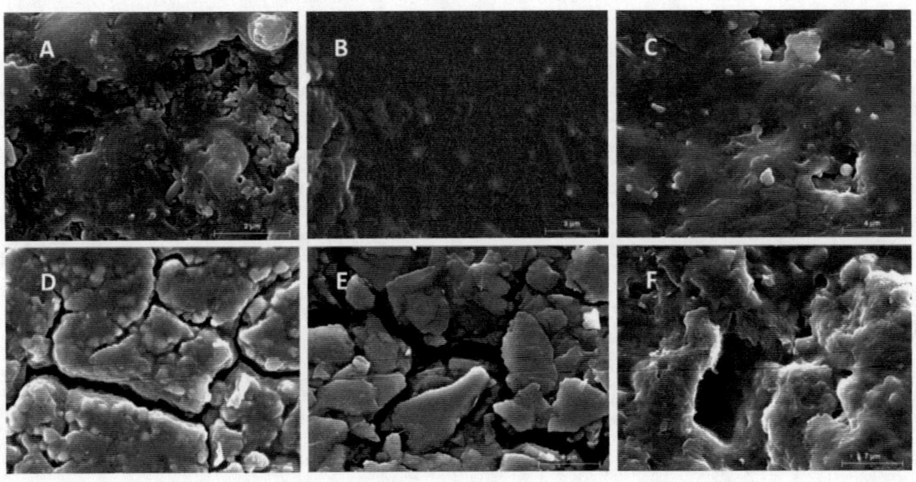

Figure 4: SEM images of the surface of protective gloves. A) Latex, B) PVC, C) Nitrile thick, D) Neoprene, E) Butyl, F) Nitrile thin.

The proposed assay assumes a high concentration of NPs in the environment close to the glove, thus it can be considered as a worst case scenario. However, the conditions of this specific test may be unrealistic, since in a real case scenario not only the NPs are important, but also the solvent or any other chemical manipulated along the live cycle of the NM. Although some of the gloves tested are specifically or recommended for liquid handling, the high penetration shown for airborne NPs brings to recommend to wear a disposable glove under the reusable one to avoid the dermal contact.

3.2. Personal protective suits

Several protective suits can be found in the market depending on the risk that we want to control. The standard UNE-EN 13982:2005 stablishes the requirements and characteristics that protective clothing must offer against solid particles. As happened with the gloves, this standard does not include the efficiency offered against NPs.

The efficiency of five distinct commercial protective clothing made of different materials was tested. The inward leakage was calculated as follows:

$$IL\% = \frac{C_1}{C_2} \cdot 100$$

Where C_2 represents the concentration of aerosol in the atmosphere and C_1 is the concentration of NPs found inside the suit (particles/cm^3). It was performed, at least, 3 replicates of each positioning. Then, the arithmetic mean of the total inward leakage was calculated for the whole exercise program and for all test subjects. This result is reported to classify the performance of the protective suit, given by the Nominal Protection Factor (NPF). The NPF is the concentration ratio of contaminant in the ambient atmosphere to the concentration of the contaminant inside the suit, giving an idea of how many times is protected the wearer (i.e. a NPF of 20 means that the clothing protects 20 times more versus wearing nothing). In Table 4 can be seen a summary of this classification, along with the assigned class for each PS.

Table 4: Classification of non-ventilated protective suits.

IL (%)	Cat.	NPF
20	I	2
2	II	50
0.2	III	500

As can be seen in Figure 5, best results were achieved in PS1 and PS2. In both PS, mean values of %IL sampling in the knee were below 5%. Similar results were also obtained sampling in the waist, but in PS1 the dispersion of the results was higher. However, results of %IL obtained sampling in the chest were higher and more disperse. Furthermore, in general terms, greater inward leakage was observed when subjects were asked to do squats, probably because this movement caused a compression or deformation in the suit, allowing the apparition of gaps between the suit and the skin by which the NPs could enter inside.

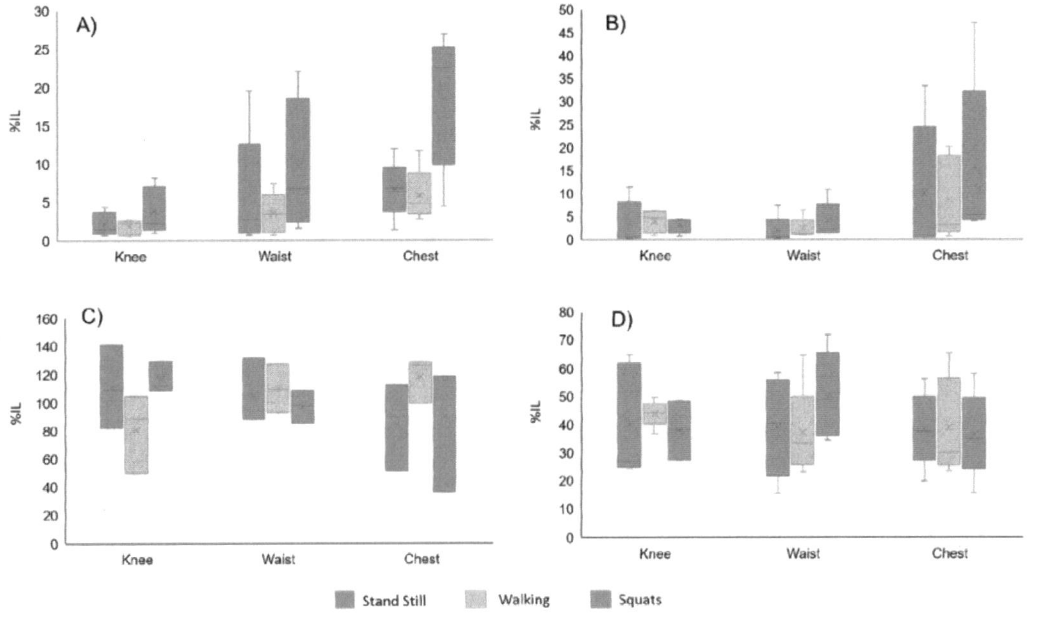

Figure 5: Values of % inward leakage obtained in the different protective suits assayed. A) Protective suit #1, B) Protective suit #2, C) Protective suit #3, D) Protective suit #4.

The PS3 offered the worst protection against NPs, due the mean values of %IL obtained in all the sampling points were practically about 100%. Moreover, the dispersion of the results achieved was high, obtaining great differences between each subject. These results might be explained because, although this coverall is classified as Cat III type 4, both the sleeves and the legs of this PS did not adjusted the body of the user, leaving spaces between the body and the suit by which the NPs could enter inside.

Results obtained for PS4 were similar in the three sampling points and for each action, with a mean value inward leakage of about 40%. This suit, although theoretically was the one which had to offer the worst protection due it was classified as Cat I type 6, improved the results offered by PS3 because, in this case, the sleeves and the legs were adjusted to the user, avoiding the penetration of the NPs by the gaps generated between the body and the suit. These results shown that the efficiency offered by the protective clothing not only depended on the type or material of suit, but also the fitting achieved with the body. Thus, the adjust between the protective clothing and the body of the user is a critical parameter to take into account at the time of choosing a PS to protect the worker against the exposition to NPs. The PS5 was not represented since the %IL was below 0.001% in every occasion, even 10 times lower, thus offering the highest protection of all.

In Figure 6 can be seen the variations of both concentrations recorded at the same time. It can be observed how the squats modified abruptly the air inside the suit, promoting the entrance of NPs by the gaps of the suit, although, immediately after stopping, the concentration decreased to lowest levels. Contrarily, concentration remained mainly constant when the subject was walking, but, during the pause for stabilization, there was an increase of NPs inside the suit due to the recovery of the shape while standing still.

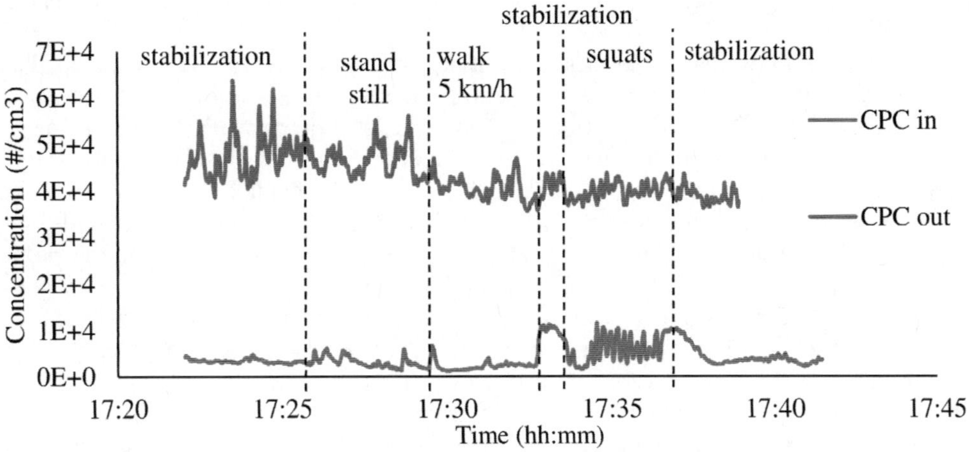

Figure 6: Concentrations in the test chamber (upper line) versus inside the suit during a series of exercises (lower line).

In order to confirm the importance of the fitting between the PS and the body, the protection offered by the fabric material was evaluated following the protocol used to study the material of the gloves. In this assay, only the efficiency of the fabric was evaluated, avoiding problems related to the formation of gaps by which the NPs could enter inside. As can be observed in Figure 7, the protection efficiency against NPs achieved by the fabrics of the PS1, PS2, PS3 and PS5 was higher than 99%. Moreover, as could be expected, fabric of the PS4 offered the lowest level of protection, although it was upper than the 95%. These results confirm that, although the fabrics offer a good level of protection against NPs, the formation of gaps between the body and the PS is a critical parameter which determines the global efficiency of the suit against NPs.

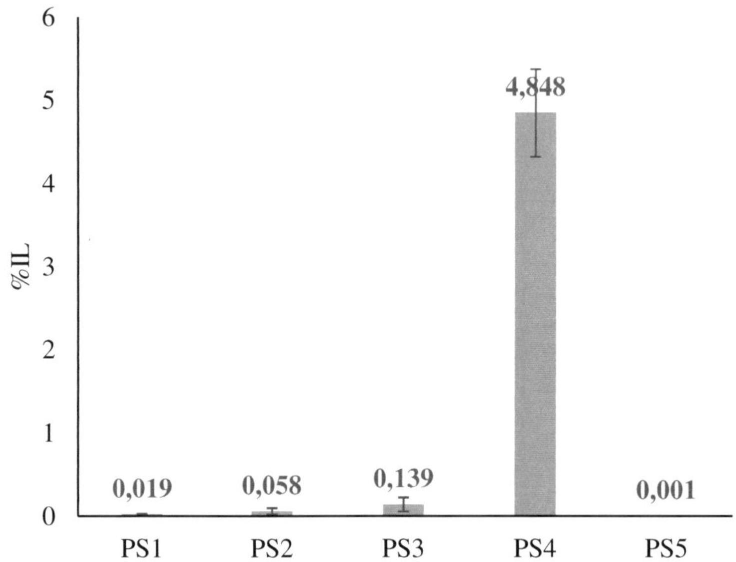

Figure 7: Results of inward leakage obtained in the different fabrics of the protective suits assayed.

4 Conclusions

In this work, the efficiency of different types of personal protective equipment against NPs has been tested and compared. Results obtained can contribute a very useful information to select the most appropriate PPE to protect workers in nanotechnology-based industries from the exposition to ENMs, offering a clear picture of the level of protection achieved by different types of personal protective clothing available in the market.

With the purpose of evaluating the efficiency of distinct types of protective clothing available in the market, different subjects were asked to do several exercises in order to simulate working conditions. Results showed that PS1 PS2 and PS5 offered the best level of protection against NPs, being upper than the 90% in most cases, close to 100% in the case of the PS5, since isolates efficiently from the exterior. The worst level of efficiency was registered by PS3 due the %IL obtained was about 100%. This result could be explained because PS3, although theoretically had to offer a good protection, the suit did not adjust the body of the subject, allowing the apparition of gaps between the suit and the skin by which the NPs might enter inside. PS4 offered a protection lower than PS1 and PS2, and higher than PS3.

On the other hand, the level of protection offered only taking into account the different fabrics was quantified. The level of efficiency obtained by the different fabrics was upper to 99% for PS1, PS2, PS3 and PS5. In any case, the lowest level of protection was still upper than 95%, achieved by the PS4. These results confirm that, although the fabrics reached a good efficiency against NPs, the fitting between the PS and the body of the user determined the global efficiency of the PS, and the small gaps between the body and the suit propitiate the leakage of NPs to the interior due to the movements of the wearer. Therefore, in order to improve the level of protection, protective suits should be sealed to the body of the user, avoiding the apparition of gaps by which the NPs can enter inside.

The efficiency of protective gloves was a function of the material, reaching values from 95% to 99% depending on the type of glove. For the same material, the ones with greater thickness achieved better results. Additionally, images about the surface of the gloves were taken by SEM, showing that some materials presented imperfections or gaps in their surface which could offer a way to the NPs to penetrate inside the glove. Thus, with the aim to achieve a better protection, the use of a double glove is recommended due it is difficult that both gloves have the imperfections in the same place.

Acknowledgements

This research was carried out as part of the European FP7 project GUIDEnano; Assessment and mitigation of nano-enabled product risks on human and environmental health: Development of new strategies and creation of a digital guidance tool for nanotech industries (FP7-NMP-2013-LARGE-7).

References

[1] Agency S C 2008 Nanotechnology - High risk with small particles?
[2] Thompson D, Leparoux M, Jaeggi C, Buha J, Pui D Y H and Wang J 2013 Aerosol emission monitoring in the production of silicon carbide nanoparticles by induction plasma synthesis *J. Nanoparticle Res.* **15**
[3] Aitken R J, Chaudhry M Q, Boxall A B A and Hull M 2006 Manufacture and use of nanomaterials: Current status in the UK and global trends *Occup. Med. (Chic. Ill).* **56** 300–6
[4] Yin J, Kang C, Li Y, Li Q, Zhang X and Li W 2014 Aerosol inhalation exposure study of respiratory toxicity induced by 20 nm anatase titanium dioxide nanoparticles *Toxicol. Res. (Camb).* **3** 367–74
[5] Jiang J, Oberdörster G, Elder A, Gelein R, Mercer P and Biswas P 2008 Does Nanoparticle Activity Depend upon Size and Crystal Phase? *Nanotoxicology* **2** 33–42
[6] Faccini M, Vaquero C and Amantia D 2012 Development of protective clothing against nanoparticle based on electrospun nanofibers *J. Nanomater.* **2012**
[7] Davies J C 2009 Oversight of next generation. Nanotechnology *Proj. Emerg. Nanotechnologies*
[8] Yeganeh B, Kull C M, Hull M S and Marr L C 2008 Characterization of airborne particles during production of carbonaceous nanomaterials *Environ. Sci. Technol.* **42** 4600–6
[9] Bergamaschi E 2009 Occupational exposure to nanomaterials: Present knowledge and future development *Nanotoxicology* **3** 194–201
[10] Evans D E, Turkevich L A, Roettgers C T, Deye G J and Baron P A 2013 Dustiness of fine and nanoscale powders *Ann. Occup. Hyg.* **57** 261–77
[11] Lee J H, Ahn K, Kim S M, Jeon K S, Lee J S and Yu I J 2012 Continuous 3-day exposure assessment of workplace manufacturing silver nanoparticles *J. Nanoparticle Res.* **14**
[12] Schulte P, Geraci C, Zumwalde R, Hoover M and Kuempel E 2008 Occupational risk management of engineered nanoparticles. *J. Occup. Environ. Hyg.* **5** 239–49
[13] Golanski L, Guiot A, Rouillon F, Pocachard J and Tardif F 2009 Experimental evaluation of personal protection devices against graphite nanoaerosols: fibrous filter media, masks, protective clothing, and gloves. *Hum. Exp. Toxicol.* **28** 353–9
[14] Demou E, Peter P and Hellweg S 2008 Exposure to manufactured nanostructured particles in an industrial pilot plant *Ann. Occup. Hyg.* **52** 695–706
[15] Yokel R a and Macphail R C 2011 Engineered nanomaterials: exposures, hazards, and risk prevention. *J. Occup. Med. Toxicol.* **6** 1–27
[16] Tsai C S J 2013 Potential inhalation exposure and containment efficiency when using hoods for handling nanoparticles *J. Nanoparticle Res.* **15**
[17] Larese F F, D'Agostin F, Crosera M, Adami G, Renzi N, Bovenzi M and Maina G 2009 Human skin penetration of silver nanoparticles through intact and damaged skin *Toxicology* **255** 33–7
[18] Ryman-Rasmussen J P, Riviere J E and Monteiro-Riviere N A 2006 Penetration of intact skin by quantum dots with diverse physicochemical properties *Toxicol. Sci.* **91** 159–65
[19] Shepard M N and Brenner S 2014 An occupational exposure assessment for engineered nanoparticles used in semiconductor fabrication *Ann. Occup. Hyg.* **58** 251–65
[20] Vinches L, Testori N, Dolez P, Perron G, Wilkinson K J and Hallé S 2013 Experimental evaluation of the penetration of TiO_2 nanoparticles through protective clothing and gloves under conditions simulating occupational use *Nanosci. Methods* **2** 1–15
[21] Park J, Kwak B K, Kim Y and Yi J 2011 Efficiency of protective dermal equipment against silver nanoparticles with water aerosol *J. Nanoparticle Res.* **13** 3043–9
[22] Dolez P, Vinches L, Perron G, Vu-Khanh T, Plamondon P, L'Espérance G, Wilkinson K, Cloutier Y, Dion C and Truchon G 2013 *Development of a Method of Measuring Nanoparticle*

Penetration through Protective Glove Materials under Conditions Simulating Workplace Use
[23] Cohen E 2012 *NTP 929: Ropa de protección contra productos químicos* vol 1
[24] INSHT 2015 *Seguridad y salud en el trabajo con nanomateriales*
[25] Gao P, Jaques P A, Hsiao T C, Shepherd A, Eimer B C, Yang M, Miller A, Gupta B and Shaffer R 2011 Evaluation of nano- and submicron particle penetration through ten nonwoven fabrics using a wind-driven approach *J Occup Env. Hyg* **8** 13–22
[26] Ben Salah M, Hallé S and Tuduri L 2016 Efficiency of five chemical protective clothing materials against nano and submicron aerosols when submitted to mechanical deformations *J. Occup. Environ. Hyg.* **13** 425–33
[27] Gómez E C 2003 NTP 748: Guantes de protección contra productos químicos
[28] Golanski L, Guiot A and Tardif F 2010 Experimental evaluation of individual protection devices against different types of nanoaerosols: Graphite, TiO2, and Pt *J. Nanoparticle Res.* **12** 83–9
[29] Dolez P, Vinches L, Wilkinson K, Plamondon P and Vu-Khanh T 2011 Development of a test method for protective gloves against nanoparticles in conditions simulating occupational use *J. Phys. Conf. Ser.* **304** 12066
[30] Liguori B, Hansen S F, Baun A and Jensen K A 2016 Control banding tools for occupational exposure assessment of nanomaterials – Ready for use in a regulatory context? *NanoImpact* **2** 1–17
[31] Brochocka A, Makowski K, Majchrzycka K and Grzybowski P 2013 Efficiency of filtering materials used in respiratory protective devices against nanoparticles *Int. J. Occup. Saf. Ergon.* **19** 285–95
[32] Brochot C, Michielsen N, Chazelet S and Thomas D 2012 Measurement of protection factor of respiratory protective devices toward nanoparticles *Ann. Occup. Hyg.* **56** 595–605

IOP Conf. Series: Journal of Physics: Conf. Series **838** (2017) 012022 doi:10.1088/1742-6596/838/1/012022

Effect of carbon black nanoparticles on methane/air explosions: Influence at low initial turbulence

David Torrado[1], Pierre-Alexandre Glaude[1,2] and Olivier Dufaud[1,2]

[1]Université de Lorraine, Laboratoire Réactions et Génie des Procédés, UMR 7274, Nancy, F-54001, France

[2]CNRS, Laboratoire Réactions et Génie des Procédés, UMR 7274, Nancy, F-54001, France

E-mail:olivier.dufaud@univ-lorraine.fr

Abstract. Nanoparticles are widely used in industrial applications as additives to modify materials properties such as resistance, surface, rheology or UV-radiation. As a consequence, the quantification and characterization of nanoparticles have become almost compulsory, including the understanding of the risks associated to their use. Since a few years ago, several studies of dust explosion properties involving nano-sized powder have been published. During the production and industrial use of nanoparticles, simultaneous presence of gas / vapor / solvents and dispersed nanoparticles mixtures might be obtained, increasing the risk of a hybrid mixture explosion. The aim of this work is to study the severity of the explosion of carbon black nanoparticles/methane mixtures and understand the influence of adding nanopowders on the behavior of the gas explosions. These results are also useful to understand the influence of soot on the efficiency of the gas combustion. Two grades of carbon black nanoparticles (ranging from 20 to 300 nm average diameter) have been mixed with methane. Tests have been performed on these mixtures in a standard 20 L explosion sphere. Regarding the scale precision, the lowest concentration of carbon black nanoparticles was set at 0.5 g.m-3. Tests were also performed at 2.5 g.m-3, which is still far below 60 g.m-3, the minimum explosive concentration of such powders previously determined in our laboratory. The influence of carbon black particles on the severity of the explosions has been compared to that of pure gas. It appears that the use of carbon black nanoparticles increases the explosion overpressure for lean methane mixtures at low initial turbulences by c. 10%. Similar results were obtained for high initial turbulent systems. Therefore, it seems that carbon black nanoparticles have an impact on the severity of the explosion even for quiescent systems, as opposed to systems involving micro-sized powders that require dispersion at high turbulence levels. Concerning the increment in the maximum rate of pressure rise, the addition of carbon black nanoparticles increased it by a factor of 1.15 in the case of lean fuel mixture. However, this behavior is only observed at high initial turbulence levels. The increment on the maximum rate of pressure rise is higher for powders with lower elementary particle diameter, which is notably due to the fragmentation phenomena that promotes the heat exchange.

1. Introduction

Nanoparticles have been progressively used in a wider variety of applications because of the different properties they have compared to those of bulk materials [1]. Nanoparticles are used to modify certain properties (particularly resistance, rheology, hardness, and magnetization) and, due to their greater surface area, they appear better suited for catalysis and biological applications [2]. The past few years have witnessed a transition in the production of certain products from traditional processes to nanoscale level manufacturing. Such materials are now widely used in cosmetics, paints, tires, catalysts and pharmaceuticals [3, 4]. The European Commission suggests that a nanomaterial should be defined as a material in which more than 50% of the particles in the number size distribution (in unbound state, in agglomerates or aggregates) have a size between 1nm and 100nm [5]. The rapid increase in nanoparticles applications have boosted the number of studies regarding the toxicity and explosion hazards of combustible nanoparticles dust [1, 3, 6]. However, there is still a limited amount of data and analysis on the explosion properties in this field [7]. Some authors focused their researches on characterizing the difference of the explosive properties between nano- and micro- dust. Dobashi [8], for instance, found that for coal, polyethylene and PVC, the explosion violence and the ignitability of powders are significantly increased when the particle size decreases until a certain size limit below which the explosion risk plateaus. The explosion severity and ignitability have also been studied for metallic dust of different sizes. Jiang et al. [9] measured the explosion severity in the standard 20L sphere for a mean aluminum particle diameter between 75 μm and 100 nm, obtaining more violent explosions for smaller particles. Similarly, Mittal [10] found that the explosion severity increases when the diameter of the particles decreases from 125 μm to 400 nm, but the violence of the explosion is lower for magnesium dust in the 30 - 200 nm range. However, the likelihood of an explosion increases significantly for the nano-range magnesium with respect to the minimum explosive energy (MIE) and the limiting oxygen concentration [10]. Similar results have been obtained for titanium, aluminum, carbon nanotubes and carbon black, powders whose ignitability increases as the particle diameter decreases and whose explosion severity increases until a plateau is reached [11-14]. In fact, this plateau and the similar explosive behavior of some micrometric and nanometric particles could be explained by the difficulty to obtain a homogeneous dispersion of elementary nanoparticles at high concentrations (greater than the minimum explosive concentration). This is mainly due to agglomeration: during the dust dispersion, the primary particles subjected to attractive forces will be rapidly transformed into larger structures [4, 15-17]. Ignition of a nano-dust cloud could be achieved before any agglomeration phenomena happens at high initial turbulence levels and at a short ignition delay time, conditions that are difficult to meet under normal industrial operations [15]. Even though nano-dust explosions have been studied in the past years, further and deeper analysis is required to understand the complex heat radiation and the behavior of the front flame. Dispersed nanometric dust may have an influence in complex solid-gas mixtures, simultaneous presence of gas/vapor/solvents and dispersed mixtures could be obtained during the production and industrial use of nanoparticles, increasing the risk of a hybrid mixture explosion. Moreover, such mixtures are systematically encountered during the combustion of organic dust, notably on the devolatilization phase, leading to a hybrid mixture of pyrolysis gases and char/unburnt powder [18].

In a previous study [19], the influence of low concentration of carbon black nanoparticles on gas mixture explosion has been analyzed. The flame front is modified from a semi-parabolic to a non-uniform shape, the latter caused by multiple perturbations occasioned by the burnt particles [19]. In the present study, the influence of the turbulence on the explosion severity and on the reaction products of carbon black/methane/air is analyzed. This work aims to document the influence of nanoparticles dispersion on hybrid gas/solid mixtures to understand the complex heat transfer, the changes on the laminar flame velocity and the possible effects of dispersed nano-carbon particles (such as soot nuclei) in the combustion chemical reaction.

2. Materials and Methods

In this study, Printex XE2 and Corax N550 (Orion) have been chosen as carbon black nanoparticles. The characteristic diameter of the powder d_{50}, the BET specific surface (using Brunauer-Emmett-Teller method) and the equivalent BET diameter are reported in Table 1. It appears that the primary nanoparticles are arranged in agglomerates of micrometric size. Moreover, Printex XE2 presents a much larger specific surface area than Corax N550 does, which may have an impact on their respective reactivity and physical properties.

Table 1. Main characteristics of the nanopowders [13]

Nanopowders	BET specific surface area (m^2g^{-1})	BET diameter (nm)	d_{50} (μm)
Printex XE2	950	3	10
Corax N550	40	75	15

The explosion severity of nanopowders/methane hybrid mixtures has been assessed by measuring the maximum overpressure P_{max} and the maximum rate of pressure rise $(dP/dt)_{max}$. The measurements of the explosion severity were performed in a 20-liter spherical vessel. In order to study the influence of an initial concentration of soot in a methane/air mixture, low concentrations of carbon black particles were chosen. With regard to the scale precision, the lowest concentration of carbon black particles was set at 0.5 $g.m^{-3}$. In addition, tests were also performed at 2.5 $g.m^{-3}$, which is far below 60 $g.m^{-3}$, the minimum explosive concentration of such powders [13]. Samples of burnt gases were taken without dilution and analyzed by micro gas-chromatography (Varian 490 MGC). Flame propagation test have been also performed but will not be detailed here.

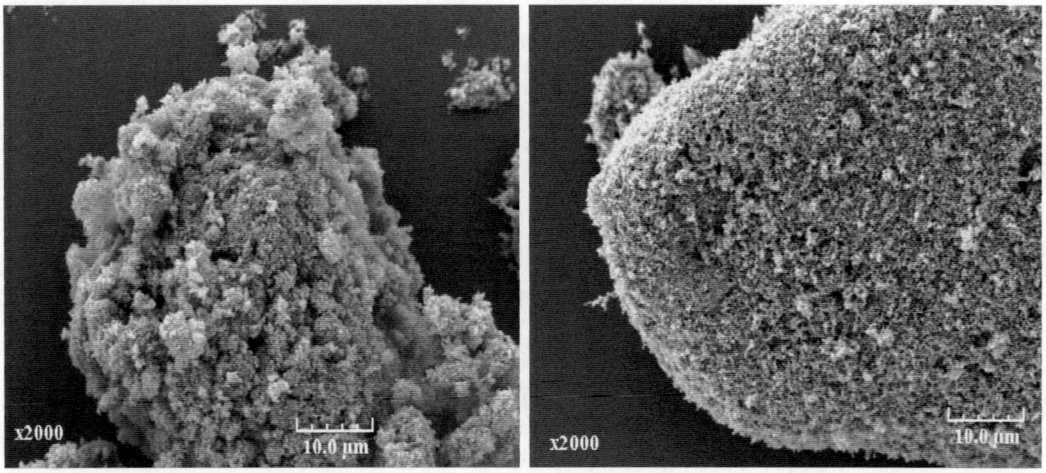

Figure 1. Scanning Electron Microscopy images of Printex XE2 (left) and Corax N550 (right)

The influence of the initial turbulence was characterized by varying the ignition delay tv (i.e. time between the beginning of the air pulse and the ignition at 100 J with chemical igniters). The severity of the explosion depends on the turbulence level, which is necessary for the dust dispersion especially for microparticles. The influence of the initial turbulence is studied by changing the delay time between the beginning of the dust dispersion and the ignition of the mixture. The intensity fluctuations estimated using the PIV method were 3.4 and 1.04 $m.s^{-1}$ at 60 and 120 ms, respectively. Due to the low settling velocity of nanoparticles, a quiescent case, which corresponds to the addition of dust without a pulse of air, was also considered.

3. Results

3.1. Dispersion Tests

Dispersion tests have been made for Printex XE2 and Corax N550 in order to analyze the stability of a dust cloud some seconds/minutes after the injection. To accomplish this, the mobility particle size was measured at different times after the powder injection by sequentially using a Differential Mobility Analyzer (Electrostatic Classifier – TSI 3080) and an Ultrafine Condensation Particle Counter (UCPC – TSI 3776). Twenty seconds after the dispersion, the estimated mean mobility diameters of the Printex XE2 and the Corax N550 were 245 nm and 346 nm, with a total concentration of $2.1x10^5$ and $2.4x10^5$ particles.cm^{-3}, respectively (Table 2). The DMA - UCPC results must be analyzed carefully because the particle size distribution analysis takes two minutes, time in which agglomeration and sedimentation phenomena occurs. However, the results obtained show that, even 3 minutes after the dispersion, the total concentration of dust is not negligible and the mean mobility diameter does not change with respect to the first measurement. It should be added that, due to the sampling process, a depression would be generated inside the reservoir, so a dilution with air is necessary to continue the DMA-UCPC tests which may explain the significant reduction of the total concentration of particles, in addition of the dust sedimentation phenomenon. These results suggest that the agglomeration phenomenon, even at high initial turbulent conditions, is limited because of the low concentration of dispersed dust. As consequence, it has be demonstrated that a stable dust cloud of agglomerates is generated and is still present at the ignition delay times of 60 and 120 milliseconds studied in this article.

Table 2. Carbon Black mean mobility diameter and total concentration after dispersion

Nanopowders	Time after dispersion (s)	Mean Mobility Diameter (nm)	Concentration (particles.cm^{-3})	Ambiance Concentration (particles.cm^{-3})
Printex XE2	20	245	$2.1x10^5$	$6.3x10^3$
	190	241	$1.5x10^5$	
Corax N550	20	346	$2.4x10^5$	
	180	350	$1.6x10^5$	

3.2. Explosion severity and burnt gases characterization

The maximum overpressure and the maximum rate of pressure rise for hybrid mixtures methane/air/Printex XE2 for different dust concentrations at quiescent conditions are shown in Figure 2. As pointed out in a previous work [19], in the case of Corax N550, the influence of the dispersion of nanoparticles agglomerates on methane/air gas explosions depends on the fuel equivalent ratio φ. For lean fuel mixtures ($\varphi < 1$), the severity of the explosion seems to increase when the concentration of Printex XE2 is augmented from 0.5 to 2.5 g.m^{-3}, but it drops for rich fuel mixtures. At low fuel concentration, dust dispersion may generate a deformation of the flame surface, causing an acceleration of the flame and an increase in the explosion severity. However, for fuel rich mixtures, heat radiation transfer appears to offset the effect of particles or agglomerates on the surface of the flame, attenuating the severity of the explosion. These results show that the risk of a gas explosion may be incremented when low concentrations of carbon black nanoparticles are present in the mixtures, even in the absence of high turbulence to guarantee the total dispersion of a nano-dust cloud.

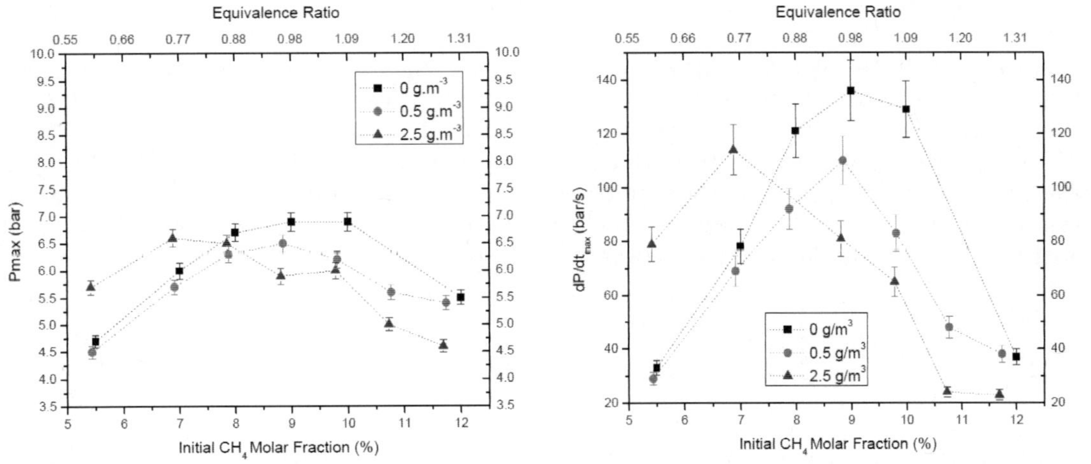

Figure 2. Influence of the carbon black concentration on the maximum overpressure - P_{max} (left) and on the maximum rate of pressure rise - dP/dt_{max} (right) for a Printex XE2/Methane/Air mixture under quiescent conditions.

For systems involving solid fuels, the initial turbulence level of the system is critical to generate an explosion and is determinant of the severity parameters. A similar behavior of the explosion severity has been observed at an initial velocity fluctuation of 1.05 $m.s^{-1}$ (see Figure 3). Nevertheless, a higher turbulence level seems to cause a higher increase in the maximal explosion overpressure for rich fuel mixtures at lower concentrations of carbon blacks. This result suggests that the initial turbulence level increases the powder deagglomeration, thereby amplifying its effects on the flame surface and on the heat radiation transfer. Even when the turbulence level was increased, no ignition of the mixture resulted at 12% v/v methane concentration due to a turbulence quenching generated by the turbulent cloud around the ignition point. The previous result was not observed for Corax N550/methane/air mixtures [19], probably because of the greater primary particle diameter.

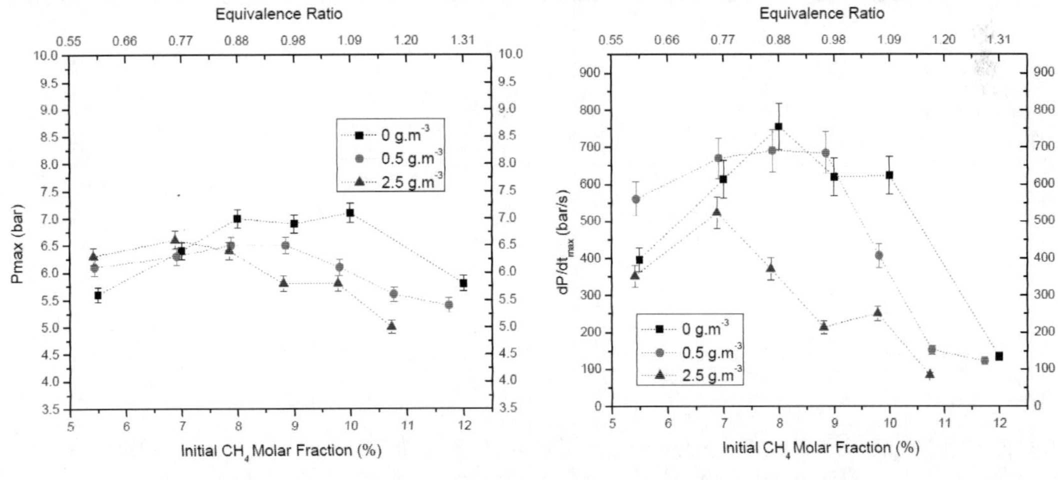

Figure 3. Influence of the carbon black concentration on the maximum overpressure – P_{max} (left) and on the maximum rate of pressure rise – dP/dt_{max} (right) for a Printex XE2/Methane/Air at an initial turbulence level u' = 1.04 $m.s^{-1}$.

Measurements of the volume fractions of the burnt gases were obtained using the micro gas-chromatography in order to analyze the effect of the carbon black and of the initial turbulence level on the chemistry of the combustion reaction. Figures 4 and 5 present the CO_2 volume fraction and the

CO/CO_2 ratio of the burnt gases for a gas mixture and hybrid mixture explosion, respectively. Under quiescent conditions, the maximum CO_2 volume fraction is obtained for stoichiometric conditions. The presence of an initial turbulence seems to increase the conversion of CO to CO_2 for lean mixtures but this conversion decreases for rich mixtures. Nevertheless, the turbulence level does not have an influence on the conversion of CO to CO_2 for the gas mixtures. For hybrid mixtures, the CO_2 volume fraction and CO/CO_2 ratio have similar gas explosions trends when the initial turbulence is augmented. However, the highest CO_2 volume fraction is obtained for the highest initial turbulence level as shown in Figure 5. The heterogeneity of the mixture increases the higher the turbulence level, generating a local equivalent ratio that may explain the difference of conversion of CO into CO_2 [19].

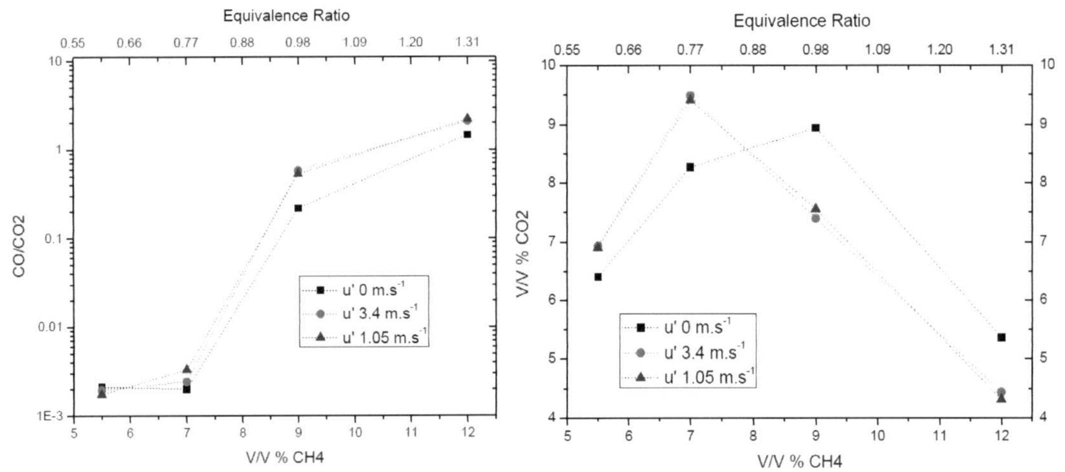

Figure 4. Influence of the turbulence level on the final CO/CO_2 ratio (left) and on the final CO_2 molar fraction (right) for a Methane/Air mixture.

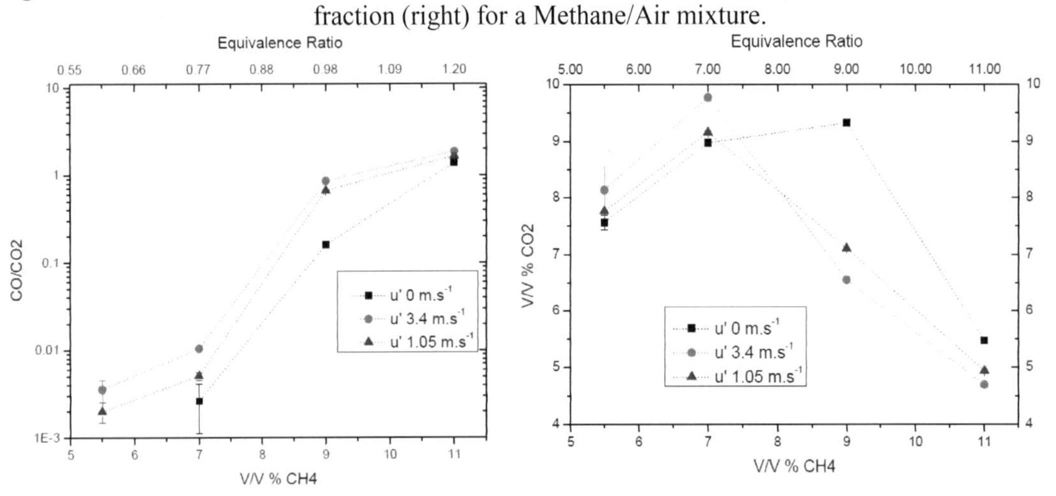

Figure 5. Influence of the turbulence level on the final CO/CO_2 ratio (left) and on the final CO_2 molar fraction (right) for a 0.5 $g.m^{-3}$ Corax N550/Methane/Air mixture.

Comparing the explosion severity for Printex XE2/Methane/Air mixture at different initial turbulence levels (Figures 2 and 3), it appears that the maximum rate of pressure rise is greatly increased at high turbulence levels, evidencing a close relationship between the turbulence level and the chemical combustion reaction. The influence of the turbulence level and combustion interactions on the explosion severity of this type of complex mixtures should be studied to design the appropriate prevention or protection devices. The micro gas-chromatography results suggest that there is a relation between the final molar CO_2 fraction and the maximum rate of pressure rise, in which the highest value of $(dP/dt)_{max}$ is obtained when the maximum conversion of CO to CO_2 is reached (Figure 6). Therefore, the conditions that promote the chemical reaction of oxidation of CO, as a higher initial turbulence level, will generate a higher release of chemical energy and, consequently, an increase on the maximum rate of pressure rise. Similar results were obtained regarding the influence of the carbon black dispersion on the maximum rate of pressure rise (Figure 2 - right), where the higher explosion severity was obtained when the maximum concentration of CO_2 was produced (Figure 7).

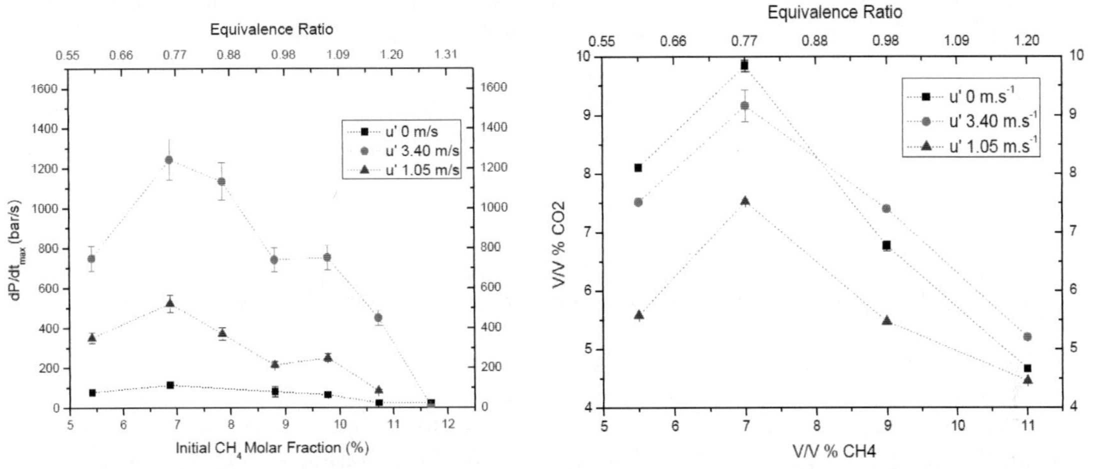

Figure 6. Influence of the turbulence level on the maximum rate of pressure rise - dP/dt_{max} (left) and on the final CO_2 molar fraction (right) for a 2.5 $g.m^{-3}$ Printex XE2/Methane/Air mixture.

The presence of a carbon black cloud affects the combustion reaction, even if the amount of solid fuel is negligible with respect to the methane volume fraction. Figure 7 shows the influence of the Printex XE2 initial concentration on the CO_2 and CO molar fraction for hybrid mixtures explosions under quiescent conditions. The explosion severity seems to increase due to the presence of dispersed agglomerates, phenomena that appears to promote the conversion of CO_2 for lean fuel mixtures. In addition, this conversion rate decreases for a fuel equivalent ratio higher than 0.8. The CO production seems to increase for lean mixtures when 2.5 $g.m^{-3}$ of Printex is added to the system, outcome that is not observed when using Corax N550 dust. Firstly, this result shows again that both powders have high stability when dispersed and influence the methane/air gas explosion. However, such impact on the explosion becomes more important for powders with a higher specific surface. The stable elementary particles and dispersed agglomerates seem to affect the flame surface (observed thanks to flame propagation tests), and, because they undergo oxidation reactions, the concentration of the final gases and the violence of the reaction vary. Secondly, the negative effect on the explosion severity for rich fuel equivalent ratio may be caused by the improvement of the soot nucleation phenomenon due to primary nuclei when initial carbon particles are added into the combustible mixture.

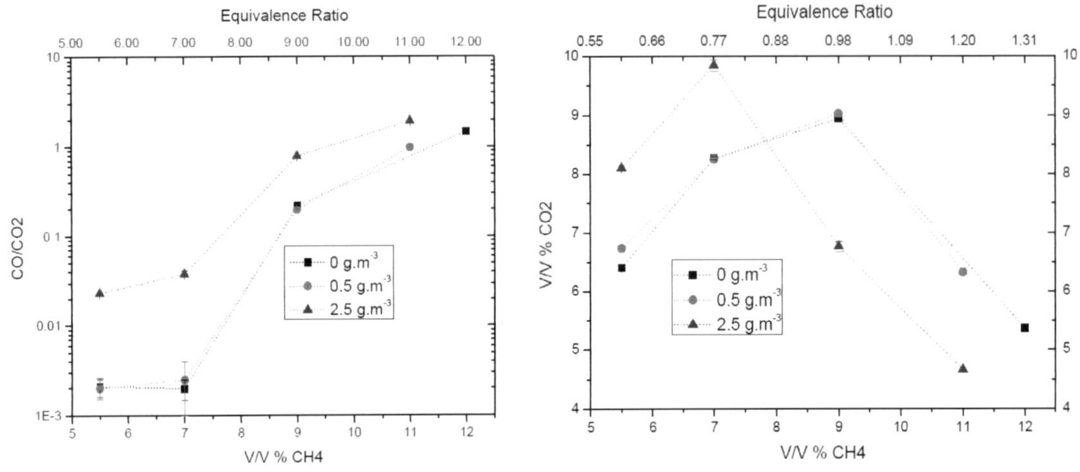

Figure 7. Influence of carbon black level on the final CO/CO_2 ratio (left) and on the final CO_2 molar fraction (right) for Printex XE2/Methane/Air mixture under quiescent conditions.

3.3. Flame propagation velocity from 20L sphere tests

The burning velocities of carbon black nano-particles and methane/air hybrid mixtures were estimated using the thin-flame model postulated by Dahoe [20] using the explosion severity results. The model assumes that the flame is an infinitely thin flame that separates the inner burnt region and the outer unburnt gas. The rate of pressure rise and the maximal overpressure are related to the burning velocity as:

$$\frac{dP}{dt} = \frac{3(P_e - P_0)}{R_{vessel}}\left[1 - \left(\frac{P_0}{P}\right)^{\frac{1}{\gamma}}\frac{P_e - P}{P_e - P_0}\right]^{\frac{2}{3}}\left(\frac{P}{P_0}\right)^{\frac{1}{\gamma}} S_u \tag{1}$$

where P_e is the maximal overpressure, P_0 is the initial vessel pressure and S_u is the burning velocity.

The estimation of the burning velocities for the hybrid mixtures Printex XE2/methane/air and Corax N550/methane/air at an initial turbulence level of u' = 1.05 $m.s^{-1}$ are presented on Figure 8. An estimation of the burning velocity is possible by equation 1. However, it should be underlined that this relationship may underestimates the burning velocity, because the flame thickness is often not negligible at high pressure and for systems involving dispersed dust clouds. Likewise, the equation 1 suppose a perfect spherical propagation of the front flame, which may not be present because of the interactions of the flame with the internal elements of the sphere, and also, because of inertial effects created by the dispersion. Shey et al. [21] reported a flame velocity of methane/air at stoichiometric concentrations of approximately 240 $cm.s^{-1}$ at an initial turbulence of velocity fluctuation u'= 1 $m.s^{-1}$, which is higher than our estimated value of 174 $cm.s^{-1}$. The addition of 0.5 $g.m^{-3}$ of Printex XE2 generates an increase of 20% on the burning velocity for fuel lean mixtures compared to gas mixtures, while a decrease is present for fuel rich mixtures. In addition, the burning velocity is considerably reduced when the carbon black concentration is augmented to 2.5 $g.m^{-3}$, even if the maximal overpressure seems to be improved with the presence of a Printex XE2 dust cloud (Figure 3). Similarly, the presence of a dust cloud of Corax N550 generates a decrease of the burning velocity with respect to gas explosions. The previous behaviour could be explained by two possible phenomena. On the one hand, the increment on the heat radiation transfer caused by the dispersed dust may generate heat losses on the propagating front and will be enhanced at a higher turbulence level. On the other hand, the cloud dust may generate changes on the flame surface, generating local flamelet regimes and reducing the overall flame velocity, which has been observed on the flame propagation effects. Ameliorations on the experimental flame velocity at different turbulence levels must be implemented in order to avoid underestimation of the protection devices at lean fuel mixtures and overestimation at rich fuel mixtures when low concentrations of dust may be present in the system.

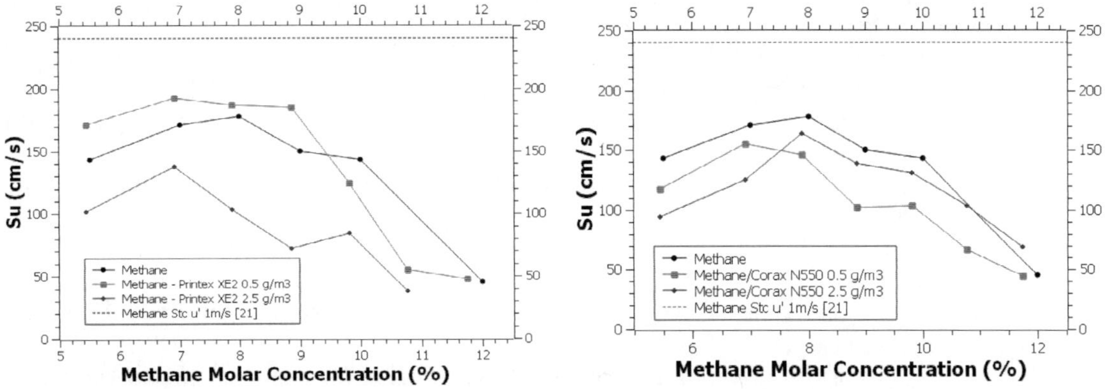

Figure 8. Burning velocity calculated by the thin flame model for Methane/ Printex XE2 /Air (left) and Methane/Corax N550/Air (right) for at an initial turbulence level u' = 1.04 m.s^{-1}.

4. Conclusions

This work studies the influence of nanoparticles on the explosion severity of gas explosion. The results obtained show that, even when adding carbon blacks at low concentration (well below the minimum explosive concentration), the heat transfer phenomena and the combustion reaction can be modified. The explosion severity tests and the burnt gas characterization suggest that:

a) The cloud of nanopowders remain highly stable some minutes after the dispersion, with a constant mean mobility diameter. The dust cloud formed, notably composed of agglomerates, has a mean diameter of 240 nm for Printex XE2 and 370 nm for Corax N550.
b) The explosion of nanoparticles can be hardly obtained under quiescent conditions, despite the high stability of the dust. Nevertheless, the explosion severity of methane/air mixtures increases when low concentrations of carbon black nanoparticles are present in the system, even at quiescent conditions. In addition, at high turbulence level, the presence of carbon black nanoparticles generates a negative effect on the ignition sensivity for fuel rich mixtures.
c) The addition of carbon black particles for fuel lean mixtures seems to promote the oxidation of CO, which will generate a higher release of chemical energy and, consequently, an increase in the maximum rate of pressure rise. The presence of carbon blacks (or as a consequence, soot) has effects on gas explosion severity which can vary as a function of the turbulence level. Such influence can lead to explosion quenching and has to be further studied in order to design appropriate prevention or protection devices as well as propose inherent safety measures.
d) The specific surface area of the dispersed particles have a great influence on the explosion severity of carbon black/ methane/ air mixtures. At high turbulence levels, the deagglomeration effects on heat transfer and flame surface are amplified for powders with higher specific surface.

References

[1] Pritchard DK. Literature review: explosion hazards associated with nanopowders. Health and Safety Laboratory; 2004
[2] Stark WJ, Stoessel PR, Wohlleben W, Hafner A. Industrial applications of nanoparticles. Chem

Soc Rev. 2015;44(16):5793–805.
[3] Bouillard J, Vignes A, Dufaud O, Perrin L, Thomas D. Explosion risks from nanomaterials. J Phys Conf Ser. 2009 May 1;170:12032.
[4] Eckhoff RK. Are enhanced dust explosion hazards to be foreseen in production, processing and handling of powders consisting of nano-size particles? J Phys Conf Ser. 2011 Jul 6;304:12075.
[5] European Comission. Commission recommendation on the definition of nanomaterial. Off J Eur Union. 2011 Oct 18.
[6] Oberdörster G, Oberdörster E, Oberdörster J. Nanotoxicology: An Emerging Discipline Evolving from Studies of Ultrafine Particles. Environ Health Perspect. 2005 Mar 22;113(7):823–39
[7] Amyotte PR. Some myths and realities about dust explosions. Process Saf Environ Prot. 2014 Jul;92(4):292–9.
[8] Dobashi R. Risk of dust explosions of combustible nanomaterials. J Phys Conf Ser. 2009 May 1;170:12029.
[9] Jiang B, Lin B, Shi S, Zhu C, Li W. Explosive characteristics of nanometer and micrometer aluminum-powder. Min Sci Technol China. 2011 Sep;21(5):661–6.
[10] Mittal M. Explosion characteristics of micron- and nano-size magnesium powders. J Loss Prev Process Ind. 2014 Jan;27:55–64.
[11] Boilard SP, Amyotte PR, Khan FI, Dastidar AG, Eckhoff RK. Explosibility of micron- and nano-size titanium powders. J Loss Prev Process Ind. 2013 Nov;26(6):1646–54.
[12] Vignes A, Dufaud O, Perrin L, Thomas D, Bouillard J, Janès A, et al. Thermal ignition and self-heating of carbon nanotubes: From thermokinetic study to process safety. Chem Eng Sci. 2009 Oct;64(20):4210–21.
[13] Bouillard J, Vignes A, Dufaud O, Perrin L, Thomas D. Ignition and explosion risks of nanopowders. J Hazard Mater. 2010 Sep;181(1–3):873–80.
[14] Dufaud O, Vignes A, Henry F, Perrin L, Bouillard J. Ignition and explosion of nanopowders: something new under the dust. J Phys Conf Ser. 2011 Jul 6;304:12076
[15] Eckhoff RK. Does the dust explosion risk increase when moving from μm-particle powders to powders of nm-particles? J Loss Prev Process Ind. 2012 May;25(3):448–59.
[16] Eckhoff RK. Influence of dispersibility and coagulation on the dust explosion risk presented by powders consisting of nm-particles. Powder Technol. 2013 May;239:223–30.
[17] Worsfold SM, Amyotte PR, Khan FI, Dastidar AG, Eckhoff RK. Review of the Explosibility of Nontraditional Dusts. Ind Eng Chem Res. 2012 Jun 6;51(22):7651–5.
[18] Cuervo N. Influences of turbulence and combustion regimes on explosions of gas- dust hybrid mixture. Université de Lorraine; 2015.
[19] Torrado D, Cuervo N, Pacault S, Dufour A, Glaude P-A, Murillo C, et al. Explosions of gas/carbon black nanoparticles mixtures: An approach to assess the role of soot formation. Chem Eng Trans. 2016;48:379–84
[20] Dahoe AE, de Goey LPH. On the determination of the laminar burning velocity from closed vessel gas explosions. J Loss Prev Process Ind. 2003 Nov;16(6):457–78.
[21] Shy SS, Lin WJ, Wei JC. An experimental correlation of turbulent burning velocities for premixed turbulent methane-air combustion. Proc R Soc Math Phys Eng Sci. 2000 Aug 8;456(2000):1997–2019.

IOP Conf. Series: Journal of Physics: Conf. Series **838** (2017) 012023 doi:10.1088/1742-6596/838/1/012023

INCINERATION OF A COMMERCIAL COATING WITH NANO CEO2

Olivier Le Bihan[1], Ghania Ounoughene[2], Laurent Meunier[1], Bruno Debray[1], Olivier Aguerre-Chariol[1]

[1] INERIS, France

[2] KU Leuven, Department of Chemical Engineering, Leuven, Belgium

olivier.le-bihan@ineris.fr

Abstract. The potential environmental risk arising from the incineration of waste containing nanomaterials is a new field which deserves further attention. Some recent studies have begun to focus on this topic but the data are incomplete. In addition, there is a need to consider real life waste.

The present study gives some insight into the fate and behavior of a commercial coating containing a commercial additive (7% w/w) based on nano-CeO_2 (aggregates of 10 to 40 nm, with elemental particles of 2-3 nm). The tests have been conducted with a system developed in the frame of the NanoFlueGas project. The test protocol was designed to respect the regulatory criteria of a good combustion in incineration plants (temperature around 850°C, highly ventilated combustion, at least 2 s residence time for the combustion gas in a post-combustion chamber at 850°C, and high oxygen/fuel contact).

Time tracking by electric low pressure impaction (ELPI) shows that the incineration produces aerosol with number concentration dominated by sub-100 nm particles. Cerium is observed by TEM and EDS analysis but as a minor compound of a sub-group of particles. No nanoCeO2 particles have been observed in the aerosol.
ICP-MS analysis indicates that the residual material consists mainly of CeO_2 (60% of the mass). Observation by TEM establishes that this material is in the form of aggregates with individual particle of 40-200 nm and suggests that sintering occurred during incineration.

As a conclusion, the lab scale incineration study led mainly to the release of nano-CeO_2 in the residual material, as the major component. Its size distribution is different than the one of the nano-CeO2 observed in the initial sample before incineration. Additional research is needed to improve the understanding of nanoCeO2 behavior, and to integrate experiments at lab and real scale.

IOP Conf. Series: Journal of Physics: Conf. Series **838** (2017) 012023 doi:10.1088/1742-6596/838/1/012023

1. Introduction

Nanomaterials (NMs) are more and more widely studied and produced. Part of these trendy materials are expected to end up in incineration waste plants due to the lack of specific recovery procedures. It seems necessary to address the issue of the fate of nano-objects during the incineration: are they destroyed, do they undergo changes in the incinerator furnace, and where are they released? To answer these questions, more and more studies investigate the behavior and the fate of nano-objects during thermal disposal at lab-scale [3, 4, 5, 6, 2, 7, 8], at pilot scale [9] and at real scale [1, 9, 10]. However, regarding the very large number of nano-objects and products, the data are far to be complete and more studies are needed to fully determine the risks related to the incineration of waste containing NMs.
The case study of nano-CeO2 appears interesting since it is the only nano-object for which the incineration has been studied at the intermediate scale [10] and real scale [2, 9, 10]. Both studies conclude that incineration plants with state-of-the art gas cleaning systems can treat efficiently nanostructured metal oxides such as Ceria. These results are also consistent with the results of the NanoFlueGas project [2].

In that context, the aim of the present study is on the one hand to enlarge the data base by considering a product available in the market, and on the other hand to support the comparison between the lab scale and the larger scales (pilot and real scale) by testing a product containing nanoCeO2.
This case study investigates the behavior and the fate of nano-CeO2 during incineration of a commercial coating containing nanoCeO2 and gives insights on potential exposure and environmental risk related to the incineration of waste containing nano-CeO2.
Incineration tests have been performed at lab-scale using a specific tubular furnace, with the purpose to characterize the combustion residues and the combustion aerosol. The results are discussed regarding the studies at intermediate and real scale.

2. Materials and methods

In this study, imagery on the sample (prior to the incineration tests), the aerosol and the residueswas carried out using transmission electronic microscope (Philips CM12 TEM 120 kV, and JEOL JEM 1400 Plus), and elemental analysis was performed using Energy Dispersive Spectroscopy (EDS).
A commercial coating (colours lasure 4ans 1l incolor code 101805 Groupe V33, Castorama – wood coating), has been enriched with 7% w/w of a commercial nanostructured additive Nanobyk 3810, which is a “Nanoparticle dispersion (cerium oxide) for long-term UV protection in aqueous system”. It has been applied on a clean and hard surface. Later the dry residue of that coating has been sampled. Incineration tests have been performed on 250 mg samples.

Laboratory scale incineration tests were performed using a specific tubular furnace fully described in a previous study [3]. The combustion conditions (within both the combustion and post-combustion zones) were controlled to satisfy the key operational parameters that govern an incineration process, *i.e.*: temperature (850 °C in the combustion and post combustion-zone), residence time (at least two seconds in the post combustion zone at 850 °C), air-excess (never below 11 % of oxygen) and turbulence (a good mix between combustible and oxygen). As illustrated in Figure 1 the “combustion chamber” is maintained at 850 °C (at least) and air cross flows penetrate and mix with the combustibles. Then, the “post-combustion chamber” is the furnace zone where the combustion aerosol stays for 2 s at 850 °C (at least). Finally, the combustion released gases and the aerosol particles were conveyed to the measurement line for on-line (an Electrical Low Pressure Impactor, Dekati (ELPI) was used downstream of a Fine Particle Sampler dilutor (FPS, Dekati)) and off-line analyses (particles from the combustion aerosol were collected on a TEM grid with a MPS ® (Mini-Particle-Sampler, Ecomesure) [11]).

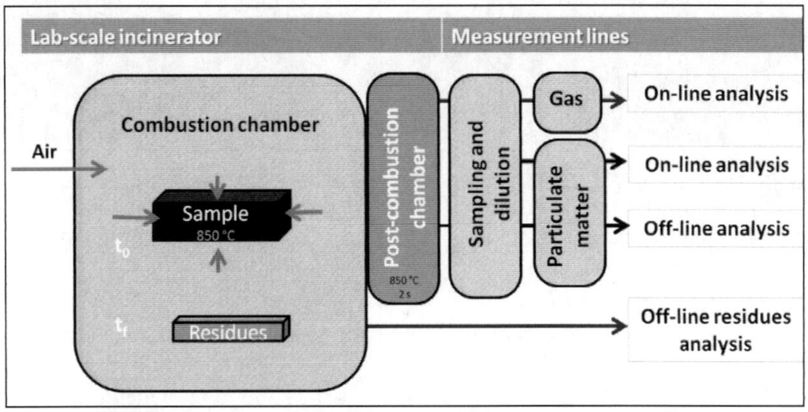

Figure 1. Lab scale incinerator and its measurement lines

A multi-gas analyzer (Portable Gas Analyzer PG-250 Horiba) was used to measure the consumption of O_2 ($\%_{vol.}$) and the production of CO_2 ($\%_{vol.}$) and CO ($ppm_{vol.}$).
The residues have been crushed, dispersed in 2 mL ultrapure water (ultra-waves). 2 µl have been deposited on a TEM grid, and dryed.

3. Results and discussions

3.1. Characterization of the sample
The sample was characterized prior to the incineration tests, by TEM imagery coupled with elemental analysis (EDS). The nanostructure of the CeO2 was observed. It is composed of aggregates of 10 to 40 nm, with elemental particles of 2-3 nm (Figure 2, 3). In addition, elemental analysis was performed and showed that the sample contains carbon, oxygen, cerium, sodium and sulfur (Figure 4).

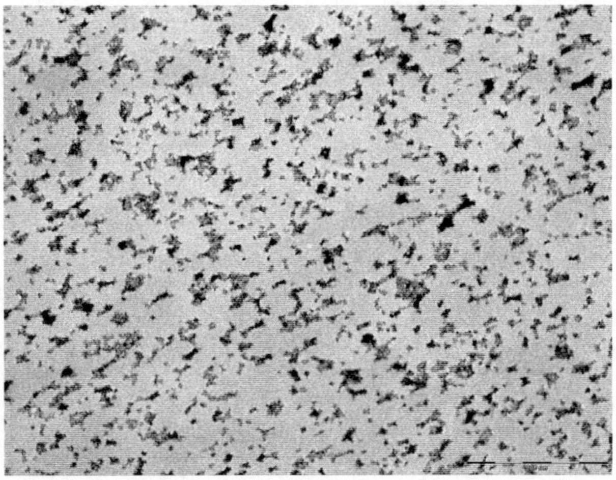

Figure 2. TEM image of the dry coating – scale 200 nm

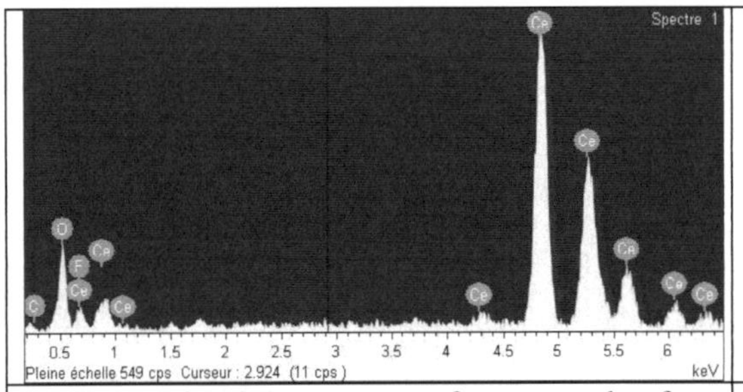

Figure 3. EDS spectrum of an example of individual particle

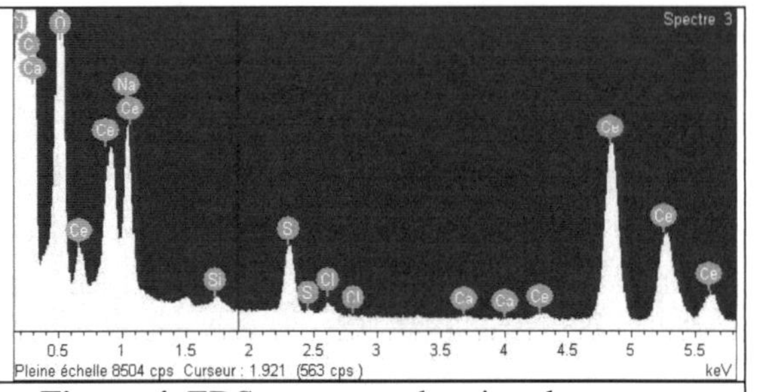

Figure 4. EDS spectrum showing the average composition of the dry coating

3.2. Time Tracking of gas and aerosol

The graphs describing the concentrations evolution during the incineration test are given in Figure 5. The two verticals delimit the time range with more than 5 % O_2 consumption from the baseline. This was named Area Of Interest "AOI" and it is considered as the time range when the combustion conditions are the closest to incineration conditions. The averages of three runs are presented with the associated standard deviation.

Ignition was observed at the beginning of the test during a period of 19 s; this could be related to the release and the combustion of volatile organic compounds (VOCs) commonly used in painting formulation.

Before the AOI, from 0 to 29 s, O_2 consumption is low, CO_2 and CO are produced, and particles are emitted at a high emission rate with a domination of PN0.1-1 (in the range of 0,1 to 1 µm) in terms of particle number. After this initial peak dominated by 0,1 – 1 µm, particles emission of particles smaller than 0,1 µm increases and becomes predominant in number concentration.

During the AOI, from 29 s to 75 s, oxygen consumption is maximal. PN0.1 becomes the most important contributor to the total number concentration.

The peak of emission of CO_2 and O_2 are not synchronized while it would be expected, as observed for instance in a previous study by Ounoughene et al. [3]. As a result, additional experiments will be needed to study further this point.

After the AOI, from 75 s to the end, O_2 reaches its baseline and gases are less and less emitted. Until the end of the test high particle concentration levels are observed. This phenomenon needs also to be studied further.

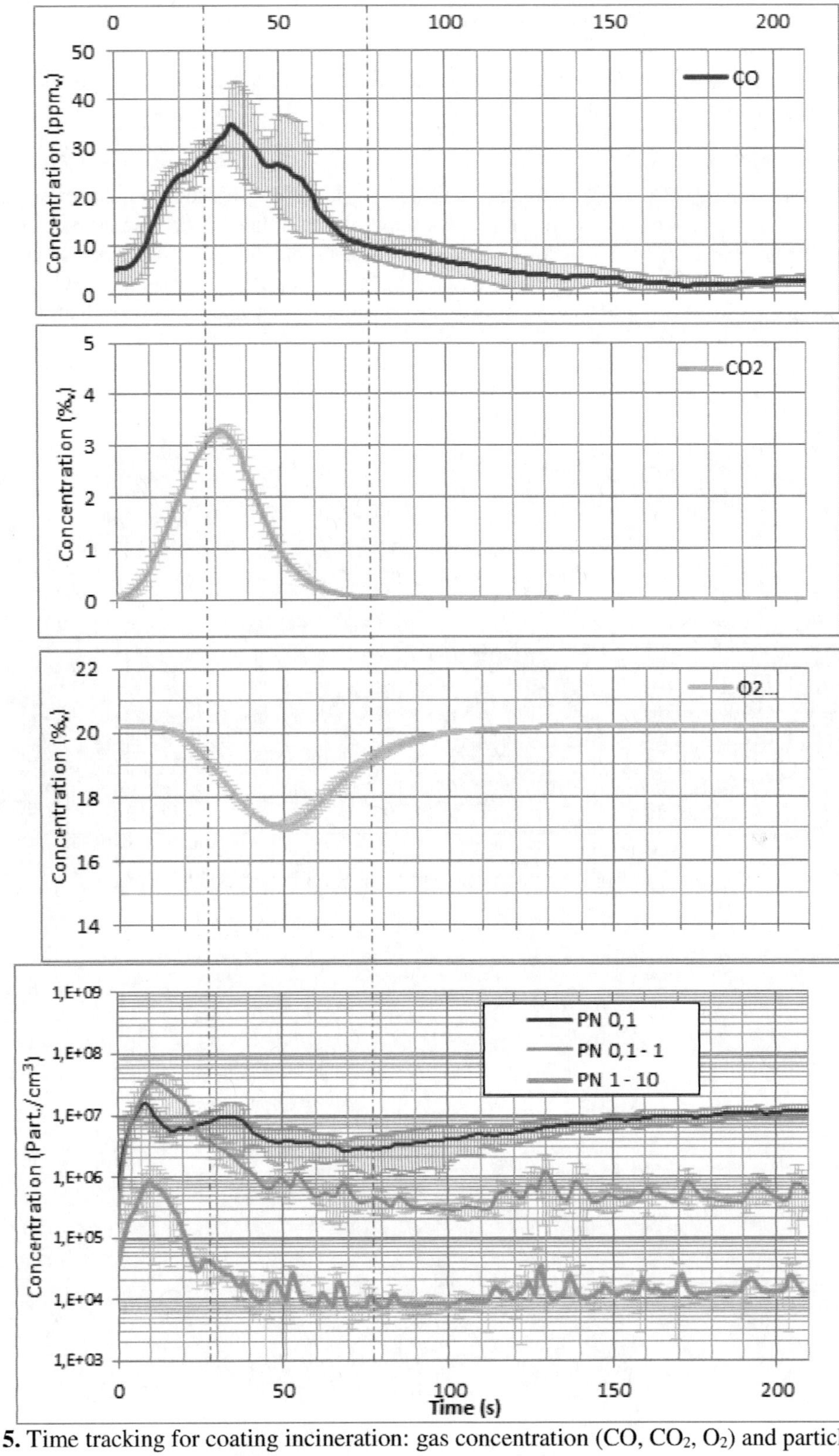

Figure 5. Time tracking for coating incineration: gas concentration (CO, CO_2, O_2) and particle number concentration. Average of three experiments.

3.3. Fate of nano-objects from the incineration of the nanowaste

The results from imagery provide qualitative information on the composition of the aerosol emitted during combustion and of the residues.

3.3.1. Analysis of the aerosol

Figure 6 shows examples of particles collected on a TEM grid during incineration. Different shapes and amount of particles have been collected (e.g soot aggregates, spherical or acicular particles, etc.), with some changes versus time. Some traces of cerium have been detected in some spherical particles mainly composed of oxygen, silica and sodium (ex. Figure 7 – a and b). Very rarely, CeO_2 particles are visible inside of some of these spherical particles (cf. Figure 7 – c and d). No free nanoCeO_2 particles have been observed.

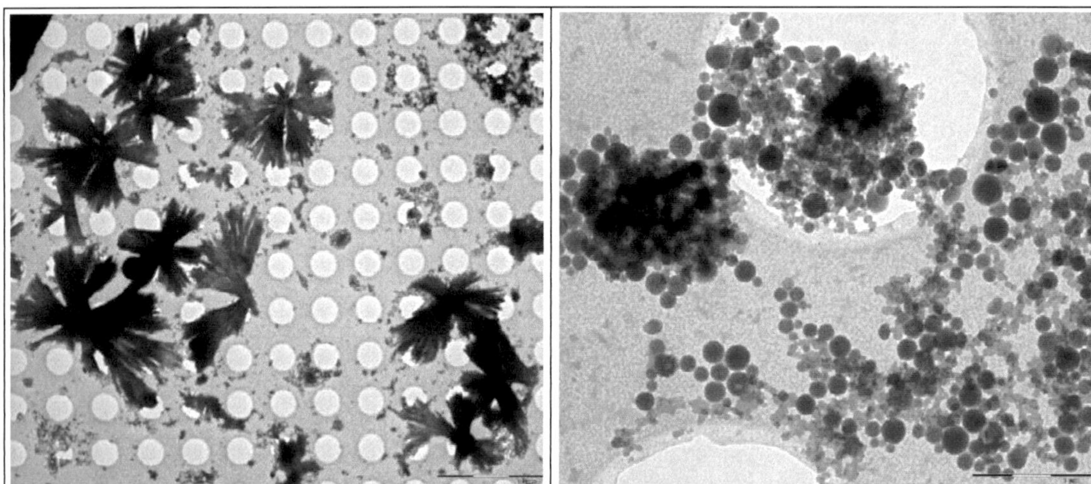

scale 10 µm	scale 1 µm

Figure 6. examples of combustion aerosol collected on TEM grid

IOP Conf. Series: Journal of Physics: Conf. Series **838** (2017) 012023 doi:10.1088/1742-6596/838/1/012023

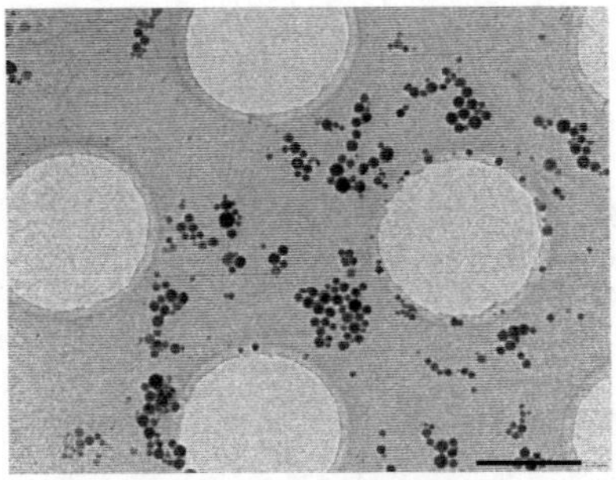

a. Scale 1 µm

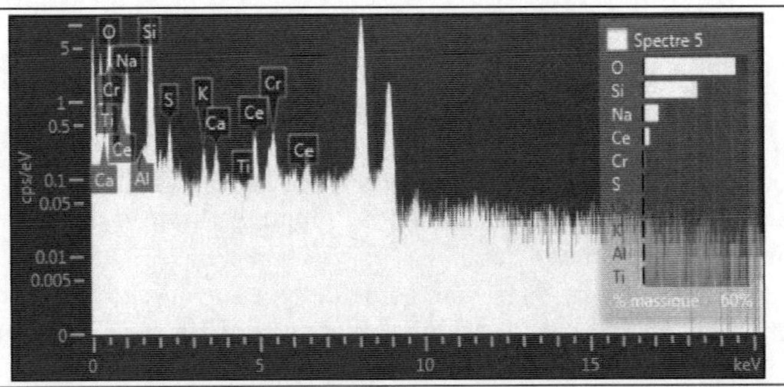

b. EDS spectrum – Log scale

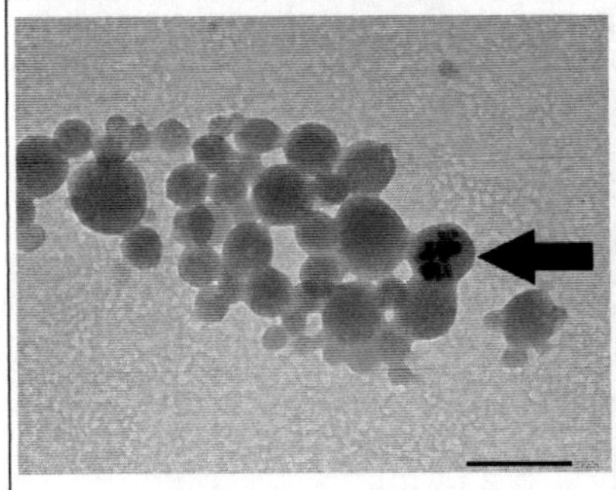

c. Scale 200 nm

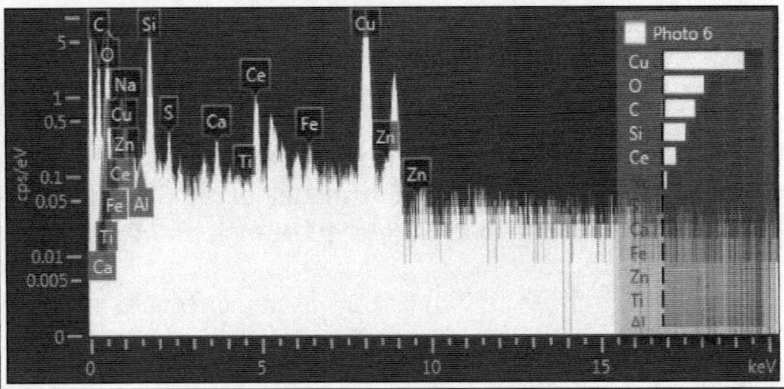

d. EDS spectrum – Log scale

Figure 7. spherical particles containing some traces of cerium

3.3.2. Analysis of the residues.

The residues represent about 6% of the sample mass before incineration. According to ICP-MS analysis, the residual material consists mainly of CeO_2 (60% of the mass).

TEM (coupled with EDS) imagery performed on combustion residues reveals that the residues consist mainly of aggregates with elemental nanoparticles of 40-200 nm (Figure 8).

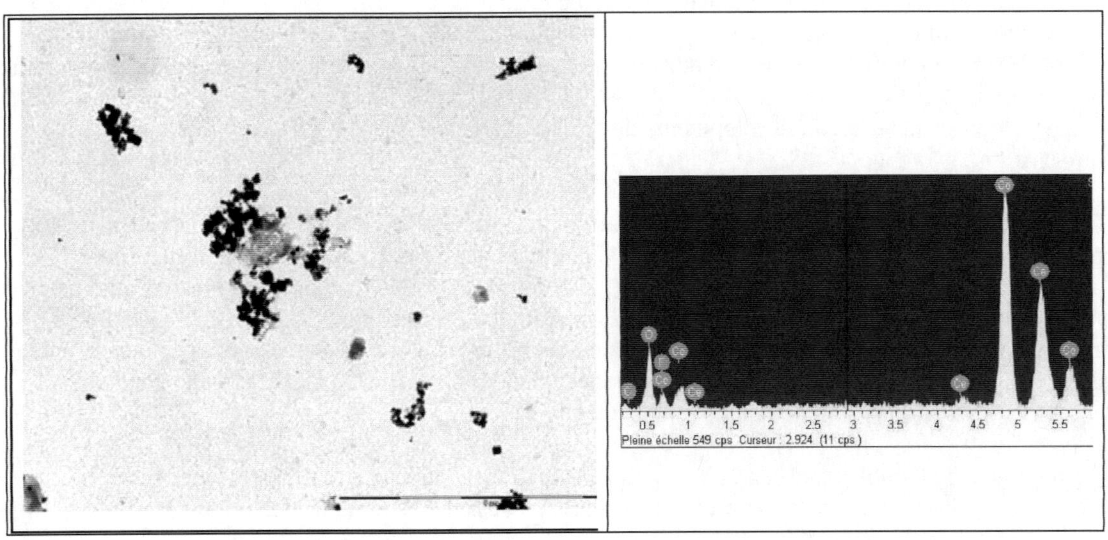

Figure 8. combustion residues analysed by TEM (scale 5 µm) and EDS

IOP Conf. Series: Journal of Physics: Conf. Series **838** (2017) 012023 doi:10.1088/1742-6596/838/1/012023

4. Discussion

4.1 Fate of nano-CeO2 (qualitative)
In the current study, a significant mass of cerium has been observed to remain in the residues, which is consistent with Walser al. 2012 who found the majority of the recovered cerium in the solid combustion residues.
The contribution of Paur el al. [9, 10] is not considered here for comparison because the ceria nanoparticles has been introduced downstream, in the post-combustion chamber.

4.2 Evolution of the size distribution of the CeO2 nanoparticles
In the frame of Paur et al. study [10], in the case of the combustion of coal, "Ceria nano-particles were mainly found in the particles size range of 1 – 10 µm", "due to the fast agglomeration with fly ash". In the case of gas as a fuel "less agglomeration was observed and the Ceria NP were in the particles size range below 1 µm." They also observed in a premixed propane flame that the nano-CeO2 of 65 nm partly evaporates forming a new particle mode at 15 nm.
These results show that the size distribution of the CeO2 particles can change very significantly, depending on various factors as the type of fuel, the temperature of combustion, the properties of the initial CeO2, etc.
In the present study, the initial CeO2 was in the form of agglomerates of very small elemental particles of 2-3 nm. Because of its high-developed surface, the NM is more reactive than the corresponding bulk material, as suggested by Luo et al. 2011 and Attarian et al. 2007 [13, 14]. So it can be supposed that, under heating at high temperature, the initial nanomaterial will undergo thermal changes, via sintering for instance which could be the case here for the residues.

5. Conclusion

Here is reported a study dedicated to the small scale incineration of a commercial coating with nano-CeO2. The results have been discussed and compared to intermediate scale and a real scale studies on incineration of nano-CeO2.
In a qualitative point of view, this leads to the conclusion that during incineration, nano-CeO2 is distributed between the combustion aerosol and the residues.
In a semi-quantitative point of view, a significant amount of cerium has been observed in the residues of combustion. Only traces of cerium have been observed in the aerosol, always as a minor compound of the particles.
Further studies and technical developments on the lab-scale pilot are needed to provide quantitative information.

Based on this study and previous works by Walser et al. and Paur et al., it is supposed that a series of phenomena can take place such as sintering, evaporation, agglomeration with fly ash, etc. These phenomena seem to be strongly influenced by the composition of the combustible fraction and the temperature at which incineration takes place. Very small elemental particles (e.g. 2-3 nm in this study) would be especially affected by these potential transformations. As a result, additional research is needed, both at lab scale and real scale in order to better understand the link between the combustion conditions and the fate of nanoCeO_2.

As suggested by Ounoughene et al. [3, 12], this leads to the conclusion that in any case, both flue gas and bottom ashes will contain nanoCeO_2. Hence, risk assessment should be performed to better estimate the risks associated with the potential release of nanoCeO2 during their thermal processing. Precautions may need to be taken to avoid exposure of workers, populations or environment to nanoCeO2 particles.

Acknowledgements

French Ministry of Environment (programme 190 and DRC 54).

References

[1] Walser, T. et al., 2012. Persistence of engineered nanoparticles in a municipal solid-waste incineration plant. Nature Nanotechnology.

[2] Le Bihan O., Joubert A., Venditti D., Tran T., Ounoughene G., Debray B., Durécu S., Le Coq L. - Characterization and reduction of particulate emissions from the incineration of wastes containing engineered NMs", Summary of the NanoFlueGas ADEME Project - 2014 - Convention n° 1181C0088 – 7p

[3] Ounoughene, G. et al., 2015. Behavior and fate of halloysite nanotubes (HNTs) when incinerating pa6/HNTs nanocomposite. *Environ. Sci. Technol.*, 2015, 49 (9), pp 5450–5457

[4] Massari, A. et al., 2014. Behavior of TiO2 nanoparticles during incineration of solid paint waste: A lab-scale test. Waste Management.

[5] Derrough, S. et al., 2013. Behaviour of nanoparticles during high temperature treatment (Incineration type). J. Phys.: Conf. Ser, 429. Available at: http://iopscience.iop.org/1742-6596/429/1/012047.

[6] Vejerano, E.P. et al., 2014. Characterization of particle emissions and fate of nanomaterials during incineration., 1(2), pp.83–192.

[7] Vejerano E. P., Holder A. L., Marr L. C., 2013, Emissions of Polycyclic Aromatic Hydrocarbons, Polychlorinated Dibenzo-p-Dioxins, and Dibenzofurans from Incineration of Nanomaterials, Environmental Science and Technology, 47, 4866−4874

[8] Vejerano E. P., Ma Y., Holder A. L., Pruden A., Elankumaran S., Marr L. C., 2013, Toxicity of particulate matter from incineration Nanowaste, Environmental Science: Nano, 2, 143-154

[9] Baumann W., M. Hauser, I.-M. Lang, H. Mätzing, N. Teuscher, H. Seifert and H.-R. Paur, Investigation of the release of engineered nanoparticles by waste incineration, Book of Abstracts of the European Aerosol Conference 2016, Tours, Sept. 2016.

[10] Paur H. R., Baumann W., Hauser M., Lang L., Teuscher N., Seifert H., Stapf W., 2016. Thermal stability and material balance of nanomaterials in waste incineration, Abstract, International Conference on Health Safety Issues related to Nanomaterials, Nanosafe 2016, Grenoble, France.

[11] R'mili, B.; Le Bihan, O.; Dutouquet, C.; Aguerre-Chariol, O.; Frejafon, E. Particle Sampling by TEM Grid Filtration, Aerosol Science and Technology 2013, 47, 7, 767-775; DOI: 10.1080/02786826.2013.789478

[12] Ounoughene G., Le Bihan O., Chivas-Joly C., Longuet C., Debray B., Joubert A., Lopez-Cuesta J-M., Le Coq L. Thermal disposal of waste containing nano-objects: first investigations on a methodology for risk management, Conference on Health Safety Issues related to Nanomaterials, Nanosafe 2016, Grenoble, France.

[13] Attarian Shandiz M., Safaei A., Sanjabi S., Barber Z.H., 2007, Modeling size dependence of melting temperature of metallic nanoparticles, Journal of Physics and Chemistry of Solids, 68, 1396–1399

[14] Luo W., Su K., Li K., Li Q., 2011, Connection between nanostructured materials' size-dependent melting and thermodynamic properties of bulk materials, Solid State Communications, 151, 229–233

IOP Conf. Series: Journal of Physics: Conf. Series **838** (2017) 012024 doi:10.1088/1742-6596/838/1/012024

Thermal disposal of waste containing nanomaterials: first investigations on a methodology for risk management

G. Ounoughene[1,2,3], O. Le Bihan[4*], B. Debray[4], C. Chivas-Joly[5], C. Longuet[2], A. Joubert[1], J-M. Lopez-Cuesta[2], L. Le Coq[1]

[1]LUNAM, Ecole des Mines de Nantes, GEPEA, CNRS, UMR 6144, 4 rue Alfred Kastler, 44307 Nantes Cedex 03, France
[2]C2MA, Ecole des Mines d'Alès, 6 Avenue de Clavières, 30319 Alès Cedex, France
[3]ADEME, 20 avenue du Grésillé, 49004 Angers Cedex 01, France
[4]INERIS, Parc Technologique Alata, 60550 Verneuil-en-Halatte, France
[5]LNE, 29 Avenue Roger Hennequin, 78197 Trappes Cedex, France
*olivier.le-bihan@ineris.fr

Abstract. Considering the wide use and production of NMs since last two decades, these trendy nanomaterials (NMs) are expected to end up in thermal disposal and waste incineration plants (WIP). It seems relevant to assess the risks related to the thermal disposal and incineration of waste containing NMs (WCNMs). The objective of this work is to present a first approach to develop a preliminary methodology for risk management in order (1) to give insights on nanosafety of exposed operators and on potential environmental risks related to the incineration and thermal disposal of WCNMs, and (2) to eventually support decision-makers and incineration plant managers. Therefore, the main challenge is to find (a) key parameter(s) which would govern the decision related to risk management of NMs thermal disposal. On the one hand, we focused on the relevant literature studies about experimental works on incineration of NMs. On the other hand, we conducted an introductory discussion with a group of experts. The review of this literature highlights that the nano-object's nanostructure destruction appears as a relevant indicator of the risks related to the NMs incineration. As a consequence, we defined a "temperature of nanostructure destruction" (TND) which would be the temperature from which the nanostructure will be destroyed. This parameter has been assumed to be a consistent indicator to develop a preliminary methodology. If the combustion chamber temperature is higher than the TND of the NM (or if they are close to each other), then the nanostructure will be destroyed and no risks related to NMs remain. If the TND of the NMs is higher than the combustion chamber temperature, then the nanostructure will not be destroyed and risks related to NMs have to be considered. As a result, five groups of NMs have been identified. WCNMs including carbonic NMs appear to be in good position to be destroyed safely in WIP. On the other hand, based on this criterion, there would be no available thermal disposal plants to safely manage WCNMs including CeO2 and ZrO2. Finally, a decision tree has been designed. TND is used as criteria to assess if a waste can be managed safely or not by a specific thermal disposal and which safety measures have to be taken.

IOP Conf. Series: Journal of Physics: Conf. Series **838** (2017) 012024 doi:10.1088/1742-6596/838/1/012024

1. Introduction

Since the last few years, the disposal via incineration of waste containing nanomaterials (WCNMs) has increasingly drawn attention of the authorities, the scientists and the waste incineration plant (WIP) managers.

Toxic effects of different types of NMs have been demonstrated (Kühnel & Nickel 2014; Wang et al. 2014). Nevertheless, the data are incomplete and more detailed studies are required. Their potential releases in the environment during their life cycle and their subsequent impacts on health have been an increasing concern for many years: firstly, the workers exposure to NMs during production (Oberdorster et al. 2010, Wohlleben et al. 2014, Bressot et al. 2015), then the emissions related to the wear and tear of products containing NMs (Shandilya et al. 2015), their end of life (storage or recycling) and finally their elimination via thermal disposal like incineration (which is the most widespread thermal disposal treatment technology in Europe). Indeed, the behavior and the fate of NMs in incineration treatment facilities have been investigated, lately, by some studies at lab-scale (Ounoughene et al. 2015; Massari et al. 2014; Derrough et al. 2013; Vejerano et al. 2014, Le Bihan et al. 2014, Vejerano et al., 2013a, 2013b;) and at real scale (Walser et al. 2012). Moreover, Roes et al. (2012) have evaluated the risks related to waste incineration of polymer nanocomposites without experimental works.

Regarding the risks related to the incineration of WCNMs, it seems necessary and relevant to develop strategies and tools to manage them. A risk is defined as function of exposure and hazard. However, in this present work, only the exposure is considered for assessing the risk because, taking account of the precautionary principle, it is assumed that the hazard is maximum as toxicity data of a lot of NMs is not yet available, so it is prudent here to assume the NMs may be toxic. In the future, when sufficient toxicological data will be available, it will be relevant to take into account the specific toxicity of each NM.

The objective of this work is to present a first approach to develop a preliminary methodology for risk management in order (1) to give insights on nanosafety of exposed workers and on potential environmental risks related to the incineration and thermal disposal of WCNMs and (2) to eventually support decision-makers and incineration plant managers. Therefore, the main challenge is to find a key parameter which would govern the decision related to risk management of NMs thermal disposal.

The aim of the present study is to provide a first insight into risks (exposure) regarding the incineration of WCNMs through the review of available data from literature concerning exclusively experimental works. This concise overview allows to identify and to define a key parameter which can determine the risks. Thus, a preliminary methodology for managing the risks related to the NMs thermal disposal can be proposed through the design of a decision tree.

2. Literature review related to experimental works on WCNMs incineration

The review of the bibliography related to incineration of WCNMs allowed to identify eight relevant studies exclusively related to experimental incineration tests performed on different NMs. The critical results of these studies are presented and their limitations are discussed. The table hereinafter (Table 1) describes succinctly those experimental studies.

Table 1: Literature review: Experimental studies on NMs incineration

References	Type of study	NMs	Results
Walser et al. 2012	Real scale: NMs injection in different WIP spot	CeO_2	CeO_2 found in residues and it has been efficiently filtered by electro-filters
Vejerano et al. 2013a	Lab-scale: mix of waste tested in a non-modified tubular furnace – 850 °C – Study of NMs influence on	Paper, plastics and NMs: Ag, NiO, TiO_2, CeO_2, C_{60}, Fe_2O_3, QD	Some NMs induce PAH concentration increase

	PAH production		
Vejerano et al. 2013b	Lab-scale: mix of waste tested in a non-modified tubular furnace – 850 °C – Study of NMs fate	Paper, plastics and NMs: Ag, NiO, TiO_2, CeO_2, C_{60}, Fe_2O_3, QD	Persistence of some NMs in fumes and slags
Derrough et al. 2013	Lab-scale: NMs tested in a non-modified tubular furnace – 850°C and 1100 °C – Study of NMs fate	Sn, Ni, Ag	At 850 °C and 1100 °C, the NMs have reached their melting point
Le Bihan et al. 2014	Lab-scale: NMs tested in a modified tubular furnace – 850°C – Study of NMs fate	Black carbon, Silicon/Silica, Paint withTiO_2	Persistence of NMs in fumes and slags
Massari et al. 2014	Lab-scale: paint tested in a non-modified electrical furnace – 950°C – Study of NMs fate	Paint withTiO_2	The TiO_2 has been found in residues without any change
Ounoughene et al. 2015	Lab-scale: nanocomposite tested in a modified tubular furnace – 850°C – Study of NMs fate and behavior	Nanocomposite (polymer/nanoclay) Halloysite nanotubes (HNTs) incorporated in a polyamide matrix	HNTs have been found in fumes and slags ; they induce a two step mechanism in the thermal decomposition of the nanocomposite
Vejerano et al. 2015	Lab-scale: mix of waste tested in a non-modified tubular furnace – 850 °C –Toxicity study of NMs and particulate matter	Paper, plastics and NMs: Ag, NiO, TiO_2, CeO_2, C_{60}, Fe_2O_3, QD	Some incinerated NMs have oxidative potential

Among those studies, the only real scale study has been performed by Walser et al. (2012). The authors have shown that nano-CeO2 is found in residues and fumes and it has been efficiently filtered by electro-filters.

The other studies have been carried out at lab-scale. Ounoughene et al. (2015) and Le Bihan et al. (2014) have highlighted that, in order to perform lab combustion tests representative of the industrial incinerator furnace, it is important to control the combustion conditions in the lab-scale furnace. Indeed, in a real scale incineration process, combustion key parameters are controlled: temperature (850 °C in the combustion and post combustion-zone), residence time (at least two seconds in the post combustion zone at 850 °C), air-excess (never below 11 % of oxygen) and turbulence (a good mix between combustible and oxygen). Apart from Ounoughene et al. (2015) and Le Bihan et al. (2014), the lab scale studies have been performed using a furnace which doesn't strictly apply incineration combustion conditions but approaches them as much as possible. However, despite of this lack concerning representativity of combustion conditions, the experimental studies give interesting insights on the incineration of WCNMs, as described below.

Through three experimental studies, Vejerano et al. (2013a, 2013b, 2015) have investigated at lab-scale the effect of NMs on PAH concentration, their fate and their toxicity. They used a tubular furnace which has not been modified (and thus which doesn't strictly respect the real incineration conditions). Furthermore, they mixed their NMs samples with paper and plastics samples (Paper, plastics and NMs: Ag, NiO, TiO2, CeO2, C60, Fe2O3, QD). They conclude that some NMs induce PAH concentration increase; some incinerated NMs have oxidative potential, and the persistence of some NMs in fumes and slags.

Massari et al., 2014 have investigated the fate of nano-TiO2 from paint at lab-scale, using a non-modified electrical furnace at 950 °C. The nano-TiO2 has been found in residues without any change. Derrough et al., 2013 have performed lab-scale tests on Sn, Ni and Ag using a non-modified tubular furnace at 850 °C and 1100°C. At 850 °C and 1100 °C, the NMs have reached their melting point. The results of Ounoughene et al. (2015) and Le Bihan et al. (2014) show that the nanostructure of the NMs may or may not evolve or may get destroyed. In fact, the black carbon particles have been destroyed, the nanoclays (HNTs) have aggregated but have kept their nanostructure; the nano-silica can undergo sintering effects or remain the same like the nano-CeO2 and the nano-TiO2. Furthermore, the polymer matrices emit aggregated/agglomerated nanoparticles during the incineration i.e.: an

IOP Conf. Series: Journal of Physics: Conf. Series **838** (2017) 012024 doi:10.1088/1742-6596/838/1/012024

organic matrix emits nanometric soot and a hybrid organic/inorganic matrix (silicone type) emits nanometric silicon oxycarbides. The presence of NMs doesn't seem to increase the soot production during the matrix thermal decomposition.

In order to clarify the definition of the nanostructure, the reader should consider that the nanostructure is related to the nanoscopic character (size) of a nanomaterial and more precisely to its specific surface area. According to the European Commision (Commission recommendation on the definition of nanomaterial, 2011/696/EU), a material should be considered as falling under the definition of a nanomaterial when its specific surface area by volume (SSV) is greater than 60 m^2/cm^3. When the nanomaterial's SSV is lower than 60 m^2/cm^3, then regarding that definition the material "loses" its nanostructure. Besides, "waste containing nanomaterials" has not been defined from a quantitative point of view.

These results lead to the conclusion that both flue gas and slag treatments will be affected by concerns about NMs risks and hence, potential precautions will need to be taken in their subsequent treatment steps. Furthermore, those studies didn't quantify the nanostructure of the NMs found in the fumes and in the residues (in terms of mass concentration in the residues or size distribution, for example); they use only SEM and TEM (scanning and transmission electronic microscopes) analysis in order to evaluate qualitatively the presence NMs nanostructure.

3. Design of a preliminary method

On the basis of the literature review focused on experimental works related to NMs incineration, and of an introductory discussion with a group of French experts (experts on nanosafety and on waste management), a key parameter which governs the risks related to WCNMs incineration has been identified and a preliminary risk management methodology concerning incineration of WCNMs could be proposed.

3.1. The risk could be determined by the fate of the nanostructure

Previously, we have highlighted that the nanostructure can be destroyed, or transformed (i.e only deteriorated, without being destroyed) during the incineration process, or it can remain the same.
In this way, the investigation of the fate of the nanostructure seems to be a good way to determine the risk. If the nanostructure is destroyed, then no ("nano"-)risks remain after the incineration. But, if the nanostructure is not destroyed (it remains the same or it is deteriorated), then a risk exists. In order to manage the risk, it is thus important to identify an indicator of the nanostructure destruction.

3.2. The temperature as an indicator of the nanostructure destruction

Actually, we could define a "Temperature of Nanostructure Destruction" (TND) which would be the temperature from which the nanostructure will be destroyed. The nanostructure destruction would be reached when the nanomaterial loses all its nanoscopic dimensions which could happen after a decomposition (via combustion of organic materials), or sintering or a fusion for instance. This could be measured *via* the specific surface area by volume or by mass. As the literature lacks of studies dealing with the concept of the nanostructure destruction, this topic is worth studying.
According to the NMs nature (carbonic/organic compound, metal, metal oxide, oxide…), this TND can be related to the temperature of the thermal decomposition for organic or carbonic nanomaterials; or can be related to the melting point for the inorganic nanomaterials and nano-oxides.

3.3. In a first approach, for operational reasons: the TND is considered equal to the melting point of the bulk material

Many studies have investigated the relation between the melting point of NMs (nano-oxides for example) and the melting point of the corresponding bulk materials (non-nanostructured material) (Cao et al. 2004, Qi et al. 2004, Qi et al. 2005, Luo et al. 2011, Attarian et al. 2007). The melting point of a NM depends on its size. Because of its high-developed surface, the NM is more reactive than the corresponding bulk material, so the melting point of the NM is lower than the melting point of the corresponding bulk material. Yet the actual values of melting points are not available for most NM because the literature lacks of data.

In order to facilitate the use of the TND in an operational way, in the context of this paper and in the following part: the TND is considered equal to the melting point of the corresponding bulk material. The Figure 1 shows the TND of various NMs. The list of NMs is far to be exhaustive, and should be extended in the future. The NMs have been selected in order to present the concept of TND. In this study, it is considered that the NM is not embedded in a matrix; or if the NM is embedded in a matrix, then the matrix's decomposition temperature is lower than the TND (which is the case for a polymer matrix for example). This is an interesting issue to address later in order to improve the methodology. When NM's incineration has been studied experimentally in literature, a blue triangle positioned next to the NM mentions the study which deals with the fate of the NM (Figure 1). This figure shows that relevant literature data on the incineration of some NMs does exist, however it highlights that further investigations are needed to make it complete. The NMs are classified following the distance of their TND from the combustion (and post-combustion chambers) temperatures of different thermal disposal plants. Five groups have been identified as described hereinafter.

If the TND of the NM is close to the combustion chamber temperature of a WIP (850 °C), then the NM would lose its nanostructure during its stay in the furnace. This is the case of the NMs from the Group 1. Indeed, in the Group 1 which consists of organic NMs like CNTs, fullerene, and black carbon, the NMs would be consumed in the furnace as observed in few studies (Vejerano et al., 2013a; Le Bihan et al., 2014). The study of Schlagenhauf et al. 2014 has highlighted that CNTs are destroyed when the temperature is higher than 600 °C and when the atmosphere is oxidative. Then the nanostructure of a NM from Group 1 is destroyed in a WIP.

With regard to the Group 2, which consists of nanosilver, nano-tin oxide and nano-gold: on the one hand, their nanostructure could change (deterioration or destruction) in a WIP since their melting point is slightly higher than the WIP combustion chamber temperature; on the other hand, their nanostructure could be destroyed in a specialized WIP because their melting point would be close to the temperature of the post-combustion chamber from a specialized WIP (1100 °C). The nanosilver has been tested by Derrough et al., 2013 in a tubular furnace at 850 °C and 1100 °C. It has melted under the two temperatures and thus has lost its nanostructure. This demonstrates that the NMs can have melting temperatures lower than the actual melting point of the bulk material. It also suggests that WCNMs could be destroyed in either standard municipal solid waste incinerators (at 850°C) or specialized industrial waste incinerators at 1100°C. However, further investigations in a furnace which applies the same conditions than an industrial incineration furnace would be necessary (as explained previously).

The melting points of the NMs from the Group 3 ($Fe2O3$, Nanoclays, Ni, $Cu2O$, $TiO2$) are much higher than the temperature of the WIP combustion chambers, then, it is expected that the nanostructure of these materials will not change in a WIP (nor in a specialized WIP). This is the case of the nanoclays studied by Ounoughene et al. (2015), and the Ni studied by Derrough et al. (2013), or NiO and $Fe2O3$ studied by Vejerano et al. (2013a, 2013b, 2015). In order to destroy the nanostructure of these NMs from the Group 3, one possibility would be to send them to steelmaking or glassmaking plants which work with furnace reaching 1600 °C. However, it is assumed that the NMs could enter the mentioned industrial thermal processes without affecting their effectiveness. This point needs obviously further investigations.

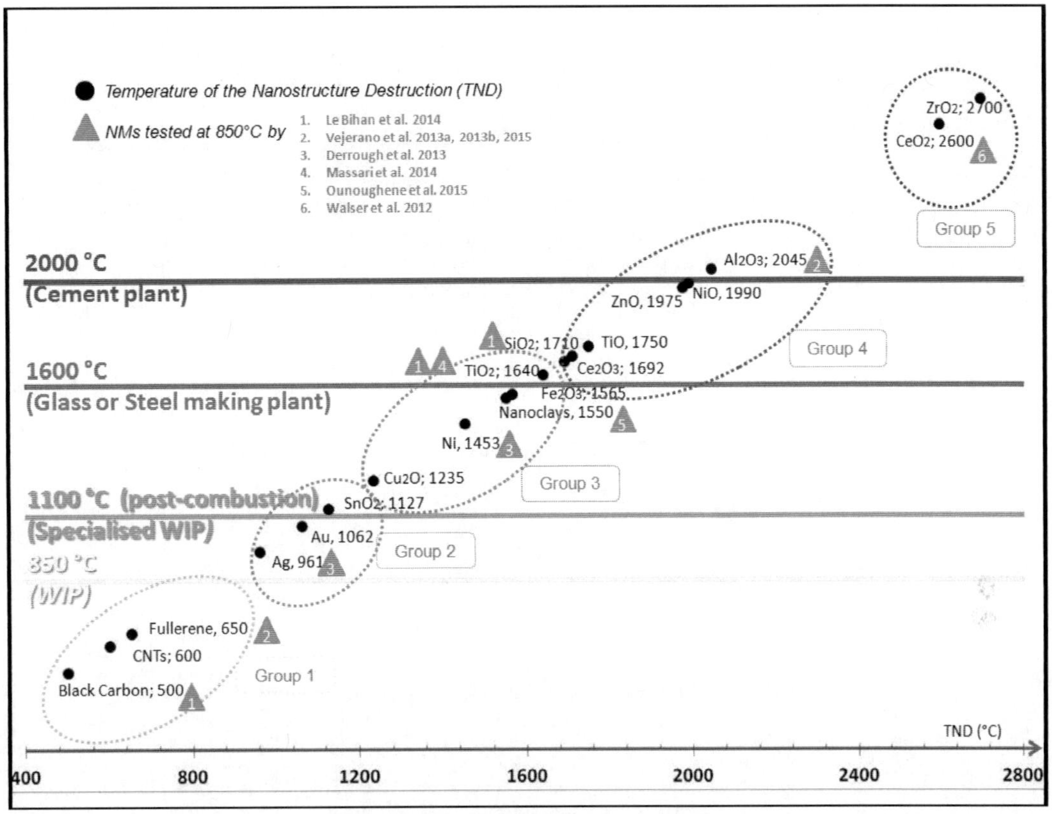

Figure 1: NMs classification according to their TND

The Group 4 consists of many nano-oxides like Al2O3, NiO, ZnO, SiO2, Ce2O3 and TiO. The only way to destroy their nanostructure would be by using a furnace from the cement industry operating at 2000 °C.

As for the materials from Group 5 (ZrO2 and CeO2), their high melting point (>2000 °C) makes their nanostructure non-destructible in a WIP combustion chamber of 850 °C as highlighted by Walser et al. 2012 and also non-destructible in any existing furnace from a thermal plant. NMs of this group will be found in the incineration residues and provisions will need to be taken to manage the risk associated with their potential dissemination.

Besides, we have considered that no NMs vaporization occurs in the furnace. In fact, the vaporization temperature of NMs shouldn't be reached in order to avoid any re-crystallization or nucleation process which would lead to the formation of new nanoparticles which could represent a supplementary risk. Here again, further studies are necessary to determine the risks.

At this point, it clearly appears that more investigations are necessary to fully study the thermal disposal of WCNMs and the nanostructure destruction.

3.4. Decision tree

It is now possible to gather all this knowledge and use it at an operational level. A decision tree can be designed. It is a first attempt and its final objective is to eventually support waste managers for the selection of suitable thermal disposal treatment for WCNMs (Figure 3).

IOP Conf. Series: Journal of Physics: Conf. Series **838** (2017) 012024 doi:10.1088/1742-6596/838/1/012024

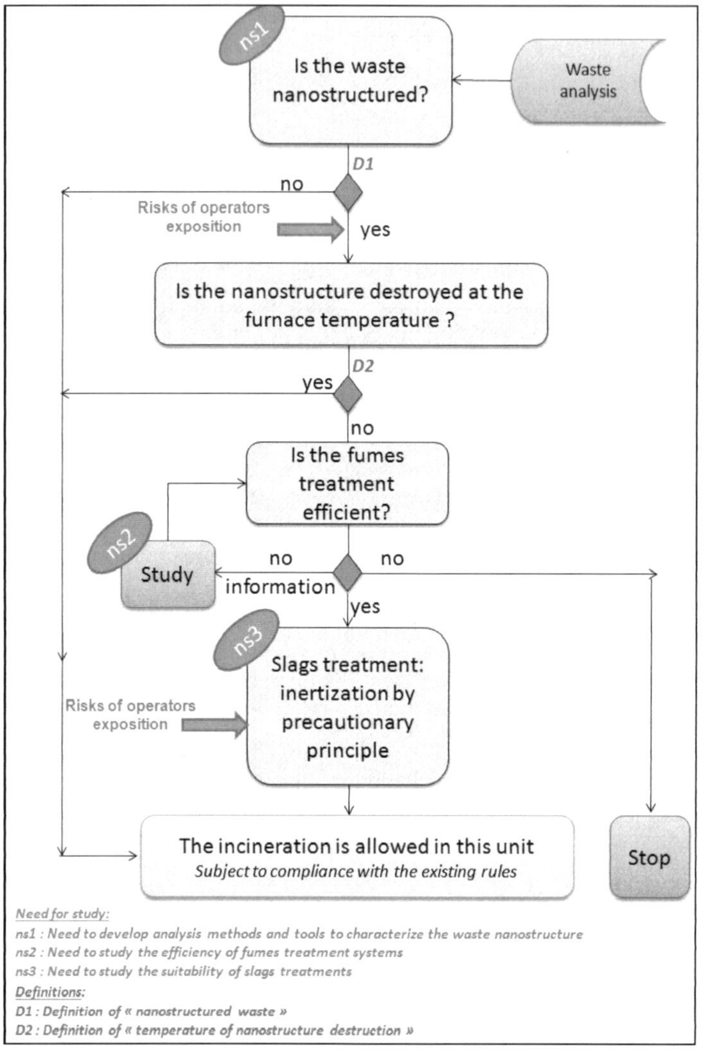

Figure 2: Decision tree: first approach for developing a risk management methodology

The decision tree takes into account the possible release of the NMs during a thermal treatment. It is considered that when WCNMs enter a thermal disposal plant: the operators (or workers) could be exposed to the NMs and the NMs could be released in the environment. Then, in order to prevent any NMs release, a series of questions should be answered.

The first step consists in analyzing the waste by investigating whether the waste is nanostructured related to the first question: "is the waste nanostructured?". This issue is associated to the first need for study (ns1): "need to develop analysis methods and tools to characterize the waste nanostructure". Indeed, there is an identified lack in the nanostructure characterization of NMs and new methods should be developed to evaluate the nanostructure of a NM (for example, through the measurement of specific surface area by volume). If the waste is not nanostructured, then there is no risk related to the size of NMs and the incineration is allowed in this unit (subject to compliance with the existing rules regarding other acceptance criteria). If the waste is nanostructured, the next question is: "is the nanostructure destroyed at the combustion chamber temperature?". This question refers to the temperature of the nanostructure destruction (TND). So, if the TND of the NM is reached in the furnace, the nanostructure is destroyed and the waste can be safely incinerated in this unit. But if the TND is not reached, then another question should be raised: “is the fumes treatment efficient?”. If the answer is “no”, then the incineration shouldn’t be done in this unit and another solution could be envisioned: for example the NM could be sent to another kind of thermal disposal plant working at a temperature higher than TND of the NMs which could destroy the nanostructure of the NM. As mentioned above, those kinds of thermal disposal units could be: specialized WIP (post-combustion chamber at 1100°C), glass industry and steelmaking plant (around 1600°C), or cement plant (around 2000°C). However, as mentioned previously, further studies need to investigate how the NMs waste stream would affect those industrial thermal processes in order to assess the feasibility. Finally, NMs with TND higher than 2000°C (group 5), should be stored as no available thermal disposal plant works at temperature high enough to destroy their nanostructure.

If the unit has no information about the efficiency of the fumes treatment regarding NMs, then a study has to be launched that refers to “ns2: need to study the efficiency of fumes treatment systems”. If the answer is “yes”, then by precautionary principle, the slags have to undergo inertisation treatments. This action leads to the “ns3: need to study slags treatments suitability”.
Moreover, because the waste is nanostructured, operators can be exposed to the NMs and thus they have to be protected and equipped with adequate personal protective equipments.

4. Conclusion

In order to provide a first insight into nanosafety regarding the incineration of WCNMs, the present study proposes an introductory approach for developing a preliminary methodology for managing the risks through the design of a decision tree.

This work presents a review of existing data about experimental works related to incineration of NMs and identifies a key parameter which could determine how to manage the risks related to the incineration of WCNMs. Only a few experimental data is currently available. It has been highlighted that the risk can be determined by the fate of the nanostructure. A temperature of nanostructure destruction (TND) has been defined. By default and for operational reasons, this TND can be equal to the melting point of the corresponding bulk material.
If the WIP combustion chamber temperature is higher than the TND of the NM or if they are close to each other, then the nanostructure will be destroyed and no risks related to the nanostructure of NMs remain (other risks still need to be considered). If the TND of the NMs is higher than the WIP combustion chamber temperature then the nanostructure will not be destroyed and another way to destroy the nanostructure would be to send the NM in a other kind of thermal disposal plant working at a temperature high enough to destroy the nanostructure of the NM, namely equal or higher than the TND (specialized WIP, glass industry, steelmaking plant, or cement plant), assuming that the effectiveness of the thermal industrial processes would not be affected by the NMs waste stream.

Finally, this preliminary study highlights the necessity to develop analytic tools in order to quantify the nanostructure in the waste, in the residues and in the fumes and to further investigate the thermal disposal of WCNMs.

It is important to underline that this method is preliminary: some gaps have to be considered in its future developments, for example, the effect of the matrix (where the NM is embedded) on the TND, and the effect of NM waste stream on the efficacy of the other kinds of thermal processes. In addition, some specific experiments are needed to support the concept of TND and a larger variety of NM should be considered.

However, despite its limits, the present approach could be a useful starting point that highlights issues which are worth studying in order to build a proper methodology for managing the risks related to the incineration of WCNMs.

Acknowledgements

Financial support for this work was provided by ADEME, INERIS (prog. 190, DRC 54), and LNE. We are grateful to D. Venditti from TREDI (Séché Environnement), E. Fiani (ADEME) and B. Schnuriger (INERIS) for participating to the group discussion.

References

Attarian Shandiz M., Safaei A., Sanjabi S., Barber Z.H., 2007, Modeling size dependence of melting temperature of metallic nanoparticles, Journal of Physics and Chemistry of Solids, 68, 1396–1399

Bressot, C. et al., 2015. Exposure Assessment Based Recommendations to Improve Nanosafety at Nanoliposome Production Sites, 2015.

Cao G., 2004, Chapter 8: Characterization and Properties of Nanomaterials, Nanostructures And Nanomaterials: Synthesis, Properties & Applications, 2nd Edition, Imperial College Press

Derrough, S. et al., 2013. Behaviour of nanoparticles during high temperature treatment (Incineration type). J. Phys.: Conf. Ser, 429. Available at: http://iopscience.iop.org/1742-6596/429/1/012047.

Kühnel, D. & Nickel, C., 2014. The OECD expert meeting on ecotoxicology and environmental fate - Towards the development of improved OECD guidelines for the testing of nanomaterials. Science of the Total Environment, 472, pp.347–353. Available at: http://dx.doi.org/10.1016/j.scitotenv.2013.11.055.

Le Bihan O., Joubert A., Venditti D., Tran T., Ounoughene G., Debray B., Durécu S., Le Coq L. - Characterization and reduction of particulate emissions from the incineration of wastes containing engineered NMs", Summary of the NanoFlueGas ADEME Project - 2014 - Convention n° 1181C0088 – 7p

Luo W., Su K., Li K., Li Q., 2011, Connection between nanostructured materials' size-dependent melting and thermodynamic properties of bulk materials, Solid State Communications, 151, 229–233

Massari, A. et al., 2014. Behavior of TiO2 nanoparticles during incineration of solid paint waste: A lab-scale test. Waste Management.

Oberdorster G., "Safety assessment for nanotechnology and nanomedicine: concepts of nanotoxicology," Journal of Internal Medicine, vol. 267, no. 1, 89–105, 2010

Ounoughene G. et al., 2015. Behavior and fate of halloysite nanotubes (HNTs) when incinerating pa6/HNTs nanocomposite. Environmental Science and Technology.

Qi W.H., 2005, Size effect on melting temperature of nanosolids, Physica B, 368, 46–50

Qi W.H., Wang M.P., 2004, Size and shape dependent melting temperature of metallic, Materials Chemistry and Physics, 88, 280–284

Roes L., Patel M. K., Worrell E., Ludwig C., 2012, Preliminary evaluation of risks related to waste incineration of polymer nanocomposites, Science of the Total Environment, 417, 76-86.

IOP Conf. Series: Journal of Physics: Conf. Series **838** (2017) 012024 doi:10.1088/1742-6596/838/1/012024

Schlagenhauf L., Nuesch F., Wang J., 2014, Release of Carbon Nanotubes from Polymer Nanocomposites, Fibers, 2, 108-127

Shandilya, N. et al., 2015. Emission of titanium dioxide nanoparticles from building materials to the environment by wear and weather. Environmental Science and Technology, 49(4), pp.2163–2170.

Vejerano, E.P. et al., 2014. Characterization of particle emissions and fate of nanomaterials during incineration. , 1(2), pp.83–192.

Walser, T. et al., 2012. Persistence of engineered nanoparticles in a municipal solid-waste incineration plant. Nature Nanotechnology.

Wang, Q. et al., 2014. A bibliometric analysis of research on the risk of engineering nanomaterials during 1999-2012. Science of the Total Environment, 473-474, pp.483–489. Available at: http://dx.doi.org/10.1016/j.scitotenv.2013.12.066.

Wohlleben W., Kuhlbusch T. A. J., Schnekenburger J., and Lehr C. M., Eds., Safety of Nanomaterials alongTheir Lifecycle: Release,Exposure, and Human Hazards, CRC Press, 2014.

In vivo evaluation of the potential neurotoxicity of aerosols released from mechanical stress of nano-TiO_2 additived paints in mice chronically exposed by inhalation

S Manixay[1], S Delaby[2], F Gaie-Levrel[3], M Wiart[4], C Motzkus[5] and A Bencsik[1]

[1] Université de Lyon 1, ANSES, Unité Maladies Neurodégénératives, Lyon, France
[2] Centre scientifique et technique du bâtiment (CSTB), Saint-Martin d'Hères, France
[3] Laboratoire national de métrologie et d'essais (LNE), Paris, France
[4] Univ-Lyon, CarMeN laboratory, Inserm U1060, INRA U1397, Université Claude Bernard Lyon 1, INSA Lyon, Charles Mérieux Medical School, Oullins, France
[5] Centre scientifique et technique du bâtiment (CSTB), Marne-la-Vallée, France

French Agency for Food, Environmental and Occupational Health Safety (ANSES)

E-mail: anna.bencsik@anses.fr

Abstract. Engineered Nanomaterials (ENM) provide technical and specific benefits due to their physical-chemical properties at the nanometer scale. For instance, many ENM are used to improve products in the building industry. Nanoscaled titanium dioxide (TiO_2) is one of the most used ENM in this industry. Incorporated in different matrix, cement, glass, paints… TiO_2 nanoparticles (NPs) provide the final product with anti-UV, air purification and self-cleaning properties, thanks to their photocatalytic activity. However, ageing processes of such products, as photocatalytic paints, during a mechanical stress have been shown to release TiO_2 NPs from this matrix associated with sanding dust. Thus, workers who sand painted walls could be exposed to TiO_2 NPs through inhalation. As inhalation may lead to a translocation of particulate matter to the brain via olfactory or trigeminal nerves, there is an urgent need for evaluating a potential neurotoxicity. In order to provide new knowledge on this topic, we developed a dedicated experimental set-up using a rodent model exposed *via* inhalation. The aerosol released from a mechanical stress of photocatalytic paints containing TiO_2 NPs was characterized and coupled to an exposition chamber containing group of mice free to move and chronically exposed (2 hours per day for 5 days a week during 8 weeks).

1. Introduction

Engineered Nanomaterials such as titanium dioxide (TiO_2), carbon nanotubes, silica dioxide (SiO_2) or silver (Ag), are increasingly used in the construction products, e.g. cement, wet mortar and concrete, coatings and paints [1-3]. It is assumed that the majority of nanomaterials are most likely safe, at least at concentrations for which workers are exposed. However, many questions are still open concerning the impact of these nanomaterials on workers' health when handling these products. As an example, the degradations induced by the abrasion mechanisms of nanocomposite materials could induce the release of nanoparticles into the environment for which chemical nature, size and concentrations are mostly unknown [4-6]. In this context, the most common route of exposure to nanoparticles is inhalation and the nanoparticle sources are aerosols of which nano-aerosols generated by activities in the construction industry [7]. Nanoscaled TiO_2 is one of the most used ENM in this industry and an emblematic use concerns photocatalytic paints [8]. Incorporated in the paints matrix, TiO_2 nanoparticles (NPs) provide the final product with anti-UV, air purification and self-cleaning effects properties, thanks to their photocatalytic activity [9]. However, ageing processes of these products during a mechanical stress have been shown to release TiO_2 NPs from their matrix associated with sanding dust [5, 10]. Thus, workers who sand painted walls could be exposed to TiO_2 NPs in

IOP Conf. Series: Journal of Physics: Conf. Series **838** (2017) 012025 doi:10.1088/1742-6596/838/1/012025

combination to paint matrix through inhalation. Although some experimental studies underlined the potential hazardous effects of inhaled nanoparticles like lung inflammation or affection of the cardiovascular system, to the best of our knowledge, none investigated a potential impact on the nervous system [11-13]. However, since nanoparticles inhaled through the nose can translocate to the brain directly *via* the nerve endings of olfactory neurons, there is a crucial need to study their potential neurotoxicity, especially in case of chronic exposure [14, 15].

We thus developed a dedicated experimental set-up using a rodent model exposed *via* inhalation, with the aim of determining if chronic exposure through inhalation to sanding dusts containing TiO_2 NPs may impact brain functions. The objective of this paper is to report the study strategy and the key elements.

2. Materials and Methods

2.1. Nanocomposite materials

Homemade nano-TiO_2 enriched paint (5% in mass) was used compared to nano-TiO_2 NPs alone (Degussa-P25: composed of rutile (80%) and anatase (20%) crystallites, primary particles with a mean diameter approximately of 21 nm and a specific surface of 50 +/- 15 m^2/g), as well as paint matrix alone. The paints (enriched or not) are applied on a 10x10cm plate of stainless steel (Figure 1).
The dry paint weighs 5.4 g for the homemade nano-TiO_2 enriched paint (with a standard deviation of 0.4g) and represents the nanomaterial tested in our study. A painted plate allowed generating sufficient number of particles in the aerosol form for a 30 minutes exposure. Abrasion tests were performed using a rotating abraser sander to mimic the mechanical abrasion process (Figure 2).

Figure 1: Illustration of painted plates used in the study

2.2. Physico-chemical characterization of the generated aerosol

Aerosol productions from the mechanical stress applied to these nanocomposite materials were characterized. The particle size distributions and the associated concentrations and morphologies (free or agglomerated nanoparticles or nanoparticles embedded in polymer matrix) of aerosolized particles were measured with on-line (FMPS, ELPI, APS, CPC, TEOM) and off-line measurements (electronic microscopy on TEM grid). Repeatability and reproducibility measurements were performed in accordance with the ISO 5725-2 standard in order to compare the aerosol production between the set-up characterization and the exposition phases.

2.3. Mice exposure

In order to investigate the potential neurotoxicity of these sanding dusts containing TiO_2 NPs, a dedicated experimental set-up developed by our group was used to chronically expose wild type mice *via* inhalation. The generations of aerosol during the mechanical solicitation of paint dust with or without TiO_2 NPs were characterized. The set-up developed in this project permits a secured and entirely controlled exposure to the aerosol. Coupled to an exposition chamber containing a group of 30 female mice (C57Bl6 mice, 6-week-old) free to move, it allowed performing chronic exposure during 2 hours per day for 5 days a week for 8 weeks. All procedures were carried out in complete compliance with the guidelines laid down by the French Ethical Committee (Decree 87-848) and

IOP Conf. Series: Journal of Physics: Conf. Series **838** (2017) 012025 doi:10.1088/1742-6596/838/1/012025

European Community Directive 86/609/EEC. The complete inhalation study was approved by National Committees on the Ethics of Animal Experiments (ANSES/ENVA/UPEC (n°15-068) and CELYNE). Animal experiments were performed in the ANSES animal facilities which have the relevant approval to carry out animal experiment (C 69 387 0801) by licensed people working in the animal experiment unit (license numbers AB: 69 387 531, LL: 69 387 191).
In order to investigate a time-dependent effect, groups of 5 mice were studied after 1, 2, 3, 4 and 8 weeks of exposure. Several parameters were collected *in vivo* such as weight, general appearance, evolution of locomotor performances using rotarod tests and *in vivo* brain imaging by Magnetic Resonance Imaging (MRI) at 0, 2, 4 and 8 weeks of exposure. Other parameters are collected post-mortem, such as histopathological studies, performed on brain sections. Some brains were dedicated to elemental content analysis (Ti brain mapping). Since the data collection and related analyses are still under progress, only the project strategy is presented in this paper.

3. The Release NanoTox project

The graphical summary of our study principle is given in Figure 2.

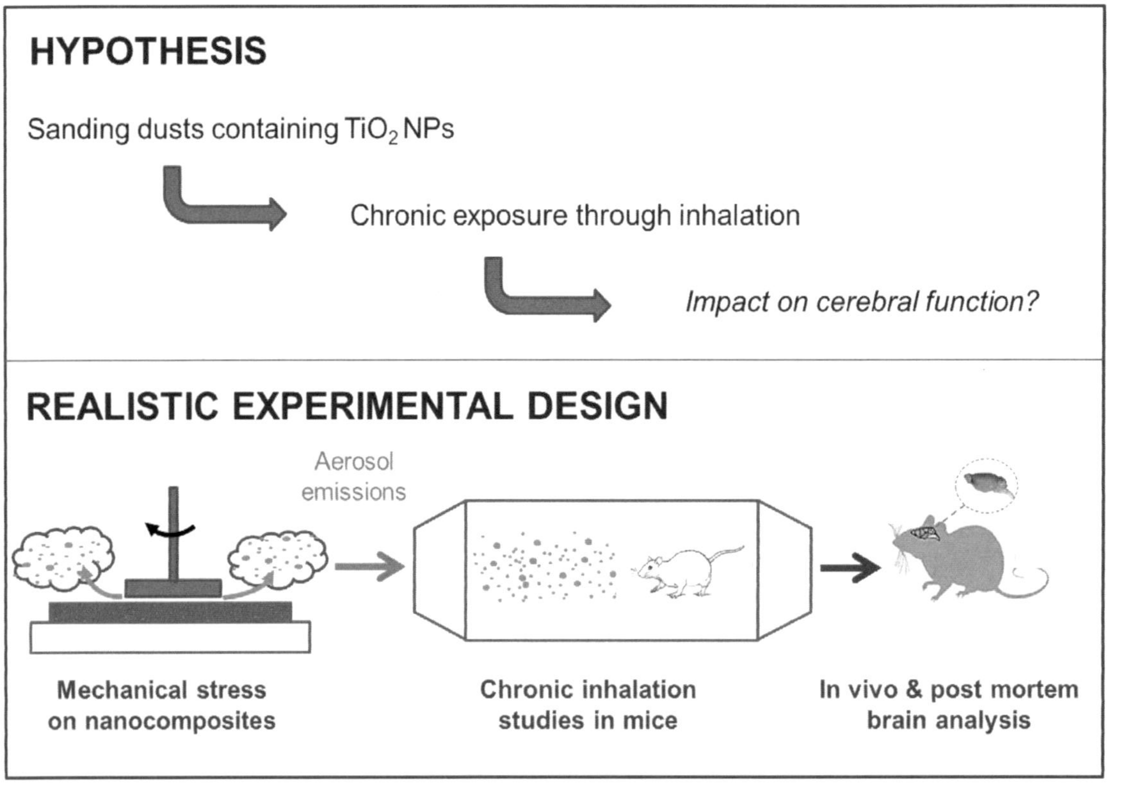

Figure 2: Principle of the Release NanoTox study

The emitted aerosols, produced under realistic conditions, were first characterized, and then used for *in vivo* studies to evaluate their potential neurotoxicity in mice after chronic inhalation exposure. This experimental design mimics a realistic scenario of workers' exposure.

MRI analysis gives access to potential morphological and functional abnormalities that can further be related to other parameters recorded *in vivo* such as the rotarod tests performances, both being suitable to detect a neurotoxic effect. *Post mortem* histological assessment of potential neuroinflammation and neuronal loss will also allow interpreting, in an appropriate manner, a potential negative effect on the brain function.

The work plan composed of several tasks and sub-tasks is presented in the Figure 3.

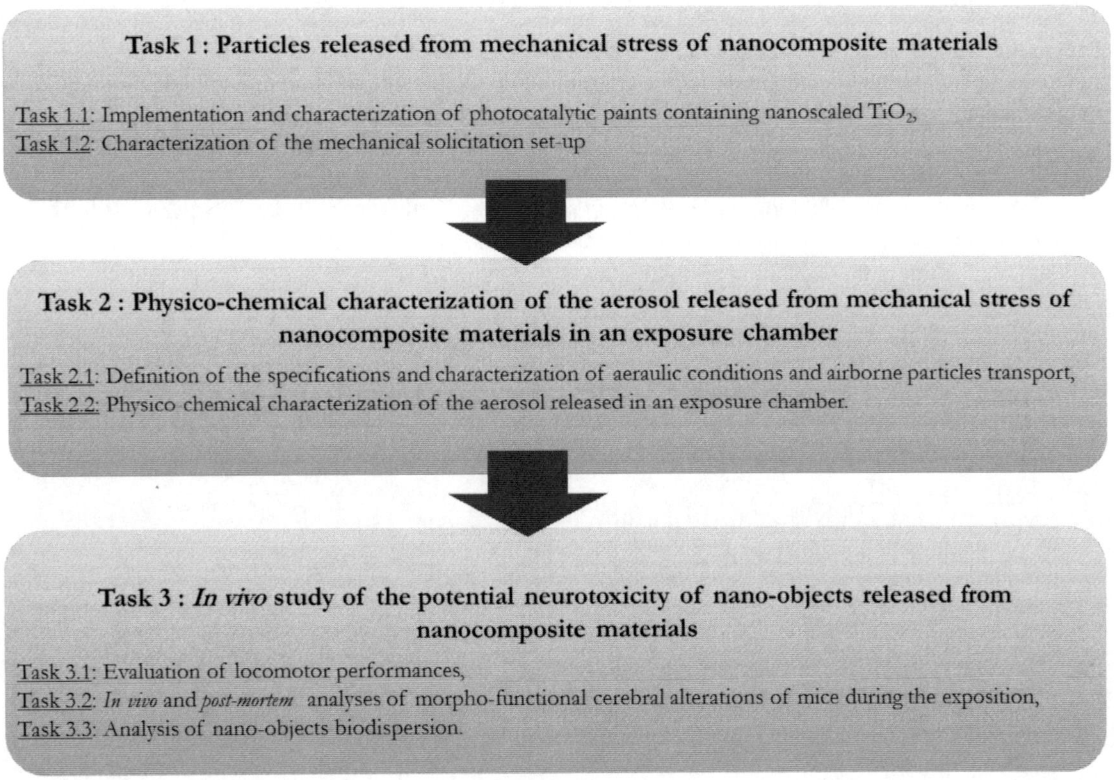

Figure 3: Workplan of the Release NanoTox project

4. Conclusion

The main originality of the Release NanoTox project consists in the convergence work on particle release from nanocomposite materials and the study of the *in vivo* neurotoxicity of nano-objects. Indeed, to date and to the best of our knowledge the work on toxicity is focused on the impact of nano-objects only. As it has been shown that the wear of a nanocomposite material leads to the production of polymorphic particles in size and composition, with a majority of particles consisting of nanoparticles encapsulated in the matrix material, it is necessary to assess their potential toxicity compared to the added NM alone. Moreover, the chemical composition of the nanocomposite material's matrix is a factor that is not taken into consideration in toxicology studies. Similarly, no study has focused on the neurotoxic effect of particles from nanocomposite materials. Our study will bring new knowledge and should contribute to clarify these important questions for the workers' health.

IOP Conf. Series: Journal of Physics: Conf. Series **838** (2017) 012025 doi:10.1088/1742-6596/838/1/012025

Acknowledgements

This work was founded by a grant from ANSES (EST-2014/1/178). Stéphane Manixay was supported by a grant from ANSES (EST-2014/1/178).

References
[1] Ricaud M, Moniteur L 2012 *INRS* **n°5651**
[2] Som C, Wick P, Krug H, Nowack B 2011 *Environ Int* **37** 1131
[3] Hincapie I, Caballero-Guzman A, Hiltbrunner D, Nowack B 2015 *Waste Manag* **43** 398
[4] Koehler A, Som C, Helland A, Gottschalk F 2008 *Journal of Cleaner Production* **16** 927
[5] Koponen IK, Jensen KA, Schneider T 2009 *Journal of Physics: Conference Series* **151** 012048
[6] Vorbau M, Hillemann L, Stintz M 2009 *Journal of Aerosol Science* **40** 209
[7] Savolainen K, Alenius H, Norppa H, Pylkkanen L, Tuomi T, Kasper G 2010 *Toxicology* **269** 92
[8] van Broekhuizen FA, van Broekhuizen JC 2009 *European Commission*
[9] Maggos T, Bartzis JG, Liakou M, Gobin C 2007 *J Hazard Mater* **146** 668
[10] Koponen IK, Jensen KA, Schneider T 2011 *J Expo Sci Environ Epidemiol* **21** 408
[11] Larsen ST, Jackson P, Poulsen SS, Levin M, Jensen KA, Wallin H, Nielsen GD, Koponen IK 2016 *Nanotoxicology* **10** 1254
[12] Smulders S, Luyts K, Brabants G, Golanski L, Martens J, Vanoirbeek J, Hoet PH 2015 *Toxicol Lett* **232** 333
[13] Smulders S, Luyts K, Brabants G, Landuyt KV, Kirschhock C, Smolders E, Golanski L, Vanoirbeek J, Hoet PH 2014 *Toxicol Sci* **141** 132
[14] Dhuria SV, Hanson LR, Frey WH, 2nd 2010 *J Pharm Sci* **99** 1654
[15] Oberdörster G, Elder A, Rinderknecht A 2009 *J Nanosci Nanotechnol* **9** 4996

IOP Conf. Series: Journal of Physics: Conf. Series **838** (2017) 012026 doi:10.1088/1742-6596/838/1/012026

In vivo nanotoxicological profile of graphene oxide

Marcela Durán[1,2*], Nelson Durán[2,3,4,*], Wagner J. Fávaro[1,2]

[1]Lab.Urogenital Carcinogen. and Immunother. UNICAMP, Campinas-SP, Brazil.
[2]NanoBioss, Inst. Chem. UNICAMP, Campinas-SP, Brazil
[3]Chem. Inst., UNICAMP, Campinas, SP, Brazil.
[4]LNNano (CNPEM), Campinas, SP, Brazil.

*duran@iqm.unicamp.br, marceladuranduran@gmail.com

Abstract. Graphene oxide, a widely studied nano materials, exhibits numerous beneficial effects in medical devices. The graphene sample Graphene oxide (GO):Single-layer graphene oxide, purity 99%, thickness 0.7-1.2 nm (AFM); ~300-800nm X&Y dimensions is the standard size <450 nm & 1-20 µm lateral dimensions from Cheap Tubes Inc., Bratleboro, USA was selected for this study. Characterization of GO and stability were previously studied. All the assays were carried out at a concentration of 1 mg/mL (stock suspension by ultrasonication)(stable 10 days). In this study the first parameter evaluated was in animal toxicity in acute and chronic responses. The second one was to observe the morphology and histological changes on major organs, like signs of inflammatory areas. A total of 25 rats (Fischer 344) were divide in 5 groups (n = 5 animals): negative control group (NCT); Positive cancer control group (PCCT); GO group-1 (GO 1); GO group-2 (GO 2) and GO group-3 (GO 3), with different concentrations of graphene oxide. All the groups were treated by via intra peritoneal (i.p) administration. Biochemical and histopathology results of acute toxicity showed no alterations. In the chronic toxicity was found deposit sections and morphology alterations in different tissues (inflammations). Genotoxicity was dose dependent of GO. This study gave us the limit concentrations of GO free for any biological study.

1. Introduction

Graphene oxide one of derivatives of graphite, in the last years has become a promising candidates for important biomedical applications. Due at its characteristics it can be used at base to material which will be used with biological functionality [1-3]. The comprehension of the toxicity of these materials has been highlighted in numerous studies [4,5].

One of the concerns in the use of these nanomaterials is how these structures could influence the cellular environment where they were inserted, There are studies that indicate a correlation between the methodology used in its synthesis and chemical composition with the cytotoxic effects [6,7]. Study by Paul and Sharma [8], assessed the platelet action or hemolysis of graphene during growth and differentiation in stem cells. The authors concluded that the graphene does not produce platelet aggregation and induction of hemolysis when did not exceed 0.1%, being permitted in the case of medical materials is at most 1%. In another study performed in 2011[7] using normal human lung cells (BEAS-2B) could be observed toxicological aspects cytotoxicity and apoptosis induced by

IOP Conf. Series: Journal of Physics: Conf. Series **838** (2017) 012026 doi:10.1088/1742-6596/838/1/012026

evaluating graphene by 3- (4,5- dimethylthiazol-2-yl) - 2,5-diphenyltetrazolium bromide (MTT) after 24 and 48 h of exposure. There was a significant decrease in cell viability dependent on the concentration and exposure time and increase of early and late apoptotic cells compared to control cells. Direct responses in hemolytic action or platelet in the presence of aqueous graphene concentrations 1.0 mg/mL, 5.0 mg/mL and 7.0 mg/mL were observed [6], showing the importance of knowing the nanomaterials and their concentrations without toxic effect when they are being used in biological systems. Then, the aim of this work is to study the *in vivo* toxicity of graphene oxide.

2. Methods

2.1. Preparation of graphene oxide suspensions (GO)
Preliminary studies have shown that it is possible to obtain stable suspensions of GO in deionized water (1mg/mL) for up to 10 days. The suspensions of graphene oxide (1.0 mg/mL) was prepared in sterile deionized water and conditioned in 15 mL Falcon tubes (sterile) and placed in ultrasonic bath for 30 minutes.

2.2. Analysis of genotoxicity of GO sample: micronucleus test
The micronucleus test is a widely used method for monitoring of genotoxic damage in populations exposed to mutagenic and carcinogenic substances. The frequency of micronuclei observed at a given time can be considered a complex response between genotoxic activity and efficiency of the physiological defence mechanism of the test organism [9,10].

A total of 25 rats of the Fischer 344 strain were divided into 5 groups (5 animals per group): Group Negative Control (NC): received a dose of 0.3 mL of 0.9% saline intraperitoneally (i.p); Positive Control Group (PC): received a dose of 1.5 mg/mL of N-methyl-N-Nitrosourea (MNU - Sigma, St. Louis, MO, USA)(i.p.); Graphene Oxide Group 1 (GO 1): received a dose of 0.05 μg/g of GO (i.p.); 2 graphene oxide (GO 2): received a dose of 0.5 μg/g of GO (i.p.); 3 graphene oxide (GO 3): received a dose of 5.0 μg/g of GO (i.p.).

The animals were killed 24 h after the administration of MNU. Femurs were removed and cleaned; and epiphysis was sectioned to expose the spinal canal. The bone marrow was aspirated with a 1 ml of foetal bovine serum (FBS) and placed in Falcon tube containing 2 mL of FBS. The bone marrow was resuspended several times to form a suspension with FBS. Subsequently, the samples were centrifuged at 1000 rpm for 5 min. The supernatant was removed and the precipitate was carefully aspirated with a pipette. Then 100 μL of the precipitate of each sample were applied in histological slide and smear done with the aid coverslip. The slides were smears with drying for 24 h at room temperature. After this period, the samples were fixed in absolute methanol for 15 min and stained with Giemsa (1:10 dilution in pH 6.8 sodium phosphate buffer).

The samples were analysed in light microscope with 100X immersion objective and the number of micronuclei will be counted in 3.000 polychromatic erythrocytes (PCEs) per animal. The micronucleus identification followed the criteria described by Fenech [11]:
a) micronucleus should be one third smaller than the core;
b) a micronucleus should not touch the main core and;
c) a micronucleus should not refract, or should have the same colour and intensity of the main core.

2.3. Histopathological analysis
For the analysis of systemic toxicity after 30 days of treatment, all animals will be anesthetized with xylazine / ketamine (1: 1) and euthanized. The bodies will be collected (spleen, liver and peritoneum) and fixed with Bouin liquid for 12 hours. Tissues follow standard fixation procedures, dehydration, and inclusion diafanization using Paraplast Plus (ST. Louis, MO, USA). The materials will be sectioned on rotary microtome CUT5062 (SLEE MAINZ, Munich, Germany) with a thickness of 5 microns, stained with hematoxylin and eosin (HE) and Masson, and photographed the light

microscope Leica DM2500 (Leica, Munich, Germany) equipped with DFC295 camera (Leica, Munich, Germany) [12].

2.4. Biochemical analyses

The animals were anesthetized and before the euthanasia the blood samples were collected by cardiac puncture (left ventricle). Serum was separated by centrifugation at 6000 rpm, 5 min, and used to investigate the possible hepatotoxic effects, nephrotoxic, cardiotoxic of different GO concentrations through analysis: TGO (transaminase glutamic oxaloacetic) and TGP (transaminase glutamic pyruvic transaminase), a specific marker for injury hepatic parenchyma. Circulating levels of urea and creatinine were performed to check kidney function. The measurement were performed by the microplate reader Multi-Mode Microplate Reader Synergy H1M (Bio-Tek Instruments, USA), with temperature control and analyzed by Kcjúnior software (Bio-Tech Instruments, Inc., Winooski, VT, USA).

3. Results and Discussion

3.1. Deposition of GO in application sites

During the sacrifice were found sites of GO deposition (regions where the doses were applied) and GO concentration in the peritoneum, especially in the group received GO suspension at higher dose (5.0 µg /g) (Figures 1 A and B).

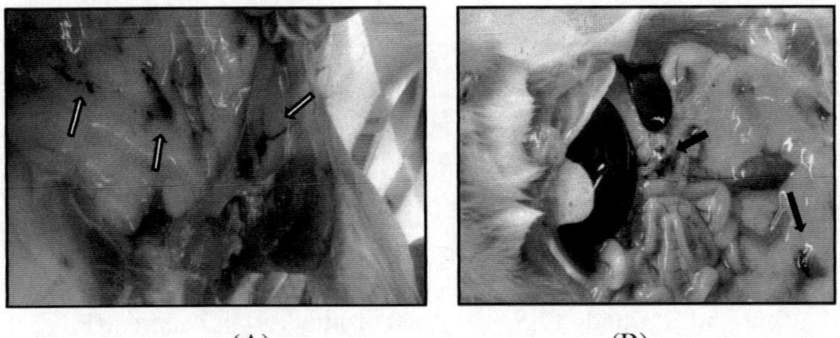

(A) (B)

Figure 1 AB. Exposure of abdominopelvic cavity in the animals from GO 5.0 µg/g group. Presence of dark agglomerates (arrows) in the peritoneum.

3.2. Systemic toxicity

The systemic toxicity was correlated with the degree of inflammation in organs: the peritoneum, liver and spleen. The degree of inflammation was assessed by a semi-quantitative scale. The inflammation was scored as: 0, absent inflammation; 1, weak inflammation (less than five lymphocytes within a grid 6200 to 10610, an area of 0.25 mm^2; 2, moderate inflammation (inflammatory mononuclear cells scattered throughout the tissue, but the stroma still clearly visible); 3, severe inflammation (mononuclear inflammatory cells infiltrating the stroma thick) (modified from [13] (Table 1).

3.2.1. Systemic Toxicity - Histophatologycal Results

Spleens, livers and peritoneum of the control groups showed no inflammation, hemosiderin deposits and GO deposits. In the group which received 0.05 µg/g GO can be seen that spleens inflammation was absent in 40% and 40% of the animals showed mild inflammation with hemosiderin deposits. At the peritoneum, although there specific areas of GO deposition, 80% of the animals showed mild inflammation, but there was no evidence of deposition in this group. Similar to the control group, the

IOP Conf. Series: Journal of Physics: Conf. Series **838** (2017) 012026 doi:10.1088/1742-6596/838/1/012026

liver of did not show inflammation and deposition of hemosiderin, but 20% of the animals showed GO deposition in the liver parenchyma. In the group receiving 0.5 μg/g GO can be observed intense inflammation hemosiderin deposits in the spleen and GO 40% of the animals. At the peritoneum, 80% of the animals showed mild inflammation, 60% of the animals showed deposition of GO. In the liver of the animals 20% of this group exhibited mild inflammation, and a deposition GO occurred in 60% of animals. The worst result occurred in the group receiving 5.0 μg/g of GO, as expected. The spleen was exhibited intense inflammation in 100% of animals, and 60% hemosiderin deposits and deposition of GO in 80% of animals. They presented peritoneum intense inflammation and deposition of GO in 100% of animals. In the liver, 10% of the animals showed mild inflammation and in 80% of animals showing deposition of GO. Some pictures of these results can be seen in figure 3.

Table 1: Degree of inflammation in the spleen, liver and peritoneum of the animals from different experimental groups (n=5).

Tissue	Groups																			
	Control					GO 0.05 μg/g					GO 0.5 μg/g					GO 5.0 μg/g				
Number of animals	1	2	3	4	5	1	2	3	4	5	1	2	3	4	5	1	2	3	4	5
Spleen	0	0	0	0	0	1	0	2[a]	2[a]	0	3[a]	3[a,b]	2	2[b]	1	3[b]	3[b]	3[a]	3[a,b]	3[a,b]
Liver	0	0	0	0	0	1	1	1[b]	1	1	1[b]	1[b]	1	1[b]	1	1	1[b]	1[b]	1[b]	1[b]
Peritoneum	0	0	0	0	0	1	1	1	1	0	2[b]	2[b]	2	2[b]	1	3[b]	3[b]	3[b]	3[b]	3[b]

Note: 0, absent inflammation; 1, weak inflammation; 2, moderate inflammation; 3, intense inflammation. a hemosiderin agglomerate. b GO agglomerate.

3.3. Genotoxicity

The number of micronucleus was significantly higher in the group of animals Positive Control (induced using a single dose of 1.5 mg/mL/rat body (7.5 μg/g) of N-methyl-N-nitrosourea (MNU)) compared to other experimental groups. The micronucleus number of animals in the GO 5.0 μg/g was significantly higher compared to GO groups 0.5 μg/g, GO 0.05 μg/g and Negative Control (without any treatment), which were not statistically different (Table 2). Still, they were observed in all concentrations, GO agglomerates from polychromatic erythrocytes and micronucleated polychromatic erythrocytes, indicating the migration of the peritoneum compounds (application site) to the bloodstream (Figure 4). The results indicate that the higher the concentration of GO, the greater the genotoxicity.

3.4. Biochemical analysis

The results showed that serum levels of TGO, TGP, urea and creatinine in treated animals 24 h before with GO 0.05μg/g, GO 0.5 μg/g and GO 5.0 μg/g did not differ significantly between itself, as well as the control group (Table 3). These results demonstrated that the GO solutions in concentrations showed no acute toxicity in the liver and kidneys.

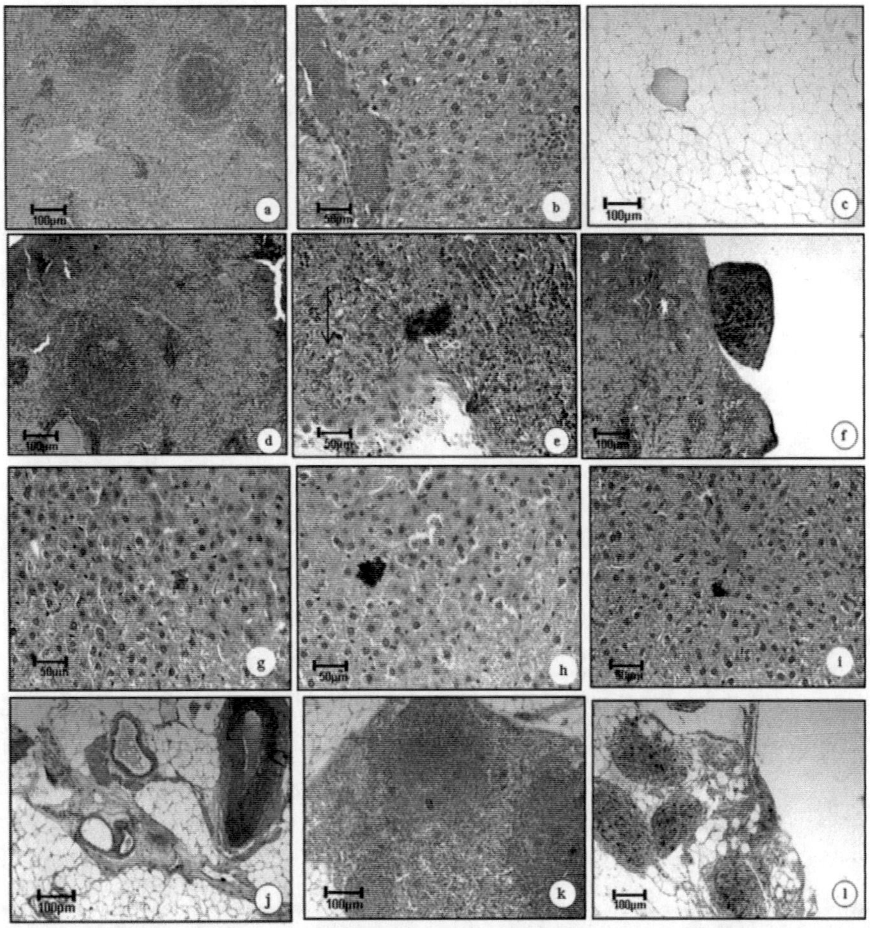

Figure 3. Photomicrographs of the spleen, liver and peritoneum of the animals from Control (a, b, c), GO 0.05 µg/g (d, g, j), GO 0.1 µg/µL (e, h, k) and GO 5.0 µg/g (f, i, l) groups. (a), (b), (c) Spleen (a), liver (b) and peritoneum (c) showed no signs of inflammation, as well as hemosiderin and GO agglomerates. (d) Spleen showed moderate inflammation. (e), (f) Intense inflammation in the spleen with GO and and hemosiderin agglomerates. (g), (h), (i) Weak inflammation was observed in the liver. (j) Peritoneum showed moderate inflammation. (k), (l) Intense inflammation in the peritoneum with GO agglomerates.

Table 2: Average of micronuclei number in relation to 3.000 polychromatic erythrocytes in the different experimental groups.

Groups	Number of Polychromatic erythrocytes	Average Micronucleus
Negative Control (n=5)	3,000	10 a
Positive Control (n=5)	3,000	283 b
GO 0.05 µg/g (n=5)	3,000	10 a
GO 0.5 µg/g (n=5)	3,000	26 a
GO 5.0 µg/g (n=5)	3,000	34 c

Different lowercase letters indicate statistically significant difference (P <0.01) between groups after the Tukey test.

Figure 4. Photomicrographs of polychromatic erythrocytes and micronucleated polychromatic erythrocytes of the animals from Negative Control (a), Positive Control (b, c), GO 0.05 μg/g (d, e), GO 0.5 μg/g (f, g) and GO 5.0 μg/g (h, i) groups.

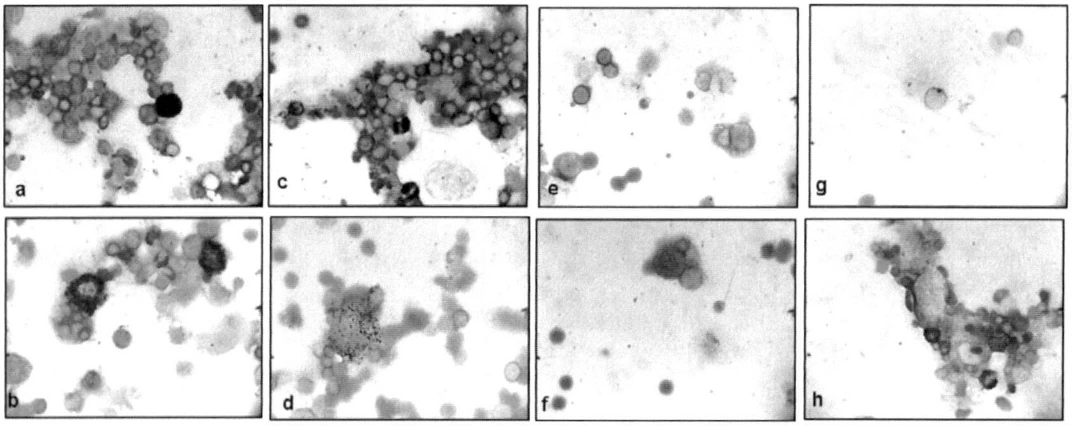

3.5. Biochemical analysis

The results showed that serum levels of TGO, TGP, urea and creatinine in treated animals 24 h before with GO 0.05 μg/g, GO 0.5 μg/g and GO 5.0 μg/g did not differ significantly between itself, as well as the control group (Table 3). These results demonstrated that the GO solutions in concentrations showed no acute toxicity in the liver and kidneys.

Table 3. Biochemical parameters of liver and renal functions 24 hours after intraperitoneal administration of GO 0.05 μg/g, GO 0.5 μg/g and GO 5.0 μg/g.

Samples	**TGP** ($U\ L^{-1}$)	**TGO** ($U\ L^{-1}$)	**Urea** ($mg\ dL^{-1}$)	**Creatinine** ($mg\ dL^{-1}$)	***P value***
Control	28.7 ± 2.1	51.7 ± 2.1	29.7 ± 1.5	0.27 ± 0.06	$P < 0.01$
GO (0.05 μg/g)	31.0 ± 2.8	53.5 ± 2.1	31.5 ± 0.7	0.31 ± 0.06	$P < 0.01$
GO (0.5 μg/g)	32.5 ±3.5	53.0 ± 2.8	27.0 ± 2.8	0.26 ± 0.03	$P < 0.01$
GO (5.0 μg/g)	27.7 ± 2.0	55.2 ± 1.6	32.06 ± 0.6	0.28 ± 0.04	$P < 0.01$

TGP (transaminase glutamic pyruvic transaminase) and TGO (transaminase glutamic oxaloacetic).
Data were expressed as mean ± standard deviation (n = 05)
None of the average values differ by Tukey test.

4. Conclusions

Evaluation of the systemic toxicity of GO solutions showed signs of inflammation in the spleen at the lowest concentration of GO (0.05 μg/g); intense inflammation, concomitant with hemosiderin regions were detected in the spleens of animals receiving doses from 0.5 μg/g and 100% of the animals was detected intense inflammation and 60% hemosiderin group recioving 5.0 μg/g. Also in 80% of the animals of this group deposition of GO was observed.

In relation to the liver, there were no signs of inflammation or deposition GO in the control group and GO of 0.05 µg/g; although, in GO groups 0.5 µg/g and 5.0 µg/g mild inflammation was observed. Deposition of GO was observed from the GO 0.5 µg/g in a considerable number of animals (60%) and even more significant in the GO group of 5.0 µg/g (80%). When in the peritoneum higher concentration of GO was applied, greater inflammation was observed. 80% of animals which received 0.05 µg/g showed mild inflammation, and 60% GO deposition in animals When they received GO 0.5 µg/g a 100% of animals exhibeted inflammation and GIO deposition. At GO 5.0 µg/g a 100% exibited high inflammation and large GO deposition.
Regarding the genotoxicity of GO solutions it was verified that the counting of micronucleus of the positive control group (MNU) was significantly higher than the other experimental groups, and the group received GO 5.0 µg/g, compared to other groups (GO 0.5 µg/g, and negative control) also showed a greater number of micronucleus. It was also possible to notice the presence of GO between erythrocytes indicating migration from GO peritoneum (application site) to the bloodstream.
Through biochemical tests can be verified that there was no change in hepatic or renal functionality of animals that received different doses of GO, 24 h after this administration. These data demonstrated that no impairment in this study was observed. Then, it could be concluded that GO of dose less than 0.5 µg/g failed to exhibit systemic toxicity and mutagenesis, while the dose of more than 5.0 µg/g exhibited obvious toxicity and showed potential to cause mutagenesis.

Acknowledgements: FAPESP (Proc. 2014/08349-5), the Laboratory of Nanostructure Synthesis and Biosystem Interactions-NANOBIOSS (MCTI), Brazilian Network on Nanotoxicology and Laboratory of Urogenital Carcinogenesis and Immunotherapy, UNICAMP.

References

[1] Shen H, Liu M, He H, Zhang L, Huang J, Chong Y, Dai J and Zhang Z 2012 *Appl. Mater. Interface,* **4** 6317
[2] Qin XC1, Guo ZY, Liu ZM, Zhang W, Wan MM and Yang BW 2013 *J. Photochem. Photobiol.* **120** 156
[3] Maity AR1, Chakraborty A, Mondal A and Jana NR 2014 *Nanoscale* **6** 2752
[4] Durán N, Guterres SS and Alves OL 2014) (Eds). *Springer* **pp 412** (ISBN 978-1-4614-8992-4)
[5] Seabra AB, Paula AJ, de Lima R, Alves OL and Durán N 2014 *Chem. Res. Toxicol.* **27** 159
[6] Chowdhury SM1, Kanakia S, Toussaint JD, Frame MD, Dewar AM, Shroyer KR, Moore W and Sitharaman B 2013 *Nature Sci. Rep.* **3 2584** 1
[7] Vallabani, NVS, Mittal S, Shukla RK, Pandey AK, Dhakate SR, Pasricha R and Dhawan A 2011 *J. Biomed. Nanotechnol.* 7 106
[8] Paul W and Sharma CP 2011.*Trends Biomater. Artif. Organs.* **25** 91
[9] Schmid W 1975 *Mutat. Res.* **31** 9
[10] Mersch, J, Beauvais MN and Nagel P 1996 *Mutat Res.* **371** 47
[11] Fenech M 2000 *Mutat. Res.* **455** 81
[12] Fávaro W, Billis A, Nunes IS and Durán N 2012 *J. Urol.* **187** E231
[13] Offersen BV, Knap MM, Marcussen N, Horsman MR, Hamilton-Dutoit S and Overgaard J 2002 *Br. J. Cancer.* **87** 1422

Natural lipids in nanostructured lipid carriers and its cytotoxicity

Paula A. Lima[1], Caroline A.D. Rampazo[1], Amanda F. Costa[1], Tiago Rodrigues[2], Carolina M. Watashi[2], Nelson Durán[1,3,4,*]

[1]Institute of Chemistry, Biological Chemistry Laboratory, University of Campinas (UNICAMP) - Campinas - SP, Brazil
[2]Center of Natural and Human Sciences, Federal University of ABC, Santo André, SP, Brazil, [3]NanoBioss (UNICAMP), SP. Brazil
[4]National Nanotechnology Laboratory (LNNano-CNPEM), Campinas, SP, Brazil

*duran@iqm.unicamp.br

Abstract. Nanostructured lipid carriers (NLCs) are active carrier systems which modulate the sustained release of actives and protect unstable compounds against degradation. NLCs can also protect skin from sun light, due to its particulates nature, which gives them intrinsic scattering properties. In this work, we present the preparation of NLCs using natural lipids and its cytotoxicity profile. It was used a vegetal butter with melting point (m.p.) ~32-40°C, an animal wax (m.p. 35-40°C) and a vegetal oil (boiling point ~120-150°C). NLCs were prepared by hot high pressure homogenization method and particles were characterized by average size (Z_{ave}), polydispersity index (PDI) and zeta potential (PZ) (Fig.1). The thermal behavior of the NLCs was studied using Differential Scanning Calorimetry (DSC). All the formulations were followed up for 60 days in order to evaluate their stability. NLCs exhibited a Z_{ave} around 150-200 nm, PDI less than 0.2 and PZ varying from -25 to -40 mV. The m.p. for the lyophilized NLCs was about 40-56°C. Cytotoxicity of the formulations were evaluated for human keratinocytes (HaCaT) and melanocytes (Melan-A) in the exponential growth phase. Cell viability was used as indicator of cytotoxicity and determined after 4 days of culture by MTT assay. It was found that the NLC formulations were not toxic against HaCaT and Melan-A cells. Results showed that the NLCs produced are potential carriers for nanocosmetics and sunscreen products.

1. Introduction

Although ultraviolet (UV) radiation is only around 5% of the total radiation emitted by the sun, UV radiation is mainly responsible for causing damage to human health [1], increasing the risk of various skin diseases such as sunburn, photo-aging and cancer [2]. The biological effects of UV radiation vary widely with wavelength and therefore, the UV spectrum is divided into three regions: UVA (320-400 nm), UVB (290-320 nm) and UVC events (100 - 290 nm) [1]. While UVC radiation is absorbed by the ozone layer in the stratosphere, the UVA and UVB radiation can reach the Earth's surface, affecting humans and ecosystems [3].

At sea level, the UVA radiation comprises about 95% of the UV radiation emitted by the sun and is mainly responsible for the persistent darkening of the skin, besides being related to premature aging and disease [3-5].

The use of TiO_2 as an active sunscreen ingredient (loading around 2-15%) for a long time has raised concerns about potential risks from the generation of free radical formation. To date, remediation attempts have concentrated on reducing the yield of free radical generation by TiO_2 upon sunlight exposure. Many new strategies are continuously developed, such as nontoxic, biocompatible shell that

neutralizes the free radicals by scavenging them with natural antioxidants before they exit the particle. This is interesting but is costly to the cosmetic industry [5´].
When Ti_2O compared with other common solar protector, as benzophenone-3 (BZ-3), the latter is biodegradable, Ti2O is not.

Although benzophenone-3 or oxybenzone (BZ-3) is used as an active ingredient in sunscreens for over 40 years [3], it has a high incidence of photoallergies' effects [4,5]. Some studies relate to BZ-3 absorption of endocrine problems and experiments using *in vitro* and on animal showed that BZ-3 influences the reproduction and the action of sex hormones. Furthermore, it is suspected that the BZ-3 influence the appearance of hormone-dependent diseases [3].
The toxicity of BZ-3 is related to the degree of penetration into the skin, which depend strongly on the nature of the vehicle in which the formulation is prepared. Properties such as solvent polarity, particle size and type of vehicle used are important in sunscreen behavior on the skin surface [2]. Thus, the development of formulations that prevent the penetration of the skin BZ-3 is a challenge for the cosmetics manufacture.
A strategy to prevent the penetration of active ingredients on the skin can be encapsulation in nanometric particles. Among them, the so-called solid lipid nanoparticles (SLNs) were developed in the early 1990s as an alternative to the system such as, emulsions, liposomes and polymeric nanoparticles. The SLNs are produced by substituting the liquid lipid (oil) of an oil / water emulsion of a solid lipid or a mixture lipids, so that the lipid matrix is solid at room temperature and body. Lipids are stabilized by 0.5 to 5% (w / w) of surfactant [6]. While these have good stability and low toxicity [7], the main disadvantage of SLNs is its low loading capacity, ranging from 25-50% depending on the solubility of the active lipid matrix method used in the preparation and polymorphic state of the lipid matrix [8].

The second generation of such material is called nanostructured lipid carriers (NLCs) and they improved the loading capacity. Lipid nanoparticles made from nontoxic lipids are potentially good systems for the carrying of sunscreens. Furthermore, the lipid nanoparticles act as physical sunscreens, due to their nanoparticulated characteristic. This effect is important because it is possible to reduce the amount of active encapsulated without compromising the effectiveness of the sunscreen, but reducing the risk of side effects [9].

Vegetable lipids are ingredients commonly found in cosmetics because of its moisturizing properties and therefore, are strong candidates in the production of NLCs. One is the cupuaçu (*Theobroma grandiflorum*) from Amazon. The seeds contain several fats and can be utilized in the extraction of butter, which is used in food and cosmetics [10]. The cupuaçu butter has a balanced composition of saturated and unsaturated fatty acids, which gives it a low melting point, facilitating the contact with skin. Moreover, it has high power of water absorption, an important feature in the preparation of cosmetics [11]. Lanolin (m.p. 36.1°C) is also another solid fat that are usually used in NLC [12].

The Buriti oil from *Mauritia flexuosa* palm native of South America exhibits rich composition in unsaturated fatty acids such as omega-6 (linoleic acid), known by the potential to reduce the effects of oxidative stress. Furthermore, it shows high β-carotene content and α-tocopherol (antioxidant) and a large oxidative stability [13].
This work aims to produce and characterize nanostructured lipid carriers for the encapsulation of benzophenone-3, using natural lipids cupuaçu butter, lanolin and oil buriti, aimed at use mainly as sunscreens, exhibiting less toxicity than the BZ3 alone.

2. Methods

2.1. Reagents

The materials were obtained from different sources (previosly being notified to PatGen-Brazil): Cupuaçu butter (courtesy from Inovam Brazil); - Buriti Oil (Beraca); - Lanolin Anhydrous U.S.P.

(Courtesy from Croda Brazil); - Pluronic F68 (Aldrich); - Benzophenone-3 (Aldrich); - Methanol HPLC grade (Merck); - Potassium chloride (USB); - Carbopol 940 (Synth).

2.2. Preparation of nanostructured lipid carriers (NLCs) by high pressure homogenization (HPH)

The preparation of NLC was performed by high pressure homogenization (HPH). It was used a methodology adapted from Durán et al. [14] and Durán and Deni [12]. Briefly : A mixture of 3-4 g of cupuaçu butter, 1-2 g of lanolin and 1-3 g of Buriti oil was melted at about 45°C. In the samples was added benzophenone-3 and waited until complete solubilization. The mixture was then poured into 300 ml of an aqueous solution of 0.3-0,7% of the stabilizer Pluronic F68 at the same temperature, under stirring of 10,000 rpm in an Ultra-Turrax T18. The mixture was kept under stirring for about 2 min. The obtained pre-emulsion was immediately added to the high-pressure homogenizer (Panda 2K, Niro Soavi). Then, 2-5 homogenization cycles were performed, with pressures of 400-800 bar. Next, the obtained formulation was cooled in an ice bath to room temperature to give the NLCs. Part of the dispersion was stored in a refrigerator (temperature around 4°C) and the other at room temperature for comparison of the effects of storage temperature stability.

2.3. NLCs characterization and stability study

2.3.1. Size of particles, polydispersity index and zeta potential

The average particle size (Z_{ave}) and polydispersity index (PDI) of NLCs were analyzed by dynamic light scattering technique (DLS) using a Nano ZS Zetasizer device Malvern, at 25°C. The dispersions CLNs were diluted in deionized water at 1: 100 and analyzed in polystyrene cuvettes. The zeta potential of the dispersions was obtained using equipment Nano ZS Malvern Zetasizer, at 25°C. Samples were diluted at 1: 100 with 1 mM KCl. All analyzes were performed in triplicate and the formulations were followed for up to 75 days.

2.3.2. Dispersion pH monitoring

The pH of CLNs dispersions were measured using the pH meter Qualxtron QX 1500 Plus, which was calibrated using commercial buffer pH 4 and 7. The formulations were followed for up to 75 days.

2.3.3. Measurement of encapsulation efficiency (EE%) and loading capacity (LC%)

The encapsulation efficiency and loading capacity of NLCs were estimated by spectrophotometer in the UV and visible regions, employing a AJ Micronal AJX-6100PC equipment. Initially, it was constructed a calibration curve of benzophenone-3 in calibration solution of methanol / water 90:10, and analyzed at wavelength 324 nm. The analyses of the patterns were carried out in triplicate.
For the preparation of the samples, it was used an adapted methodology from Luan et al. [15]. For % EE and LC% calculations, two types of sample were prepared in order to calculate the total mass of the suspensions of BZ-3 (M_T) and the mass encapsulated in nanoparticles (M_E). The total mass of the suspensions BZ-3 was estimated as follows: a known volume of the formulation was transferred to a 25 ml volumetric flask, where the volume was measured so that the ratio methanol / water was equal to 90:10 . The resulting suspension was brought to a Branson 1510 ultrasonic bath for 30 min to cause breakage of the particles. Then the samples were centrifuged in a Eppendorf 5804R centrifuge for 30 min at 5000 rpm to cause the sedimentation of the lipid fraction. The obtained solutions were analyzed by UV-vis spectrophotometer at a λ 324 nm. The analyses were performed in duplicate.
The BZ 3-encapsulated mass was determined as follows: A known volume of formulation was filtered using a microfilter centrifuge with 10 kDa cut mass in an Eppendorf centrifuge 5424-14000 rpm for 40 min. Next, the filtered lipid fraction was removed from the microfilter inverting it in a vacuum

microfilter and subjecting to centrifugation at 14000 rpm for 40 min. The solid was transferred to a 25 ml volumetric flask, which was measured using a solution of methanol / water 90:10. The resulting suspension was brought to a Branson 1510 ultrasonic bath for 30 min to cause solubilization of BZ-3 in methanol solution. Then the samples were centrifuged in a Eppendorf 5804R centrifuge for 30 min at 5000 rpm to cause the sedimentation of the lipid fraction. The obtained solutions were analyzed by UV-Vis spectrophotometer at λ 324 nm. The analyses were performed in duplicate.
The encapsulation efficiency calculations (EE%) and loading capacity (LC%) equations were performed by using equations 1 and 2. Where T_M is total mass of BZ-3 in suspensions, M_E is the mass encapsulated in nanoparticles and $M_{lipidic}$ is the mass of lipids used.

$$EE\% = \frac{M_E}{M_T} 100\% \qquad (1)$$

$$LC\% = \frac{M_E}{M_{lipídica}} \qquad (2)$$

2.3.4. Lyophilization-redispersion assays

Assays were performed by freeze (lyophilization)-re-dispersion with NLC samples without BZ-3 (CLN0) in the absence and presence of trehalose (2 to 5%), and NLC samples containing 10% BZ-3 (CLN10) in the absence and presence of glucose 5-30%. Samples were lyophilized and re-dispersed subsequently using ultrasound bath for 30 min. To assess the conservation of the nanoparticles after a freeze-re-dispersion cycle, average particle size measures and zeta potential were performed.

2.3.5. Differential Scanning Calorimetry (DSC) measurements

DSC analyzes were conducted using previously lyophilized samples. It was used a TA Instruments Q10 DSC equipment, operating in the temperature range 5°C to 55°C at a heating rate of 5°C min^{-1} under Nitrogen.

2.4. Cytotoxicity

The tetrazolium reduction assay was done as described by Denizot and Lang [16]. Briefly, 0.1 ml of serum–free medium containing 3-(4,5-dimethylthiazol-2-yl)-2,5-diphenyltetrazolium bromide (MTT) (1 mg/ml) was added to each well. After incubation of NLCs for 4 h, the supernatant was removed and the blue formazan product obtained was dissolved in 0.1 ml of ethanol with stirring for 15 min on a microplate shaker after which the absorbance was read at 570 nm, using for human keratinocytes (HaCaT)(CLS Heidelberg, Germany) and melanocytes (Melan-A) (obtained from Dr. Dorothy Bennett, St. George's Hospital, UK) in the exponential growth phase.

3. Results and discussions

The high pressure homogenization (HPH) is a technique used for years for the production of nanoemulsions for parenteral nutrition. For the production of LSNs and NLC, HPH is a reliable and powerful technique, and easily scalable for large quantities [17]. In addition, HPL leads to homogeneous products, with low tendency to aggregation and therefore more stable [18].

NLCs produced showed on the day of production, average diameter of 173.8 ± 1.9 nm with no noticeable agglomerates of particles in the micrometer size range. The NLC0 (without active) presented PDI less than 0.2 in all measures, indicating that the system has low polydispersity. Due to the low polydispersity of the produced system, the aging Ostwald can be avoided in the long term [19]. On the day that it was produced, the NLC0 presented PZ = -38.3 ± 0.9 mV, and a high value may indicate good stability related to the aggregation processes.

The physicochemical stability of the formulation in the long term is an important parameter to assess the feasibility of application of carrier systems. The NLC0 was analyzed over 154 days. It was observed subtle changes in Z_{ave} values, indicating that no significant aggregation of the nanoparticles in this period. However, for the zeta potential, there was a gradual decrease in the zeta potential module, reaching a peak of -26.6 ± 5.4 mV after 64 days of storage for samples kept at 4°C, and -31, 8 ± 5.6 mV for the specimen kept at room temperature. Next, there was an increased zeta potential module followed by a stabilization at values close to -35 mV. Zeta potential changes may be related to changes in the crystalline materials network without, however, causing the particle average size changes.

Monitoring of the pH of the NLC0 formulation showed a small variation in pH during the period of 154 days (4.77 to 4.95). A low variation in the pH is related to the chemical stability of the system, so that the NLC0 shows good chemical stability during the analysis period that is an important feature for the application of cosmetic formulations.

Moreover, it is observed that the dispersions stored at 4°C and at room temperature showed the same trends during the storage period at all parameters. Thus, the NLCs produced were stable in the long term (about 5 months) even when stored at room temperature.

The NLC2,5-15,0 (5-15% percentage of BZ-3) had, after their preparation, Z_{ave} = 175.8 ± 2.0, 178.5 ± 3.8, 166.7 ± 0.62 and 163.5 ± 3.0 nm, respectively (Table 1). It was observed that there were no significant differences in average particle size when NLC2,5 and NLC5,0 containing BZ-3 were compared to NLC0. The samples showed NLC10,0 and NLC15,0 Z_{ave} about 10 nm and smaller than NLC2,5 and NLC5,0, possibly due to the greater amount of encapsulated BZ-3, which should lead to changes in the crystalline structure of the nanoparticles. The polydispersity indexes obtained for the NLCs were lower than 0.2 on the preparation of these formulations, which indicates to obtain monodisperse samples with low tendency to Ostwald aging [19].

It was observed that, during the storage periods, there were few changes in average size of particle, which indicates that there was no formation of aggregates. This feature is important for the development of cosmetics, since the storage at room temperature brings less handling difficulties with products.

The zeta potential obtained for the NLC2.5 to 15.0 on the day of preparation were -21.1 ± 1.5, -20.0 ± 2.7, 26.8 ± 2.6 and -34, 3 ± 1.3 mV, respectively (Table 1). All values obtained for the NLCs containing BZ-3 were less negative than those obtained for the CLN0, due to possible changes in the external surface of the nanoparticles. However, steric effects are also responsible for the stability of NLCs. For systems in which there is a combination of steric and electrostatic effects, as in the case of NLCs produced in this work a zeta potential of ± 20 mV may indicate be stable [20].

The monitoring of NLCs during the storage time shows that small changes occur in Zeta potential of the formulations. In general, the NLCs stored at 4°C and at room temperature showed similar trends of zeta potential and pH in the studied periods. For all formulations containing BZ-3, Zeta potential remained in the range between -17 and -43 mV. The observed changes were due possibly to structural changes, such as the expulsion of BZ-3 to the outer regions of nanoparticles or penetration into the innermost regions. The BZ-3 interaction with the surfactant Pluronic F68, the surface of the nanoparticles could also be a factor causing changes in the surface charge and thus the zeta potential.

During the storage period, the zeta potential of most NLCs presented to more negative than -20 mV, indicating that the formulations exhibit good stability.

On the first day, NLC2,5-15,0 had pH 5.03-5.36, respectively. Then, no significant changes were observed in the NLCs pH during storage period, which can indirectly indicate that the formulations are chemically stable.

The encapsulation efficiency and entrainment capacity are important characteristics to assess the implementation feasibility of a carrier system for certain targets. In this study, both were determined using an indirect method consisting of the comparison between the amount of BZ-3 encapsulated in lipid nanoparticles, and the total amount of the colloidal system, in accordance with equations 1 and 2.

The BZ-3 is an active desirable as sunscreen exhibiting a protection factor (SPF) greater than 15, due to their strong absorption in the regions of UVA (320-400 nm) and UVB (290 - 320 nm) [21].

Quantification of BZ-3 in NLCs the samples were performed through rupture of nanoparticles by sonication using methanol, which is capable of solubilizing BZ-3 without any significant solubilization of the lipids present in the sample.

Table 1 shows the results of EE% and LC% obtained for all the NLCs2.5-15 after production. High encapsulation efficiencies were obtained for all NLCs produced, varying between 87.2 and 91.0%. The high efficiency of encapsulation can be attributed to the high concentration of liquid lipids and the complex mixture of solid lipids, which generate large imperfections in the crystal lattice of lipids, which serve as spaces for encapsulating the asset. Furthermore, BZ-3 should have high solubility in liquid lipids [22].

Table 1. Encapsulation efficiency (EE%) and loading capacity (LC%) of NLCs 2.5-15.0 and their physicochemical properties.

Formulation	% of BZ-3	EE%	LC%	Z_{av} (nm)	Zeta Potential (mV)
NLC2.5	2.5	90.8	2.4	175.8 ± 2.0	-21.1 ± 1.5
NLC5.0	5.0	91.0	4.8	178.5 ± 3.8	-20.0 ± 2.7
NLC10.0	10.0	87.2	9.1	166.7 ± 0.6	-26.8 ± 2.6
NLC15.0	15.0	87.5	14.4	163.5 ± 3.0	-34.3 ± 1.3

For the viability of developed formulations, it is important that the active is not released during storage, i.e. the %EE should not change by this time. The encapsulation efficiency and loading capability of NLC2.5 and NLC5.0 formulations were reanalyzed after 48 days storage at 4°C or at room temperature.

According to the data, there was a small increase in EE% of NLC2.5 and NLC5.0 formulations during storage. Furthermore, there was no significant difference between the NLCs maintained for 48 days at 4°C and room temperature, which indicates that storage of the formulation can be done at room temperature. This feature is very important in the development of products with possible applications cosmetic due to the good conservation of the cosmetic characteristics of the ambient conditions.

The results are consistent with other systems containing 3-BZ. Lacerda et al. [23] obtained NLCs from carnauba wax and isodecyl oleate by high pressure homogenization with 91% encapsulation efficiency for formulations containing 48 µg BZ-3 per mg of lipids. However, after 30 days of storage EE% of these systems decreased to 58.8%. The complexity of the mixture of lipids contained in NLC2.5 and NLC5.0 produced in this work, it is probably responsible for the increased stability of the encapsulation, due to increased disorganization of the crystalline lattice.

Lyophilization-re-dispersion tests were performed in order to evaluate the behavior of nanoparticulated systems after dehydration by lyophilization. For this analysis was performed using NLC0 and NCL15.0 and in the presence or absence of cryoprotectant, which would function preventing agglomeration of the nanostructured carriers. For all samples, after the addition of water and sonication, it was observed the formation of a gel lumpy appearance, unlike the prior lyophilization homogeneous dispersion. Particle size measurement and polydispersity index performed under the same conditions used in the NLCs analysis does not freeze, showed results which indicated the presence of large agglomerates and high polydispersity, which indicates that lyophilization-re-dispersion does not produce equivalent systems these early. Therefore, the storage of NLCNs tested in lyophilized form was not suitable due to not ensure the maintenance of the characteristics of systems.

IOP Conf. Series: Journal of Physics: Conf. Series **838** (2017) 012027 doi:10.1088/1742-6596/838/1/012027

The results obtained in DSC analyses show that the melting temperature ranges for NLCs from 2 to 40% liquid lipid varied very little from each other. Furthermore, the fusion of NLCs occurred within a wide temperature range, which was consistent with the fact that the NLCs were composed of a complex mixture of lipids, and not to a pure substance.

Subtle differences were observed between the melting points of the different NLCs, which indicates the possibility of using formulations with up to 40% liquid lipid. It is assumed that, the larger the amount of the lipid, the greater the benefits the NLC presented at the end, since this oil has a high composition linoleic and palmitic acids, giving it the character of a non-drying oil, e.g., prevents dryness [24]. However, as a greater amount of oil in the formulation generates particles with a larger average diameter.

Cytotoxicity of the formulations was evaluated for human keratinocytes (HaCaT) and melanocytes (Melan-A) in the exponential growth phase. Cell viability was used the indicator of cytotoxicity and determined after 4 days of culture by MTT assay (Fig.1 ab). It was found que the NLC formulations were not toxic against HaCaT and Melan-A cells. However, the free BZ-3 was toxic to HaCaT cells (IC_{50} of ~ 25 μM) and to Melan-A was less toxic (IC_{50} of ~ 200 μM).

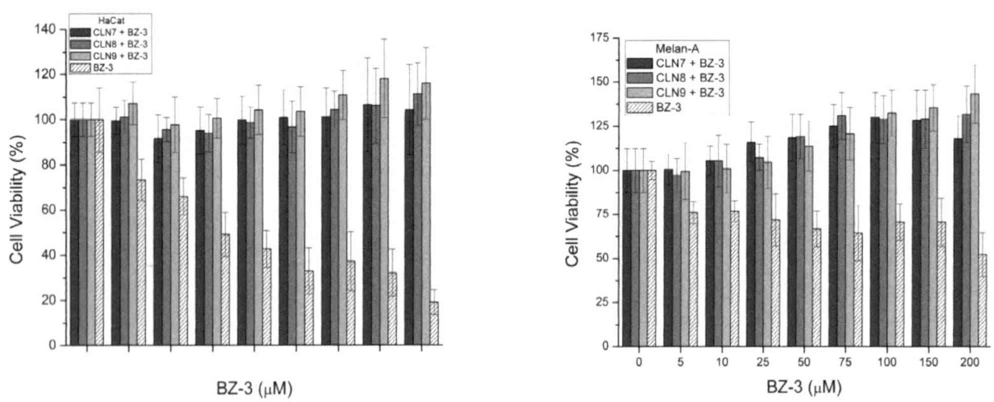

(a) (b)

Figure 1. Cell viability of HaCat cells (a) and Melan-A cells (b) treated with encapsulated BZ-3 and its free form.

4. Final remarks

Nanostructured lipid carrier's formulations were produced containing BZ-3 through the high pressure homogenization technique. The NLCs produced had average particle sizes around 175 nm and low polydispersity index, which indicates that the samples were approximately monodisperse. The zeta potential of the produced NLC was compatible with systems steric and electrostatic stabilized. The monitoring of NCLs produced at 4°C or at room temperature showed that the formulation remained stable in storage. It was observed the presence of aggregates and relatively small changes were observed in the pH and zeta potential of nanoparticles, which should be related to structural changes in particles that may have occurred during the storage period.

It was observed that there were no significant differences between samples stored at 4°C and room temperature, so that the storage of NLCs at room temperature is possible. This feature is desirable in the development of cosmetics. Finally, this nanostructured sunscreen appears biodegradable and more stable than the common sunscreen, as TiO_2, that is free radical producer and non biodegradable.

IOP Conf. Series: Journal of Physics: Conf. Series **838** (2017) 012027 doi:10.1088/1742-6596/838/1/012027

Acknowledgements

Support from FAPESP, CNPq, INOMAT (MCTI/CNPq), NanoBioss (MCTI) and Brazilian Network on Nanotoxicology (MCTI/CNPq) are acknowledged.

References

[1] Diffey BL 2002 *Methods* **28** 4
[2] Gulbake A, Jain A, Khare P and Jain SK 2010 *J. Microencapsul.* **27** 226
[3] Kim S and Choi K. *Environ. Inter.* 70, 143–157 (2014).
[4] Sambandan DR and Ratner D 2011 *J. Amer. Acad. Dermatol.* **64** 748
[5] Lautenschlager S, Wulf HC and Pittelkow MR 2007 *Lancet* **370,** 528
[5′] Morsella M, D'Alessandro N, Lanterna AE, Scaioano JC 2016. *ACS Omega* **1**, 464
[6] Pardeike J, Hommoss A and Aiman RH 2009 J. *Inter. J. Pharm.* **366** 170
[7] Poulsen HS, Kristensen NHG, Heydenreich AV and Pedersen N 2006 *Eur. J. Pharm. Biopharm.* **62** 155
[8] Weber S, Zimmer A and Pardeike J 2014 *Eur. J. Pharm. Biopharm..* **86** 7
[9] Puglia C, Bonina F, Rizza L, Blasi P, Shoubben A, Perrotta R, Tarico MS and Damiani E 2012 *J. Pharm. Sci.* **101** 301
[10] Genovese MI, Trunchado P, Tomar-Barberan FA and Pugliese AG 2013 *J. Agric. Food Chem.* **61** 2720
[11] Mancini-Filho J, Torres RP, Solis VS, Alvarado LF and Vasquez-Ocmin PG 2010 *Grasas y Aceites* **61** 390
[12] Durán N and Deni, AXP 2015 *Brazilian Patent PIBr.* 102014028002 2; PCT/BR2015/000121
[13] Moura A, Oliveira MM, Rangel JHG and Aroucha I 2010 Congresso de pesquisa e inovação da rede norte nordeste de educação tecnológica, N. **5**, Maceió. Artigos. Maceió: CONNEPI.
[14] Durán N, Caverzan J and Marcato PD 2013 *Brazilian Patente. PIBr.* 102013033412-0
[15] Luan J, Zheng F, Yang X, Yu A and Zhai G 2015 *Colloids Surf. A: Physicochem. Eng. Aspects* **466** 154
[16] Denizot F and Lang R 1986 *J. Immunol. Methods* **89** 271
[17] Mehnert W and Mäder K 2012 *Advan. Drug Deliv. Rev.* **64** 83
[18] Nalwa HS 2004 *Encyclopedia of Nanoscience and Nanotechnology* **10** 43
[19] Fan H, Liu G, Huang Y, Li Y and Xia Q 2014 *Appl. Surf. Sci.* **288** 193
[20] Tamjidi F, Shahedi M, Varshosaz J and Nasirpour A 2013 *Innovative Food Sci. Emerging Technol.* **19** 29
[21] Baughman BM, Stennett E, Lipner RE, Rudawsky AC and Schmidtke SJ 2009 *J. Phys. Chem. A* **113** 8011
[22] Badawi AA, Elbayoomy TS, Abdelmalak NS and Sanad RA 2010 *Amer. Assoc. Pharm. Sci.* **11** 1684
[23] Lacerda SP, Cerize NNP and Ré MI 2011 *Inter. J. Cosmetic Sci.* **33** 312
[24] Rincón MAD 2011 *Ph.D. Thesis*, Chemical Engineering Faculty, UNICAMP, Campinas,Brazil

IOP Conf. Series: Journal of Physics: Conf. Series **838** (2017) 012028 doi:10.1088/1742-6596/838/1/012028

Additive interaction of carbon dots extracted from soluble coffee and biogenic silver nanoparticles against bacteria

Patricia F. Andrade[1], Gerson Nakazato[2], Nelson Durán[1,3,4,*]

[1]Institute of Chemistry, University of Campinas (UNICAMP), Campinas, SP, Brazil.
[2]Biology Sciences Center, Londrina State University (UEL), Londrina, Brazil.
[3]NanoBioss, Institute of Chemistry, UNICAMP, Campinas, SP, Brazil.
[4]Brazil. Nat. Nanotechnol. Lab. (LNNano-CNPEM), Campinas, SP, Brazil.

*duran@iqm.unicamp.br

Abstract. It is known the presence of carbon dots (CDs) in carbohydrate based foods. CDs extracted from coffee grounds and instant coffee was also published. CDs from soluble coffee revealed an average size of 4.4 nm. CDs were well-dispersed in water, fluorescent and we have characterized by XPS, XRD analysis, fluorescence and by FTIR spectra. The MIC value by serial micro-dilution assays for CDs on *S. aureus* ATCC 25923 was 250 μg/mL and *E. coli* ATCC 25922 >1000 ug/mL. For silver nanoparticles biogenically synthesized was 6.7 μg/mL. Following the checkerboard assay with combining ½ MIC values of the MICs of 125 μg/mL of carbon dots and 3.4 μg/mL of silver nanoparticles, following the fractionated inhibitory concentration (FIC) index methodology, on *S. aureus* gave a fractionated inhibitory concentration (FIC) value of 1.0, meaning additive interaction. In general, the unfunctionalized CDs showed to be inefficient as antibacterial compounds, however the CDs extracted from Coffee powder and together silver nanoparticles appeared interesting as antibacterial association.

1. Introduction

Carbon dots (CDs) are carbon based photoluminescent structures, which may be applied in a wide range of areas. Photoelectronic, and electronic properties, photoacoustic imaging, bioimaging application as agents for targeted drug delivery, photodynamic therapy and in light emitting diodes, biocompatibility and biomedical applications, as gene and drug delivery nanocarriers, in particular as anticancer agents in nanomedicine. Also de CDs are non toxic compared with quantum dots (QDs) [1]. Various synthetic methods have been developed for the preparation or extraction of carbon dots, including natural products extraction [1]. Moreover, Durán et al. [1] showed that it is possible to extract CDs, under mild condition, from complex matrices. In this way, processed foods have a high potential as CDs sources.

Sk et al. [2] were the first to verify this hypothesis by extracting CDs from carbohydrate based foods, such as bread, jaggery, sugar caramel, corn flakes and biscuits. The CDs extracted were amorphous and fluorescent. Amongst the processed food, coffee grounds [3] and instant coffee [4], such as Nescafe®, are an interesting source. Coffee based carbon dots are usually attracted considerable attention because of their fascinating merits, including low cytotoxicity, good biocompatibility, excellent solubility and nitrogen present in its surface which strongly enhances the quantum yield.

CDs, in spite of its photoluminescence, in general exhibit very low cytotoxicity in several cell cultures[5], and even in *in vivo* assays. For this reason, there has been an increasing interest in using it as levels for images in many diseases [6,7]. However, some reports deals with their antibacterial activities.

Water soluble CDs extracted from rice straw [8] did not exhibit any toxicity against sewage water bacteria up to 0.5 mg mL^{-1}. In this work, a hydrothermal method was used for the preparation of water dispersible CDs by using *S. officinarum* juice as carbon source. The growth curve of *E. coli* showed that CDs were non-toxic up to 400 mg mL^{-1} for this bacterium [9].

Raw soot CDs passivated with a PEG2000N solution (CDs-NH_2) were applied to a *Salmonella typhimurium* mutagenicity test and showed no detectable mutagenicity up to 0.1 mg/plate under the experimental conditions[6]. Amino-functionalized carbon dots were synthesized in the presence of glycerin and PEG diamine and this material over 3.000 μg mL^{-1} showed no inhibition of carbon dots on *E. coli* culture [10]. Bare CDs (0.3 mg mL^{-1}) showed no zone of inhibition against Gram negative *Pseudomonas aeruginosa* or Gram positive *Bacillus subtilis* [11]. However, multifunctional CDs with glucose and poly(ethyleneimine) (PEI), which were further quaternized with benzyl bromide exhibited a significant antibacterial activity against Gram positive and Gram negative bacteria with a MIC around 16 μg mL^{-1} [12].

Considering the potential application of CDs as antibacterial agents and the simple processing to achieve CDs from processed foods, the aim of our work was to extract the different fractions (separated by size) of non-functionalized CDs from the soluble coffee Nescafe®, characterize and determine its antibacterial activity.

Herein, CDs from soluble coffee revealed an average size of 5.7 nm. They were well-dispersed in water, photoluminescent and characterized by X-ray photoelectron spectroscopy (XPS), X-ray diffraction (XRD) and Fourier-transform infrared (FTIR) spectra. The cytotoxicity study showed that the CDs from coffee did not cause any toxicity to cells at concentrations up to 20 mg mL^{-1} and may applied to cells and fish imaging[4].

2. Methods

2.1.Materials

Nescafe® Original instant coffee powder was purchased from at the supermarket of Brazil. Quinine sulfate and sulfuric acid were purchased from Sigma Aldrich and was synth, respectively. The aqueous solutions were prepared in deionized water.

2.2. Carbon dots extraction from instant coffee

The preparation process for carbon dots was carried out following a modified method of Jiang et al. [4]. Briefly, 4 g of Nescafe® Original instant coffee powder were added to 20 mL of distilled water at 90°C and the mixture vigorously stirred for 10 minutes. Afterwards, the mixture was centrifuged at 14,000 rpm for 20 min 3 time to remove particles in suspension. The resulting supernatant was further filtered through a 0.22 μm membrane to remove agglomerated particles. Finally the filtrate was purified by Sephadex G-25-40 gel filtration chromatography with deionized water as eluent. All fractions, containing carbon dots with different mean diameter, were accompanied by fluorescence measurements. The fractions obtained (~120 mL) were then lyophilized and stored at 4 °C for further characterization and use.

2.3.Biosynthesis of silver nanoparticles

The *Fusarium oxysporum (F. oxysporum)* strain used was the following: 07 SD, from ESALQ-USP Genetic and Molecular Biology Laboratory-Piracicaba, S.P., Brazil. The fungal inoculates were prepared in a malt extract 2% and yeast extract 0.5% at 28 °C in Petri dishes. The liquid fungal growth was carried out in the presence of yeast extract 0.5% at 28 °C for 6 days. The biomass was filtrated and resuspended in sterile water. Approximately 10 g of *F. oxysporum* biomass was taken in a conical flask containing 100 mL of distilled water, kept for 72 h at 28 °C and then the aqueous solution components were separated by filtration and/or centrifugation. To this solution, $AgNO_3$ (10^{-2} M) was added and kept for 72 hours at 28 °C [13].

2.4. Association of silver nanoparticles and carbon dots

To evaluate the antibacterial effects and interactions of CD nanoparticles combined with AgNPs produced by *F. oxyporum* against *Methicillin-resistant Staphylococcus aureus* (MRSA), assays of microdilution (described below) in double-antimicrobial gradient were performed.

To perform these experiments, solutions containing the minimum inhibitory concentration (MIC) of CDs and various concentrations of AgNPs (including MIC) were prepared.

2.5.Characterization

The fluorescence measurements were obtained by a fluorescence spectrophotometer (Varian, Palo Alto, USA) equipped with a 1 cm quartz cell at room temperature, used a wavelength 370 nm of the sample with excitation and emission slits of 5 mm, acquisition of emission spectra in the interval 300-700 nm.

The Attenuated Total Reflection Fourier Transform Infrared (ATR-FTIR) spectra was obtained from solid using an Agilent Spectrometer, operating at a 4 cm^{-1} resolution, with 16 scans between 500 and 4000 cm^{-1}.

CDs morphology, size and size distribution was investigated in a JEOL JEM-2100F URP transmission electron microscope operating at 200 kV, equipped with a 2k x 2k Gatan CCD camera. For that, fractions 12 and 14 were dripped on top of carbon coated copper grids and left to dry.

Carbon dots X-ray diffraction (XRD) measurements were carried out on an X-ray diffractometer (Shimadzu XRD 7000). The patterns with Cu K_α radiation ($\lambda = 0.15406$ nm) at a 40 kV voltage, 30 mA current with a 2° min^{-1} scanning rate were recorded in the 5°- 90° 2θ region.

The morphology and particle size of the AgNPs were investigated in a Carl Zeiss Libra 120 Plus (with a Ω-filter in column) transmission electron microscope, operated at an 80 kV acceleration voltage and using a tungsten thermionic source. An Olympus camera with iTEM software was used for image.

2.6.Determination of the minimum inhibitory concentration (MIC)

MIC was determined by serial micro-dilution assays in 96-well microplates, as suggested by the CLSI 15]. In brief, single colonies of bacterial cultures (on *Staphylococcus aureus* ATCC 25923 – *S. aureus* and *Escherichia coli* ATCC 25922 – *E. coli*) grown in Mueller-Hinton agar (Sigma-Aldrich, USA) media were diluted in saline solution and adjusted to 0.5 on the MacFarland scale, which corresponds to 1.5 x 10^8 CFU (colony forming units)/ml. Then, the bacterial suspensions were diluted in Mueller-Hinton broth (Difco, USA) and plated in 96-well microplates at a density of 5.0 x 10^5 CFU/well. Finally, different concentrations of the analyzed compounds were added to each well to determine the MIC values. The plates were incubated at 37ºC for 18 h, and then the optical density values at 600 nm were determined using a Bio-Rad Microplate Reader model 3550. MIC was determined at total inhibition of growth after 24 h incubation. All assays were carried out in triplicate [14].

IOP Conf. Series: Journal of Physics: Conf. Series **838** (2017) 012028 doi:10.1088/1742-6596/838/1/012028

To evaluate the MIC values of the association between both antimicrobials, the fractionated inhibitory concentration (FIC) index was used as described by Chin et al.[15].

$$FIC = MIC(Cc)/MIC(Ca) + MIC(Sc)/MIC(Sa)$$

where MIC(Cc) is the MIC of CDs nanoparticles used combined with the AgNPs, MIC(Ca) is the MIC of carbon dot nanoparticles used alone, MIC(Sc) is the MIC of the AgNPs used combined with carbon dots and MIC(Sa) is the MIC of the AgNPs used alone. FIC indexes were interpreted as follows: FIC ≤ 0.5 = synergic interaction; $0.5 < FIC \leq 1.0$ = additive interaction; $1.0 < FIC \leq 4.0$ = no interaction; FIC > 4.0 = antagonist interaction.

3. Results and discussions

Carbon dots (CDs) extracted from the soluble coffee Nescafé® were further purified by Gel Permeation chromatography (GPC) using deionized water as eluent. Every 120 mL fraction of the eluted material was accompanied by fluorescence spectroscopy. Figure 1a shows excitation and emission CDs spectra in water. The strongest emission peak was located at 460 nm with an excitation wavelength in 370 nm. Therefore, the GPC fractions analyses were taken at the optimal excitation wavelength of 370 nm. Figure 1b shows the fluorescence spectra of fraction 12 to 21, in which CDs are present. The highest concentration of CDs in fraction 12 is verified by the fluorescence intensity at 460 nm. The above results indicate that not only carbon dots were successfully obtained but also that it possess stable and strong fluorescence in water, which may serve as a biosensor. Prior to investigating its fluorescence and phosphorescent properties, it was necessary to determine the carbon dots size and composition.

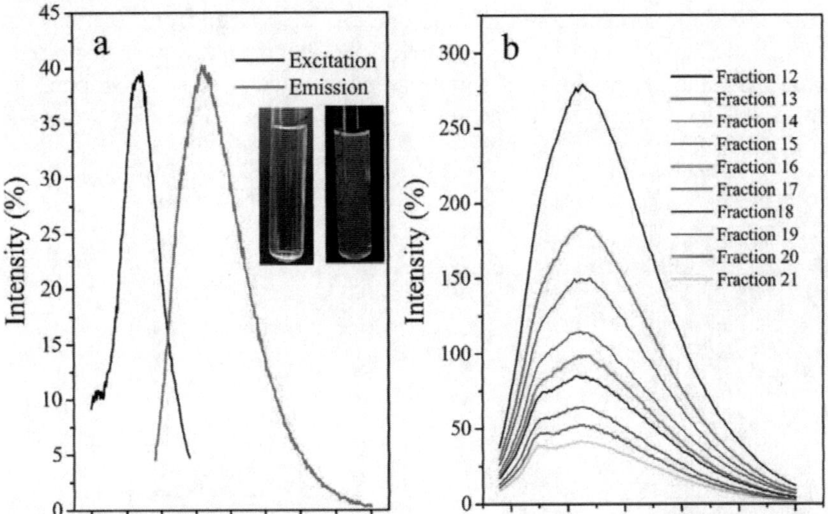

Figure 1. (a) Fluorescence emission spectra of carbon dots with excitation wavelengths in aqueous solution. Inset: C-dot containing solution in room (left) and ultraviolet (right) light, (b) Fluorescence emission spectra at room temperature of carbon dots at excitation wavelength 370 nm, obtained of different fractions.

Figure 2 shows the FTIR spectra of the Fraction 12 CDs, which showed bands centered at 3457 cm^{-1}, 3260 cm^{-1}, 1637 cm^{-1}, 1371 cm^{-1} and 1023 cm^{-1}. These bands reveal the surface functional groups present at the CDs. The band centered at 3457 cm^{-1} are assigned to stretching vibrations of O-H and N-H, while the appearance of a band at 3260 cm^{-1} can be assigned to the bending vibrations of C-N groups. The bands at 1637 cm^{-1}, 1371 cm^{-1} and 1023 cm^{-1} correspond to the vibrations of C-O-C bonds, CH_2 and C-N, respectively. Therefore, the FTIR results indicated that the surfaces of carbon dots were full hydrophilic, and presented hydroxyl, carbonyl and amine groups. The hydrophilic surface is probably the reason for the reason for the good dispersibility in water.

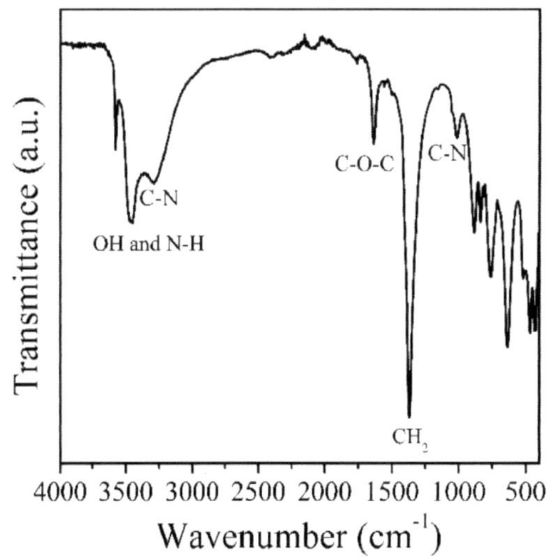

Figure 2. Carbon dots FTIR spectrum.

To investigate the CDs size and morphology, TEM micrographs were taken from fractions 12 and 14. Figure 3a shows that the CDs, indicated by white arrows, are spherical and well dispersed. Fractions 12 and 14 presented the same morphology, distribution and average size. The histogram in Figure 3b was calculated from 360 measurements of fractions 12 and 14 CDs. These measurements yielded an average diameter of 5.7 ± 2.5 nm.

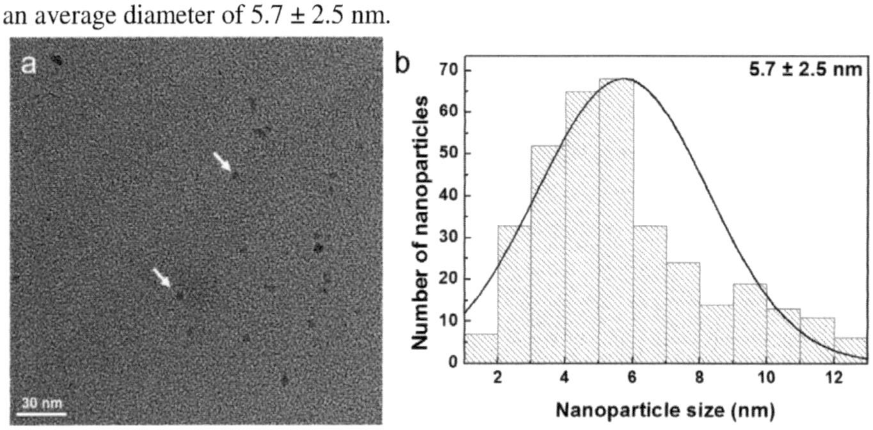

Figure 3. Carbon dots TEM micrograph (a), diameter histogram (b) The total number of particles counted for the histogram was 358.

High resolution TEM (HR-TEM) from a single fraction 14 carbon dot is shown in Figure 4a. The nanoparticle boundaries are highlighted by the white lines. It can be observed at the micrograph the atomic planes characteristic of crystalline subjects. The fast Fourier transforms (FFT), shown in Figure 4b, presents four white spots, which confirms the nanoparticle crystallinity. On the other hand, the XRD measurements of fraction 12 CDs (Figure 4c) showed only an amorphous halo with maxima at 2θ ≈ 22°. The lack of crystalline peaks at the XRD, in spite of the HR-TEM crystallinity confirmation, is due to the small particle size. Since XRD only detects long-range order, the few nanometers diameter of the CDs is insufficient to produce detectable X-ray diffraction.

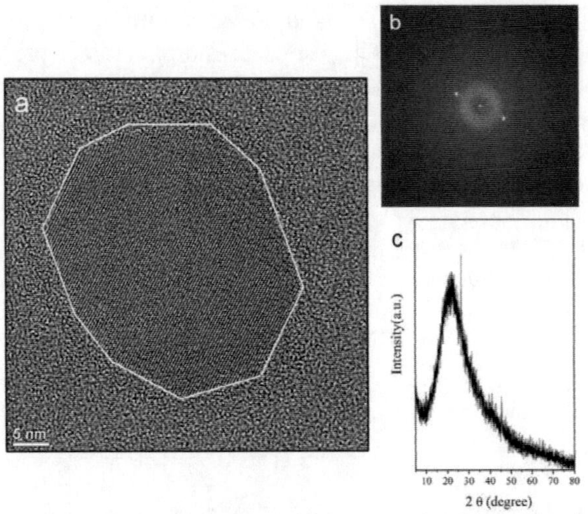

Figure 4. HR-TEM micrograph (a) and its respective FFT (b) of a single fraction 14 carbon dot. XRD diffractogram of fraction 14 CDs (c).

Fluorescence quantum yield (QY) of as prepared CDs is estimated by comparing the wavelength integrated intensity of testing sample to that of a standard (std). A solution of quinine sulfate in 0.01M H_2SO_4 (QY of 54% at 360 nm, n = 1.33) is used as a standard. The QY is determined by the following equation:

$$Q = Q_{std}\,\frac{I}{I_{std}}\,\frac{A_{std}}{A}\,\frac{n^{2}}{n_{std}^{2}}$$

Where Q is the quantum yield, I is the integrated emission intensity. A refer to absorbance, which is measured by a UV-Vis spectrophotometer, and n is refractive index of the solvent. The subscript std denotes the standard value. Specifically, the CDs is dissolved in deionized water (n=1.33) and excited under 360 nm. To avoid inner filter effect, the optical density must be below 0.1 at the excitation wavelength. 5 nm slit width is used for both excitation and emission. According to above equation, QY of the CDs is determined to be 11% (Table 1), which is better than previous reported ones (e.g., QY of 3.8% [3], 5.5% [4]) (Table 1).

Table 1: Quantum yield of CDs

Sample	Integrated emission Intensity (I)(365-700 nm)	Absorbance at 360 nm (A)	Quantum yield (Q)	Z_{av} (nm)
Quinine sulfate	39033.20	0.04	0.540	
CDs [this paper]	34843.20	0.18	0.110	5.7 ± 2.5
CDs Coffee beans[3]			0.038	5 ± 2
CDs Coffee powder[4]			0.055	3-6

To investigate the antimicrobial activity of these CDs, the MIC against *S. aureus* ATCC 25923 and *E. coli* ATCC 25922 was calculated by the serial micro-dilution method and was found to be 250 μg mL^{-1} and >1000 ug mL^{-1}, respectively.

On the other hand, the MIC of biogenic silver nanoparticles (AgNPs) on *S. aureus* ATCC 25923 was found to be 6.7 μg mL^{-1}. The morphology, size and size distribution of these AgNPs was also investigated by TEM. Figure 5a shows AgNPs with different morphologies (pseudo-spherical triangular truncated, nanorods and platelets), while the histogram, Figure 5b, showed a mean AgNPs diameter value and standard deviation of 55 ± 18 nm. It is important to point out that AgNPs with an average size of 4 nm can be also observed. Further characterization of these nanoparticles, such as zeta potential and DLS measurements, can be found at the supplementary material.

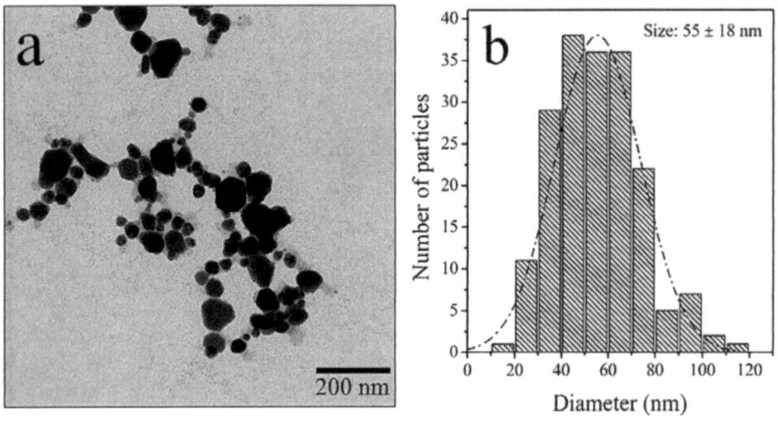

Figure 5. AgNPs TEM micrograph (a) and diameter histogram (b). The total number of particles counted for the histogram was 198.

Chin et al. [15] used the checkerboard assay methodology to combine two different antimicrobial agents. In this work, the combination of ½ MIC value of the CDs (125 μg mL^{-1}) and ½ MIC value of the AgNPs (3.4 μg mL^{-1}) through the checkerboard methodology acting on *S. aureus* gave a fractionated inhibitory concentration value of 1.0 which showed a remarkable synergy between these antimicrobial agents.

The only CDs that showed a better antibacterial activity was from a complex CDs produced by glucose and poly(ethyleneimine) (PEI), and posterior quaternized with benzyl bromide. This CDs exhibited a significant antibacterial activities against Gram (+) and Gram (-) bacteria with a MIC around 16 μg mL^{-1} [12].

4.Conclusions

CDs nanoparticles were isolated and characterized from soluble coffee (~6 nm). The CDs exerted antibacterial activity alone (MIC of 250 ug mL^{-1} on *S. aureus* ATCC 25923) and exhibited an additive effect using a checkerboard assay (combination of ½ MIC) with AgNPs biogenically synthesized. These data of the presence of CDs in human drinks may help their potential applications in cell tracking and more practical biological applications. This is a relevant result, since the CDs from different sources, such as rice straw sources [8], from hydrothermal method of CDs synthesis from sugar cane source [9], from CDs from raw soot [6], from CDs synthesized from glycerin and PEG-diamine [10] or from Gum Arabic by microwaves method [11] did not exhibited any toxicity against different kinds of bacteria as shown in this manuscript.

These data of the presence of CDs from soluble coffee, apparently without any toxic effect on humans, could be an important nanomaterial for applications in cell tracking and more practical biological applications.

Acknowledgements

Support from FAPESP, CNPq, INOMAT (MCTI/CNPq), NanoBioss (MCTI) and Brazilian Network on Nanotoxicology (MCTI/CNPq) are acknowledged. The authors also acknowledged the Brazilian National Nanotechnology Laboratory (LNNano- CNPEM) and the Electron Microscopy Laboratory (LME-LNNano) staff for TEM technical support and to Prof. T.D.Z. Atvars from Institute of Chemistry, University of Campinas, Brazil for the preliminary results on fluorescence measurements.

References

[1] Durán N, Simões MB, Moraes ACM, Fávaro WJ and Seabra AB 2016 *J. Biomed. Nanotechnol.* **12** 1323
[2] Sk MP, Jaiswal A, Paul A, Ghosh SS and Chattopadhyay A 2012 *Sci Rep* **2** 383
[3] Hsu PC, Shih ZY, Lee CH and Chang HT 2012 *Green Chem.* **14 917**
[4] Jiang C, Wu H, Song X, Ma X, Wang J and Tan M 2014 *Talanta* **127** 68
[5] Wang Y, Anilkumar P, Cao L, Liu JH, Luo PG, Tackett II, KN., Sahu S, Wang P, Wang X and Sun YP 2011 *Exper. Biol. Med.* **236** 1231
[6] Wang K, Gao Z, Gao G, Wo Y, Wang Y, Shen G and Cui D 2013 *Nanoscale Res. Lett.* **8** 122
[7] Zheng ZT, Ananthanarayanan A, Luo KQ and Chen P 2015 *Small* **11** 1620
[8] Mandal TK and Parvi N. 2011 *J. Biomed. Nanotechnol.* 7 846
[9] Mehta VN, Jha S and Kailasa SK 2014 *Mat. Sci, Eng. C* **38** 20
[10] Krishna AS, Radhakumary C, Antony M and Sreenivasan K 2014 *J. Mater. Chem. B* **2** 8626
[11] Thakur M, Pandey S, Mewada A, Patil V., Khade M, Goshi E and Sharon M. 2014 *J. Drug Deliv.* **2014** Article ID 282193.
[12] Dou Q, Fang X, Jiang S, Chee PL, Lee, TC and Loh XJ 2015 *RSC Adv.* **5** 46817
[13] Durán N, Marcato PD, Alves OL, De Souza GIH and Esposito E 2005 *J. Nanobiotechnol.* **3** 1
[14] CLSI (Clinical and Laboratory Standards Institute) 2013 Methods for dilution antimicrobial susceptibility tests for bacteria that grow aerobically. Approved Standard, M07-A8, eighth edition. CLSI, Wayne, PA, USA
[15] Chin NX, Weitzman I and Della-Lata P 1997 *Antimicrob. Agents Chemother.* **41** 850

IOP Conf. Series: Journal of Physics: Conf. Series **838** (2017) 012029 doi:10.1088/1742-6596/838/1/012029

Combined assessment of cyto-, geno-, and embryotoxicity of silica nanoparticles in experiments in vitro

G A Protasova, V B Popov, L V Shabasheva, I V Strekalovskii, Y A Panferova, N S Khlebnikova, and A A Radilov

Research Institute of Hygiene, Occupational Pathology and Human Ecology, Federal Medical Biological Agency, G/P Kuzmolovsky, Vsevolozhsky District, Leningrad Region, 199663 Russian Federation

A complex of express methods for cyto-, geno-, and embryotoxicity assessment of nanoparticles (NPs), based on the simultaneous use of three models: the cultures of pre- and post-implantation embryos of mice (egg cleavage, blastocyst formation, 1–5 day of development) and rats (head fold–neurulation stages, 30 somite pairs, 9.5–11.5 day of development), as well as human peripheral blood lymphocytes, is developed. Pathogenic (dysmorphogenic) effects of SiO_2 NPs in concentrations of 100 and 200 μg/mL on early embryo development processes were established. The cytogenetic methods used for cyto-, geno-, and mutagenicity assessment (micronuclear test under cytokinesis-block conditions, chromosomal aberrations test, and DNA comet assay) revealed no effects of SiO_2 NPs in concentrations 200, 100, and 20 μg/mL.

Introduction

The broad-spectrum applications of SiO_2 NPs in different fields, including medicine, food industry, optoelectronics, pharmacology, cosmetic industry, etc., make such NPs priority objects for detailed safety assessment [1, 2]. The toxic effects of SiO_2 NPs are first of all associated with their ability to react with cell membranes and thus stimulate excess formation of reactive oxygen species (ROSs) [3, 4]. Some research on the effect of SiO_2 NPs revealed damage of cell membranes, resulting in increased levels of reduced glutathione and malondialdehyde [5] on the background of disturbed cellular calcium homeostasis [6]. According to [7], SiO_2 NPs enter the cell nucleus and cytoplasm, react with DNA, enhance topoisomerase I-mediated DNA cleavage, inhibiting replication, transcription, and proliferation processes. Yang et al. [6] suggested that the cytotoxic effect of SiO_2 NPs can also be realized via inhibition of the mitochondrial function [6]. An important factor responsible for the potential impact of SiO_2 NPs on human health is their size which is directly related to cytostatic activity. Napierska et al. [7] showed that 70-nm silica nanoparticles exhibit a stronger cytotoxic effect compared to their microsized analogs.

Thus, analysis of published data literature showed that amorphous SiO_2 NPs have been insufficiently studied in terms of general toxicity and scarcely studied in terms of their effect on mammalian embryogenesis. Therefore, in the present work we set ourselves the aim to develop a complex approach to express in vitro cyto-, geno-, and embryotoxicity assessment of SiO_2 NPs. Previously we developed a test system including rodent pre- and postimplantation embryo cultures and the human peripheral blood lymphocyte culture for complex assessment of the effect of chemical substances and unfavorable environmental factors on the genetic apparatus of the cell and human and animal embryogenesis. The complex assessment of the pathogenic impact on embryo development and human peripheral blood lymphocytes is based on an analysis of the embryotoxic, embryolethal, and teratogenic (dysmorphogenic) effects, dynamic morphofunctional disorders, as well as cytotoxic (cytostatic) and genotoxic (mutagenic) effects [8, 9, 10, 11]. The test system was many times applied in toxicological, environmental hygienic, and clinical research, and in the present work it was applied for the first time for the express toxicity assessment of silica nanoparticles.

Materials and methods

Silica nanoparticles 12 nm in diameter were purchased from Sigma Aldrich (Aldrich-718483). The stock aqueous suspension of SiO_2 NPs (2 mg/mL) was prepared by ultrasonication of a mixture of the NPs and water. To prevent agglomeration, following the protocol developed by Bihari et al. [12], the suspension was ultrasonicated and then BSA (15 mg/mL) was added as a stabilizer. The nanoparticles were introduced into the culture medium in concentrations of 20, 100, and 200 μg/mL. The size of SiO_2 NPs in the culture solutions was estimated by means of laser correlation spectroscopy (LCS) [13]. The LCS spectra were measured at an INTOX LCS-03 laser correlation spectrometer (INTOKS OOO, St. Petersburg, Russia). The processed LCS spectra of the prepared suspensions showed the Gaussian distributions of the hydrodynamic linear particles sizes peaking at 15 ± 2 nm for the concentrations 20 and 100 μg/mL and 16± 2 nm for the concentration 200 μg/mL.

SPF animals: F1 female mice (C57BL x CBA) and outbred Wistar female rats, were obtained from the Pushchino Animal Breeding Facility (Moscow Region, Russia). The experimental animals were handled and used in accordance with the "Rules of Laboratory Practice (GLP) in the Russian Federation" (approved by the Order no. 267 of the Ministry of Health of the Russian Federation of 19.06.2003).

Culturing preimplantation mouse embryos. For dated pregnancy female mice were stimulated by injections of Puregon (10 ME) and Pregnyl (5 ME) (MSD Pharmaceuticals, The Netherlands) and placed together with male rats in a 2 : 1 proportion. The day of vaginal plug detection was designates as the gestation day 1 (GD 1). On GD 2 the female mice were sacrificed by cervical dislocation, after which 2-cell embryos were washed out from the oviducts using the EmbryoAssist medium (Origio, Denmark). Morphologically normal embryos were transferred to 40-μL droplets (15−20 embryos per a droplet) of the culture medium preloaded with SiO_2 NPs, and the cultured under paraffin oil in Petri dishes (35 mm, Denmark) by the Brinster method. For the control we used embryos cultured in a straight EmbryoAssist medium. Culturing was performed in a $CO_2/O_2/N_2$ incubator (Sanyo MCO-18M, Japan) for 72 h at 37 °C under a humid air (5% CO_2, 5% O_2, and 90% N_2). After culturing had been complete, live, dead, and abnormal embryos were counted, and the development stages were determined. Embryos were fixed in a methanol–acetic acid mixture (3:1), and cytological preparations were prepared [14] to assess the cellular mass, proliferative activity, and cell death (as determined by pyknotic nuclei score).

Culturing early rat embryos. To obtain postimplantation rat embryos, pregnant female rats were sacrificed by cervical dislocation at GD 9.5, the embryos (2–4 somite pairs) were taken out from the uterus and transferred to the culture flasks containing 6–8 mL of inactivated rat blood serum (ca. 1 mL per 1 embryo) spiked with SiO_2 NPs at concentrations of 20, 100, and 200 μg/mL. Culturing was performed for 46–48 h in rollers (10–40 rpm, rolling diameter 15 cm) at 37ºC in a $CO_2/O_2/N_2$ incubator (Sanyo MCO-18M, Japan). after which the embryos were examined for the presence of heart beating and circulation in yolk sac vessels and chorioallantois to determine the number of live and dead embryos. In live embryos, a number of main development parameters which characterize the growth processes, degree of development of structures and organs, degree of establishment and completion of morphological processes were estimated, as well as the number of embryos with congenital abnormalities and the character of the pathologies.

Assessment of gene expression. Assessment of the expression of genes involved in embryonic neurulation was performed in rat embryos (E 11.5) in control (blank) cultural medium and the cultural medium spiked with a suspension of SiO_2 NPs (100 and 200 μg/mL). Quantitative PCR using RT2 SYBR Green I qPCR primer assay kit (Sabioscience, Germany) with normalization by the 18S rRNA gene was applied. Two- or higher-fold changes in gene expression, as calculated by delta-delta-Ct method, were considered significant. The markers were genes involved in primary neurulation in rat embryos: bone morphogenetic proteins 2 and 4 (BMP2 and BMP4), sonic hedgehog (SSH), tubby-like protein 3 (TULP3, SSH inhibitor, signaling regulator), and NEUROG1. Disorders in these gene expression lead to craniofacial abnormalities.

IOP Conf. Series: Journal of Physics: Conf. Series **838** (2017) 012029 doi:10.1088/1742-6596/838/1/012029

The cytotoxicity and genotoxicity (mutagenicity) of amorphous SiO_2 NPs in the human peripheral blood lymphocyte culture was assessed using (a) micronuclear test under cytokinesis-block conditions (score of binuclear cells with micronuclei, proliferative activity, frequency of pyknosis and apoptosis) [15]; (b) chromosome aberration assay [16]; and (c) alkaline DNA comet assay to reveal primary DNA damage in cell nuclei [17].

Micronclear test under under cytokinesis-block conditions. For lymphocyte culturing, blood of healthy donors (0.5 mL) was taken from a finger with a micropipette filled with heparin and transferred to sterile tubes with the culture medium composed of 5 mL of RPMI-1640 (Sigma, USA) containing 15% of inactivated fetal bovine serum, a mixture of penicillin and streptomycin, and PHA-M (10 µg/mL). After 24-h culturing at 37°C, suspensions of SiO_2 NPs at concentrations of 100 and 200 µg/mL were added to the PHA-stimulated lymphocyte culture.

To inhibit cytokinesis and obtain polynuclear cells, cytochalasin B (6 µg/mL) was added to the medium 44 h after initiation of culturing, and in an additional 28 h the cells were separated by centrifugation (1000 rpm), resuspended in a cold (4°C) hypotonic solution (0.56% KCl) for 5 min, fixed in duplicate in a 4:1 methanol:glacial acetic acid, and air-dried using a cytocentrifuge. The preparations (20 object plates per concentration) were Giemsa stained and analyzed on an Axiostar plus microscope (Carl Zeiss, Germany); no less than 100 cells in each of 20 object plates for each concentration were analyzed [15, 18].

The cytotoxic effect was assessed using the entire cell pool, where the total scores of mono-, bi-, tri-, tetranuclear, and higher nuclear lymphocytes and their percentages were determined. The frequencies of pyknosis, apoptosis, and mitosis were calculated in per mille with respect to the absolute value, and the cytokinesis block proliferation index (CBPI) was calculated by the formula CBPI = .(M1+2M2+3(M3+M4))/N, where M1–M4 are the scores of mono-, bi-, tri-, and tetranuclear cells, multiplied by the number of cell cycles that have passed and N is the total cell score (except for pyknotic cells) [19]. The mutagenic effect was assessed by the score of micronuclei per 1000 binuclear cells.

Chromosome aberrations test. The structural alterations in the human peripheral blood lymphocyte chromosomes were identified, as well as grouping of chromosome abnormalities was performed using the unstable chromosome aberration assay [20]. The blood samples obtained as described above were introduced into sterile tubes (0.6 mL per tube) with the culture medium composed of 6.0 mL of the RPMI 1640 medium, 1.0 mL of inactivated fetal bovine serum; 50 µL of a mixture of penicillin and streptomycin, and 0.2 mL of PHA-M. After 48-h culturing at 37°C, a suspension of SiO_2 NPs at a concentration of 200 µg/mL was added to the lymphocyte culture. The known mutagen Mitomycin-C (positive control) was introduced at a concentration of 0.1 µg/mL to the lymphocyte culture at the 48th hour of culturing. The control was the straight lymphocyte culture. To block mitosis at the metaphase stage, a solution of colcemid (Sigma, USA) was added to the culture medium 1 h before the end of culturing to the final concentration of 0.6 µg/mL. After culturing had been complete, the cells were precipitated by centrifuging at 1000 rpm for 10 min. The supernatant was discarded, and the residue was poured with a warm (37°C) hypotonic solution (0.75 M KCl), the mixture was thermostated at 37°C for 20 min, centrifuged under the above conditions, and the supernatant was discarded. A freshly prepared fixer (3:1 methanol–glacial acetic acid) was added to the precipitate. The fixing procedure was repeated three times. The cytological preparations were obtained immediately after the last fixing run. The cell suspension was applied by drops on cold object glasses prewetted with distilled water, and Giemsa stained. In each experimental group, 100 metaphase plates containing no less than 44 chromosomes having no less than 3 chromosome overlaps in one metaphase were analyzed. Chromosome-type (paired acentric fragments) and chromatid-type (single fragments) aberrations on metaphase plates were scored.

Alkaline DNA comet assay. Lymphocytes isolated from the whole blood of human donors by the Ficoll method were incubated for 3 h at 37°C in the L-glutamin-free RPMI-1640 medium spiked with a suspension of SiO_2 NPs (100 and 200 µg/mL) and then applied in an agarose gel (mp $<$42°C) on slides of agarose gel (mp $<$ 65°C). Cell lysis, alkaline denaturation, electrophoresis, fixation, and staining were performed. The resulting preparations were analyzed on an Axioskop 40 luminescent microscope equipped with a digital photo camera. Analysis of the resulting images was performed using the CometScore sortware (http://tritekcorp.com). The total level of DNA damage was estimated by the percentage of cells with tail DNAs.

Statistical analysis was performed using the Bartlett's test with preliminary assessment of the homogeneity of variances by means of univariate dispersion analysis (ANOVA) with the subsequent adjustment of the level of difference between control and test groups by the Bonferroni *t*-test or Dannett q-test. In the absence of homogeneity, the group medians were compared using the Kruskal–Wallis test with Dunn's adjustment. Chi-squared test or Fisher's exact test [21, 22] were used for comparison of differences between discrete variables (embryolethal effect, pathological effects). The differences were considered significant at $p < 0.05$. Data processing was performed using the Prizm 5 software (GraphPad Software, USA).

Results and Discussion

For the assessment of cyto-, geno-, and embryotoxicity of silica nanoparticles in the present work we employed our developed combined approach based on a combined use of three models: pre- and post-implantation embryos of SiO2 NPs was performed using three models: in vitro cultured pre- and post-implantation embryos of laboratory animals (mice and rats) and human peripheral blood lymphocyte culture. Early embryos cultures were used to reveal the embryolethal and teratogenic effects (dysmorphogenesis), dynamic morphofunctional disorders, and gene expression disorders, caused by SiO2 NPs, and human peripheral blood lymphocyte cultures, to reveal their cytotoxic (cytostatic), genotoxic, and mutagenic effects.

The advantage of embryonic models is that embryonic development occurs at stages most sensitive to chemical hazards: the stage of embryo cleavage (from fertilization to formation of blastocyst capable for implantation) and initial organogenesis stages involving the most important morphogenetic processes (neurulation, body axis formation, somitogenesis, etc.), active initiation of organs (brain, vision and acoustic organs, limb rudiments, etc.), development of the embryo's support system (establishment of trophic functions, change of the respiration modes, yolk sac and liver hematopoiesis, cardiac function activation, circulation initiation, etc.).

The advantage of human peripheral blood lymphocytes is their high proliferation rate in the culture, which makes them an appropriate model for direct human cytotoxicity and genotoxicity (mutagenicity) assessment.

In vitro assessment of the effect of SiO_2 *NPs on preimplantation mouse embryos at the stages cleavage and blastocyst hatching.*

Silica NPs at concentrations of 200 and 100 μg/mL caused single embryo deaths (8.7–4.2 %) and decreased the rate of hatching, which resulted in a decrease of the total cell mass by the end of culturing by a factor of 1.3 compared to control ($P < 0.01$, $P < 0.001$). After 72-h culturing 14.3 and 17.4 % of developing embryos stayed at the early blastocyst stage and showed inhibited cavitation (Table 1, Fig. 1). As little as 85.7 and 82.6 % (control 100%) of embryos reached the formed blastocyst stage and had a well-defined internal cell mass (ICM). The number of hatching blastocysts (Table 1) was lower than in control ($P \geq 0.05$) with both concentrations of SiO_2 NPs. Hatching (the process when an embryo frees itself from the enveloping pellucid zone) is quite an important stage of embryogenesis, which creates conditions for embryo implantation within uterine crypts. The changes in the in vitro cultured preimplantation mouse embryos in response to exposure to SiO_2 NPs at concentrations of 200 and 100 μg/mL were similar both in character and in strength.

The nanoparticles at 20 μg/mL had no effect on the development of preimplantation mouse embryos. All explanted embryos reached the formed blastocyst stage by the end of culturing, and 38.1% had already hatched.

The cytological analysis of air-dried embryo preparations after exposure to SiO_2 NPs at 200, 100, and 20 µg/mL revealed no changes in the proliferative activity of blastomers (mitotic score) and cytotoxic effect (pyknotic score) with respect to control.

Table 1. Effect of SiO_2 NPs on in vitro development of preimplantation mouse embryos (direct exposure)

Concentration of SiO_2 NPs, µg/mL	Control	200	100	20
Total embryo score	17	23	24	21
Dead embryo score, total/ %	-	2/8.7	1/4.2	-
Score of embryos formed by the end of culturing: 9–16 blastomers total/ %	-	-	-	-
17–32 blastomers total/ %	-	-	1/4.3	-
33–64 blastomers total/ %	-	7/33.3	5/21.7	2/9.5
65 and more blastomers total/ %		14/66.7	17/74	19/90.5
Score of embryos at the stage of morula total/ %	-	-	-	-
early blastocyst total/ %	-	3/14.3	4/17.4	-
blastocysts total/%	17/100	18/85.7	19/82.6	21/100
including hatched blastocysts total /%	7 / 41.2	5/29.8	6/31.6	8/ 38.1
Mean nuclei score total ± m	94.88 3.95	74.29** 4.12	72.61*** 3.61	91.90 4,00
Mitotic score (per embryo) ±m	2,3 0.4	2.8 0.4	2.9 0,4	2.8 0.5
Pyknotic score (per embryo) ±m	0.8 0.2	0.8 0.2	0.8 0.2	0.9 0.2
Abnormal embryo score total /%	-	-	-	-
Net embryotoxic effect total /%	-	-	-	-

(*, **, ***) Statistically significant result with $P < 0.05$, $P < 0.01$, and $P < 0.001$, respectively, relative to control.

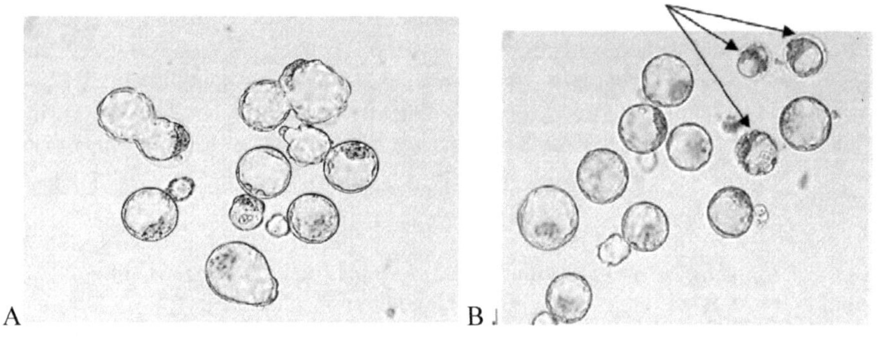

Fig. 1. Culturing preimplantation embryos for 72 h: (A) control: all blastocysts are hatching and (B) exposure to SiO_2 NPs (200 µg/mL): the arrows point to early blastocysts (few-celled blastocysts with inhibited cavitation process). Magnification 100x.

Thus, amorphous SiO_2 NPs at concentrations of 200 and 100 μg/mL showed an embryotoxic effect on preimplantation mouse embryos developing in vitro, as evidenced by slowed down cleavage entailing a decrease in the total cell mass, appearance of few-celled blastocysts with inhibited cavitation process ($P < 0.01$, $P < 0.001$), and decrease in the hatching blastocyst score (by a factor of 1.4–1.3 compared to control). Silica nanoparticles at a concentration of 20 μg/mL had no effect on the preimplantation mouse embryo development.

In vitro assessment of the effect of SiO_2 NPs on early rat organogenesis.

Culturing postimplantation rate embryos (E 9.5–11.5) in the medium containing SiO_2 NPs at concentrations of 20, 100, and 200 μg/mL caused no embryolethality, but at the highest concentration some structural (neural tube dysgraphia, brain deformation, reduction of cerebral hemispheres, micro- and anophthalmia; disturbance of axial rotation) and dynamic (brain and pericardial edemas, brain and caudal artery hemorrhages) disorders were observed (Fig. 2, A–C). General morphogenesis inhibition, specifically retardation of somitogenesis, reduction of craniocaudal dimension, and reduction of total protein (Table 2) was observed. The yolk sac growth and vascularization rates did not differ from those in control.

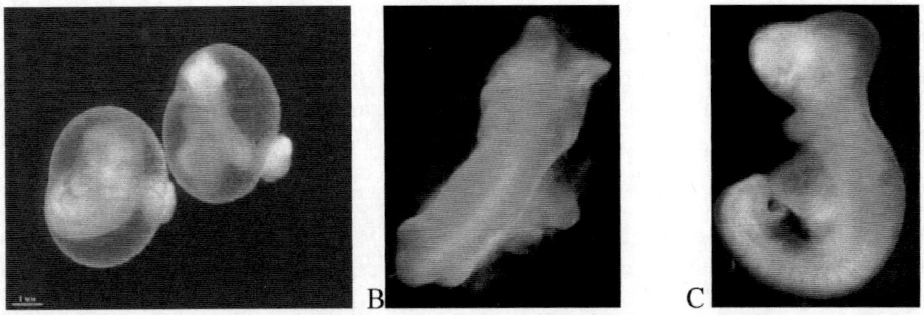
.A. B C

Fig. 2. (A) Control: rat embryos in the course of culturing (the embryo is sheathed in the visceral wall of the yolk sac, containing blood vessels and abnormal embryos after 48-h after culturing with 200 μg/mL SiO_2 NPs: (B) disturbed neurulation at the head end and (C) reduction of brain hemispheres and brain deformation. Magnification 10x.

Sister chromatid exchange test revealed no genotoxicity in embryos growing in the medium containing 200 μg/mL SiO_2 NPs.

Table 2. Effect of SiO_2 NPs on in vitro development of postimplantation rat embryos

Agent	Control	SiO_2		
Concentration, μg/mL		20	100	200
Total embryos score	42	30	22	28
Dead embryos score, abs/%	0/0	0/0	1/4.8 ±4.7	2/7.1 ± 4.1
Live embryos score	42	30	21	26
Yolk sac dimensions:				
longitudinal, mm	4.21± 0.06	4.51 ± 0.08	3.50± 0.18***	3.49± 0.11***
transversal, mm	3.83 ± 0.05	3.92 ± 0.07	3.23 ± 0.17***	3.23 ± 0.09***
vertical, mm	3.68 ± 0.05	3.79 ± 0.06	3.18 ± 0.16***	3.23 ± 0.10***
volume, mm^3	31.69 ± 1.26	35.78 ± 1.61	21.53±2.83***	20.03 ± 1.49***
Total somite score	25.50 ± 0.24	25.97 ± 0.21	20.52±1.07***	18.54 ± 0.92***
Craniocaudal size, mm	3.37 ± 0.04	3.63 ± 0.06	3.02 ± 0.11 **	2.46 ± 0.11***

Head size, mm a	1.70 ± 0.03	1.81 ±0.04	1.44 ± 0.07***	1.17 ± 0.05 ***
b	0.67 ± 0.01	0.74 ± 0.02	0.58 ± 0.03**	0.47 ± 0.02***
c	0.89 ± 0.02	0.91 ± 0.02	0.77 ± 0.04***	0.60 ± 0.02***
d	1.25 ± 0.03	1.32 ± 0.03	1.09 ± 0.05 **	0.88 ± 0.04***
e	0.94 ± 0.02	1.00 ± 0.02	0.83 ± 0.04**	0.71 ± 0.03***
Total protein, μg/embryo	375 ± 2.3	370 ± 6.6	316 ± 27**	227 ± 4.7***
Dysmorphogenesis				
total	5	5	13	23
%	10.60 ± 5.56	18.18 ± 2.34	63.70±20.62**	87.5 ± 7.97***
- structural defects	3	4	12	23
- dynamic disorders	0	2	13	16
- retarded development	4	1	11	18
SCE frequency/cell	4.10 ± 0.22	—	—	3.83 ± 0.27

(*, **, ***) Statistically significant result with $P < 0.05$, $P < 0.01$, and $P < 0.001$, respectively, relative to control.

Assessment of gene expression (quantitative PCR).

Analysis of the expression of genes involved in neurulation showed that SiO_2 NPs at concentrations of 100 and 200 μg/mL activated Neurog1 gene expression at the transcription level (6.77 and 5.1 times, respectively), and the expression levels of the other genes associated with neuronal differentiation and neurulation did not significantly differed from control (Table 3). The activation of neurogenin can be explained by either disorders in morphogenesis or neurulation processes.

Table 3. Effect of SiO_2 NPs on gene expression in rat embryos (E 11.5)

Gene	Control	SiO_2 (100 μg/mL)	SiO_2 (200 μg/mL)	Effect
BMP2	1	1.33	1.03	-
BMP4	1	1.67	1.20	-
NEUROG1	1	6.77	5.10	Enhanced expression
TULP3	1	1.61	1.26	-
SHH	1	1.22	1.30	-

Thus, high concentrations of SiO_2 NPs exert dysmorphogenic effects at early embryogenesis stages (cleavage division, blastocyst formation, early organogenesis).

Cytotoxicity and genotoxicity (mutagenicity) assessment of SiO_2 NPs in the human peripheral blood lymphocyte culture.

Micronuclear test. In the micronuclear test under cytokinesis-block conditions, the addition of SiO_2 NPs to the lymphocyte culture (200 and 100 μg/mL) caused cell clustering but without signs of aggregation. The scores of pyknotic and apoptotic cells did not change (Table 4). Slightly more expressed mitotic activity of lymphocytes was noted ($P \geq 0.05$) but statistically significant differences were not obtained. Changes were observed in the scores of polynuclear cells and their proliferative activity. The data in Table 4 show that exposure to SiO_2 NPs at concentrations of 200 and 100 μg/mL increase the scores of bi-, tri-, and tetranuclear cells with respect to control ($P<0.05$, $P<0.01$, and $P<0.001$, respectively). The cytokinesis block proliferation index (CBPI), too, revealed a statistically significant increase in the cell proliferation rate ($P<0.05$, $P<0.01$).

The micronuclear test for mutagenicity revealed no increase in the number of micronuclei in binuclear cells in the cultures containing SiO_2 NPs in all the three studied concentrations.

IOP Conf. Series: Journal of Physics: Conf. Series **838** (2017) 012029 doi:10.1088/1742-6596/838/1/012029

Table 4. Cyto- and genotoxicity (mutagenicity) assessment of SiO_2 NPs in the human peripheral blood lymphocyte culture

Agent		Control with BSA	SiO_2	
Concentration			100 μg/mL	200 μg/mL
Cytological analysis				
Total cell score,	abs	2373	2541	2242
•Mean per 100 cells	n± m	113.0±2.5	121.0±3.1	124.6±4.9*
Mononuclear cells	abs/ %	1160/48,9	1163/45.8	956/42.6
•Mean per 100 cells	n± m	55.2±1.6	55.4±2.0	53.1±3.0
Binuclear cells	abs/ %	1018/42.9	1050/41.3	1025/45.7
•Mean per 100 cells	n± m	48.5±1.2	50.0±1.4	56.9±2.3**
Trinuclear cells	abs/ %	107/4,5	164/6.5	121/5.4
•Mean per 100 cells	n± m	5.1±0.4	7.8±0.4***	6.7±0.6*
Tetranuclear cells	abs/ %	88/3.7	164/6.5	140/6.2
•Mean per 100 cells	n± m	4.2±0.5	7.8±0.5***	7.8±0.6***
CBPI	n ± m	1.6±0.02	1.7±0.02**	1.7±0.02*
#Pyknotic score	total	12	12	10
	‰	5.1	4.7	4.5
#Apoptotic score	total	9	6	3
	‰	3.8	2.4	1.3
# Mitotic score	total	49	63	57
	‰	20.6	24.8	25.4
Micronuclear test				
Binuclear cells (BC) without micronucleus (MN)		1013	1045	1021
BC with 1 MN		4	4	3
BC with 2 MN		1	1	1
BC with 3 MN		0	0	0
##Micronuclei	total	6	6	5
BC with MN	total	5	5	4

(##) Calculated per total BC score; (#) calculated per total cell score; (•) averaged over 20 object plates, with no less than 100 cells scored at each plate; (*, **, ***) statistically significant result with $P < 0.05$, $P < 0.01$, and $P < 0.001$, respectively, relative to control.

Thus, the mutagenicity assessment of SiO_2 NPs using the micronuclear test revealed no increased scores of micronuclei in binuclear cells. The addition of SiO_2 NPs at concentrations of 200 and 100 μg/mL to the lymphocyte culture produced no other effect than proliferation activation as evidenced by an increased score of polynuclear cells, which can potentially lead to tumor cell transformation. This suggestion agrees with the data in [23, 24] and calls for further research.

DNA comet assay was used to assess DNA damage in peripheral human blood lymphocytes, caused by exposure to SiO_2 NPs. The DNA damage was measured in terms of the percentage of DNA in the comet tail. After 3-h incubation of the human peripheral blood lymphocyte culture containing 200 and 100 μg/mL silica NPs, no statistically significant difference with control was found (Table 5).

Table 5. Levels of DNA damage in human peripheral blood lymphocytes after exposure to SiO_2 NPs

Method		Alkaline DNA comet assay		
Agent (concentration)		Control with BSA	SiO_2 (100 μg/mL)	SiO_2 (100 μg/mL)
Total level of cellular DNA damage				
Total comet score		136	85	176
Tail DNA %	mean ±m	5.7±0.3	7.1±0.5	6.5±0.3
Fractions of cells with damaged (tail) DNA				
0–5%	total	59	31	73
	mean ±m	1.9±0.2	2.4±0.3	2.3±0.2
5–10%	total	55	33	62
	mean ±m	7.1±0.2	7.2±0.3	7.6±0.2
10–15%	total	20	13	39
	mean ±m	12.0±0.3	11.5±0.3	12.2±0.2
15–20%	total	2	7	2
	mean ±m	15.5±0.3	16.8±0.6	15.9±0.6
> 20%	total	0	1	0
	mean ±m			

Chromosome aberrations test. The cytogenetic analysis of the human peripheral blood lymphocytes exposed to SiO_2 NPs at a concentration of 200 μg/mL revealed no statistically significant increase in the frequency of unstable chromosome aberrations, and the numbers of double and single fragments did not differ from control (Table 6).

Table 6. Frequency of metaphase chromosome aberrations in the human peripheral blood lymphocyte culture on direct exposure to SiO_2 NPs

Agent (concentration)	Negative control	Mitomycin-C (0.1 μg/mL)	SiO_2 (200 μg/mL)
Number of analyzed metaphase plates	400	400	400
Total number of aberrations, abs/%	9/2.25	71/17.75***	11/2.75
of them:			
single fragments	8/2.0	57/14.25	9/2.25
double fragments	1/0.25	14/3.5	2/0.5
Number of aberrations per one metaphase plate	0.02±0.07	0.2±0.01***	0.03±0.008

(*, **, ***) Statistically significant result with $P < 0.05$, $P < 0.01$, and $P < 0.001$, respectively, relative to control.

Thus, the cyto- and genotoxicity (mutagenicity) assessment of amorphous SiO2 NPs in the human peripheral blood lymphocyte culture by three cytogenetic methods (chromosome aberration assay, DNA comet assay, and micronuclear test) did not reveal increased scores of chromatid- and chromosome-type aberrations, single- and double-strand DNA breaks, and micronuclei in binuclear cells.

Conclusions

IOP Conf. Series: Journal of Physics: Conf. Series **838** (2017) 012029 doi:10.1088/1742-6596/838/1/012029

The developed combined approach to the express assessment of the effect of amorphous SiO_2 nanoparticles using three models (pre- and post-implantation rodent embryos, and human peripheral blood lymphocyte culture) allowed us to show that SiO2 NPs at concentrations of 200 and 100 µg/mL impairs early embryogenesis (cleavage division, blastocyst formation, early organogenesis). The cytogenetic tests in the human peripheral lymphocyte culture (micronuclear test under cytokinesis-block conditions, chromosomal aberrations assay, and alkaline DNA comet assay) revealed no cytotoxic, genotoxic, and mutagenic effects, except that SiO_2 NPs at concentrations of 200 and 100 µg/mL caused more active proliferation as evidenced by the increased score of polynuclear cells, which may result in the subsequent tumorigenic cell transformation.

Acknowledgements

The authors are grateful to A. Nikolaev, Leading Researcher, Research Institute of Hygiene, Occupational Pathology and Human Ecology, for LCS measurements and data treatment.

References

[1] Salata O 2004 Applications of nanoparticles in biology and medicine *J. Nanobiotechnol.* pp 2–3

[2] Trifonova E A, Shirkin L A 2009 *Environmental safety of nanoparticles, nanomaterials an nanotechnologies* (Vladimir: Vladimir. Gos. Univ.) pp 1–64

[3] O'Farrell N, Houlton A, Horrocks B R 2006 Silicon nanoparticles: applications in cell biology and medicine *Int. J. Nanomed.* vol 1 chapter 4 pp 451–472

[4] Unfried K, Albrecht C, Klotz L O, von Mikecz A, Grether-Beck S, Schins RPF 2007 Cellular responses to nanoparticles: Target structures and mechanisms *J. Nanotoxicol.* vol 1 pp 52–71

[5] Lin W, Huang Y, Zhou X-D, Ma Y 2006 In vitro toxicity of silica nanoparticles in human lung cancer cells *J. Toxicol.Applied Pharmacol.* pp 252–259

[6] Yang H, Wu Q, Tang M, Kong L, Lu Z 2009 Cell membrane injury induced by silica nanoparticles in mouse macrophage *J, Biomed. Nanotechnol.* vol 5 chapter 5 pp 528–35

[7] Napierska D, Leen C, Thomassen J, Lison D, Martens J A, Hoet P H 2010 The nanosilica hazard: another variable entity *J. Particle Fibre Toxicol.* pp 7–39

[8] Popov V B, Protasova G A, Shabasheva L V, Vasilyuk and Gusev R V 2008 Acute Poisoning by Chemical Toxins: A Rapid Assessment of Embryo -, Cyto -, and Genotoxic factors in the Blood of Patients *J Vestnik Rossiiskoi Voenno-Medicinskoi Akademii* vol 1 chapter 21 pp 169–174 (in Russian)

[9] Popov V B, Protasova G A 2009 Experimental emryotoxicology of chemical substances in vitro. *Toxicology, hygiene, occupational pathology in the work with highly hazardous chemicals* (St. Petersburg) pp 1–372

[10] Popov V B, Protasova G A, Protasova O V, Maksimova I A 2010 Elicitation and Evaluation of Embryopathogenic Factors in Pregnant Women Blood using the Culture of Early Embryos of Laboratory Animals *J. Molek. Meditsiny* vol 3 pp 47–52

[11] Popov V, Protasova G 2011 *In vitro Embryotoxicity Testing in Reproductive and Developmental Toxicology* (R. C. Gupta, Ed., New York: Elsevier) chapter 12 pp 147–157

[12] Bihari P, Vippola M, Schultes S, Praetner M, Khandoga A G, Reichel C A, Coester C, Tuomi T, Rehberg M and Krombach F 2008 *Particle Fibre Toxicol.* pp 1–14

[13] Lopatin V N, Priezzhaev A V, Aponasenko A.D. 2004 *Light scattering techniques in the analysis of disperse biological media* (Moscow: Izd. Fizikal i Matematikal Literatury).

[14] Tarkowski A K 1966 An air-drying method for chromosome preparations from mouse eggs *J. Cytogenet.* vol 5 pp 394–400

[15] Fenech M 2007 Cytokinesis-block micronucleus cytome assay *Nat. Protoc.* vol 2 chapter 5 pp 1084–1104

[16] Froland A A 1962 Micromethod for chromosome analysis on peripheral blood-cultures *Lancet* pp 1281–2

[17] Dhawan A, Anderson D Eds 2009 *The Comet Assay in Toxicology* (The Royal Society of Chemistry) pp 1–477

[18] Fenech M 2000 The in vitro micronucleus technique *J. Mutat. Res.* vol 455 pp 81–89

[19] Abou-Eisha A, Marcos R, Creus A 2004 Genotoxicity studies on the antimicrobial drug sulfamethoxazole in cultured human lymphocytes *J. Mutat. Res.* vol 564 chapter 1 pp 51–66

[20] Hungerford P A 1965 Leukocytes cultured from small inocula of whole blood and the preparation of metaphase chromosomes by treatment with hypotonic KCl *J. Stain. Techn.* vol 40 pp 333–338

[21] Glantz S A 1992 *Primer of Biostatistics* (3d ed., New York: McGraw-Hill) pp 1–440

[22] Zar J H 1984 *Biostatistical Analysis* (2d ed., Englewood Cliffs, NJ: Prentice Hall) pp 224–246

[23] Stavropoulou V, Xie J, Henriksson M, Tomkinson B, Imreh S and Masucci M G 2005 Mitotic infidelity and centrosome duplication errors in cells overexpressing tripeptidyl-peptidase II *J Cancer Res* vol 65 chapter 4 pp 1361–8

[24] Ingel' F I 2006 Potential uses of the micronuclear test in human peripheral blood lymphocytes cultured under cytokinesis-block conditions *Ekologicheskaya Genetika* vol 4 chapter 3 pp 11–19

IOP Conf. Series: Journal of Physics: Conf. Series **838** (2017) 012030 doi:10.1088/1742-6596/838/1/012030

Graphistrength© C100 MultiWalled Carbon Nanotubes (MWCNT): thirteen-week inhalation toxicity study in rats with 13- and 52-week recovery periods combined with comet and micronucleus assays

Jean-François Régnier[1], Daniela Pothmann-Krings[2], Sophie Simar[3], Eva Dony[4], Jean-Loïc Le Net[5] and Julien Beausoleil[6]

[1]Arkema France, Département Toxicologie et Environnement, 420 rue d'Estienne d'Orves, 92705 Colombes, France
[2]Harlan Laboratories Ltd, Zelgliweg 1, 4452 Itingen, Switzerland
[3]Institut Pasteur de Lille, 1 rue du Pr. Calmette, 59019 Lille, Cedex, France
[4]Harlan Cytotest Cell Research GmbH, In den Leppsteinwiesen 19, 64380 Rossdorf, Germany
[5]Le Net Pathology Consulting, 18 rue Henry Dunant, 37400 Amboise, France
[6]Arkema France, Groupement de Recherches de Lacq (GRL), 64170 Lacq, France.

E-mail: Jean-francois.regnier@arkema.com

Abstract. Graphistrength© C100 provides superior electrical and mechanical properties for various applications and is one of the industrial MWCNT referenced in the OECD sponsorship program for the safety testing of nanomaterials. Graphistrength© C100 is formed of MWCNT (ca. 12 walls, outer mean diameter ca. 12 nm, length ca. 1 µm) agglomerated in particles with a granulometry centered on 400 µm. A general feature of MWCNT after inhalation or intratracheal exposures is the induction of an inflammatory reaction in the lungs sometimes associated with local genotoxic effects. Most of the *in vitro* and *in vivo* genotoxicity data available on Graphistrength© C100 are negative. However, a weak DNA damage activity in the *in vitro* and *in vivo* FPG-modified Comet assays and a weak clastogenic effect in the *in vitro* micronucleus test were reported. After investigating different parameters for the aerosol generation, male and female Wistar rats were exposed by nose-only inhalation (6h/day, 5d/week) to target concentrations of 0.05, 0.25 and 5.0 mg/m^3 air of a respirable aerosol (MMAD < 3 µm) and sacrificed immediately after 4 and 13 weeks of exposure and 13 and 52 weeks of recovery after the 13-week exposure. Clinical, biological and histological evaluations were performed according to the OECD TG 413. Broncho-alveolar lavage fluid (BALF) was collected and analysed for cytokines and inflammatory parameters. Immediately after 13 weeks of exposure, chromosomal aberrations in the bone marrow cells of males and females were evaluated by the micronucleus test (OECD TG 474) and DNA damage in the lung, kidney and liver cells of males were assessed by both the standard and the human 8-oxoguanine DNA N-glycosylase 1 (hOGG1)-modified comet assay (OECD TG 489). Concentration-related deposition of black particles (MWCNT) was observed in lungs. At all sacrifice periods, an inflammatory lung reaction was observed in rats exposed to 5.0 mg/m^3 associated with changes in the differential white blood cells counts. The lung inflammation was characterized by changes in the cytological, biochemical and cytokine parameters of the BALF, an increase of the lung weight, an interstitial inflammation mainly around the alveolar ducts at the bronchiole-alveolar junction and a cell hypertrophy/hyperplasia in the terminal and respiratory bronchioles. The slight changes in BALF parameters observed at 0.25 mg/m^3 recovered after the 13-week treatment-free period and were not associated with any of the histological changes observed in lungs at 5.0 mg/m^3. Signs of lung clearance of the MWCNT were observed at 0.05 and 0.25 mg/m^3. After a one year treatment-free period, the inflammatory lung reaction was slight and of similar intensity that at the earlier sacrifice periods. Additional findings were minimal/slight bronchiolar/alveolar cell hypertrophy/hyperplasia and focally extensive alveolar septal fibrosis. No other pathological change was observed, nor was there any brain translocation via the olfactory bulb. The microscopic observations of the pleura were unremarkable. Neither increase in the number of micronucleated polychromatic erythrocytes nor increase in percent DNA damage were observed at any concentration. In conclusion, a lung inflammation characteristic of an overload with insoluble particles was observed after a 13-week inhalation exposure to 5.0 mg/m^3 of Graphistrength© C100. A No-Observed Adverse Effect Concentration (NOAEC) of 0.25 mg/m^3 was established for the repeated-dose toxicity and Graphistrength© C100 appears of low concern in term of local and systemic genotoxicity.

1. Introduction

Graphistrength© C100 provides superior electrical and mechanical properties for various applications. Graphistrength© C100 is formed of entangled MWCNT agglomerated in particles with a granulometry centered on 400 µm and contains a residual amount (<0.23 %) of small agglomerates (<15 µm) [1]. These small particles are comparable to those observed in the atmosphere of our high safety laboratory dedicated to scientific experiments with MWCNT [2], indicating a possible inhalation exposure to these small particles in case of an insufficient workers protection.

A general feature of MWCNT after inhalation or intratracheal exposures is the induction of an inflammatory reaction in the lungs sometimes associated with local genotoxic effects [3]. Therefore, conducting an inhalation subchronic toxicity study combined to genotoxicity assays was judged to be a key feature in the safety assessment of Graphistrength© C100.

An aerosol generation procedure was developed [4] in order to perform a valid study which fulfils the requirement of the inhalation OECD test guideline no. 413 [5]. The micronisation process used is an enrichment of the small particle fraction and is allowing the worst case material to be used as expected by the regulatory authorities. Another important criterion was ensuring that the administered aerosol has physico-chemical properties similar to the original material. After a careful evaluation, the defined technical conditions for the generation of respirable aerosols from Graphistrength© C100 were assessed in a 5-day range finding inhalation toxicity study in rats with a 28-day recovery period [4]. Then, a 13-week inhalation toxicity study was performed in rats, including an interim sacrifice after 4 weeks of exposure, 13- and 52-week recovery periods and the evaluation of the pulmonary inflammation parameters. A part of this study have been previously published by the same authors [4], in addition, the results obtained after the 4-week exposure and the 52-week recovery periods are presented in this article for completeness.

There are a number of publications reporting *in vitro* and *in vivo* genotoxicity studies on Graphistrength© C100 [4] under the coded named NM 402 or JRCNM04002a [6]. A weak DNA damage activity was reported in the *in vitro* (C3A and HK-2 cells) [7][8] and the *in vivo* (lung cells) [9] FPG-modified Comet assay. A weak clastogenic effect was observed in the *in vitro* micronucleus test in human lymphocytes [10] and A549 cells [11][12]. On the other hand, negative results were also reported using the same *in vitro* and *in vivo* assays [11][12][13][14].

Therefore, the genotoxic potential of Graphistrength© C100 was evaluated in cells directly in contact, and at a distance in case material was translocated from the lungs. A standard and hOGG1-modified comet assay was performed on the lung, liver and kidney cells of the rats exposed by inhalation for 13 weeks to Graphistrength© C100. The hOGG1-modified comet assay [15] was chosen because it is more specific than the FPG (formamidopyrimidine glycosilase) comet assay for the identification of oxidative DNA damage. A micronucleus assay was performed on the bone marrow cells of the rats exposed to Graphistrength© C100 in case of translocation from the lungs to the bone marrow.

2. Methods

The study was conducted in an AAALAC-accredited laboratory according to the OECD GLP Principles, the OECD test guidelines no. 413 [5], 474 [16] and 489 [17] and the OECD recommendations [18] for the revision of the tests guidelines applicable to the inhalation toxicity testing of nanomaterials.

2.1. Test materials

Graphistrength© C100 (Figure 1) is formed of tightly bound spherical, ovoid or irregular shaped agglomerates (granulometry centred on 400 μm) of entangled MWCNT (diameter ≈ 12 nm, wall number ≈ 12, length ≈ 1 μm, 92% C, 3% Al, 2.7% Fe).

Figure 1. Scanning Electron Microscopic images of Graphistrength© C100

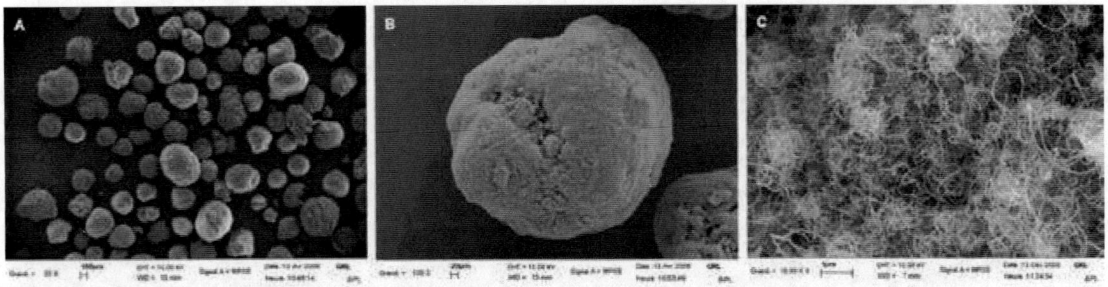

Legend: (A) Magnification: 22 fold. (B) Magnification: 120 fold. (C) Magnification: 10'000 fold.

2.2. Aerosol generation and monitoring

Graphistrength© C100 was ground in a ceramic ball mill for 12 hours under an argon atmosphere to reduce oxidation. The high aerosol concentrations was generated from the milled and sieved Graphistrength© C100 using a SAG 410 Solid Aerosol Generator connected to a micronizing jet mill and a cyclone and two elutriators thereafter (Figure 2). The low and intermediate concentrations were achieved by serial dilution of the high concentration. Aerosol concentrations were determined by gravimetric analysis. The cumulative particle size distribution was determined using Mercer cascade impactors and a Wide Range Particle Spectrometer©.

IOP Conf. Series: Journal of Physics: Conf. Series **838** (2017) 012030 doi:10.1088/1742-6596/838/1/012030

Figure 2. Aerosol generation and animal exposure

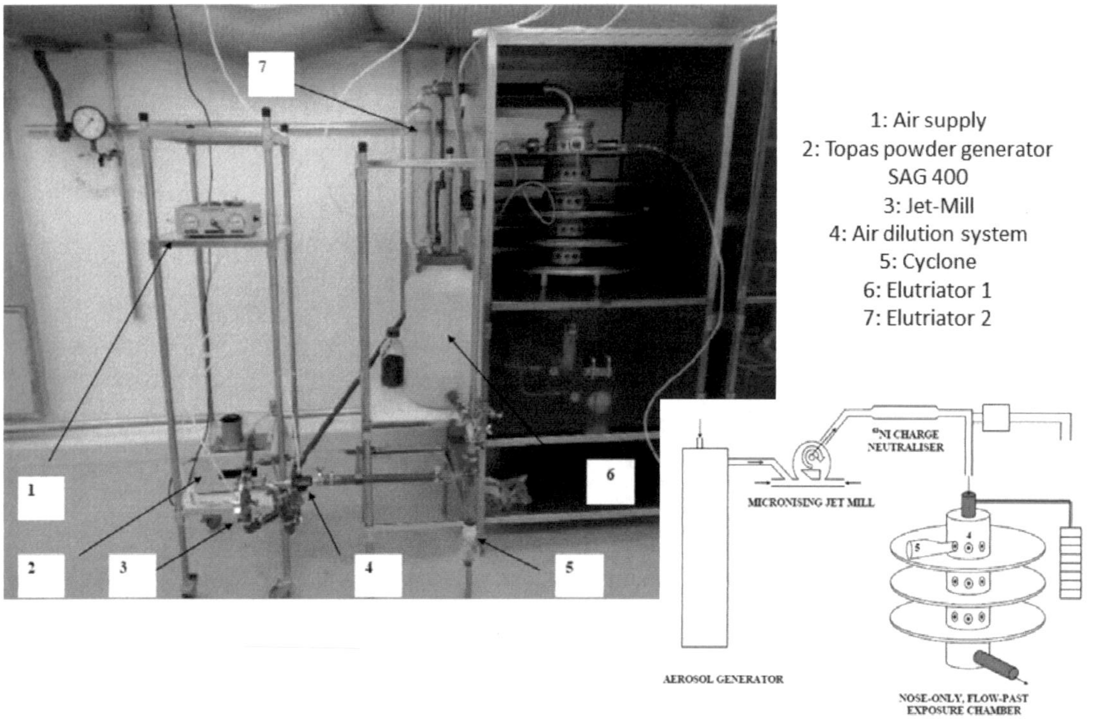

2.3. Physico-chemical characterisations
The original Graphistrength© C100 and samples taken at different steps of the aerosol generation process during the method development and during the inhalation exposure were analyzed for physico-chemical parameters (see Table 2).

2.4. Animal exposure

2.4.1. Graphistrength© C100. Four groups of 35 8-week old male and female Wistar rats were exposed by nose-only inhalation, 6 h/day, 5 d/week for 4 or 13 weeks to a respirable aerosol (MMAD 1.62-2.30 µm) at target concentrations of 0, 0.05, 0.25 and 5.0 mg/m^3 air and measured concentrations of 0, 0.06, 0.28 and 4.84 mg/m^3.

2.4.2. Positive controls for the genotoxicity studies. For the micronucleus assay, cyclophosphamide monohydrate (CPA, 20 mg/kg) was administered orally 24 h before tissue sampling. For the Comet assay, methyl methanesulfonate (MMS) was administered orally three times at 100 mg/kg bw at approximately 24 h intervals.

2.5. Study design

2.5.1. Repeated dose toxicity. Clinical, biological (haematology, biochemistry and urinalysis), broncho-alveolar lavage (BAL), sperm, oestrus cycle and histopathological examinations were performed according to the OECD TG no. 413 [5]. Sacrifices were performed after 4 and 13 weeks of exposure and after 13- and 52-week treatment-free periods (Table 1). BAL fluid (BALF) was examined for cytological and biochemical parameters and cytokine levels (TNF-α, IL-1α, IL-1β and IL-5). The olfactory bulb was examined microscopically at the end of the 13-week exposure. Microscopic

IOP Conf. Series: Journal of Physics: Conf. Series **838** (2017) 012030 doi:10.1088/1742-6596/838/1/012030

examinations on the 13- and 52-week recovery rats were limited to the respiratory tracts, pleura, heart and/or aorta.

Table 1. Design of 4/13-week exposure and 13/52-week recovery study

Examinations	4-week interim animals	Main animals	13-week recovery animals	52-week recovery animals
Clinical signs, body weight, food consumption				
Ophthalmology				
Oestrus cycle				
Functional observation battery				
Blood pressure				
Hematology, blood chemistry and urinalysis				
Bronchoalveolar lavage fluid				
Full histopathology				
Respiratory tract histopathology				
Sperm analysis				
Genotoxicity assays				

2.5.2. Micronucleus test. Femoral bone marrow was collected from the last 5 males and 5 females of each group of 10 rats, 24 h after the last day of the 13-week exposure to Graphistrength© C100 or the single administration of the positive control. Nucleated cells were separated from the erythrocytes using the method of Romagna [19]. The preparation and the scoring of the slides and the interpretation of the data were performed as recommended by the OECD TG 474 [16].

2.5.3. Comet assay. Parts of the lung, kidney and liver were collected from the first 5 males from each group of ten rats sacrificed 24 h after the last day of the 13-week exposure to Graphistrength© C100 or 6 hours after the last treatment with the positive control. The organs were minced using fine scissors in cold mincing buffer. The preparation and the scoring of the slides and the interpretation of the data were performed as recommended by the OECD TG 489 [17].

3. Results

3.1. Physico-chemical analysis
Minor changes (Table 2) were noted between the starting material and the ball milled and aerosol samples. For apparent density, surface to volume ratio, and MWCNT length, the changes observed are inherent to the process to generate the aerosol form.

IOP Conf. Series: Journal of Physics: Conf. Series **838** (2017) 012030 doi:10.1088/1742-6596/838/1/012030

Table 2. Physico-chemical characterization of Graphistrength© C100 before and after aerosol generation

	Graphistrength© C100 original	**Graphistrength© C100** 12-h ball-milled under Argon	**Graphistrength© C100** aerosol[1]
Apparent Density (g/cm^3) (mean ± sd) by porosimetry with Hg intrusion	0.106 ± 0.06 (n = 3)^B	0.2, 0.2^B	0.17, 0.18^B
Elementary organic analysis by calcination			
% C	92.0, 91.6	91.1, 90.8	90.2, 90.1
% H, N, O	< LoD	< LoD	< LoD
Ash content (%) by calcination	8.2 ± 0.0 (n = 3)	nd	nd
Particle Size Distribution (μm) by laser			
D_{10}	223	9.3	
D_{50}	418	27.0	MMAD: 1.62-2.30[2]
D_{90}	655	57.2	
Specific area (m²/g) by BET	225.6	244	242
Metal Content by lithium tetraborate fusion			
Al (% w/w)	3.0 ± 1.5 (n = 4)	2.9, 3.0	3.0, 3.0
Fe (% w/w)	2.7 ± 0.6 (n = 4)	2.2, 2.3	2.1, 2.1
Chemical Surface Analysis by XPS			
C (% w/w)	99.5 ± 0.2 (n = 14)	99.1 ± 0.2 (n = 4)	99.2 ± 0.3 (n = 4)
O (% w/w)	0.54 ± 0.20 (n = 14)	0.70 ± 0.12 (n = 4)	0.62 ± 0.22 (n = 4)
N (% w/w)	< 0.2 (n = 14)	< 0.2 (n = 4)	< 0.2 (n = 4)
Al (% w/w)	< 0.2 (n = 14)	0.17 ± 0.06 (n = 4)	0.13 ± 0.08 (n = 4)
Fe (% w/w)	< 0.2 (n = 14)	<0.1 (n = 4)	<0.1 (n = 4)
Diameters by TEM			
External Diameters (nm) (mean ± sd)	12.1 ± 3.5	12.1 ± 3.5	11.8 ± 3.0
Internal Diameters (nm) (mean ± sd)	4.4 ± 1.5		
Walls number (mean ± sd) by TEM	12 ± 4	12 ± 5	12 ± 4
Lenght (nm) by TEM			
mean ± sd	1069 ± 1102	713 ± 537	750 ± 623
D_{50}	708	569	563
Surface to Volume ratio (m^{-1})	$2.4 \cdot 10^7$	$4.9 \cdot 10^7$	$4.2 \cdot 10^7$
Ends and alignment of carbon by TEM Nanotubes (% open tips)	20	nd	25

[1]collected in the inhalation exposure system
[2]gravimetric determination

There was no apparent alteration by TEM of the MWCNT structure between the original, milled and aerosolized Graphistrength© C100 (Figure 3).

Figure 3. SEM and TEM images of Graphistrength© C100 aerosol

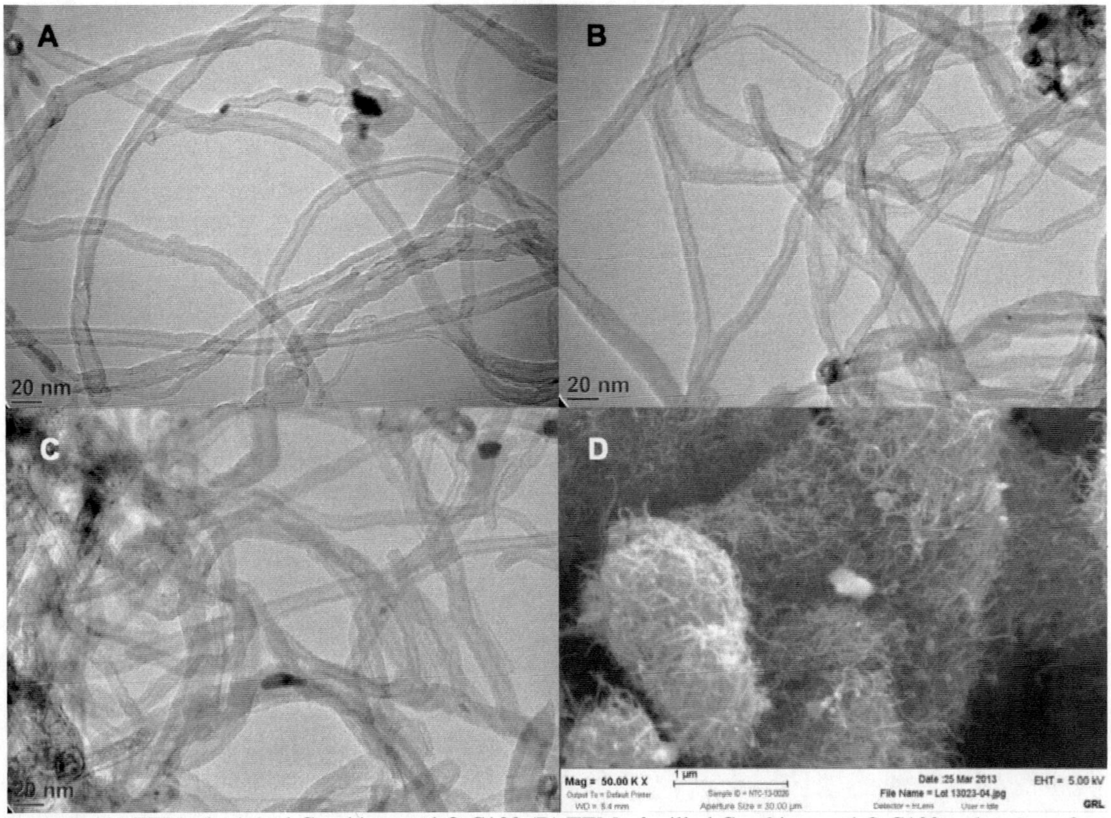

Legend: (A) TEM of original Graphistrength© C100 (B) TEM of milled Graphistrength© C100 under argon for 12 h and sieved (63 μm) (C) TEM of aerosol sample (D) SEM of aerosol sample. (A, B, and C: 350'000x, D: 50'000x)

3.2. *In life animal observations*

No mortality, specific clinical signs and exposure-related adverse effects on body weight gain, food consumption, FOB parameters, blood pressure, ophthalmoscopic examinations, blood chemistry, estrus cycle and urinalysis parameters were observed.

An increase in relative and absolute blood neutrophil counts and slight decrease of the relative (but not absolute) lymphocyte counts were noted at all sacrifice periods in both sexes of rats exposed to 5.0 mg/m^3.

3.3. *Post-mortem observations*

3.3.1. Semiology. No exposure-related adverse effects were observed on sperm counts, motility and morphology after 13 weeks of exposure and 13 and 52 weeks of recovery.

3.3.2. BALF. Presence of black particles was observed in the BALF of almost all exposed rats, from minimal at 0.05 mg/m^3 to severe at 5.0 mg/m^3.

At all sacrifice periods, a significant increase in neutrophils and lymphocytes with a concomitant decrease in macrophages were observed in BALF of rats exposed to 5.0 mg/m^3 (Figure 4). At 0.25 mg/m^3, the effect was slight after 13 weeks of exposure and reversible after 13 and 52 weeks of recovery.

From 4 weeks of exposure, significant changes in biochemical parameters were observed in rats exposed to 5.0 mg/m^3 (Figure 4), maximal after 13 weeks of exposure and slightly improved during the recovery periods. At 0.25 mg/m^3, GGT increased after 13 weeks of exposure, fully recovered in males and partially in females.

TNF-α levels were increased at 0.25 and 5.0 mg/m^3. Levels decreased after 13 weeks of recovery.

Figure 4. BALF parameters of male and female rats after 4 (A, B) and 13 weeks (C, D) of exposure to Graphistrength© C100

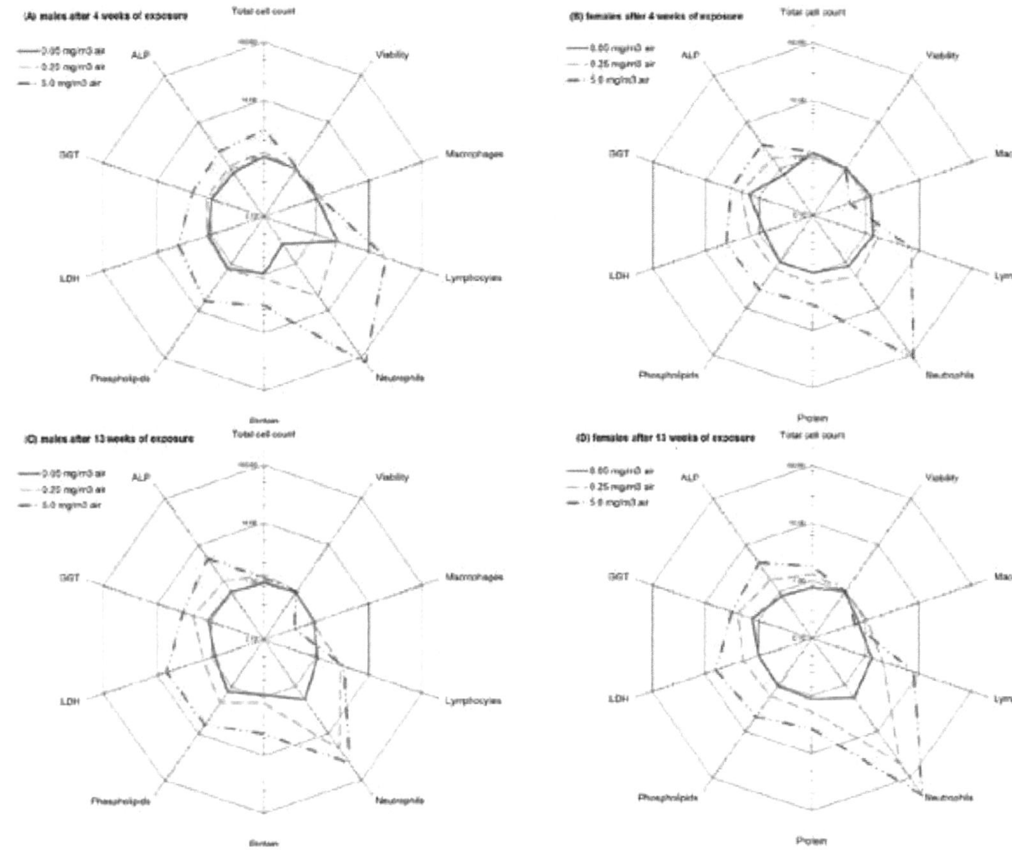

IOP Conf. Series: Journal of Physics: Conf. Series **838** (2017) 012030 doi:10.1088/1742-6596/838/1/012030

Figure 4 (continued). BALF parameters of male and female rats after 13 (E, F) and 52 (G, H) weeks of recovery

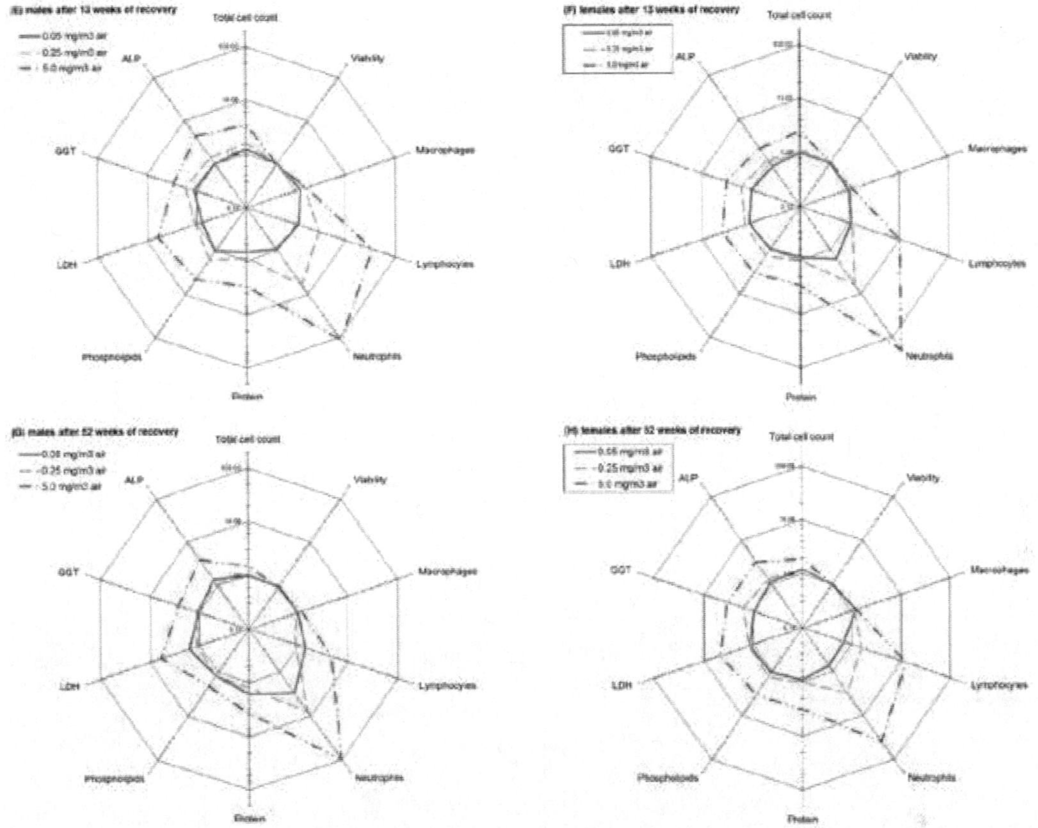

Changes are shown as x-fold differences compared to controls using a logarithmic scaling.
Abbreviations: ALP: alkaline phosphatase, GGT: γ-glutamyltransferase, LDH: lactate dehydrogenase.

3.3.3. Macroscopic findings. Dark red discoloration of the lung at the 4-week interim sacrifice, and black brown discoloration of the lung and/or greenish foci at the other sacrifice periods were seen in most of the animals exposed to 5.0 mg/m^3. Black discoloration was also recorded in the bronchial lymph nodes of most of these animals.

3.3.4. Organ weights. Increase of the absolute and relative lung weight, maximal at the 13-week recovery sacrifice, was observed at 5.0 mg/m^3 (Figure 5).

Figure 5. Changes of the relative weight lungs of male and female rats after 4 and 13 weeks of exposure to 5.0 mg/m^3 and 13 and 52 weeks of recovery.

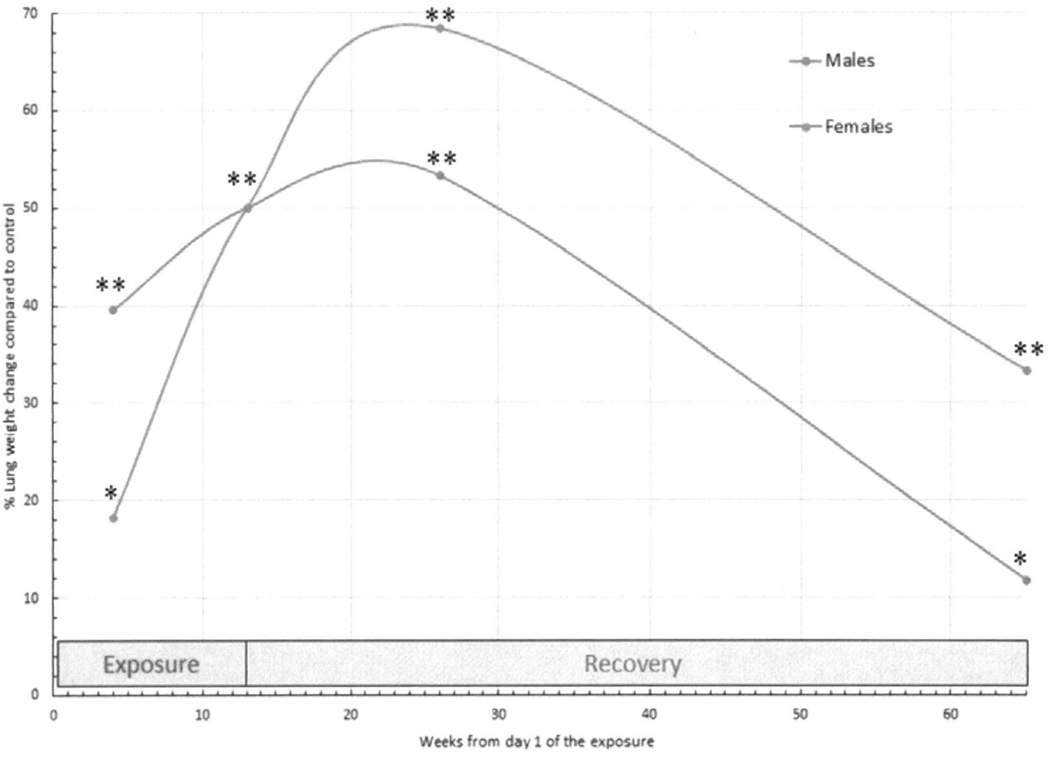

3.3.5. Microscopic findings. Deposition of variably-sized and shaped black particles, localized in the lungs, within the alveolar macrophages, were observed in all rats exposed to 0.05 and 0.25 mg/m^3 and within tissue macrophages or free within the alveolar lumen in rats exposed to 5.0 mg/m^3 (Figure 6A and 6B).

Every times the pleura was unremarkable (Figure 6A and 7A).

Adverse histological changes in the respiratory tract were only observed in rats exposed to 5.0 mg/m^3:

- minimal alveolar granulocyte infiltration in lungs (all sacrifice periods) (Figure 6B),
- minimal/moderate intra-alveolar eosinophilic material deposition in lungs, considered to be the result of macrophages membrane cell rupture (Figure 6C),
- minimal/slight interstitial inflammation in lungs (all sacrifice periods) (Figure 6A),
- minimal/slight focal/multifoca/focally extensive alveolar septa fibrosis in lungs (Figure 7B),
- minimal/slight granulomatous fibrosing inflammation in lungs (Figure 7C),
- minimal/slight bronchiolar cell hypertrophy/ hyperplasia in lungs (Figure 7D),
- Moderate/marked cytoplasmic eosinophilic globules (inclusions) in the respiratory and nasal epithelial cells (Figure 7E).

Minimal to marked concentration-related deposition of black particles in cortex/paracortex of the tracheobronchial lymph nodes was observed in rats exposed to 0.25 and 5.0 mg/m^3 (Figure 6D and 7F) consistent with continuous drainage of black particles from the lungs.

Specifically, no histological lesions were observed in aorta, heart and olfactory bulb and no deposit of MWCNT aggregate was observed in the liver, kidneys and bone marrow and other organs of the exposed animals.

IOP Conf. Series: Journal of Physics: Conf. Series **838** (2017) 012030 doi:10.1088/1742-6596/838/1/012030

Figure 6. Histological changes after 13 weeks of exposure to 5.0 mg/m^3.

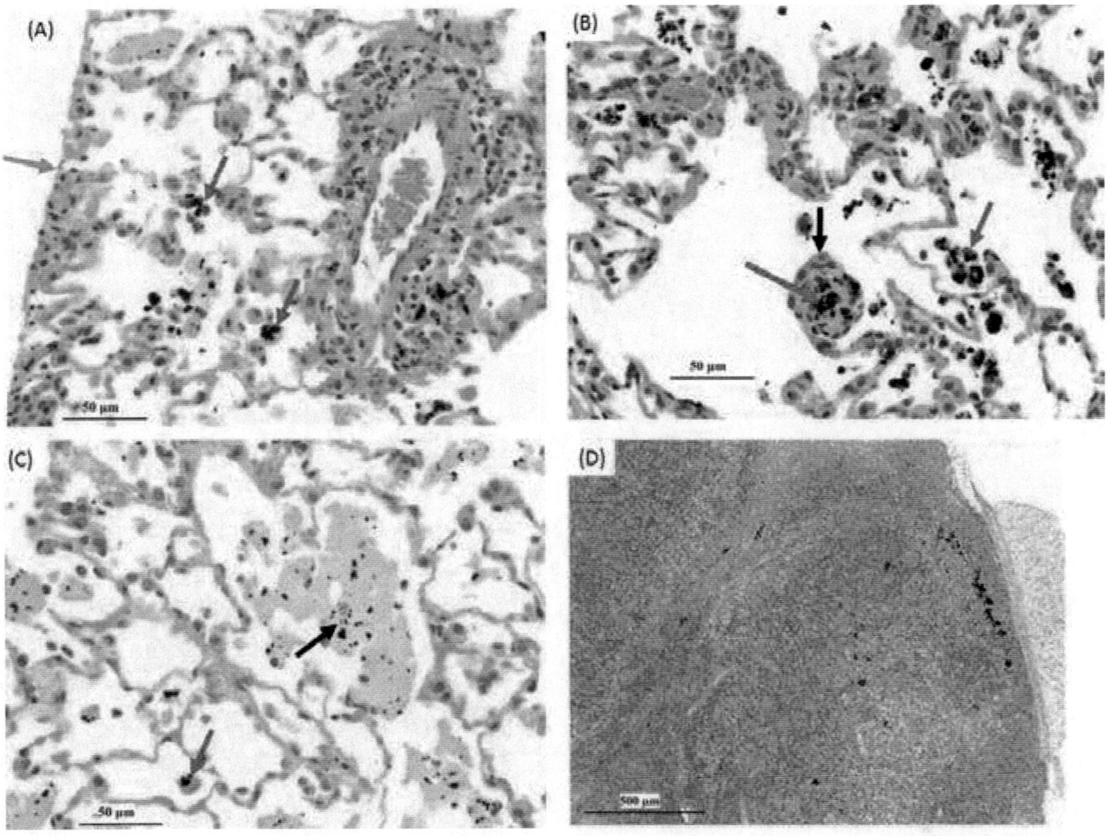

Legend: (A) Presence of black particles within the alveolar macrophages (bleu arrow) and tissue macrophages (yellow arrow) around a blood vessel. The pleura overlying the parenchymal inflammation is unremarkable (green arrow). (B) Presence of black particles within the alveolar macrophages (bleu arrow) and tissue macrophages (red arrow). Note the interstitial inflammation around the alveolar duct with macrophages arranged as a small nodule-like structure (black arrow), forming concentric layers around black particles. (C) Presence of black particles within alveolar macrophages (blue arrow) or free within the alveolar lumen, admixed with the eosinophilic material (black arrow). (D) Accumulation of black particles associated with increased lymphocytes in the cortex/paracortex of the tracheobronchial lymph nodes

Figure 7. Histological changes 52 weeks after 13 weeks of exposure to 5.0 mg/m^3.

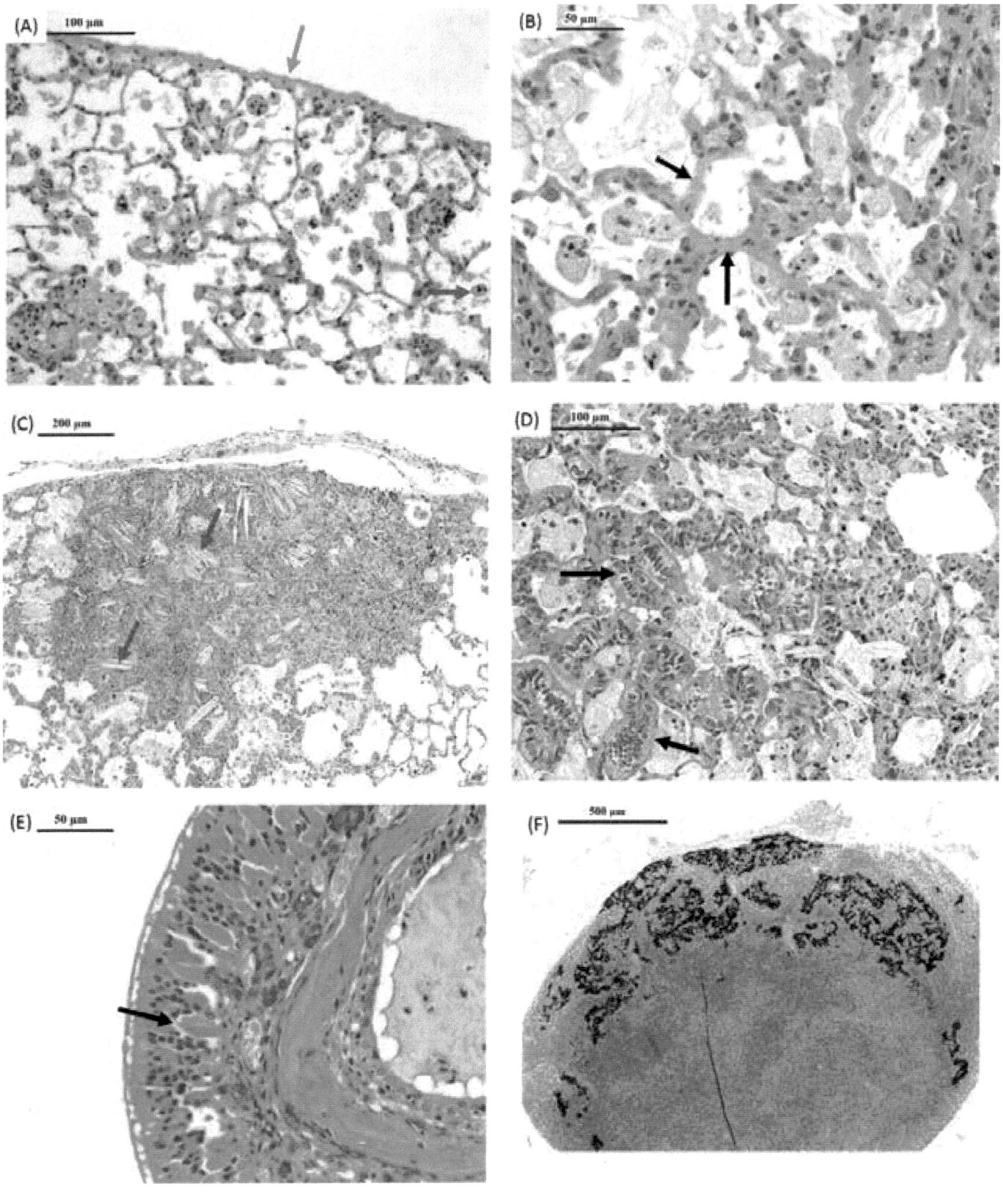

Legend: (A) Presence of black particles within the alveolar macrophages (bleu arrow). The pleura overlying the parenchymal inflammation is unremarkable (green arrow). (B) Focally extensive alveolar septal fibrosis (arrows) (C) Focal subpleural granulomatous fibrosing inflammation. Note the presence of cholesterol-like clefts (arrows). (D) Bronchiolar/alveolar cell hypertrophy/hyperplasia (arrows) associated with the presence of intra-alveolar eosinophilic material (E) Eosinophilic globules are observed within nasal epithelial cells (arrow) (F) severe accumulation of black particles in the cortex/paracortex of tracheobronchial lymph nodes in relation to the pulmonary clearance process and the black discoloration observed at necropsy.

IOP Conf. Series: Journal of Physics: Conf. Series **838** (2017) 012030 doi:10.1088/1742-6596/838/1/012030

3.4. Genotoxicity assays

3.4.1. Comet assay. No increase in the tail intensity (mean of median), in absence and presence of hOGG1 (Table 3), was observed in isolated lung, liver and kidney cells from male rats exposed to Graphistrength© C100 for 13 weeks.

Table 3. Results of the h-OGG1-modified comet assay in the lung, kidney and liver of male rats.

Test groups (n=5/group)	hOGG1	Air control	Graphistrength© C100 (mg/m³) 0.05	0.25	5.0	Positive control[3]
LUNG						
% of DNA in tail[1]	-	8.8 ± 5.5	3.4 ± 1.5*	8.2 ± 5.1	4.7 ± 3.7	48.7 ± 3.0**
	+	20.2 ± 5.2	16.7 ± 5.7	22.3 ± 11.8	17.9 ± 5.9	79.1 ± 7.9**
Relative ratio of ghost cell[2]	-	-	1.0	1.2	0.8	0.6**
	+	-	2.1**	2.0**	1.3	1.1
KIDNEYS						
% of DNA in tail[1]	-	7.4 ± 4.4	7.5± 4.2	8.2± 4.2	9.7± 3.1	72.5± 5.8**
	+	22.8 ± 7.4	22.4 ± 7.3	23.4 ± 8.4	24.3 ± 8.0	74.0 ± 5.9**
Relative ratio of ghost cell[2]	-	-	1.6	0.9	1.5	2.8**
	+	-	1.1	1.6*	1.7**	0.6*
LIVER						
% of DNA in tail[1]	-	3.8 ± 3.1	3.4 ± 1.6	1.8 ± 1.0	5.5 ± 2.9	71.2 ± 9.1**
	+	11.6 ± 6.4	7.5 ± 1.5	7.7 ± 1.1	9.6 ± 3.0	78.6 ± 3.7**
Relative ratio of ghost cell[2]	-	-	0.2	0.5	2.8**	4.9**
	+	-	0.5*	0.2**	1.1	1.7**

[1]Mean of median ± sd
[2]Mean value in treated groups/mean control value,
[3]MMS
*/**: Non-parametric Mann-Whitney test significant <0.05 5% (*), <0.01 (**)

3.4.2. Micronucleus assay. No increase in the frequency of micronucleated polychromatic erythrocytes (PCE) (Table 4) was observed in the bone marrow of male and female rats exposed to Graphistrength© C100 for 13 weeks.

Table 4. Results of the micronucleus assay in the bone marrow of male and female rats.

Test groups (n=5/group)	**Concentration (mg/m³)**	**PCEs with micronuclei (%)**	**Range**[2]	**PCE per 2000 erythrocytes**
		MALES		
Air control	0	0.340	2 - 17	1099
Graphistrength© C100	0.05	0.440	5 - 11	1073
	0.25	0.280	2 - 10	1094
	5.00	0.210	3 - 8	1081
Positive control[1]	20	0.833*	10 - 26	806
		FEMALES		
Air control	0	0.290	3 - 9	1155
Graphistrength© C100	0.05	0.210	1 - 8	1223
	0.25	0.190	1 - 5	1147
	5.00	0.220	2 - 9	1091
Positive control[1]	20	0.750**	8 - 28	813

*/**: Non-parametric Mann-Whitney test significant <0.05 (*) or <0.01 (**),
[1]CPA, 20 mg/kg
[2]per 2000 PCEs per animal

4. Discussion and conclusion

This 13-week inhalation toxicity study on MWCNT Graphistrength© C100 was performed after a careful tuning of the conditions for the generation of a respirable aerosol which respect the physicochemical properties of the MWCNT [4]. A concentration-related increase of black inclusions (regarded to be MWCNT) were observed in the alveolar space and the cytoplasm of infiltrated macrophages, indicating an adequate exposure of the lungs.

The signs of systemic effects were limited to an increase in neutrophil counts and a concomitant decrease in lymphocyte counts in blood of rats exposed to 5.0 mg/m^3. After 4 and 13 weeks of exposure to 5.0 mg/m^3 and 13 and 52 weeks post-exposure, a black brown discoloration of the lung and/or greenish foci were seen in most of the rats. This was associated with an increase of the lung weights maximal at the 13-week recovery sacrifice. The pulmonary reaction to the overload with these insoluble particles was revealed by changes in the cytological, biochemical and cytokine parameters of BALF, slight and reversible at 0.25 mg/m^3 but marked at 5.0 mg/m^3.

Histological changes were only observed in rats exposed to 5.0 mg/m^3. Inflammatory changes in the lungs and eosinophilic globules in the nasal epithelium were observed at all sacrifice periods. After 13 and 52 weeks of recovery, minimal or slight focal collagen deposition was also observed within alveolar septae. Even one year post exposure to 5.0 mg/m^3, no microscopic changes were observed in pleura, heart and aorta.

Twenty-four hours after a 13-week inhalation exposure to MWCNT Graphistrength© C100, in the presence of a clear inflammatory reaction in the lungs of the rats exposed to 5.0 mg/m^3, no primary and hOGG1-sensitive oxidative DNA damage was detected by the comet assay, either in the lung cells directly in contact with the MWCNT or systemically in the liver and kidney cells and no increase of the micronucleus frequency was detected in the bone marrow.

These results differs significantly to those reported on the thick and long fibre-like MWCNT-7 (70-170 nm x ≈ 5 μm from Hodogaya Chemical) which induced lung carcinomas in rats after a chronic inhalation exposure [20] and is classified by IARC [21] as possibly carcinogenic to humans (Group 2B). In the lung cells of ICR [22] and C57Bl/6 [23] mice instilled intratracheally with MWCNT-7, DNA damage, analysed by the comet assay, increased in a dose-dependent manner. Moreover, DNA oxidative damage, indicated by 8-oxo-7,8-dihydro-2'-deoxyguanosine and heptanone ethenodeoxyribonudeosides occurred in the lungs [22]. Increases of DNA strand breaks in lung and bronchoalveolar lavage (BAL) cells and of micronucleated alveolar type II cells were also observed in mice exposed to aerosolized MWCNT-7 (8.2 ± 1.7 mg/m^3), for 4 days (4 h/day) [23]. In contrast, thin and tangled MWCNT (8-15 nm x 0.37 μm from Cheap Tubes) like Graphistrength© C100, did not increase the DNA damages in BAL and lung cells of mice after a single pharyngeal aspiration (1-200 μg/mouse) and an inhalation exposure (17.5±2.0 mg/m^3, 4 h/day for 4 days) [23].

Therefore, MWCNT Graphistrength© C100 appeared of low concern in terms of local and systemic genotoxicity even in presence of a pulmonary inflammation. Considering the limited and reversible effects on the BALF parameters and the lack of adverse pathological changes in the lungs of rats exposed to 0.25 mg/m^3 up to one year after the end of exposure, this concentration was considered to be the No-observed Adverse Effect Concentration (NOAEC).

References

[1] Le Bihan OLC, Ustache A, Bernard D *et al.* 2014 Experimental Study of the Aerosolization from a CarbonNanotube Bulk by a Vortex Shaker *Journal of Nanomaterials* Article ID 193154 [https://www.hindawi.com/journals/jnm/2014/193154/]

[2] R'Mili B, Dutouquet C, Sirven JB *et al.* 2011 Analysis of particle release using LIBS (laser-induced breakdown spectroscopy) and TEM (transmission electron microscopy) samplers when handling CNT (carbon nanotube) powders *Journal of Nanoparticle Research* **13(2)** 563-77

[3] Pacurari M, Lowe K, Tchounwou PB, Kafoury R 2016 A Review on the Respiratory System Toxicity of Carbon Nanoparticles *Int J Environ Res Public Health* **13(3)** 325

[4] Pothmann D, Simar S, Schuler D, Dony E, Gaering S, Le Net JL, *et al.* 2015 Lung inflammation and lack of genotoxicity in the comet and micronucleus assays of industrial multiwalled carbon nanotubes Graphistrength© C100 after a 90-day nose-only inhalation exposure of rats *Particle Fibre Toxicology* **12**:21 http://particleandfibretoxicology.biomedcentral.com/articles/10.1186/s12989-015-0096-2]

[5] Organisation for Economic Cooperation and Development (OECD) 1998 OECD guidelines for testing of chemicals, Section 4: Health Effects, Test No. 413: Subchronic Inhalation Toxicity: 90-day Study. Paris: Organisation for Economic Cooperation and Development

[6] Rasmussen K, Mast J, De Temmerman P-J, Verleysen E, Waegeneers N, Van Steen J, Pizzolon J-C 2014 Multi-walled Carbon Nanotubes, NM-400, NM-401, NM-402, NM-403: Characterisation and Physico-Chemical Properties. JRC Repository: NM-series of Representative Manufactured Nanomaterials, Report EUR 26796 EN [https://ec.europa.eu/jrc/sites/default/files/mwcnt-online.pdf]

[7] Kermanizadeh A, Gaiser BK, Hutchison GR, Stone V 2012 An *in vitro* liver model-assessing oxidative stress and genotoxicity following exposure of hepatocytes to a panel of engineered nanomaterials. *Particle Fibre Toxicology* **9**:28

[http://particleandfibretoxicology.biomedcentral.com/articles/10.1186/1743-8977-9-28]

[8] Kermanizadeh A, Vranic S, Boland S, Moreau K, Baeza-Squiban A, Gaiser BK, Andrzejczuk LA, Stone V 2013 An in vitro assessment of panel of engineered nanomaterials using a human renal cell line: cytotoxicity, pro-inflammatory response, oxidative stress and genotoxicity *BMC Nephrology* **14** 96 [http://www.biomedcentral.com/1471-2369/14/96]

[9] Cao Y, Jacobsen NR, Danielsen PH, Lenz AG, Stoeger T, Loft S, Wallin H, Roursgaard M, Mikkelsen L, Møller P 2014 Vascular effects of multiwalled carbon nanotubes in dyslipidemic ApoE-/- mice and cultured endothelial cells *Toxicology Science* **138(1)** 104-16

[10] Tavares AM, Louro H, Antunes S, Quarré S, Simar S, De Temmerman PJ, Verleysen E, Mast J, Jensen KA, Norppa H, Nesslany F, Silva MJ 2014 Genotoxicity evaluation of nanosized titanium dioxide, synthetic amorphous silica and multi-walled carbon nanotubes in human lymphocytes *Toxicology in vitro* **28(1)** 60-9

[11] Louro H, Pinhão M, Santos J, Tavares A, Vital N, Silva MJ 2016 Evaluation of the cytotoxic and genotoxic effects of benchmark multi-walled carbon nanotubes in relation to their physicochemical properties *Toxicology Letters* **262** 123–134

[12] Norppa H, Siivola K, Fessard V, Tarantini A et al. 2013 WP 5: *In vitro* methods for genotoxicity. Deliverable 5: In vitro testing strategy for nanomaterials including database. Report no.: Grant Agreement n° 2009 21 01. Nanogenotox Joint Action, 2013-05-04 [http://www. nanogenotox. eu/files/PDF/Deliverables/nanogenotox_deliverable_wp5. Pdf]. Consulted on 19 October 2016

[13] Fessard V 2013 Nanogenotox Joint Action: Characterisation of manufactured nanomaterials for their clastogenic/aneugenic effects or DNA damage potentials and correlation analysis. Final report, March 2013 [http://www.nanogenotox.eu/files/PDF/DELIVRABLES2/nanogenotox%20deliverable6_bis. pdf]. Consulted on 19 October 2016

[14] Jackson P, Kling K, Jensen KA, Clausen PA, Madsen AM, Wallin H, *et al.* 2015 Characterization of genotoxic response to 15 multiwalled carbon nanotubes with variable physicochemical properties including surface functionalizations in the FE1-Muta(TM) mouse lung epithelial cell line *Environ Mol Mutagen* **56** 183–203

[15] Smith CC, O'Donovan MR, Martin EA 2006 hOGG1 recognizes oxidative damage using the comet assay with greater specificity than FPG or ENDOIII *Mutagenesis* **21(3)** 185-90

[16] Organisation for Economic Cooperation and Development (OECD) 2014 OECD guidelines for testing of chemicals, Section 4: Health Effects, Test No. 474: Mammalian Erythrocyte Micronucleus Test. Paris: Organisation for Economic Cooperation and Development

[17] Organisation for Economic Cooperation and Development (OECD) 2014 OECD guidelines for testing of chemicals, Section 4: Health Effects, Test No. 489: In Vivo Mammalian Alkaline Comet Assay. Paris: Organisation for Economic Cooperation and Development

[18] Organisation for Economic Cooperation and Development (OECD) 2012 Inhalation toxicity testing: expert meeting on potential revisions to OECD test guidelines and guidance document. Series on the Safety of Manufactured Nanomaterials **35** [ENV/JM/MONO(2012)14].

[19] Romagna F, Staniforth CD 1989 The automated bone marrow micronucleus test *Mutation Research* **213** 91–104.

[20] Kasai T, Umeda Y, Ohnishi M, Mine T, Kondo H, Takeuchi T, Matsumoto M, Fukushima S 2016 Lung carcinogenicity of inhaled multi-walled carbon nanotube in rats *Particle and Fibre Toxicology* **13** 53

[21] Grosse Y, Loomis D, Guyton KZ *et al.* 2014 Carcinogenicity of fluoro-edenite, silicon carbide fibres and whiskers, and carbon nanotubes *The Lancet Oncology* **15(13)** 1427-8

[22] Kato T, Totsuka Y, Ishino K, Matsumoto Y, Tada Y, Nakae D, Goto S, Masuda S, Ogo S, Kawanishi M, Yagi T, Matsuda T, Watanabe M, Wakabayashi K 2013 Genotoxicity of multi-walled carbon nanotubes in both *in vitro* and *in vivo* assay systems *Nanotoxicology* **7(4)** 452-61

[23] Catalán J, Siivola KM, Nymark P, Lindberg H, Suhonen S, Järventaus H, Koivisto AJ, Moreno C, Vanhala E, Wolff H, Kling KI, Jensen KA, Savolainen K, Norppa H 2016 *In vitro* and *in vivo* genotoxic effects of straight versus tangled multi-walled carbon nanotubes *Nanotoxicology* **10(6)** 794-806

Antibacterial activity of nitric oxide releasing silver nanoparticles

Amedea B. Seabra[1,2*] Nixson Manosalva[3], Bruna de Araujo Lima[4], Milena T. Pelegrino[1,2], Marcelo Brocchi[4], Olga Rubilar[3], Nelson Duran[5,6]

[1]Center of Natural and Human Sciences, Universidade Federal do ABC, Santo André, SP, Brazil.
[2]Exact and Earth Sciences Department, Universidade Federal de São Paulo, Diadema, SP, Brazil.
[3]Chemical Engineering Department,Universidad de La Frontera, Temuco, Chile.
[4]Tropical Disease Lab, Institute of Biology, Universidade Estadual de Campinas, SO, Brazil.
[5]Institute of Chemistry, Biological Chemistry Laboratory, Universidade Estadual de Campinas, SP, Brazil.
[6]LNNano (CNPEM), Campinas, SP, Brazil

*amedea.seabra@ufabc.edu.br

Abstract. Silver nanoparticles (AgNPs) are well known potent antimicrobial agents. Similarly, the free radical nitric oxide (NO) has important antibacterial activity, and due to its instability, the combination of NO and nanomaterials has been applied in several biomedical applications. The aim of this work was to synthesize, characterize and evaluate the antibacterial activity of a new NO-releasing AgNPs. Herein, AgNPs were synthesized by the reduction of silver ions (Ag^+) by catechin, a natural polyphenol and potent antioxidant agent, derived from green tea extract. Catechin acts as a reducing agent and as a capping molecule on the surface of AgNPs, minimizing particle agglomeration. The as-synthesized nanoparticles were characterized by different techniques. The results showed the formation of AgNPs with average hydrodynamic size of 44 nm, polydispersity index of 0.21, and zeta potential of -35.9 mV. X-ray diffraction and Fourier transform infrared spectroscopy revealed the presence of the AgNP core and cathecin as capping agent. The low molecular weight mercaptosuccinic acid (MSA), which contain free thiol group, was added on the surface of catechin-AgNPs, leading to the formation of MSA-catechin-AgNPs (the NO precursor nanoparticle). Free thiol groups of MSA-catechin-AgNPs were nitrosated leading to the formation of S-nitroso-mercaptosuccinic acid (S-nitroso-MSA), the NO donor. The amount of 342 ± 16 µmol of NO was released per gram of S-nitroso-MSA-catechin-AgNPs. The antibacterial activities of catechin-AgNPs, MSA-catechin-AgNPs, and S-nitroso-MSA-catechin-AgNPs were evaluated towards different resistant bacterial strains. The results demonstrated an enhanced antibacterial activity of the NO-releasing AgNP. For instance, the minimal inhibitory concentration values for *Pseudomonas aeruginosa* (ATCC 27853) incubated with AgNPs-catechin, AgNPs-catechin-MSA, and AgNPs-catechin-S-nitroso-MSA were found to be 62, 125 and 3 µg/mL, respectively. While in the case of *Klebsiella pneumoniae* (ATCC 700603) the minimum bactericidal concentration values for treatments with AgNPs-catechin, AgNPs-catechin-MSA, and AgNPs-catechin-S-nitroso-MSA were found to be 1000, 500, and 125 µg/mL, respectively. The antibacterial actions of the NO-releasing nanoparticle were superior in comparison with the antibacterial effects of AgNPs, in most of the tested antibiotic resistant bacteria strains. These results highlight the promising uses of NO-releasing AgNPs against resistant bacteria in several biomedical applications.

1. Introduction

Silver nanoparticles (AgNPs) have been attracted the attention of the scientific community in the last decades due to their significant antimicrobial properties [1-3]. Traditional methods for the synthesis of AgNPs are based on the chemical route, which uses strong reducing agents, such as sodium borohydride [4,5]. Although chemical synthesis of AgNPs has a consider control over nanoparticle size distribution, this route involves the presence of toxic chemicals, yielding hazardous byproducts, leading to the environment contamination. In addition, traditional chemical methods are significant expensive since they demand high-energy input and manufacturing [6,7]. In contrast, "green chemistry" has been considered an interesting approach to overcome the main limitations to synthesize several classes of metallic nanoparticles, including AgNPs [8,9].Biogenic synthesis of metallic nanoparticles, including AgNPs, are considered clean, cost and eco effective route and non-toxic to the environment [6-9]. The biogenic (or green synthesis) of AgNPs can be performed at room temperature and at ambient conditions.

The use of plant extract and/or compounds derived from plant extracts to obtain metallic nanoparticles have been gained considerable attention in recent years [10,11]. Catechin is a natural polyphenol and potent antioxidant molecule, which belongs to the group of flavanols, and it is the main constituent of the green tea extract [12]. In biogenic synthesis of AgNPs, catechin acts not only as powerful reducing agent of Ag^{+} to Ag° (leading to the formation of AgNPs), but also as a capping agent, stabilizing the obtained nanoparticles. In this work, AgNPs were synthesized by green chemistry by the action of catechin, leading to the formation of catechin-AgNPs. The obtained nanoparticles were characterized by different techniques. The results demonstrated the successfully formation of AgNPs. The obtained nanoparticles were stabilized by the presence of catechin as capping agent.

In a further step, mercaptosuccinic acid (MSA), a low molecular weight thiol (SH) containing molecule was conjugated on the surface of catechin-AgNPs, leading to the formation of MSA-catechin-AgNPs. MSA was maintained on the surface of catechin-AgNPs by positive electrostatic interactions. The amount of free thiol groups on the surface of MSA-catechin-AgNPs was evaluated. It should be noted that the presence of free thiol groups on the surface of a nanoparticle represents a site for nanoparticle conjugation with important biological molecules [13].

Free thiol groups on the surface of MSA-catechin-AgNPs were nitrosated leading to the formation of S-nitroso-MSA-catechin-AgNPs, which act as spontaneous nitric oxide (NO)-releasing nanoparticle. The aim of this work was to synthesize, characterize and evaluate the antibacterial activity of a new NO-releasing AgNPs. NO is an endogenous found free radical that plays several physiological and pathophysiological roles, such as the cell defense against microbes [14]. The amount of NO release from S-nitroso-MSA-catechin-AgNPs was determined. The combination of NO and AgNPs might find important applications in the combat of resistant bacteria. In this direction, the antibacterial activities of S-nitroso-MSA-catechin-AgNPs, MSA-catechin-AgNPs and catechin-AgNPs were demonstrated towards different bacterial strains (*Pseudomonas aeruginosa* (ATCC 27853), *Staphylococcus aureus* (ATCC 29213), *Klebsiella pneumoniae* (ATCC 700603), *Salmonella enterica* (ATCC 14028), and *Escherichia coli* (ATCC 35218)). The Minimal inhibitory concentration (MIC) and minimum bactericidal concentration (MBC) values were obtained. All tested nanoparticles demonstrated antibacterial effects, and in most of the cases, the NO-releasing AgNPs (S-nitroso-MSA-catechin-AgNPs) demonstrated superior antibacterial effects, compared to the other groups.

Therefore, the results demonstrated the successful biogenic synthesis of AgNPs by the polyphenol catechin, the nanoparticle functionalization with NO group and the antibacterial activities of the nanoparticles. To our best knowledge, this is the first report to demonstrate the biogenic synthesis of NO-releasing AgNPs and their antibacterial effects. More studies are required for further characterization of the antimicrobial activities of the NO-releasing biogenic AgNPs.

2. Methods

2.1. Synthesis of AgNPs by catechin

A volume of 0.5 mL of aqueous solution of catechin (0.1 mol/L) was deposited in the Erlenmeyer, followed by the addition of 96.5 mL of deionized water. The final suspension was stirred at 80°C. Then, 3 mL of a stock solution (0.1 mol/L) of $AgNO_3$ were added to obtain a final concentration of 3 mmol/L and the pH was adjusted to 10 with NaOH. Aliquots of 2 mL were removed from the Erlenmeyr flaks after 4 h of reaction and their UV-visible spectra were obtained using a spectrophotometer (Genesys 10S) at the resolution of 1 nm from 200 to 800 nm for each sample. The final mixture was further stirred for 1 h, centrifuged and washed several times with water, followed by freezer-dryer. This procedure led to the formation of catechin-AgNPs.

2.2. Caracterization of catechin-AgNPs

Catechin-AgNPs were characterized by through Fourier transformed infrared (FTIR) spectroscopy (CARY 630 FTIR Agilent Technologies) in the range 450-4000 cm^{-1} at a resolution of 4 cm^{-1}. X-ray diffraction (XRD) measurements were performed in reflection set-up, with a conventional X-ray generator, CuKα radiation of 1.5418 Å coupled to a scintillation detector. The morphology the nanoparticles was determined by Field-emission scanning electron microscopy (FEI Quanta FEG250, STEM) at 30 kV. The nanoparticle size distribution was estimated by using the software SigmaScan Pro 5.0. The hydrodynamic diameter and zeta potential were measured at 25°C by dynamic light scattering (DLS) using the Zetasizer Nano ZS90 System (Malvern Instruments, Malvern, UK). Prior to the DLS measurement, the aqueous suspensions of nanoparticles were passed through a 0.22 μm polyvinylidene fluoride (PVDF) membrane.

2.3. Functionalization of catechin-AgNPs with MSA leading to MSA-catechin-AgNPs

Catechin-AgNPs (20 mg) were suspended in 10 mL of deionized water in an ultrasound bath for 10 mim at room temperature. MSA (200 mg) is dissolved in 10 mL of deionized water. MSA solution was added to catechin-AgNPs suspension, and the final suspension was stirred for 14 h at room temperature. The MSA-catechin-AgNPs obtained were isolated by centrifugation, washed and dried.

2.4. Quantification of free thiol groups (SH) on the surface of MSA-catechin-AgNPs

The quantification of free thiol groups on the surface of MSA-catechin-AgNPs was performed by titration of SH presented in MSA with a thiol reagent 5,50-dithiobis-(2-nitrobenzoic acid) (DTNB). [13]. This quantification is based on the detection of the absorbance at 412 nm, which corresponds to the 2-nitro-5-thiobenzoate anion (TNB^{2-}) generated in the reaction of SH groups with DTNB. Appropriate amounts of thiolated nanoparticles were added to 3.0 mL of 0.01 mol L^{-1} DTNB in PBS buffer (pH 7.4) containing 1 mmol L^{-1} of ethylenediaminetetraacetic acid. After 5 min of incubation, the suspensions were filtered by centrifugal ultrafiltration using a Microcon centrifugal filter device containing ultrafiltration membranes (MWCO 10-kDa cutoff filter, Millipore). The supernatant was placed into a quartz cuvette, and the intensity of the absorption band at 412 nm was measured in an UV–vis spectrophotometer (Agilent 8453). The experiments were performed in triplicate.

2.5 Nitrosation of MSA-catechin-AgNPs leading to the formation of S-nitroso-MSA-catechin-AgNPs

MSA-catechin-AgNPs (10 mg) were dispersed in 1 mL of deionized water by using an ultrasound bath, and the pH of the suspension was adjusted to 4.0. Aliquot of 200 μL of sodium nitrite ($NaNO_2$), 60 mmol/L, was added to the MSA-catechin-AgNPs suspension under stirring for 30 min. The obtained S-nitroso-MSA-catechin-AgNPs were immediately used.

2.6. Quantification of NO release from S-nitroso-MSA-catechin-AgNPs

The amount of NO released from S-nitroso-MSA-catechin-AgNPs was measured by a NO electrode (2.0 mm ISO-NOP) connected to a TBR4100/1025 Free Radical Analyzer (World Precision Instruments). Aliquots of 100 μL of aqueous suspension of S-nitroso-MSA-catechin-AgNPs (5. mg/mL) were added to the sampling compartment, which contained 10 mL of 10 mL copper(II) chloride ($CuCl_2$) (0.1 mol L^{-1}). This condition allowed for the detection of free NO released from the NPs. The experiments were performed in duplicate with the standard error of the mean. Calibration curves were obtained with aqueous solutions of freshly prepared S-nitrosoglutathione (1–500 μmol L^{-1}) (data not shown).

2.7. Antibacterial activities of the synthesized NPs

The antibacterial activities of catechin-AgNPs, MSA-catechin-AgNPs and S-nitroso-MSA-catechin-AgNPs, at different concentrations, were evaluated against *Pseudomonas aeruginosa* (ATCC 27853), *Staphylococcus aureus* (ATCC 29213), *Klebsiella pneumoniae* (ATCC 700603), *Salmonella enterica* (ATCC 14028), and *Escherichia coli* (ATCC 35218), all samples kindly provided by Oswaldo Cruz Foundation (Fiocruz, Rio de Janeiro, Brazil). Bacterial strains were incubated with the different concentrations of nanoparticles for 24 h. Minimal inhibitory concentration (MIC) and minimum bactericidal concentration (MBC) values were obtained, using micro-dilution assays in 96- well plates, as previous described [15], and according to the Clinical and Laboratory Standards Institute (CLSI) [16].

3. Results and Discussion

The main objective of this study was to to synthesize, characterize and evaluate the antibacterial activity of a new NO-releasing AgNPs. To this end, AgNPs were synthesized by catechin, in a green synthetic route, followed by the coating of the obtained nanoparticles with S-nitroso-MSA, as a spontaneous NO releasing molecule. It is expected a superior antimicrobial activity of the NO-releasing-AgNPs. The following sections describe the obtained results.

3.1. Synthesis and characterization of catechin-AgNPs

Uv-visible spectrophotometry is a simple and direct method to confirm the formation of AgNPs from $AgNO_3$ [17]. The reduction of Ag^+ to Ag° occurs immediately upon the pH adjustment, which is accompanied by a change in the suspension color from pale yellow to brown, indicating the formation of AgNPs. Figure 1A shows the Uv-visible spectra for catechin-AgNPs at different pHs.

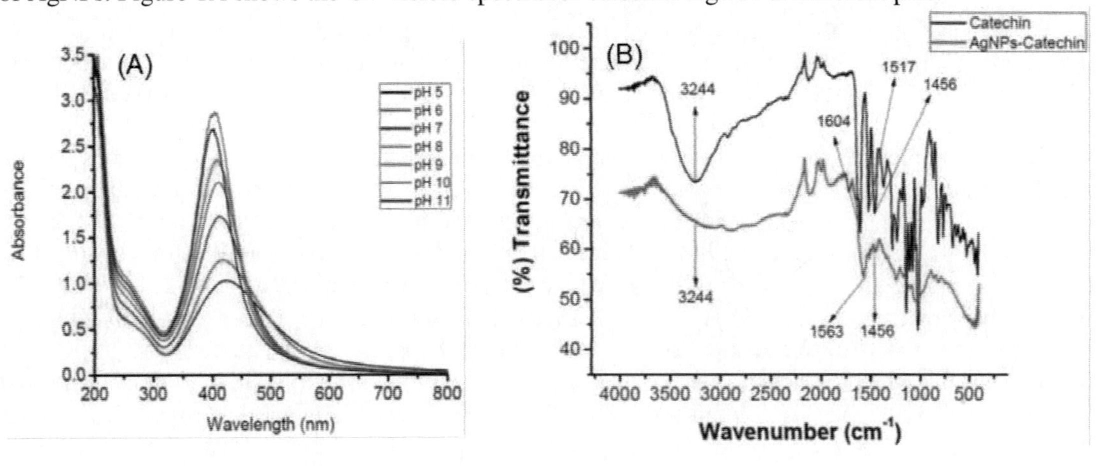

Figure 1. Plasmonic absorbantion bands of catechin-AgNPs at different pHs (A). FTIR spectra of catechin and catechin-AgNPs (B).

IOP Conf. Series: Journal of Physics: Conf. Series **838** (2017) 012031 doi:10.1088/1742-6596/838/1/012031

As can be observed, with the increase of the pH, the intensity of the plasmonic band increases, band becomes narrower, and the bands shift to lower wavelength values [20]. These peaks are due to the plasmonic band of $AgNO_3$ [18,19].

Figure 1B shows the FTIR spectra of pure catechin and catechin-AgNPs, as indicated in the Figure. Band at 3244 cm^{-1} is associated with O-H stretching vibrations of the phenolic group of catechin. Vibrations at 1604 cm^{-1} corresponds to the stretching vibration of C=C present in aromatic and aliphatic compounds, while vibrations at 1517 cm^{-1} corresponds to the vibrations of C-O of esters, ethers, and phenols, and vibrations at 1456 cm^{-1} correspond to the C-O of ethers [21,22]. Upon catechin conjugation with AgNPs, there is a decrease in the intensity of the band at 3244 cm^{-1}, which corresponds to the decrease in the OH. This can be explained since OH groups participate in the reduction of Ag^+ to Ag°. Bands at 1563 and 1456 cm^{-1} correspond to aromatic C=C in the modified catechin.

Figure 2 shows the XRD pattern of catechin-AgNPs. Peak values of around 38.10°, 44.47°, 64.63°, 77.44°, and 81.33° correspond to the XRD pattern of indexed [111], [200], [220], [311], and [222] facets of Ag° NPs [2]. These results confirm the reduction of Ag^+ to Ag° by catechin, which acts as reducing and capping agent.

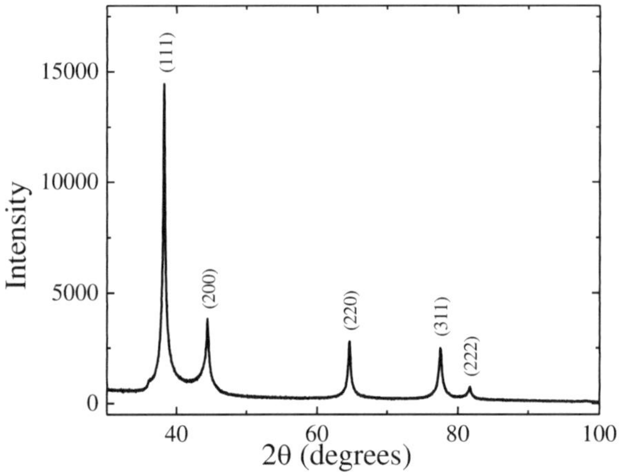

Figure 2. XDR patter of catechin-AgNPs.

Figure 3A shows the morphology of quasi spherical catechin-AgNPs with the presence of agglomerates. Figure 3B shows that the size distribution of the nanoparticles was found to be between 10 and 40 nm with average size of 23.4 ± 8.4 nm at solid state. The agglomeration observed in Figure 3A might be due to the drying process prior microscopy analysis.

DLS measurements revealed that the average hydrodynamic size of catechin-AgNPs was 44 nm, with PDI value of 0.21 and zeta potential of – 35.9 mV. The hydrodynamic size of the nanoparticles was found to be higher in comparison to the average size of the nanoparticles assayed by STEM. As expected, higher hydrodynamic sizes of NPs measured by DLS, compared with the sizes obtained by TEM, are attributed to the presence of extra hydrate layers in aqueous environments [23]. The results indicate the formation of catechin-AgNPs at the nanosize scale in aqueous suspension, and the PDI value indicates that the size distribution is moderate polydispersive. The negative value of zeta potential is due to the presence of catechin on the surface of AgNPs, since negative charge is expected for polyphenols [10]. This result indicates the presence of catechin on nanoparticle surface. Moreover, the magnitude of this zeta potential demonstrates the stability of the nanoparticles in aqueous suspension, avoiding nanoparticle agglomeration.

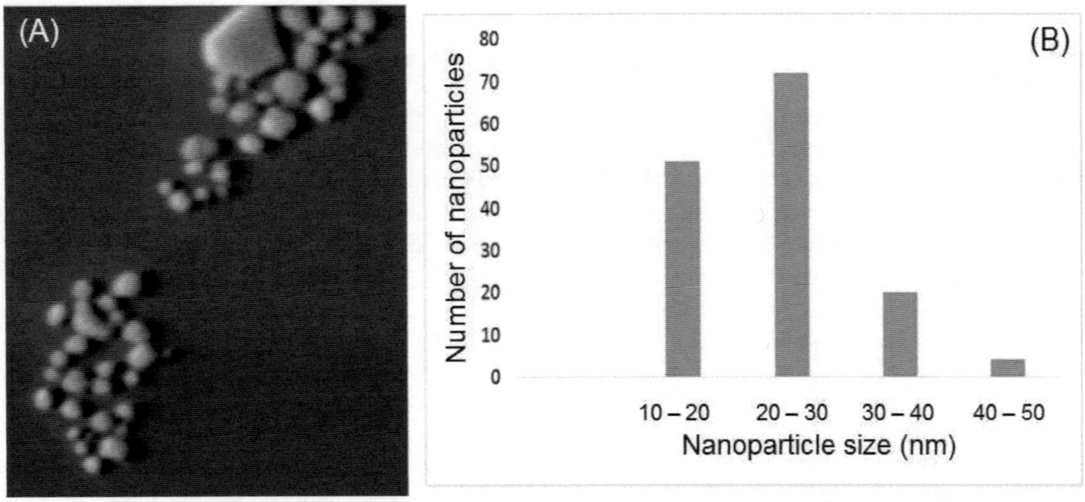

Figure 3. Field-emission scanning electron microscopy (STEM) of catechin-AgNPs (A) and their corresponding size distribution at solid state (B).

3.2. Functionalization of catechin-AgNPs with MSA leading to MSA-catechin-AgNPs

MSA, a low molecular weight thiol-containing molecule, was conjugated on the surface of catechin-AgNPs leading to the formation of MSA-catechin-AgNPs, which contain free thiol (SH) groups on their surface (Figure 4).

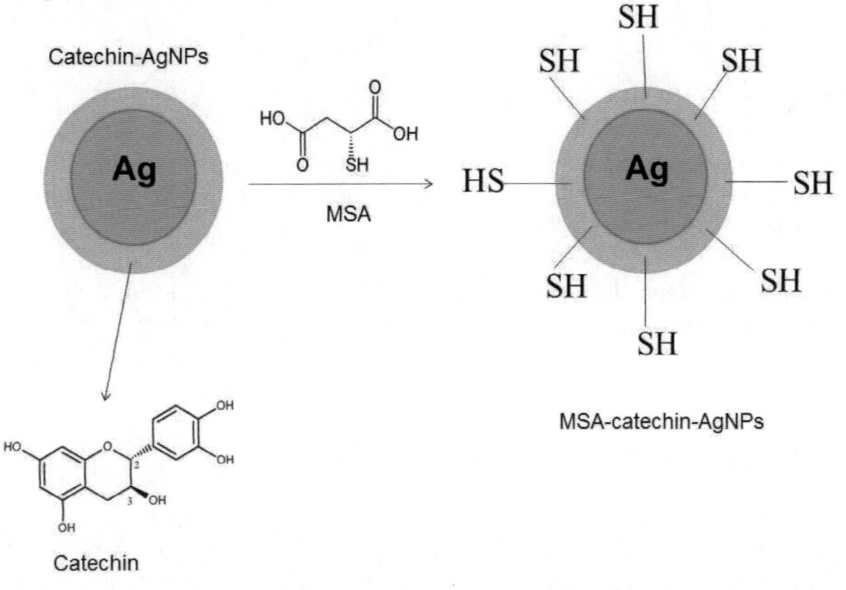

Figure 4. Schematic representation of the functionalization of catechin-AgNPs with mercaptossunic acid (MSA), a thiol containing-molecule, leading to the formation of MSA-catechin-AgNPs.

A value of 355 ± 19 µmol of free SH group per gram of MSA-catechin-AgNPs was obtained. MSA was conjugated with catechin-AgNPs by positive electrostatic interactions. The quantification of free thiol (SH) groups on the surface of MSA-catechin-AgNPs was determined by the reaction with a thiol specific reagent, DTNB, as previous described [10,23]. In drug delivery applications, the presence of free SH groups on the surface of nanoparticles represents a site for nanoparticle conjugation with important therapeutic molecules. In this present work, NO was loaded on MSA-catechin-AgNPs through SH groups.

3.3. Nitrosation of MSA-catechin-AgNPs leading to the formation of S-nitroso-MSA-catechin-AgNPs
Free thiol groups (SH) on the surface of MSA-catechin-AgNPs were nitrosated by the addition of sodium nitrite ($NaNO_2$) in slight acidified solution [10,23], leading to the formation of S-nitroso-MSA-catechin-AgNPs, which act as spontaneous NO donor due to the cleavage of S-N bound, as represented in Figure 5.

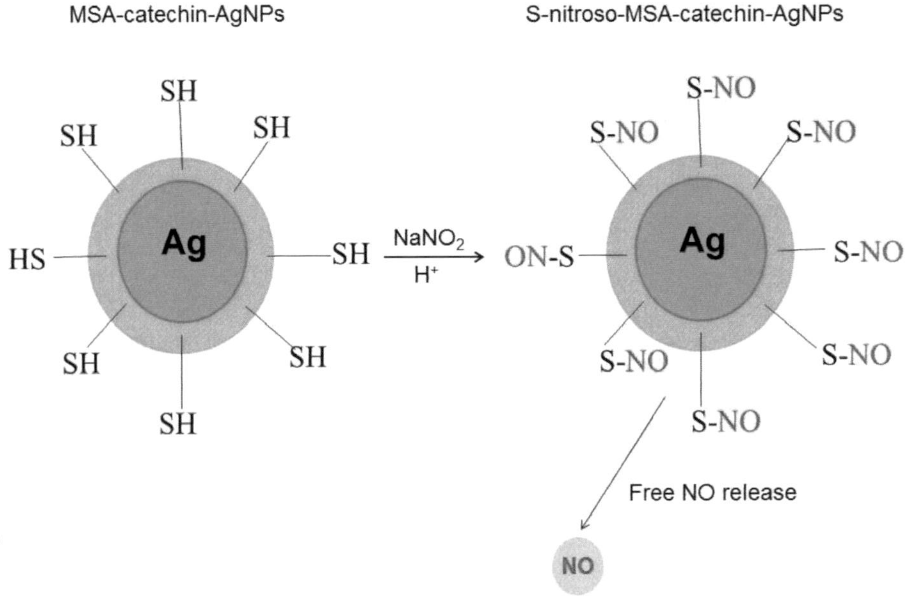

Figure 5. Schematic representation of nitrosation of free thiol groups on the surface of MSA-catechin-AgNPs by sodium nitrite ($NaNO_2$) leading to the formation of S-nitroso-MSA-AgNPs, which act as spontaneous NO donor.

The quantification of NO loading on the surface of S-nitroso-MSA-catechin-AgNPs was evaluated by electrochemical analysis with a specific NO sensor. The amount of 342 ± 16 µmol of NO was released per gram of S-nitroso-MSA-catechin-AgNP. This amount of NO release from nanoparticles is in the same range as reported for S-nitroso-MSA-Fe_3O_4 magnetic nanoparticles [24]. At this concentration range, NO is expected to have biological activities such as the antimicrobial effects [25]. To our best knowledge, this is the first report to describe the synthesis of NO-releasing AgNPs.

3.4. Antibacterial activities of the synthesized NPs
The antibacterial activities of catechin-AgNPs, MSA-catechin-AgNPs and S-nitroso-MSA-AgNPs were evaluated against different resistant bacterial strains. Table 1 and 2 show the minimal inhibitory concentration (MIC) and MBC values, respectively. As can be observed, the antibacterial effect is dependent on the bacteria strain and the nature of the AgNPs. For *Pseudomonas aeruginosa*, *Staphylococcus aureus* and *Klebsiella pneumoniae* lower values of MIC were found for the bacteria incubation with the NO-releasing nanoparticle (S-nitroso-MSA-catechin-AgNPs). For instance, MIC value of 3 µg/mL was observed for *Pseudomonas aeruginosa* treated with S-nitroso-MSA-catechin-AgNPs. In contrast, lower MIC values were found for catechin-AgNPs incubated with *Salmonella enterica* and *Escherichia coli.*

IOP Conf. Series: Journal of Physics: Conf. Series **838** (2017) 012031 doi:10.1088/1742-6596/838/1/012031

Table 1. MIC values (µg/mL) for different bacterial strains incubated for 24 h with catechin-AgNPs, MSA-catechin-AgNPs and S-nitroso-MSA-catechin-AgNPs.

Bacterial strain	Catechin-AgNPs	MSA-catechin-AgNPs	S-nitroso-MSA-catechin-AgNPs
Pseudomonas aeruginosa	62	125	3
Staphylococcus aureus	500	250	125
Klebsiella pneumoniae	1000	250	125
Salmonella enterica	62	250	125
Escherichia coli	62	250	125

Table 2. Minimum bactericidal concentration (MBC) values (µg/mL) for different bacteria strains incubated for 24 h with catechin-AgNPs, MSA-catechin-AgNPs and S-nitroso-MSA-catechin-AgNPs.

Bacterial strain	Catechin-AgNPs	MSA-catechin-AgNPs	S-nitroso-MSA-catechin-AgNPs
Pseudomonas aeruginosa	62	250	6
Staphylococcus aureus	500	500	125
Klebsiella pneumoniae	1000	500	125
Salmonella enterica	125	500	125
Escherichia coli	125	500	125

Table 2 shows that MBC values depend on bacterial strain and the nanoparticle. In all tested bacterial strains, MBC values decreased for S-nitroso-MSA-AgNPs in comparison with catechin-AgNPs, indicating that the NO releasing nanoparticles are more effective as antibacterial agent. Indeed, for *Pseudomonas aeruginosa* a MBC value of 6 µg/mL was found upon incubation with S-nitroso-MSA-AgNPs. Taking together the results demonstrated the all tested nanoparticles have antibacterial activity towards different bacterial strains. In most of the cases, the presence of NO on the nanoparticle surface enhanced the antibacterial effect due to a synergist effect of the NO donor and the AgNP.

4. Conclusions

This work describes the successful synthesis of AgNPs by catechin, the main product of green tea extract. Catechin acts as efficient reducing agent of Ag^+ to Ag° leading to the formation of catechin-AgNPs. Moreover, catechin acts a capping agent on the surface of AgNP, avoiding nanoparticle oxidation and/or aggregation. The obtained nanoparticles were characterized by different techniques, which indicate the formation of AgNP core coated with catechin. The surface of catechin-AgNPs was functionalized with MSA, a low molecular weight thiol containing molecule, leading to the formation of MSA-catechin-AgNPs. Free thiol groups on the surface of MSA-catechin-AgNPs were nitrosated by the addition of sodium nitrite leading to the formation of S-nitroso-MSA-catechin-AgNPs, which act as spontaneous NO donor. The antibacterial activities of catechin-AgNPs, MSA-catechin-AgNPs and S-nitroso-MSA-catechin-AgNPs were demonstrated towards different bacterial strains. All tested nanoparticles demonstrated antibacterial effects, as assayed by the determination of MIC and MBC values. In most of the cases, NO-releasing nanoparticles enhanced the antibacterial effect of catechin-AgNPs. These results highlight the promising uses of NO-releasing AgNPs against resistant bacteria in several biomedical applications.

Acknowledgements: FAPESP (Proc. 2016/10347-6), the Brazilian Network on Nanotoxicology (Grant number: 552120/2011-1) (MCTI/CNPq), the Laboratory of Nanostructure Synthesis and

Biosystem Interactions-NANOBIOSS (MCTI) (Grant number: 402280-2013), FONDECYT (Grant number 1130854) and CONICYT REDES (Grant number 140053).

References

[1] Durán N, Durán M, de Jesus MB, Seabra AB, Fávaro WJ and Nakazato G 2016 *Nanomedicine* **12** 789
[2] Durán N, Nakazato G and Seabra AB 2016 *Appl. Microbiol. Biotechnol.* **100** 6555
[3] Lai CY, Cheong CF, Mandeep JS, Abdullah HB, Amin N, and Lai KW 2014 *J. Mater. Eng. Perform.* **23** 3541
[4] Duan J, Yin H, Wei R, and Wang W 2014 *Biosens. Bioelectron.* **57** 139
[5] Sung HK, Oh SY, Park C and Kim Y 2013 *Langmuir* **29** 8978
[6] Terenteva EA, Apyari VV, Dmitrienko SG, and Zolotov YA 2015 *Spectrochim. Acta. A. Mol. Biomol. Spectrosc.* **151** 89
[7] Seabra AB and Durán N 2015 *Metals* **5** 934
[8] Lima R, Feitosa LO, Ballottin D, Marcato PD, Tasic L and Durán N 2013 *JPCS* **429** 012020
[9] de Lima R, Seabra AB and Durán N 2012 *J. Appl. Toxicol.* **32** 867
[10] Silva BSO and Seabra AB 2016 *Biointerface Res. Appl. Chem.* **6** 1280
[11] Herlekar M, Barve S and Kumar R 2014 *J. Nanopart. Res.* 140614
[12] Menendez C, Jimenez R, Moreno L, Galindo P, Cogolludo A, Duarte J, and Perez-Vizcaino F 2011 *Br. J. Nutr.* **105** 1287
[13] Seabra AB, Pasquoto T, Ferrarini ACF, Santos MD, Haddad PS and de Lima R 2014 *Chem. Res. Toxicol.* **27** 1207
[14] Seabra AB and Durán N 2010 *J. Mater. Chem.* **20** 1624
[15] Cardozo VF, Lancheros CA, Narciso AM, Valereto EC, Kobayashi RK, Seabra AB and Nakazato G 2014 *Int. J. Pharm.* **473** 20
[16] CSLI Methods for dilution antimicrobial susceptibility tests for bacteria that grow aerobically. Wayne, PA: CLSI; 2003:6th ed
[17] He Y, Du, Tang, Zheng, Zhang, Zhao, Lv, and Qianfa J 2013 *Int. J. Nanomedicine* 1809
[18] Patil RS, Kokate MR, and Kolekar SS 2012 *Spectrochim. Acta. A. Mol. Biomol. Spectrosc.* **91** 234
[19] Velmurugan P, Lee S-M, Iydroose M, Lee K-J and Oh B-T 2013 *Appl. Microbiol. Biotechnol.* **97** 361
[20] Alqadi MK, O. Noqtah OAA, AlzoubiJ FY and Aljarrah K 2014 *Mater. Sci.-Pol.* **32** 107
[21] Chen Z, Yu T, Zhou B, Wei J, Fang Y, Lu J, Guo L, Chen W, Liu Z-P and Luo J 2016 *Biomaterials* **81** 125
[22] Ramos-Tejada MM, Duran JDG, Ontiveros-Ortega A, Espinosa-Jimenez M, Perea-Carpio R and Chibowski E 2002 *Colloids Surf. B Biointerfaces* **24** 297
[23] Santos MC, Seabra AB, Pelegrino MT and Haddad PS 2016 *Appl. Surf. Sci.* **367** 26
[24] Molina MM, Seabra AB, de Oliveira MG, Itri R and Haddad PS 2013 *Mater. Sci. Eng. C* **33** 746.
[25] Seabra AB, Martins D, Simões MM, da Silva R, Brocchi M and de Oliveira MG 2010 *Artif. Organs.* **34** E204

IOP Conf. Series: Journal of Physics: Conf. Series **838** (2017) 012032 doi:10.1088/1742-6596/838/1/012032

Synthesis, characterization and cytotoxicity of S-nitroso-mercaptosuccinic acid-containing alginate/chitosan nanoparticles

Amedea B. Seabra[1,2*] Giulia K. Fabbri[2], Milena T. Pelegrino[1,2], Letícia C. Silva[1], Tiago Rodrigues[1]

[1]Center of Natural and Human Sciences, Universidade Federal do ABC, Av. dos Estados 5001, CEP 09210-580, Santo André, SP, Brazil.
[2]Exact and Earth Sciences Department, Universidade Federal de São Paulo, Rua São Nicolau, 210, CEP 09913-030, Diadema, SP, Brazil

*amedea.seabra@ufabc.edu.br

Abstract. Nitric oxide (NO) is an endogenous free radical, which plays key roles in several biological processes including vasodilation, neurotransmission, inhibition of platelet adhesion, cytotoxicity against pathogens, wound healing, and defense against cancer. Due to the relative instability of NO *in vivo* (half-life of *ca.* 0.5 seconds), there is an increasing interest in the development of low molecular weight NO donors, such as S-nitrosothiols (RSNOs), which are able to prolong and preserve the biological activities of NO *in vivo*. In order to enhance the sustained NO release in several biomedical applications, RSNOs have been successfully allied to nanomaterials. In this context, this work describes the synthesis and characterization of the NO donor S-nitroso-mercaptosuccinic acid (S-nitroso-MSA), which belongs to the class of RSNOs, and its incorporation in polymeric biodegradable nanoparticles composed by alginate/chitosan. First, chitosan nanoparticles were obtained by gelation process with sodium tripolyphosphate (TPP), followed by the addition of the alginate layer, to enhance the nanoparticle protection. The obtained nanoparticles presented a hydrodynamic diameter of 343 ± 38 nm, polydispersity index (PDI) of 0.36 ± 0.1, and zeta potential of – 30.3 ± 0.4 mV, indicating their thermal stability in aqueous suspension. The negative zeta potential value was assigned to the presence of alginate chains on the surface of chitosan/TPP nanoparticles. The encapsulation efficiency of the NO donor into the polymeric nanoparticles was found to be 98 ± 0.2%. The high encapsulation efficiency value was attributed to the positive interactions between the NO donor and the polymeric content of the nanoparticles. Kinetics of NO release from the nanoparticles revealed a spontaneous and sustained release of therapeutic amounts of NO, for several hours under physiological temperature. The incubation of NO-releasing alginate/chitosan nanoparticles with human hepatocellular carcinoma (HepG2) cell line revealed a concentration-dependent toxicity. These results point to the promising uses of NO-releasing alginate/chitosan nanoparticles for anti-cancer chemotherapy.

1. Introduction

Polymeric nanoparticles composed by polysaccharides have been extensively studied due to their interesting properties such as biocompatibility and biodegradability [1]. Chitosan (CS) is on the most employed biopolymer in pharmacological and biomedical applications, in the forms of films, gels, micro and nanoparticles [2]. Alginate (ALG) is naturally occurring polysaccharide extracted from brown seaweed. It is composed of two types of uronic acids, alfa-*L*-gluronic acid and beta-*D*-manuronic acid, which contributes for the bioadhesiveness and drug-binding properties of this polymer [3]. ALG can interact with calcium inducing gel formation [4], and may also form complexes with polycations, CS or peptides [5].

Nitric oxide (NO) is an important endogenous molecule, which plays several physiological and pathophysiological functions [1,6-8]. Among them, NO has toxicity towards cancer cells, depending on its concentration [9]. NO is a free radical and relatively instable in the biological medium. S-nitrosothiols (RSNOs) are an important class of NO donors, acting as NO carriers able to spontaneous release free NO [10]. Recently, RSNOs have been incorporated into nanomaterials to enhance the sustained NO release in biomedical applications [1].

The objective of this study was to describe the synthesis, characterization and cytotoxicity of CS nanoparticles containing the NO donor S-nitroso-mercaptosuccinic acid (S-nitroso-MSA), which belongs to the class of RSNOs [1,8]. ALG was added onto the surface of CS NPs, leading to the formation of S-nitroso-MSA-CS/ALG NPs (Figure 1).

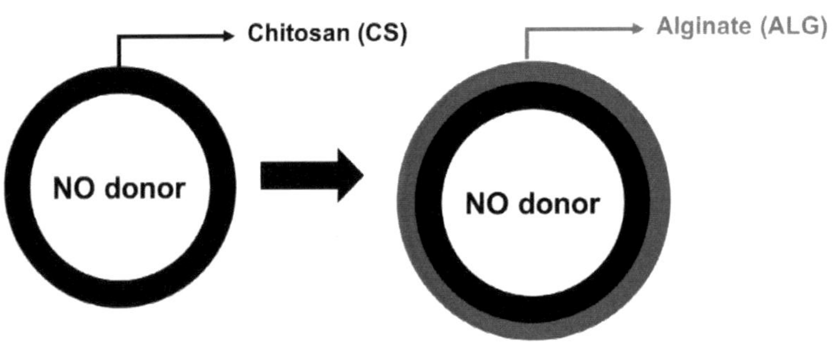

Figure 1. Schematic representation of the synthesized nanoparticles. ALG layer (grey circle) was added on the surface of CS NPs (black circle) containing the NO donor.

ALG is known to increase the mechanical properties of a biomaterial, and also to enhance the stability of encapsulated drugs decreasing the nanoparticle degradation in acid environment or minimizing the degradation actions of hydrolytic enzymes [11]. Moreover, both CS and ALG are mucoadhesive polymers, enhancing the delivery of therapeutic drugs to the mucus layers [12].

In this study, polymeric nanoparticles made of CS and ALG were evaluated regarding their size, stability and potential as drug delivery system, further, the cytotoxicity of the NO-releasing polymeric nanoparticles was demonstrated towards human hepatocellular carcinoma (HepG2) cell line. The results suggest the potential uses of this nanomaterial against cancer cells.

2. Methods

2.1. Synthesis of MSA-CS/ALG NPs

Firstly, CS NPs were synthesized by ionotropic gelation with sodium tripolyphosphate (TPP), as previous described [13]. Briefly, CS (2.6 $mg \cdot mL^{-1}$) and MSA (133.3 $mmol \cdot L^{-1}$) were mixed through magnetic stirring in an aqueous solution of acetic acid (0.175 $mol \cdot L^{-1}$) for 90 min. TPP aqueous solution (0.6 $mg \cdot mL^{-1}$) was added dropwise to the CS/MSA suspension, following the volumetric proportion of 3CS/MSA : 1TPP. The mixture was stirred for further 45 minutes at room temperature leading to the formation of aqueous suspension of CS NPs containing MSA, henceforth, referred as MSA-CS NPs. In a second step, the ALG layer was inserted onto the MSA-CS NPs, as previous described [14]. Under magnetic stirring and at room temperature, ALG solution (1 $mg \cdot mL^{-1}$) was dropwise in the previous prepared MSA-CS NPs dispersion, following the volumetric proportion of 1ALG: 1 MSA-CS NPs. The final mixture was stirred for further 30 min at room temperature forming a dispersion of MSA-CS/ALG NPs in which the MSA concentration is 50 $mmol \cdot L^{-1}$, showed in Figure 1, and referred as MSA-CS/ALG NPs.

2.2. Dynamic light scattering measurements (DLS)
The hydrodynamic size diameter, polydispersity index (PDI), and zeta potential of MSA-CS/ALG NPs were evaluated by DLS (Nano ZS Zetasizer, Malvern Instruments Co, UK.) [15]. Measurements were performed at a fixed angle of 173°, using a disposable capillary cuvette (DTS1070) after a dilution of 1:10. The results were reported as an average of three independent experiments with the error bar values expressed by their standard error of the mean (SEM).

2.3. Encapsulation efficiency of MSA into CS/ALC NPs
The encapsulation efficiency of MSA in CS/ALG NPs was measured by the titration of free thiols groups in the MSA structure with the thiol reagent 5,50-dithiobis-(2-nitrobenzoic acid) (DTNB),
as already described [16]. Non-encapsulated (free) MSA was separated from encapsulated MSA by using a Microcon centrifugal filter device (MWCO 10,000, Millipore) and titrated with DTNB. The absorption band at 412 nm ($\mathcal{E}$ = 14.15 $mmol^{-1}Lcm^{-1}$), assigned to the formation of 2-nitro-5-thiobenzoate anion that is generated in the reaction of DTNB with MSA, was measured in the ultra violet-visible spectrophotometer (Agilent 8453, Palo Alto, CA, USA). The analysis was performed in duplicates. The percentage of MSA encapsulation was determined by the described equation:

$$(\%) = (\text{mass of MSA encapsulated/mass of MSA total}) \times 100 \quad \text{(Eq. 1)}$$

2.4. Nitrosation of MSA incorporated into CS/ALG NPs
Free thiol group of MSA encapsulated into CS/ALG NPs was nitrosated by the addition of equimolar amount of sodium nitrite ($NaNO_2$, 50 $mmol \cdot L^{-1}$) to the aqueous suspension of MSA-CS/ALG NPs (pH ~4.0). The final suspension was stirred for 30 min in an ice bath (5°C), protected from the ambient light, leading to the formation of S-nitroso-MSA-CS/ALG NPs. The confirmation of the nitrosation of MSA yielding S-nitroso-MSA was performed by the appearance of the characteristic S-NO group absorption bands at 336 nm ($\varepsilon = 980.0\ mol^{-1} \cdot L \cdot cm^{-1}$) or at 545 nm ($\varepsilon = 18.4\ mol^{-1} \cdot L \cdot cm^{-1}$) by using the UV–vis spectrophotometer (Agilent 8453, Palo Alto, CA, USA).

2.5. NO release profile from S-nitroso-MSA-CS/ALG NPs at physiological temperature
The kinetics of NO release from S-nitroso-MSA-CS/ALG NPs was determined by monitoring the spectral changes at 545 nm (nN → π* transition) [15,17]. The decay of this absorption band is associated with the decomposition of S-nitroso-MSA with free NO release. The initial concentration of S-nitroso-MSA in the CS/ALG NPs was 50 $mmol \cdot L^{-1}$ and the kinetic data were collected in 30 min intervals at 37°C for 6 h of monitoring. The amount of S-nitroso-MSA decomposed is related to the amount of NO released, since the decay of the absorption band at 545 nm over time is solely assigned to the cleavage of the S–N bond and NO release. Each point in the kinetic curves represents the average of two independent experiments, with the error bar values expressed by their standard error of the mean (SEM). In order to investigate the mechanism of NO release from S-nitroso-MSA-CS/ALG NPs, the Higuchi model was applied to kinetic curve [18-20]. According to the Higuchi model, the drug release is dependent of the square-root of time, as shown in Eq. 2, where Q_T is the amount of drug released at time t, K_H is the Higuchi dissolution constant and t is time. The results were adjusted through linear regression and the correlation coefficient (R^2) was analyzed [21].

$$Q_t = K_H t^{-0.5} \qquad \text{(Eq. 2)}$$

2.6. Cytotoxicity of S-nitroso-MSA-CS/ALG NPs

The cytotoxicity of S-nitroso-MSA-CS/ALG NPs was evaluated towards human hepatocellular carcinoma (HepG2) cell line. Different concentrations of S-nitroso-MSA-CS/ALG NPs and MSA-CS/ALG NPs (control group) were added to each well containing the cells followed by incubation for 24 h at 37°C in a 5% CO_2 atmosphere. After, 3-(4,5-dimethylthiazol-2-Yl)-2,5-diphenyltetrazolium bromide (MTT) solution (0.25 mg·mL^{-1}) was added to each well followed by 4 h incubation. Then, 0.1 mL of 10% SDS (prepared in 10 mmol·L^{-1} HCl) was added and incubated overnight to dissolve the formazan crystals, and the plates were read at 570 nm with 630 nm (BiochromAsys Expert Plus Microplate Reader, Biochrom Ltd., UK). The control was performed in the absence of NPs and considered as 100%. Each point represents the average of two independent experiments, with the error bar values expressed by their standard error of the mean (SEM).

3. Results and Discussion

This main objective of this study was to obtain a polymeric NP able to carry and to deliver therapeutic amounts of NO for biomedical applications. To this end, this study reports the preparation of NO-releasing CS based NPs, their characterization, the evaluation of the NO release profile from the NPs and the cytotoxicity of the NPs towards HepG2 cell line. The following sub-sections summarize the main results.

3.1. Synthesis and characterization of MSA-CS/ALG NPs

Polymeric NPs based on CS and/or ALG have been broadly studied and employed as drug delivery system due to their unique chemical and biological properties, such as, mucoadhesiveness, biocompatibility, biodegradability, and non-toxicity [14,19,22]. Indeed, synthesis of CS NPs followed by the addition of ALG layer is able to promote sustained release of encapsulated drugs in a more efficacy manner in comparison with NPs comprised by only CS or ALG [22]. In this study, CS NPs were initially synthesized by the ionotropic gelation method, in which cationic chitosan chains readily interacts with the counter negatively charge TPP ions leading to the formation of the CS NP. This process is based on the interaction of cationic CS chains with multivalent counter ions resulting in the intermolecular and/or intramolecular network structure maintained by ionic interactions between NH_3^+ protonated groups of CS and negatively charged counter ions of TPP. This cross linking leads to the formation of CS NPs. Encapsulation of MSA into CS NPs is performed by the strong electrostatic interactions between MSA and the CS chains. In a second step, addition ALG layers were added on the surface of MSA-CS NPs, leading to the formation of MSA-CS/ALG NPs. The presence of ALG is responsive to confer protection to MSA-CS NPs, which might decrease the degradation rates of NP in biological medium, and therefore protecting the NO donor.

DLS measurements revealed that the hydrodynamic diameter of MSA-CS/ALG NPs was found to be 343 ± 38 nm with a moderate polydispersity index (PDI) of 0.36 ± 0.10. These results are in accordance Biswas and co-workers who reported a hydrodynamic diameter of 432 ± 52 nm and PDI value of 0.368 ± 0.06 for CS/ALG NPs containing antigens [14]. In addition, Costa and co-works reported the hydrodynamic diameter in the rage of 380-420 nm for CS/ALG NPs containing antibiotics [22].

The zeta potential value for MSA-CS/ALG NPs was found to be -30.3 ± 0.4 mV, indicating a moderate stability of the NPs in aqueous environment, which is necessary for biomedical applications. The negative value of the zeta potential confirms the addition of ALG layers on the surface of MSA-CS NPs, since ALG has negative charge due to the presence carboxyl groups. In contrast, a positive zeta potential value of 21.8 ± 1.1 mV was observed for MSA-CS NPs, prior the addition of ALG layers. This positive value is assigned to the presence of protonated amino groups found in CS chains. Our results are in accordance with published papers which report negative values for zeta potential of CS NPs coated with ALG layers, such as -16 to -26 mV for CS/ALG NPs containing paclitaxel [23],

IOP Conf. Series: Journal of Physics: Conf. Series **838** (2017) 012032 doi:10.1088/1742-6596/838/1/012032

and a zeta potential of *ca.* -38.2 ± 0.3 mV for antibiotic delivery, as reported by Costa and co-workers [22].

The encapsulation efficiency of MSA in CS/ALG NPs was found to be 98.05 ± 0.20 %, indicating a high affinity of MSA to the ALG/CS NPs, as previous described [16]. MSA has positive charge interactions with CS and ALG chains.

3.2 Nitrosation of MSA-CS/ALG NPs leading to the formation of S-nitroso-MSA-CS/ALG NPs

Free thiol groups of MSA-CS/ALG NPs were nitrosated by reacting by nitrous acid (HNO_2) leading to the formation of S-nitroso-MSA-CS/ALG NPs, as shown in Figure 2A. HNO_2 was generated by the presence of sodium nitrite ($NaNO_2$) is acid solution.

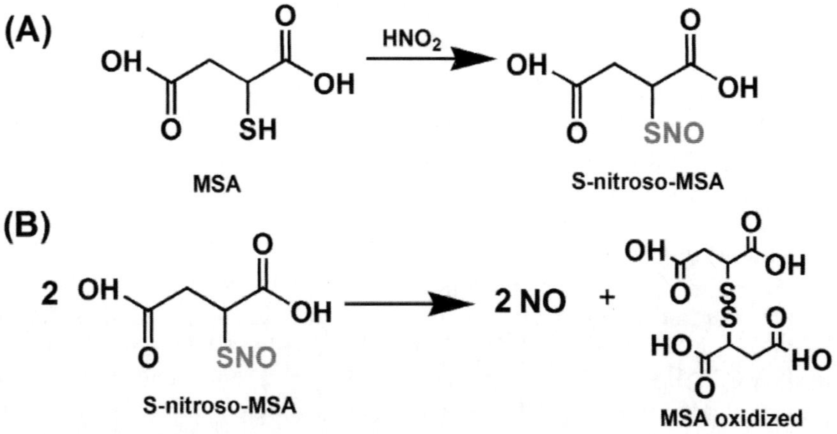

Figure 2. (A) Nitrosation of MSA forming the S-nitroso-MSA which acts as NO donor, (B) S-nitroso-MSA decomposition releasing the free NO yielding oxidized MSA.

3.3. NO release profile from S-nitroso-MSA-CS/ALG NPs at physiological temperature

S-nitroso-MSA is a spontaneous NO donor as shown in Figure 2B. The kinetics of NO release from S-nitroso-MSA-CS/ALG NPs in aqueous solution were monitored by following the spectral changes at 545 nm absorption band, associated the S-N bound cleavage with free NO release (Figure 3) [15, 16,19]. Figure 3 shows that the kinetic curve shows two distinct phases: an initial burst, in the first 2 h, of NO release with the dissolution constant (K_H) of 39.05 ± 1.8 mmol·Lh$^{-0.5}$, followed by a decrease in the rates of NO release from 2 to 6 h. Moreover, the concentration of NO release is in the mmol/L, at this concentration, NO is known to have antitumor and antibacterial activities [24-27]. The NO release from CS/ALG NPs is limited by several mechanisms, such as, diffusion, desorption, particle erosion or the combination of these factors. In order to evaluate the mechanism of NO release from the S-nitroso-MSA-CS/ALG NPs, the Higuchi model was applied to the kinetic curve. The correlation coefficient (R^2) obtained was of 0.981, which indicates that the main mechanism controlling the NO release is the diffusion process over the pores in the polymeric walls. Similar results were reported for the release of NO from CS/ALG NPs [19].

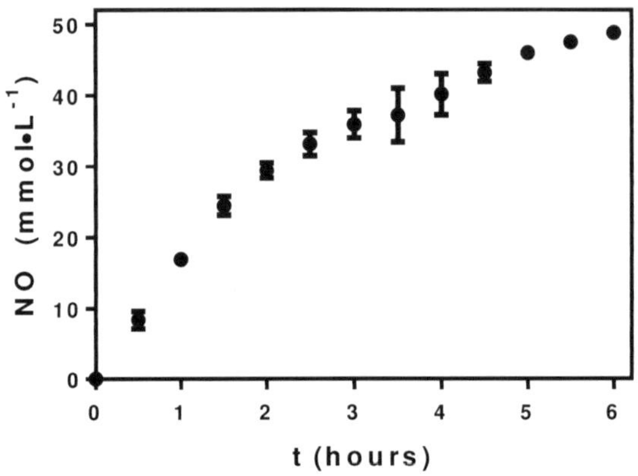

Figure 3. Kinetics of NO release from S-nitroso-MSA-CS/ALG NPs (S-nitroso-MSA concentration of 50 mmol·L^{-1}), at 37°C for 6 h. The results are presented as mean ± standard error of two independent experiments.

3.4. *Cytotoxicity of S-nitroso-MSA-CS/ALG NPs*

Figure 4 shows the viability of HepG2 cells upon treatment with S-nitroso-MSA-CS/ALG NPs and MSA-CS/ALG NPs. For both cases, a decrease in cell viability in a concentration-dependent manner was observed. At 40 μg·mL^{-1}, viability of HepG2 cells decreased to 64.771 ± 7.928% upon incubation with S-nitroso-MSA-CS/ALG NPs. This decrease is *ca.* 22% lower compared to the cell viability upon HepG2 incubation with MSA-CS/ALG NPs at the same concentration.

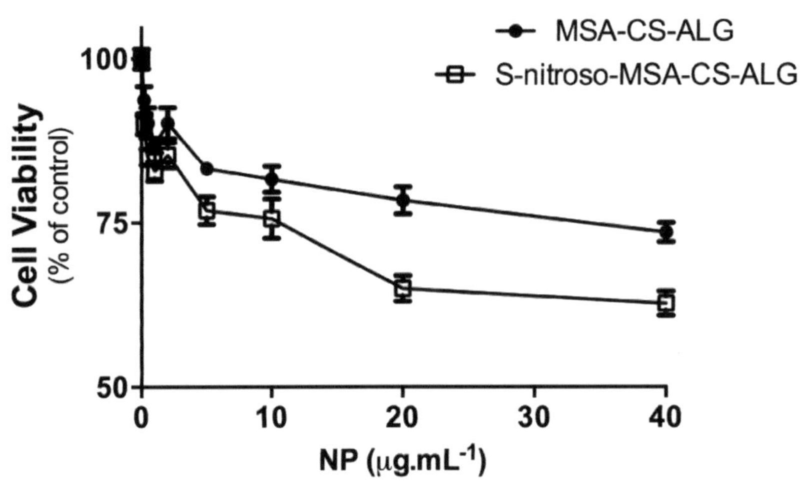

Figure 4. Viability of HepG2 tumor cells upon incubation with S-nitroso-MSA-CS/ALG NPs (empty squares) and MSA-CS/ALG NPs (black circles), at increasing concentrations. Cell viability was estimated by a tetrazolium-based (MTT) reduction assay. The results are presented in percentage of control (absence of NPs) as mean ± standard error of two independent experiments.

It is noteworthy that the polymeric nanoparticle without NO also presented toxicity in HepG2 cells and this toxicity was further increased for the nanoparticle containing the NO donor. This result indicates that NO-releasing CS/ALG NPs presents potential to use against cancer cells.

4. Conclusions

This work describes the synthesis and characterization of NO-releasing CS/ALG NPs. The NO donor S-nitroso-MSA was encapsulated into CS NPs, followed by the addition of ALG layer. This nanoparticle was able to spontaneously release free NO, following the Higuchi mathematical model. The cytotoxicity of S-nitroso-MSA CS/ALG NPs were demonstrated towards HepG2 tumor cell. Further studies to understand the molecular pathways involved in the cytotoxicity of this nanomaterial is under investigation.

Acknowledgements: FAPESP (Proc. 2016/10347-6 and 2016/07367-5), CNPq (Proc. 486067/2013-0) the Brazilian Network on Nanotoxicology (Grant number: 552120/2011-1) (MCTI/CNPq), the Laboratory of Nanostructure Synthesis and Biosystem Interactions-NANOBIOSS (MCTI) (Grant number: 402280-2013), Newton Advanced Fellowship (The Royal Society NA140046).

References

[1] Seabra AB, de Lima R and Calderón M 2015 *Curr. Top. Med. Chem.* **15** 298
[2] Yin L, Su T, Chang J, Liu R, He B and Gu Z 2014 *NANO* **9** 1450075
[3] Markeb AA, El-Maali NA, Sayed DM, Osama A, Abdel-Malek MAY, Zaki AH and Elwanis MEA, Driscoll JJ 2016 *Int. J. Breast Cancer* **2016**
[4] Manuja A, Kumara B, Chopraa M, Bajaja A, Kumara R, Dilbaghib N, Kumarb S, Singha S, Riyesha T and Yadavaa SC 2016 *Int. J. Biol.Macromolec.* **88** 146
[5] Sundar S, Kundu J and Kundu SC 2010 *Sci. Technol. Adv. Mater.***11** 014104
[6] Basudhar D, Cheng RC, Bharadwaj G, Ridnour LA, Wink DA and Miranda KM 2015 *Free Radic. Biol. Med.* **83** 101
[7] Bonavida B and Garban 2015 *Redox Biol.* **6** 486
[8] Seabra AB, Justo GZ and Haddad PS 2015 *Biotechnol. Adv.* **33** 1370
[9] Burke AJ, Sullivan FJ, Giles FJ and Glynn SA 2013 *Carcinogenesis* **34** 503
[10] de Oliveira MG 2016 *Basic Clin. Pharmacol. Toxicol.* doi: 10.1111/bcpt.12588
[11] Borges O, Cordeiro-da-Silva A, Romeijn SG, Amid M, de Sousa A, Borchard G and Junginger HE 2013 *J. Control. Release***114** 348
[12] Wu W, Perrin-Sarrado C, Ming H, Lartaud I, Maincent P, Hu X, Sapin-Minet A and Gaucher C 2016 *Nanomedicine***12** 1795
[13] Pereira AES, Narciso AM, Seabra AB and Fraceto LF 2015 *JPCS* **617** 012025
[14] Biswas S, Chattopadhyay M, Sen KK and Saha MK 2015 *Carbohyd.Polym.* **121** 403
[15] Seabra AB, Kitice NA, Pelegrino MT, Lancheros CAC, Yamauchi LM, Pinge-Filho P and Yamada-Ogatta SF 2015 *JPCS* **617** 012020
[16] Cardozo VF, Lancheros CA, Narciso AM, Valereto EC, Kobayashi RKT, Seabra AB and Nakazato G 2014 *Int. J. Pharm.* **473** 20
[17] de Oliveira MG, Shishido SM, Seabra AB and Morgon NH 2002 *J. Phys. Chem. A* **106** 8963
[18] Grillo R, de Melo NFS, de Araújo SR, de Paula E, Rosa AH, Fraceto LH 2010 *J. Drug Target* **18** 688
[19] Marcato PD, Adami LF, Barbosa RM, Melo PS, Ferreira IS, de Paula L, Durán N and Seabra AB 2013 *Curr. Nanosci.* **9** 1
[20] Kim JO, Noh JK, Thapa RK, Hasan N, Choi M, Kim JH, Lee JH, Ku SK and Yoo JW 2015 *Int. J. Biol. Macromol.* **79** 217
[21] Costa and Lobo 2001 *Eur. J. Pharm. Sci.* **13** 123
[22] Costa JR, Silva NC, Sarmento B and Pintado M 2015 *Eur. J. Clin. Microbio.l Infect. Dis.* **34** 1255
[23] Wang F, Yang S, Yuan J, Gao Q and Huang C 2016 *J. Biomater. Appl.* **23** 619
[24] Quinn JF, Whittaker MR and Davis TP 2015 *J. Control. Release* **205** 190

[25] Backlund CJ, Worley BV and Schoenfisch MH 2015 *Acta Biomater.* **29** 198
[26] de la Rosa AJ, Rodríguez-Hernández A, González R, Romero-Brufau S, Navarro-Villarán E, Barrera-Pulido L and Pereira S 2016 *Gene Ther.* **23** 67
[27] Seabra AB, Pelegrino MT and Haddad PS 2016 Antibiotic Resistance *Can nitric oxide overcome bacterial resistance to antibiotics?* Ed 1 (Elsevier) pp 187-201

IOP Conf. Series: Journal of Physics: Conf. Series **838** (2017) 012033 doi:10.1088/1742-6596/838/1/012033

Surface reactivity measurements as required for grouping and read-across: An advanced FRAS protocol

Arnaud Gandon[†,‡], Kai Werle[‡], Nicole Neubauer, Wendel Wohlleben*

BASF SE, Dept. Material Physics, 67056 Ludwigshafen, Germany
[†] present address: Laval University, Department of Chemistry, Québec, Canada
[‡] equal contribution

* corresponding author email: wendel.wohlleben@basf.com

Abstract. Oxidative stress is a widely accepted paradigm associated with different adverse outcomes of particulate matter, including nanomaterials. It has frequently been identified in in vitro and in vivo studies and different assays have been developed for this purpose. Here we describe a newly developed multi-dose protocol of the FRAS assay (Ferric Reduction Ability of Serum). The purpose of this SOP is the measurement of the surface reactivity of nanomaterials under physiological conditions. Antioxidative components as present in human blood serum (HBS) serve as reporter molecules. The assay separates the oxidative damage from the read-out of the reporter molecules. The results show significantly enhanced repeatability with better sensitivity towards low reactivity, enabling application of FRAS both to a rough grouping by reactive vs. passive nanomaterials and further to substantiation of read-across by enhanced resolution of the similarity between different nanoforms of the same substance.

1. Introduction

The French nano-inventory highlights the diversity of substances in nanoform, which are dominated by conventional fillers and pigments produced in multi-k-ton/y of various nanoforms per substance.[1] Therefore, alternative testing approaches with reduced costs and high predictivity suitable for high throughput screening and prioritization are urgently needed. In parallel, extensive research efforts are targeted to understanding modes of action of Engineered Nano-Materials (ENMs), which may also support the development of new predictive assays and which are also needed to establish NM grouping.[2] Oxidative stress is a widely accepted paradigm associated with different adverse outcomes of ENMs. It has frequently been identified in in vitro and in vivo studies [2,3] and different assays have been developed for this purpose.[4-11]

Previously, we have investigated a representative panel of 32 ENMs including functionalized amorphous silica (6 nanoforms), zirconium dioxide (4), silver (4), titanium dioxide (3), zinc oxide (2), multiwalled carbon nanotubes (3), barium sulfate and AlOOH. Surface reactivities of all were studied in a cell-free system by electron spin resonance (ESR) spectroscopy. NRK-52E cells were treated with all NP, analyzed for viability (WST-1 assay) and intracellular ROS production (DCFDA assay). Carbonylated proteins were analyzed on 1D and 2D immunoblots and identified by matrix assisted laser desorption time-of-flight mass spectrometry.[3] We found overall an excellent match of ESR vs. in vitro carbonylation.

To complement the grouping concepts by several complementary protocols to assess the extrinsic property of surface reactivity, we explore here an optimization of the FRAS assay (Ferric Reduction

IOP Conf. Series: Journal of Physics: Conf. Series **838** (2017) 012033 doi:10.1088/1742-6596/838/1/012033

Ability of Serum). The principle of the FRAS assay is well established to assess oxidative damage [4] including its application to ENM surface reactivity [5], and has shown potential to separate active from passive ENMs [6-8].

2. Methods and Materials

2.1. *FRAS assay principle*

In short, the antioxidative components as present in human blood serum (HBS) serve as reporter molecules.[4,5] The assay separates the oxidative damage from the read-out of the reporter molecules, as schematically shown in Figure 1: (1), HBS is incubated with nanomaterials, which are then (2) removed by centrifugation. Afterwards (3) an aliquot of that pre-incubated HBS is mixed with a solution containing a Fe3+ complex and is again incubated. The ability of the pre-incubated HBS to reduce Fe3+ to Fe2+ is detected optically by a color change from transparent to blue. Any damage to anti-oxidative species in HBS by nanomaterials during step (1) will reduce the blue color in the final step (3).

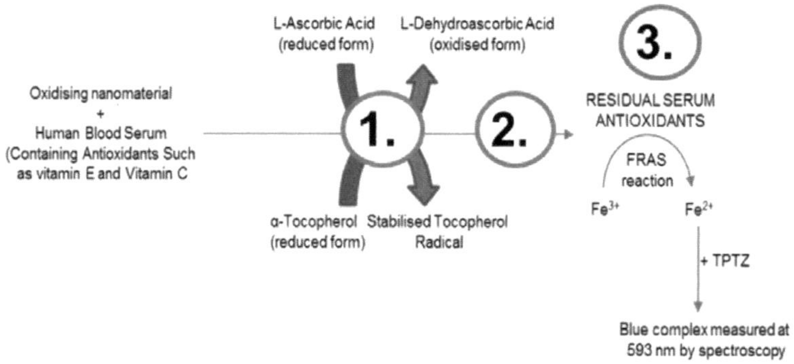

Figure 1: Generic FRAS schematic

2.2. *FRAS assay optimization summary*

The individual steps of the protocol were not / not all / not sufficiently described in earlier publications.[5-8] We optimized them for high sensitivity & increased statistical significance, and found the following elements to be essential at each of the respective steps:

1. Optimized step ENM@serum incubation time is 3 h at 37°C. The use of stock solutions is strongly disadvised because the dilution of serum by de-ionized water create a substantial signal. Instead, HBS is given directly on ENM powder. Despite potential trade-offs by the denaturation of proteins, we found short (1 min bath) sonication highly beneficial to enhance reproducibility, presumably by making the ENM surface accessible to the antioxidants
    a. Tested incubation time from 60 to 180 min – preferred 180 min for enhanced sensitivity towards low reactive ENM.
    b. Sonication preferred, otherwise visibly large (>μm) agglomerates can remain even after the incubation time, leaving doubts on the biologically accessible surface area.
    c. For screening purposes established ENM mass dose range 0.75 g/L, 2 g/L, 5.5 g/L, 15 g/L, 40 g/L (which extends to both sides from the previous default of 10 g/L)
    d. Triplicate testing of dose response, then averaging of fit parameters.
2. Centrifugal extraction to remove ENM and yet to retain antioxidants works fine with 150 minutes at 14,000 g. When transferring ENM-free serum from centrifugation tubes into the FRAS reagent, the reproducibility of the volume of 100 μL translates directly into the final reproducibility of the assay.
    a. Tested 2000 rpm, found no increase of antioxidants, hence no issues with inadvertent antioxidant depletion by ENM separation.

IOP Conf. Series: Journal of Physics: Conf. Series **838** (2017) 012033 doi:10.1088/1742-6596/838/1/012033

   b. Pipettes tested: plastics, coated plastic "Ultra Low Retention" tips, glass with "positive displacement" pipettes using a Teflon plug (strongly recommended).
   c. Tested fresh pipette for each vial – no effect.
3. Handling of the FRAS reagent in the dark is absolutely essential. The doubled dilution of 0.1 mL HBS into 2 mL FRAS reagent enhances the sensitivity to low reactive ENM, because it ensures that the final state of Fe oxidation is limited by the antioxidant concentration and not by the Fe-TPTZ concentration.
   a. Optimized serum@FRAS reaction time at room temperature is 60 min with ± 1 min precision.
   b. Tested under ambient lab conditions vs. no light during handling.
   c. Tested pre-wetting of pipette tips in water / in HBS, found no influence, tested glass vs. plastic vials and pipette tips, found strong influence.
   d. Tested range of reaction times: from 60 to 120 min, with 5 minute deliberate variations, or in reversed order – only reaction time counts.

The following sections 2.2.1 to 2.2.3 document the exact materials, disposables, equipment used in our labs, marking these as "recommended" and **highlighting in bold** those that we now know to be critical.

2.2.1. Recommended materials:
- sodium acetic trihydrate: Sigma Aldrich BioUltra, purity: ≥99.5%, CAS number: 6131-90-4, EC number: 204-823-8, Product number: 71188
- glacial acetic acid: Alfa Aesar, purity ≥99,7%, CAS number: 64-19-7, EC number: 200-580-7, Product number: 36289
- 2,4,6-tri(2-pyridyl)-s-triazine (TPTZ): Sigma Aldrich, purity: ≥98%, CAS number: 3682-35-7, RC number: 222-965-9, Product number: T1253
- 1M HCl: Riedel-De Haen, concentration: 1M, CAS number: 7647-01-0, Product number: 35328
- $FeCl_3.6H_2O$: Sigma Aldrich, purity: ≥98-102%, CAS number: 10025-77-1, EC number: 231-729-4, Product number: 44944
- human blood serum (HBS): Sigma Aldrich, P2918-100ml (frozen in 2 g aliquots)
  Note that HBS is of human origin and requires appropriate risk management. It is anticipated that the source of human blood serum has an influence on the absolute FRAS

2.2.2. Recommended disposables:
- Pipette: Eppendorf Reference, 100-1000 μL, with plastic tips
- Disposal **glass Pasteur pipettes** with pipetting aid: volac, (product number D810, 150 mm); Brand accu-jet pro
- 100 mL glass bottles: Kavalier Simax, product number: 2070 M, EAN: 8593419604089, code: 1632414321100, diameter: 56 mm, height: 100,00 mm, GL: 45
- 30 mL **glass vials**: chemoline, product number 50-2030
- 10 mL **glass vials**: Max Wiegand, product number 4382810)
- PTFE stirrers: Polygon, length:15 mm diameter: 6 mm [glass stirrers would be preferable but could not be obtained at this small dimensions required for low volumes]
- Centrifuge tubes: Beckman Coulter, Centrifuge Tubes Ultra-Clear, 11 x 60 mm, product number 344062
- Brand Transferpettor **with 100 μL glass tips**
- Quartz cuvettes: Helma Analytics, High Precision Cell made of Quartz SUPRASIL, light path 10 mm, 104-10-40

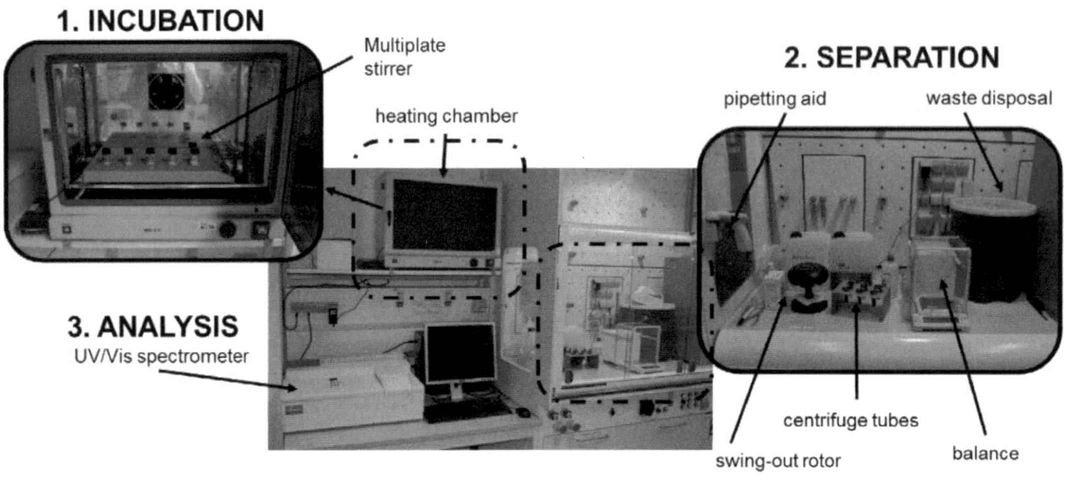

Figure 2: Overview of FRAS implementation

2.2.3. Recommended equipment:

- Millipore Advantage A10 → DI water
- Balance: Mettler AT250
- Ultrasonication bath: Sonorex Digital 10p (320 W electrical power)
- Shaker: IKA Vibrax VXR (set to level 200), equipped with home-made extension for 10 mL vials (see photo below)
- Heating chamber: mytron WB 55
- Stirrer: Thermo Scientific, Variomag Poly 15
- Centrifuge: Beckman Coulter OPTIMA-XL80K with swing-out rotor SW60-TI
- UV/Vis spectrometer: Perkin Elmer Lambda 35 equipped with **automatic linear cell changer for 8 samples** (the automatic changer is not ultimately required, but makes it much easier to keep the timing identical for all doses.)

2.3. Optimized multi-dose protocol of the FRAS assay

2.3.1. Preparation of FRAS reagent:

As the following solutions cannot be stored, they have to be prepared freshly before every experiment.

Prepare S1:

- Weigh 0.2021 g of sodium acetic trihydrate to a 100 mL glass bottle.
- Add 50 g DI water.
- Add 1.060 mL (2 x 0.53 ml) of glacial acetic using a pipette.
- Add DI water to achieve a total of 100 g.

Prepare S2:

- Weigh 0.0946 g of TPTZ to a 30 mL glass vial.
- Add 15 g DI water.
- Add 1.2 mL (2 x 0.6 mL) of 1M HCl using a pipette.
- Add DI water to achieve a total of 30 g.
- **Put S2 to an ultrasonication bath and sonicate for 30 minutes with maximum power at room temperature.**

Prepare S3:

- Weigh 0.1635 g of $FeCl_3.6H_2O$ to a 30 mL glass vial.

- Add DI water to achieve a total of 30 g.

Prepare FRAS reagent:

- Weigh 50 g of S1, 5 g of S2, and 5 g of S3 to a 100 mL glass bottle.
- **Store FRAS reagent in the dark while shaking slowly.**

2.3.2. Test protocol:

Make sure to only use glass equipment to avoid interferences with plastic parts.

1. Unfreeze 6 x 2 g human blood serum (HBS) at 37°C for 30 minutes.

STEP (1) INCUBATION:

2. Prepare powder nanomaterial dispersions in concentrations of 0 (blank), 0.75, 2, 5.5, 15, and 40 g/L by first weighing the powder nanomaterial to six 10 mL glass vials and then adding 1.5 g of unfrozen HBS to each glass vial by using glass Pasteur pipettes.
3. Add a PTFE stirrer to each bottle, close bottles with caps.
4. Put the nanomaterial dispersions to an ultrasonication bath and sonicate for 1 minute with maximum power at room temperature.
5. Incubate dispersions **at 37°C for 3 hours in the dark** while stirring constantly at 400 rpm.

STEP (2) SEPARATION:

6. Transfer nanomaterial dispersions completely to six centrifuge tubes.
7. Centrifuge at **14,000 g** (11,900 rpm in our centrifuge) **for 150 minutes**. Earlier implementations of the FRAS assay [8] specify the same g-number for twice 20 minutes. We found this procedure less effective due to convection and re-diffusion issues. We checked that this condition does not deplete the antioxidants from serum (FRAS signal remains the same with only 2000rpm)
8. During centrifugation, label 18 10 ml glass vials with *blank A, C1A, C2A, C3A, C4A, C5A, blank B, C1B, C2B, C3B, C4B, C5B, blank C, C1C, C2C, C3C, C4C, C5C.*
9. Add 2.00 g ± 0.002 g of FRAS reagent to each glass vial by using a glass Pasteur pipette. **Make sure to avoid any direct light during aliquotation of FRAS reagent**. Put each glass vial to the dark directly after the addition of FRAS reagent.
10. Label six 10 ml glass bottles with *blank, C1, C2, C3, C4, C5.*
11. **After centrifugation, carefully recover the liquid part of samples (approximately 0.5 mL) by using disposal glass Pasteur pipettes. Make sure to recover no solids: neither particles from the sediment nor floating components. Add the removed supernatants to the six labelled 10 ml glass bottles by disposing the first and the last drop.**

STEP (3) ANALYSIS:

12. Take glass vials *blank A, C1A, C2A, C3A, C4A, C5A* containing the FRAS reagent out of the dark and add 100 µl of the supernatant of blank to blank A, C1 to C1A, C2 to C2A, C3 to C3A, C4 to C4A, C5 to C5A by using a specific pipette (Brand Transferpettor, **100 µL glass tip) according to a strict time schedule depending on the time required for UV/Vis detection (absorption at 500-700 nm). Such a time schedule is crucial to ensure that a reaction time of 60 ± 1 min. for each sample is achieved.** In our case detection by UV/Vis takes approximately 45 seconds per sample (approximately 4 minutes in total) for a wavelength range of 500-700 nm resulting in the time schedule below. If using another UV/Vis detector, the time schedule has to be adapted accordingly in dependence on the detection duration of the used spectrometer. Start timer directly after adding the supernatant of blank to FRAS reagent of blank A and afterwards carefully follow the time schedule. Put each sample to the shaker in the dark, directly after adding the supernatant to the vial.
13. Accordingly, take glass vials filled with FRAS reagent *(blank B, C1B, C2B, C3B, C4B, C5B)* out of the dark and add 100 µL of the supernatant blank to blank B, C1 to C1B, C2 to C2B, C3 to C3B, C4 to C4B, C5 to C5B by strictly following the time schedule.
14. Accordingly, take glass vials filled with FRAS reagent (*blank C, C1C, C2C, C3C, C4C, C5C*) out of the dark and add 100 µL of the supernatant blank to blank C, C1 to C1C, C2 to C2C, C3 to C3C, C4 to C4C, C5 to C5C by strictly following the time schedule.

Table 1 Recommended timing scheme

sample	time A in min	time B in min	time C in min
blank	0:00	4:45	9:28
C1	0:45	5:25	10:09
C2	1:26	6:05	10:50
C3	2:07	6:46	11:30
C4	2:47	7:28	12:11
C5	3:27	8:08	12:51

15. Take glass vials blank A, C1A, C2A, C3A, C4A, C5A out of the dark and transfer each sample to a quartz cuvette by using glass Pasteur pipettes. Put these 6 cuvettes to the automatic linear cell changer of the UV/Vis detector and measure absorption of samples at 500-700 nm. Make sure that this step is performed accurately timed ensuring that all samples are measured exactly after 60 minutes of reaction time.
16. Prepare and measure samples of series B and C accordingly.

2.3.3. Calibration to Trolox-equivalent units (TEU):

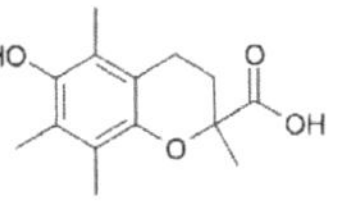

Trolox ($C_{14}H_{18}O_4$ or 6-hydroxy-2,5,7,8-tetramethylchroman-2-carboxylic acid, M_{Trolox} = 250.29 g·mol^{-1}) is a water-soluble analog of vitamin E. It is an antioxidant like vitamin E and it is routinely used in biological or biochemical applications to reduce oxidative stress or damage – and to measure antioxidant capacity of foods, beverages and supplements.[7,9] We introduce the quantities:

Abs = Absorption [-]
l = light path of cuvette [cm]
kE = extinction coefficient of TPTZ induced by 1 Mol antioxidant [TEU]
d = dilution factor 0.048 (0.1 ml HBS in 2 ml FRAS reagent)

Performing the FRAS protocol step (3 ANALYSIS), replacing the HBS serum by a concentration series of Trolox from 0.001 to 0.1 g/L, we obtain a FRAS absorption signal that can be linearly fitted with Abs_{Trolox} = kE * l * d * c_{Trolox} [mM] + b (R^2 = 0.9998). The offset absorption b = 0.075 at zero antioxidants determines the maximum detectable oxidative damage. The linear fit factor kE * l * d = 2.654 /(mM Trolox) calibrates the absorption signal per „Trolox-Equivalent Unit" (TEU) antioxidant molarity (mM TEU) for our specific implementation of the FRAS assay. Removing the dilution factor and the optical path length, the extinction coefficient of TPTZ per Trolox calibration is kE = 55.76 mM TEU /cm for our FRAS reaction time of 60 minutes.

The earlier study by Hsieh et al [6] found a value of 4.18 / mM/L TEU at unspecified factors l and d. Assuming that the protocol in [6], [7] and [8] is identical, the description in [8] "1-10 mg typical quantities; 1.0 mL serum at 37 °C for 90 min, separation of the ENM, transfer of 100 μL of the supernatant to 1.0 mL of working FRAS reagent." indicates a dilution factor of 0.091. Assuming further identical values of cuvette depth l = 1 cm and 60 min FRAS reaction time, considering our slightly increased damage by longer incubation but halfed signal by dilution, their value of 4.18 / mM TEU is very consistent with our value of 2.65 / mM TEU.

2.3.4. Data analysis:

1. Calculate Biological Oxidative Damage (BOD) for A1-A5, B1-B5, C1-C5, according to the following equation, shown as an example for sample A. in mM TEU: (x= 1 to 5)

$$BOD = \frac{Abs(blank\ A) - Abs\ (Ax)}{l*kE*d}\ \text{[in mM TEU]}$$

   The denominator is determined by calibration with Trolox, see section 2.3.3.
2. Convert the nanomaterial doses to surface metrics

   $Dose_x$ [m²/L] = c_x [g/L] * BET [m²/g]
   BET = BET surface of nanomaterial [m^2/g]

c_x = concentration of nanomaterial [mg/mL] [note: mg/mL = g/L, no conversion required]

3. Convert to the maximum oxidative damage in the two relevant metrics (both max at same x)
   a. Mass-based max damage mBOD = max {BOD (c_x) / c_x } [nmol TEU /mg]
   b. Surface-based max damage sBOD = max {BOD ($dose_x$) / $dose_x$ } [nmol TEU /m²]
4. Plot each series A, B, C independently. Use log (Dose) on x-axis and log (BOD) on y-axis.
5. Fit each series A, B, C independently with a linear regression curve, giving slope and offset for series A, B, C. [note: for reactive materials, the linear fit is not appropriate and should be replaced by sigmoidal fit, but most ENM remain at very low damage, hence benchmark dose approaches or related sigmoidal metrics are not easily applicable]
6. Calculate from these three values the average and standard deviation for:
   a. Slope [mM TEU / m²]
   b. Offset [mM TEU]

3. Results

3.1. Repeatability of blank measurements (= in the absence of ENM)

By performing the complete protocol as described in section 2.3, but without adding ENM in step 2.3.2-2, we assess the repeatability of the protocol and eliminate sources of variability. As demonstrated by Figure 3, the scatter is reduced by nearly an order of magnitude (from 8 to 1 µM TEU), and the average remaining signal from blanks is reduced from 6 to 2 µM TEU.

In contrast, the HBS used here has a total antioxidant capacity of 366 µM TEU, as derived from the Trolox calibration and the absorption signal induced by the blank measurements (0.97). For comparison, a value of 530 mM TEU is given by Hsieh et al. for their HBS.[7] We believe the unit mM to be a typo (correct: µM), as all other values and units are consistent with ours. The difference thus represents the natural spread of HBS sources. As the FRAS assay detects only relative changes, this difference is not critical to the comparison of oxidative damage results.

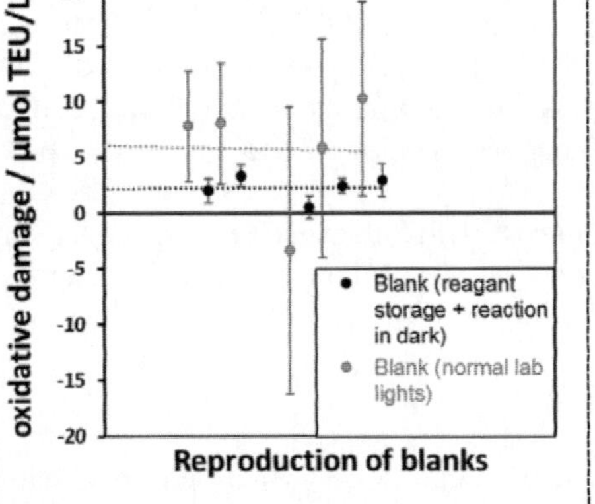

Figure 3. Repeatability of blank measurements by handling FRAS reagent in normal lab light and by handling without direct light, reacting in the dark.

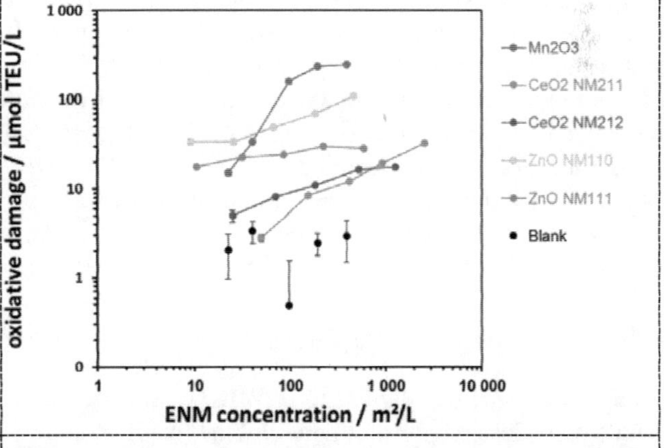

Figure 4. Dose-dependent results for positive control Mn_2O_3 and for each two nanoforms of ZnO and CeO_2 from the OECD sponsorship program. Error bars (1 SD) from triplicate testing are shown on every data point, but are smaller than the symbol in most cases.

Note the logarithmic scale of oxidative damage.

IOP Conf. Series: Journal of Physics: Conf. Series **838** (2017) 012033 doi:10.1088/1742-6596/838/1/012033

3.2. Positive control: Mn2O3

Based on earlier report of both FRAS assay [7] and cytochrome c assay [10] and band gap analysis [11], Mn_2O_3 is recurring as the most reactive, yet insoluble material. Hence, we chose to establish Mn_2O_3 as positive control. We use the same batch as in [10], sourced from Skyspring Nanomaterials, with primary particle diameters ranging from 20 to 100 nm (TEM) and a specific surface area of 19.9 m^2/g (BET).[10] As expected, Mn_2O_3 induces maximum antioxidant damage already at low dose (Figure 4, red line), and levels off into a plateau at higher doses. The maximum sBOD of 1610 ± 4 nmol TEU/ m^2 (equivalent to mBOD of 32.0 ± 0.09 nmol TEU/ mg) is reached at a surface dose of 244.5 m^2/L (corresponding to mass concentration of 4.9 g/L), generating a BOD of $156{,}680 \pm 435$ mM TEU.

The same substance in a different grade (BET = 7.4 m^2/g) was studied earlier. The simple BOD, dose and standard deviation were not reported, but the mBOD was reported as 104 ± 5 nmol TEU / mg and sBOD as 20,555 nmol TEU / m^2.[7,8] The mBOD value is a factor 5 higher than ours (the sBOD value a factor 13, but marked by Hsieh et al. as potentially influenced by uncertainty of the BET value). As we observe a plateau at high dose, interpreted as maximal damage of all antioxidants that are susceptible to the specific reaction pathway induced by Mn_2O_3, dose is probably not the origin of the quantitative discrepancy.

Despite the quantitative disagreement between the two implementations of FRAS, both agree that Mn_2O_3 is among the most reactive materials, of which it is the only one with vanishingly low solubility. Both also agree on excellent reproducibility of Mn_2O_3 reactivity (<1% standard deviation in our triplicates). Mn_2O_3 is thus chosen as our positive control, as established also for the cytochrome-c [10] and bandgap assays.[11]

3.2.1. Limit of Detection. At low reactivity, Hsieh et al. note their limit of statistical relevance for mBOD at 1.7 nmol TEU/mg, probably determined at their concentration of 10 g/L (corresponding to 17 μM TEU). In comparison, our blank signal of 2 ± 1 μM TEU allows an estimate for the limit of detection at blank + three standard deviations, resulting in 5 μM TEU. Together with the extended concentration range up to 40 g/L, our limit of statistical relevance for mBOD is around 0.12 nmol TEU/mg. This is more than an order of magnitude improvement already for single-dose triplicates. Additionally, the dose-dependence provides highly relevant information.

At high reactivity, the limit is ultimately the complete consumption of the HBS antioxidant capacity of 366 μM TEU, which can be converted into an upper detection limit of mBOD, assuming a material that damages all antioxidants at the lowest routine dose of 0.75 g/L. This results in a maximum mBOD of 488 nmol TEU /mg detectable by screening ENMs with the routine dose range, which is well above the value of the positive control Mn_2O_3. Thus, the reactivity range from below 1% to above 1000% of the control material is detectable.

This consideration confirms again that the FRAS assay is more sensitive than DCFH, especially for ENMs with low to moderate oxidative damage potential, and may serve as a more biologically relevant substitute for acellular ROS measurements of ENMs.[6,8]

3.3. First application to compare nanoforms of the same substance

3.3.1. Comparing CeO_2, OECD NM211 vs. NM212, we find that both have a significant reactivity with a clear dose response (Figure 4, blue lines). Both remain far below the positive control, and would not have been detected without the improved protocol. Within the statistically relevant range of BOD > 5μM TEU, the curves a very close, so that for each individual dose, the sBOD values are very similar, as expected for surface-induced reactions. (In contrast, the NM211 has higher mBOD values than NM212 due to the higher BET surface of NM211). The slope of the dose response is the same or steeper for NM211, consistent with the increased share of unsaturated Ce^{III} by the smaller particles. All this is very plausible, and any further interpretation is beyond the scope of the present protocol-centered contribution.

We note that the same CeO_2 NM212 shows only borderline reactivity in a macrophage assay,[13] and that it has not been possible to determine significant reactivity by the cytochrome-c assay, ascorbic acid depletion, glutathione depletion, whereas Mn_2O_3 was significantly reactive on all of these.[10] This evidences the sensitivity of the advanced FRAS assay to weakly reactive ENM.

3.3.2. Comparing ZnO, OECD NM110 vs. NM111, we find significant oxidative damage. This was expected from several other assays.[2,7,8,10,11,13] For the surface-coated NM111 we find significant suppression of reactivity as compared to the naked NM110. Most notably, the oxidative damage at lowest doses is above Mn_2O_3 , but remains below Mn_2O_3 at higher doses. A crossing of dose-response curves points to different reaction pathways – in this case attributed to the contribution of Zn^{2+} ions, which will remain relevant at doses down to the steady-state solubility far below our lowest dose of 0.75 g/L. We note that ZnO NM110 is used as benchmark "dissolving ENM" in the ECETOC grouping scheme.[3] All this is very plausible, and any further interpretation is beyond the scope of the present protocol-centered contribution.

Any dose can be formally converted to a mBOD, with values for the naked ZnO NM110 ranging from 3 to 44 nmol TEU /mg (sBOD ranging from 239 to 3700 nmol TEU /m²). This is fully consistent with the single-dose value of mBOD of 12 ± 3 nmol TEU /mg (sBOD 340 ± 40 nmol TEU /m²) determined by Hsieh et al..[7]

4. Conclusion and Outlook

We developed the FRAS assay further to become a protocol to determine dose-dependent oxidative damage by nanomaterials dispersed in physiological medium, with statistical relevance also for intermediate and weakly reactive materials. Already for single-dose determination of oxidative damage, our advanced protocol achieves a limit of statistical relevance for mBOD around 0.1 nmol TEU/mg. This constitutes an improvement by more than an order of magnitude. Further, we propose to establish Mn_2O_3 as positive control material, in accord with other recent assays of oxidative stress. Thus, the range from below 1% to above 1000% of the control material is measurable with our routine dose range. Thus, single-dose screenings can group materials by properties of concern (here: surface reactivity), which we currently define as "sBOD more than 10% of Mn_2O_3 reactivity".[3] Additionally, the dose-response relationship may point to differences of oxidative stress mechanisms, such as the contributions by dissolved ions, and serve to substantiate read-across.

To benchmark FRAS against other measures of surface reactivity and against toxicological studies, the same silica, titania, zincoxide families as before[2] and families of organic DPP pigments, metal-organic CuPhthalocyanins and iron oxide will be investigated by the present multi-dose protocol of the FRAS assay. For most of these materials, in vivo inhalation and macrophage assay results are available for benchmarking the proposed groupings/screenings.[3,12,13]

Based on the outcome of this study we currently develop a comprehensive testing strategy for assessing the oxidative stress potential of ENMs, which combines acellular methods (ESR, FRAS) and fast in vitro screening approaches, as well as a more involved detailed redox proteomics approach. This allows for screening and prioritization in a tier of "system-depending properties" and, if required, also for unraveling mechanistic details down to compromised signaling pathways. This approach appears highly useful within an ENM grouping strategy.

Acknowledgement

Partial funding by BMBF NanoGRAVUR is gratefully acknowledged (FKZ 03XP0002B). We sincerely thank Christian Riebeling and Andrea Haase (BfR Berlin), Martin Wiemann (IBE Münster), Thomas A. J. Kuhlbusch (BAuA Dortmund), Bryan Hellack (IUTA Duisburg), Jürgen Schnekenburger (WWU Münster) and Robert Landsiedel (BASF Ludwigshafen) for excellent discussions on surface reactivity and its positioning in grouping concepts.

References

[1] Ministère de l'Environnement, de l'Énergie et de la Mer (**2015**) Éléments issus des déclarations des substances à l'état nanoparticulaire: Exercice 2015.

[2] Riebeling C, Wiemann M, Schnekenburger J, Kuhlbusch TAJ, Wohlleben W, Luch A and Haase A (**2016**). "A redox proteomics approach to investigate the mode of action of nanomaterials." *Toxicology and applied pharmacology* 299: 24-29.

[3] Arts JHE, Irfan M-A, Keene AM, Kreiling R, Lyon D, Maier M, Michel K, Neubauer N, Petry T, Sauer UG, Warheit D, Wiench K, Wohlleben W and Landsiedel R, (**2016**). "Case studies putting the decision-making framework for the grouping and testing of nanomaterials (DF4nanoGrouping) into practice." *Regulatory Toxicology and Pharmacology* 76: 234-261.

[4] Mahecha L, Nuernberg K, Nuernberg G, Martin J, Hubbermann EM, Knoeller S, Claeys E, De Smet S, Dannenberger D. (**2011**) Antioxidant enzyme activities and antioxidant capacity in longissimus muscle from bulls fed diets rich in polyunsaturated fatty acids. *Food Chemistry* 127: 379–386 O. I. Aruoma. (**1994**) Nutrition and health aspects of free radicals and antioxidants. *Food and Chemical Toxicology* 32 (7) 671-683

[5] Rogers EJ, Hsieh SF, Organti N, Schmidt D, & Bello D (**2008**) A high throughput in vitro analytical approach to screen for oxidative stress potential exerted by nanomaterials using a biologically relevant matrix: Human blood serum. *Toxicology in Vitro* 22(6):1639-1647.

[6] Pal A, Hsieh S-F, Khatri M, Isaacs J, Demokritou P, Gaines P, Schmidt D, Rogers E, & Bello D (**2014**) Screening for oxidative damage by engineered nanomaterials: A comparative evaluation of FRAS and DCFH. *Journal of Nanoparticle Research* 16(2):1-20.

[7] Hsieh S-F, Bello D, Schmidt DF, Pal AK, Stella A, Isaacs JA, & Rogers EJ (**2013**) Mapping the biological oxidative damage of engineered nanomaterials. *Small* 9(9-10):1853-1865.

[8] Pal AK, (**2013**) Multi-tiered toxicity assessment of engineered nanomaterials : exploring the utility of oxidative stress, PhD-Thesis, University Of Massachusetts Lowell

[9] Benzie IFF & Strain JJ (**1996**) The ferric reducing ability of plasma (FRAP) as a measure of “antioxidant power”: The FRAP assay. *Analytical Biochemistry* 239(1):70-76.

[10] Delaval M, Wohlleben W, Landsiedel R, Baeza-Squiban A, & Boland S (**2016**) Assessment of the oxidative potential of nanoparticles by the cytochrome c assay: Assay improvement and development of a high-throughput method to predict the toxicity of nanoparticles. *Archives of Toxicology:* 91(1): 163–177

[11] Zhang H, Ji Z, Xia T, Meng H, Low-Kam C, Liu R, Pokhrel S, Lin S, Wang X, Liao YP, Wang M, Li L, Rallo R, Damoiseaux R, Telesca D, Mädler L, Cohen Y, Zink JI, & Nel AE (**2012**) Use of metal oxide nanoparticle band gap to develop a predictive paradigm for oxidative stress and acute pulmonary inflammation. *ACS Nano* 6(5):4349-4368.

[12] Hofmann T, Ma-Hock L, Strauss V, Treumann S, Rey Moreno M, Neubauer N, Wohlleben W, Gröters S, Wiench K, Veith U, Teubner W, van Ravenzwaay B, & Landsiedel R (**2016**) Comparative short-term inhalation toxicity of five organic diketopyrrolopyrrole pigments and two inorganic iron-oxide-based pigments. *Inhalation Toxicology* : 28(10): 463–479.

[13] Wiemann M, Vennemann A, Sauer UG, Wiench K, Ma-Hock L, & Landsiedel R (**2016**) An in vitro alveolar macrophage assay for predicting the short-term inhalation toxicity of nanomaterials. *Journal of Nanobiotechnology* 14(1):1-27.

IOP Conf. Series: Journal of Physics: Conf. Series **838** (2017) 012034 doi:10.1088/1742-6596/838/1/012034

Degradation of bis-*p*-nitrophenyl phosphate using zero-valent iron nanoparticles

Maiby Valle-Orta[1], David Díaz[2], Inti Zumeta Dubé[2], José Luis Ortiz Quiñonez[2] and Rubén Saldivar Guerrero[1]

[1]Centro de Investigación en Química Aplicada, Blvd. Enrique Reyna Hermosillo No. 140 C.P. 25294 Saltillo, Coahuila/ México and [2]Universidad Nacional Autónoma de México, Facultad de Química, Ciudad Universitaria, Ciudad de México, Coyoacán, C.P.04510/ México.

E-mail: maiby.valle@ciqa.edu.mx

Abstract. Phosphate esters are employed in some agrochemical formulations and have long life time in the Environment. They are neurotoxic to mammals and it is very difficult to hydrolyze them. It is easy to find papers in the literature dealing with transition metal complexes used in the hydrolysis processes of organophosphorous compounds. However, there are few reports related with degradation of phosphate esters with inorganic nanoparticles. In this work bis-4-nitrophenyl phosphate (BNPP) was used as an agrochemical agent model. The BNPP interaction with zero-valent iron nanoparticles (ZVI NPs), in aqueous media, was searched. The concentration of BNPP was 1000 times higher than the ZVI NPs concentration. The average size of the used iron nanoparticles was 10.2 ± 3.2 nm. The BNPP degradation process was monitored by means of UV-visible method. Initially, the BNPP hydrolysis happens through the P-O bonds breaking-off under the action of the ZVI NPs. Subsequently, the nitro groups were reduced to amine groups. The overall process takes place in 10 minutes. The reaction products were identified employing standard substances in adequate concentrations. The iron by-products were isolated and characterized by X-RD. These iron derivatives were identified as magnetite (Fe_3O_4) and/or maghemite (γ-Fe_2O_3) and lepidocrocite (γ-FeOOH). A suggested BNPP degradation mechanism will be discussed.

1. Introduction

Hydrolysis of phosphoester bonds is involved in many biological processes. They have close relation to information store and conduction, energy transfer and protein phosphoacylation [1]. Phosphoesters have also been the most widely used as pesticides for many decades. They are a large source of potential hazard to humans [2,3]. These compounds are extremely stable and are highly resistant toward hydrolytic processes [4]. Bis-p-nitrophenyl phosphate (BNPP) is a phosphodiester compound that can be used as a model to study the hydrolysis of P-O bonds by catalytic interaction with metal nanoparticles. The stability of this molecule is relatively high; its $t_{1/2}$ is 2000 years in water at 20 °C and 53 years in water at 50 °C [5]. The hydrolytic cleavage of this molecule assisted by transition metal complexes is well-known [6-9]. However, there are few reports related with degradation of phosphate esters with inorganic nanoparticles.

In the other hand, the use of zero-valent iron (ZVI) as reductant to degrade organic contaminant compounds has received considerable attention. Compared with zero-valent iron, ZVI nanoparticles (NPs) have the most promising applications in environmental remediation [10-15]. This effect is due to the decrease of the iron particle size and the increase of the surface area and, thereby, the rate of reaction. The surface area of the ZVI NPs has a direct influence on the number of active surface sites. These sites are called Hot Spots, which are sites for catalytic transformation [16]. The increase of the specific surface area also results in an increase of the fraction of iron atoms that are on the particle surface, thereby, a greater reductive capacity per gram. In other words, these hot spots are places where the electronic hardness is very pronounced (Lewis acid sites or Pearson sites) and they are able to polarize bonds like P-O.

The reductive capacity of the metallic iron is increased when the size of particles is reduced due to the high surface/volume ratio, having greater number of active sites to be carried out this process. Furthermore electron transfer of these nanoparticles is explained by their particle size dependent redox properties.

In this paper, a simple method of BNPP degradation employing ZVI NPs is reported and a mechanism is proposed. In order to reinforce this mechanism, the metallic iron NPs also reacts with standard intermediate products. The degradations were monitored by UV-visible characterization.

2. Materials and Methods

2.1. Materials

Iron(II) bromide, FeBr (98% Aldrich); sodium borohydride, $NaBH_4$, (98% Aldrich); triethylamina, TEA (99.5% Aldrich); ethylene glycol, EG (99.9% J. T. Baker); acetone, $(CH_3)_2CO$, (99.6% J. T. Baker); bis(*p*-nitrophenyl)-phosphate, BNPP (99% Aldrich); *p*-nitrophenol, *p*-NP (99% Aldrich); *p*-nitrophenyl-phosphate, *p*-NPP (99% Aldrich); p-aminophenol, *p*-AP (99% Aldrich); argon, Ar (99.998% Praxair Inc.)

2.2. Instrumentation

High-resolution transmission electron microphotographs (HR-TEM) were obtained in a JEOL 2010 FasTEM analytical microscope, operating at 400 kV, by deposition of a drop of the corresponding dispersion onto 200 mesh copper grids coated with carbon/collodion layers, and processed with Digital Micrograph 1.2 software. The HR-TEM images were used to obtain the particle size distributions. The electronic absorption spectra were collected in a Hewlett-Packard 8452A UV-visible spectrophotometer.

2.3. Synthesis of ZVI NPs

The synthesis of ZV iron NPs in a previous work was published [17] and was carried out as follow: A 0.01078 g portion of $FeBr_2$ was dissolved in 15 mL of ethylene glycol under vigorous stirring and argon bubbling, at room temperature. Then, 0.01890 g of $NaBH_4$ was *in situ* dissolved in 8.75 mL of EG and simultaneously added to the reaction mixture. Instantaneously, a black dispersion of ZVI NPs with gas evolution (molecular hydrogen) was obtained. In order to increase the stability of these NPs, 1.25 mL of 1 M TEA in EG was added to the ZVI dispersion. These stabilized NPs were separated by centrifugation and washed three times with acetone (previously dried in 4 Å zeolites and bubbled with argon for 10 min in order to eliminate the dissolved molecular oxygen). Finally, the resulting black powder was vacuum-dried.

2.4. BNPP degradation studies

The interaction between ZVI NPs and BNPP was carried out as following. 8.65 mg of powder zero-valent iron nanoparticles were suspended in 25 mL of deionized water and argon bubbled, previously. Subsequently, 0.85 mg BNPP were added to this dispersion of nanoparticles under vigorous stirring and allowed to interact for 10 min, the interaction was followed for 1 day. At this time an aliquot of 8 mL was taken of the reaction mixture, which remains stirring. This portion is centrifuged and the supernatant was taken to measure the UV-visible spectrophotometer. Fe nanoparticles final concentration in this mixture was equal to $1x10^{-7}$ M while the final concentration of BNPP was $1x10^{-4}$ M. Similarly, the interaction between Fe^0 NPs and *p*-nitrophenol (*p*-NP) and *p*-nitrophenyl phosphate (*p*-NPP) were carried out. Afterwards, 0.85 mg BNPP were also added to this dispersion of nanoparticles and allowed to interact for a time range under vigorous stirring.

In order to verify if active iron species were released in the aqueous medium, another experiment was carried out in the same experimental conditions. The zero-valent iron nanoparticles were previously suspended in 25 mL of deionized water and argon bubbled. These nanoparticles were kept under magnetic stirring for 30 minutes. Subsequently, they were centrifuged and separated from the dispersing medium. The volume of total water was used to dissolve 0.85 mg of BNPP and it was kept under stirring for 30 minutes. The electronic absorption spectra were recorded at 10 and 30 minutes.

3. Results

Figure 1 shows a representative ZVI NPs Z-contrast TEM micrograph. The nanoparticles appear agglomerated, forming chain-like structures due to magnetic and electrostatic interactions that exist between them. The average diameter of 10.2 nm (SD ± 3.3 nm) was obtained from 111 counted crystallites from this Z-contrast TEM image.

It is important to mention that the fuzzy appearance of the Z-Contrast image is due to the fact that the magnetic particles are able to deflect the flow of electrons through the observed sample, which results in the presence of astigmatism in the micrograph presented. Other characterizations of ZV iron NPs were published [17].

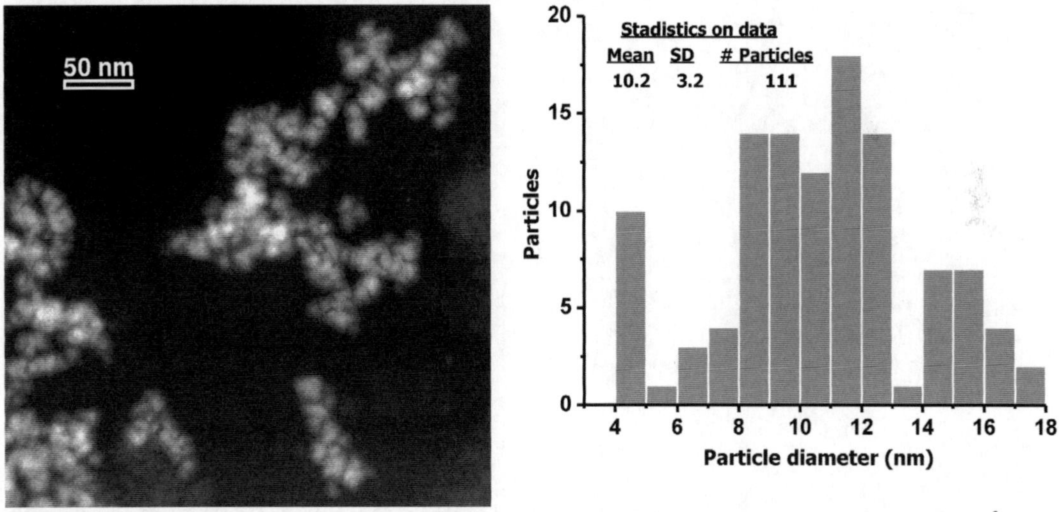

Figure 1. Z-Contrast of ZVI nanocrystallites (after 1 day) prepared from $FeBr_2$ (2×10^{-3} M), $NaBH_4$ (2×10^{-2} M), and TEA (2×10^{-2} M) in ethylene glycol, the corresponding particle size distribution histogram.

Table 1 summarizes the calculated parameters of the ZVI dispersion. These parameters were calculated considering the particle size distribution, assuming a completed reaction (excess of $NaBH_4$ was used) and spherical particles, see reference [17] for details. Parameters in Table 1 correspond to the ZVI NPs obtained in a typical synthesis as described in the Materials and Methods section.

IOP Conf. Series: Journal of Physics: Conf. Series **838** (2017) 012034 doi:10.1088/1742-6596/838/1/012034

The precursor salt concentration was 2 x 10^{-3} M. The organic contaminant concentration was calculated considering that BNPP is 1000 times more concentrated than the iron salt concentration for their interaction.

Table 1. Parameters estimated of zero-valent Fe NPs dispersions from data of size distribution.

Metal	Concentration of nanoparticles (particles/mL)	Total surface (m^2)	Relation Surface/volume (m^{-1})
iron	2.04 x 10^{13}	0.18	5.5 x 10^8

Figure 2 show the electronic absorption spectra of BNPP during 5 days. As it can be see there was no significant change in band positions during this time. The BNPP provides resistance to natural degradation processes.

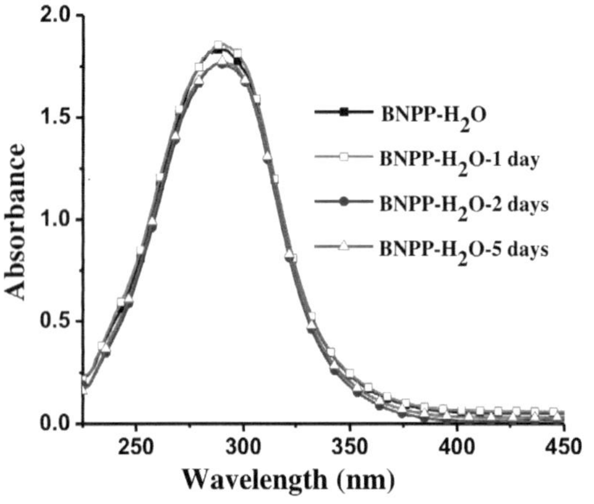

Figure 2. Electronic absorption spectra recorded at room temperature of a BNPP aqueous dissolution (1x10^{-4} M), the final pH was 7.

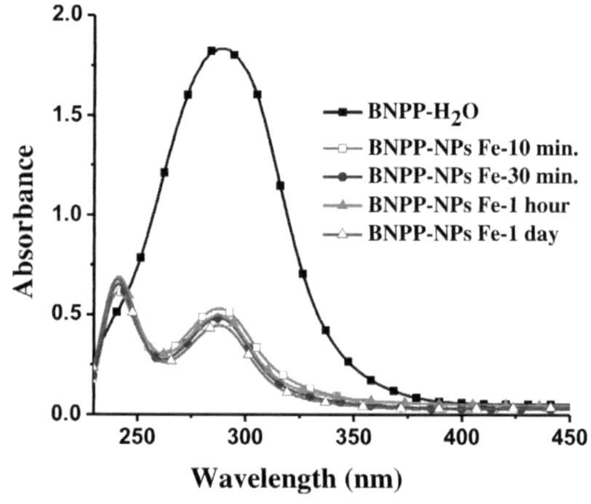

Figure 3. Time-dependent changes in the electronic absorption spectra of an aqueous reaction mixture, BNPP (1x10^{-4} M) and ZVI NPs (1x10^{-7} M). This process takes place at room temperature, and the final pH was 7.

In the Figure 3, the evolution in time of electronic absorption spectra for BNPP degradation by the interacting with the ZVI nanoparticles is displayed. The BNPP spectrum initially exhibits a single band at 289 nm, with an approximate 1.83 absorbance units. After 10 minutes of reaction a drastic change in the spectrum is observed. Two bands, the first is located at 239 nm, with 0.67 units and the second at 288 nm with 0.54 absorbance units are obtained. After 30 minutes and until one day of interaction of the BNPP with the ZVI NPs, the electronic absorption spectrum of reaction mixture remains without appreciable changes. This new spectrum were assigned to the formation of *p*-aminophenol (*p*-AP) and *p*-aminophenyl phosphate (*p* -APP), which are products that result from two consecutive processes; first occurs the hydrolysis of BNPP and subsequently happens reduction of the nitro group [18].

The mechanism proposed for the degradation of BNPP assisted by zero-valent iron NPs is shown in Scheme 1. Initially, the BNPP hydrolysis occurs through the P-O bonds breaking-off under action of the ZVI NPs. Thus, the molecule is divided into two parts forming the *p*-nitrophenol and *p*-nitrophenyl phosphate. Subsequently, due to the great reducing power of Fe nanoparticles, nitro groups are reduced to amines, in this case it is proposed that the water is the source of H^+. The effectiveness of the reduction of nitroaromatic compounds by zero-valent iron gives as primary products aromatic amines, which is well documented in the literature [19-22]. Further, the mechanisms that have been published suggest that reduction of pollutant compounds in the presence of zero-valent iron occurs on the active sites of the surface of this metal [23].

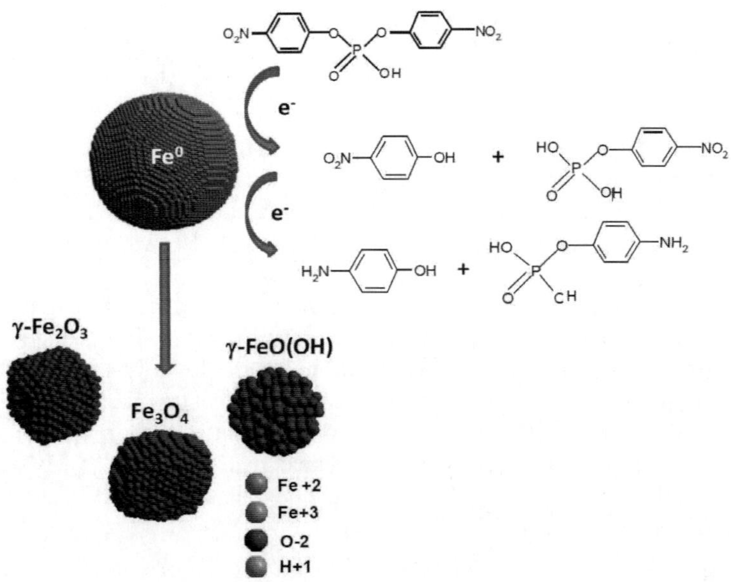

Scheme 1. Suggested Mechanism of BNPP Degradation by interaction of this pollutant with the Fe^0 NPs surface to form *p*-aminophenol and *p*-aminophenyl phosphate.

In order to corroborate the second process of the proposed mechanism in the degradation of BNPP, *p*-NP and *p*-NPP were used as standards. These compounds were subjected to the same experimental conditions used in the degradation of BNPP and interacted with the Fe NPs. Firstly, the UV-visible spectra of *p*-NP and *p*-NPP versus time are recorded to determine their stability when they remains in aqueous solution.

Subsequently, the interaction of Fe NPs with *p*-NP was performed. In the Figure 4, the change of the electronic absorption spectrum of *p*-NP after 30 minutes of interaction with the iron particles is observed. After 1h, the spectra remained almost unchanged and were identified with the *p*-aminophenol spectrum. Also, the absorbance values are close to those of *p*-aminophenol (*p*-AP) prepared at the same concentration ($1x10^{-4}$ M).

The Figure 5 shows the same process for monitoring by UV-vis spectroscopy in the interaction of the *p*-NPP with Fe^0 NPs. Again a drastic change occurs in the spectrum of this standard, evidencing a reaction by the reducing action of the Fe^0 NPs. In this case, it was not possible to compare it with the *p*-aminophenyl phosphate spectrum.

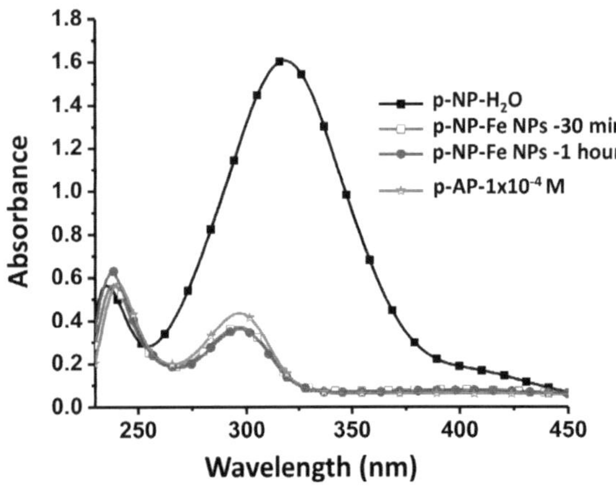

Figure 4. Time-dependent changes in the electronic absorption spectra of an aqueous reaction mixture, *p*- NP ($1x10^{-4}$ M) and ZVI NPs ($1x10^{-7}$ M), the final pH was 7.

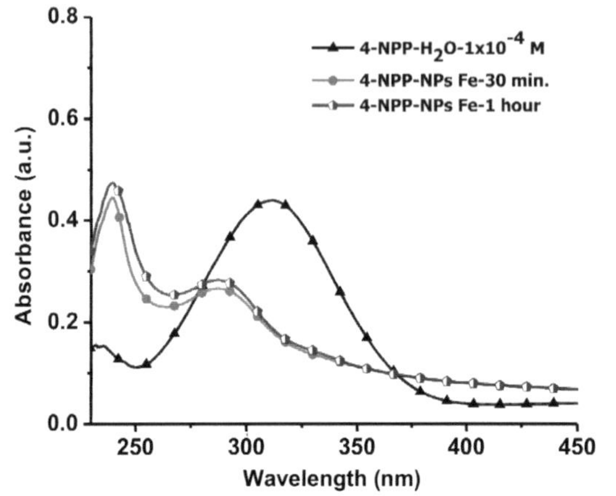

Figure 5. Time-dependent changes in the electronic absorption spectra of an aqueous reaction mixture, *p*- NPP ($1x10^{-4}$ M) and ZVI NPs ($1x10^{-7}$ M), the final pH was 7.

It is important to mention that the dispersion of the nanoparticles in water was also studied by UV-vis spectroscopy and no absorption bands corresponding to iron species were obtained. This result was corroborated with a new experiment, which is shown in Figure 6. The indicated figure exhibits the evolution in time of the electronic absorption spectra of BNPP solubilized in water. The water used to solubilize the BNPP previously for 30 minutes was in contact with the metallic Fe nanoparticles. The results of this experiment reveal that there are no Fe species soluble in the reaction medium that contribute significantly to the degradation of BNPP, at least under the reaction conditions in which the experiments were performed.

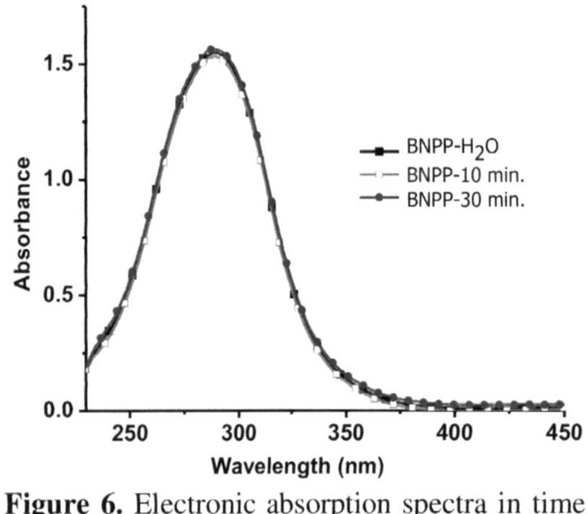

Figure 6. Electronic absorption spectra in time of an aqueous solution of BNPP ($1x10^{-4}$ M). The water previously was in contact by 30 minutes with ZVI NPs under stirring. This process takes place at room temperature, and the final pH was 7.

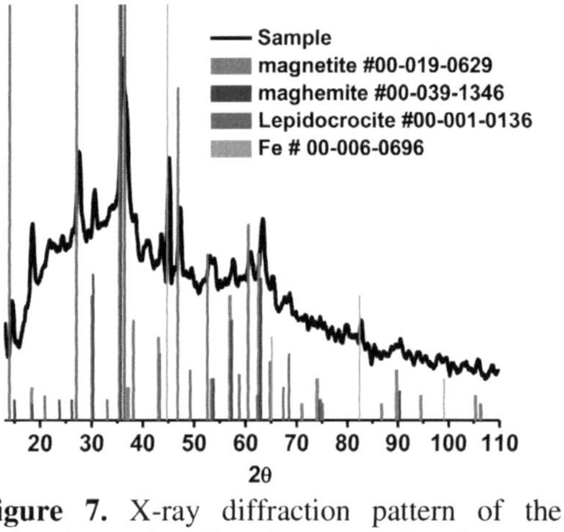

Figure 7. X-ray diffraction pattern of the isolated powder after the interaction between ZVI nanoparticles and BNPP in aqueous solution.

The degradation products of the Fe NPs was washed with water and acetone and characterized by X-Ray diffraction. The diffractogram obtained is represented in the Figure 7. It was detected magnetite (Fe_3O_4) and/or maghemite (γ-Fe_2O_3) which have similar crystalline structure. Also, was identified lepidocrocite which is an oxohydroxo of iron (γ-FeOOH).

4. Concluding Remarks

Zerovalent iron nanoparticles with average diameter 10.2 ± 3.2 nm were synthesized by a reliable method. The ZVI nanocrystallines are able to degrade BNPP, under normal reaction conditions. The overall degradation process takes place in approximately 10 minutes. It was proposed a simple mechanism based in the reduction of organic contaminant compound of active sites on the surface of Fe^0 NPs. The by-products of BNPP are aromatic amines, which are more susceptible to biodegration than the original nitro compounds. The by-products of ZVI NPs were identified as magnetite (Fe_3O_4) and/or maghemite (γ-Fe_2O_3) and lepidocrocite which is an oxohydroxo of iron (γ-FeOOH).

5. References

[1] Westheimer F H, Why nature chose phosphates, 1987, *Science*, **235**, 1173.

[2] Bazzicalupi C, Benchini A, Berni E, Bianchi A, Fedi V, Fusi V, Giorgio C, Paoletti P and Valtancoli B, Carboxy and Diphosphate Ester Hydrolysis by a Dizinc Complex with a New Alcohol-Pendant Macrocycle, 1999, *Inorg. Chem.*, **38**, 4115.

[3] Spiro T G and Stigliani W M, *Química Medioambiental*, segunda edición, **2004**, Pearson Educación.

[4] Jia-Qing X, Ci L., Min W, Bing-Ying J., Preparation of a new metallomicelle catalyst for the hydrolysis of bis(4-nitrophenyl) phosphate, 2013, *Chemical Papers,* **67(4)**, 365.

[5] Chin J, Banaszczyk M, Jubian V and Zou X, 1989, *J. Am. Chem. Soc.*, **111**, 186.

[6] Yuanqin Z, Fan L, Qingxiang X and Junru X, A study on the hydrolysis of bis(nitrophenyl)phosphate catalyzed by N-methyldiethanolamine-Ce(III) complex, 2004, *Int. J. Chem. Kinet.*, **36**(12), 687.

[7] Jiang W, Xu B, Lin Q, Li J, Liu F, Zeng X and Chen H, Metal-promoted hydrolysis of bis(*p*nitrophenyl) phosphate by trivalent manganese complexes with Schiff base ligands in Gemini micellar solution, 2008, *Colloid. Surf. A: Phys. Eng. Aspects*, **315**, 103.

[8] Kou D, Meng X-G, Liu Y, Du J, Kou X-M and Zeng X-C, Comparative studies on hydrolysis of bis(*p*nitrophenyl) phosphate catalyzed by short- and long-alkyl-multiamine metallomicelles, 2008, *Colloid. Surf. A:Physicochem. Eng. Aspects*, **324**, 189.

[9] Tafesse F and Eguzozie K, Efficient hydrolysis of 4-nitrophenylphosphate catalyzed by copper bipyridyl in microemulsions, 2009, *Ecotoxicol. Environ. Safety*, **72**, 954.

[10] Li F, Vipulanandan C and Mohanty K K, Microemulsion and solution approaches to nano-scale particle iron production for degradation of trichloroethylene, 2003, *Colloids Surf. A,* **223**, 103–112.

[11] Kim H, Hong H-J, Lee Y-J, Shin H-J, Yang J-W, Degradation of trichloroethylene by zero-valent iron immobilized in cationic exchange membrane, 2008, *Desalination*, **223**, 2 12.

[12] Nurmi J T, Tratnyek P G, Sarathy V, Baer D R, Amonette J E, Pecher K, Wang C, Linehan J C, Matson D W, Penn R L, Driessen M, Characterization and properties of metallic iron nanoparticles: Spectroscopy, electroscopy, electrochemistry and kinetics, 2005, *Environ. Sci. Technol.*, **39**, 1221.

[13] Li X-Q, Zhang W-X, Sequestration of metal cations with zerovalent iron nanoparticles, a study with high resolution X-ray photoelectron spectroscopy (HR-XPS), 2007, *J. Phys. Chem. C*, **111**, 6939.

[14] Poder S M, Darab J G, Bucher J, Caulder D, Craig I, Davis L, Edelstein N, Lukens W, Nitsche H, Rao L, Shuh D K, Mallouk T E, Surface Chemistry and Electrochemistry of Supported Zerovalent Iron Nanoparticles in the Remediation of Aqueous Metal Contaminants, 2001, *Chem. Mater.*, **13**, 479.

[15] Ma C, Wu Y, Dechlorination of perchloroetylene using zero-valent metal and microbial community, 2008, *Environ. Geol.,* **55**, 47.

[16] Shishido, T.; Morishima, R.; Yoshinaga, Y.; Hasegawa, S.; Tanaka, T., Effect of support on the suppression of hot spot formation in the partial oxidation of methane, 2009, *Chem. Lett.*, **38**(7), 720.

[17] Valle-Orta M, Diaz D, Vazquez-Olmos A R, Santiago P and Reguera E, Instantaneous synthesis of stable zerovalent metal nanoparticles under standard reaction conditions, 2008, *J.Phys. Chem. B*, **112**(46), 14427.

[18] Pradhan N, Pal A and Pal T, Catalytic reduction of aromatic nitro compounds by coinage metal nanoparticles, 2001, *Langmuir*, **17**(5), 1800.

[19] Agrawal A and Tratnyek P G, Reduction of nitroaromatic compounds by zero-valent iron metal, 1996 *Environ. Sci. Technol.*, **30**, 153.

[20] Devlin J F, Klausen J and Schwarzenbach R P, Kinetics of nitroaromatic reduction on granular iron in recirculating batch experiments, 1998, *Environ. Sci. Technol.*, **32**, 1941.

[21] Hayakawa K, Yoshimura T and Esumi K, Preparation of gold-dendrimer nanocomposites by laser irradiation and their catalytic reduction of 4-nitrophenol, 2003, *Langmuir*, **19**, 5517.

[22] Fleming I and Trost B M, Comprehensive organic synthesis. Vol. 8, 366, Pergamon Press, fifth impression, Elsevier, 2005.

[23] Weber E J, Iron-mediated reductive transformations: investigation of reaction mechanism, 1996, *Reser. Comm.*, **30**(2), 716.

IOP Conf. Series: Journal of Physics: Conf. Series **838** (2017) 012035 doi:10.1088/1742-6596/838/1/012035

Polymeric film of 6-arm-poly(ethylene glycol) amine graphene oxide with poly (ε-caprolactone): Adherence and growth of adipose derived mesenchymal stromal cells culture on rat bladder

Marcela Durán[1,2*], Nelson Durán[1,3,4*], Angela C.M. Luzo[5], Adriana S. S. Duarte[5], Bruno B. Volpe[5], Helder J. Ceragioli[6], Patricia F. Andrade[3], Joel G. De Souza[2], Wagner J. Fávaro[1,2].

[1]NanoBioss, Institute of Chemistry (IQ), UNICAMP, Campinas, SP, Brazil.
[2]Urogenital Carcinogenesis: Urogenital and Immunotherapy Lab., Inst. Biology, University of Campinas (UNICAMP), Campinas, SP, Brazil.
[3]IQ, Biol. Chem. Lab., UNICAMP, Campinas, SP, Brazil.
[4]Braz. Nanotechnol. Nat. Lab. (LNNano-CNPEM), Campinas, SP, Brazil.
[5]Mol. Biol. Lab., Hemocenter-Medicine Fac., UNICAMP, Campinas, SP, Brazil.
[6]Electric Eng. Comput. Fac., UNICAMP, Campinas, SP, Brazil.

*duran@iqm.unicamp.br, marceladuranduran@gmail.com

Abstract. Nanotechnology has been more present in different fields related to health. The need to find a durable material, of easy use, and which does not interfere significantly in the growth and differentiation of stem cells for the construction of a scaffold for use in urologic surgery, with the purpose of reducing infections, regeneration times and even graft rejection during reconstitution in patients with urethral stricture was conducted a broad survey of information about this and came to the consensus of this project: using graphene oxide, a widely studied nanomaterials which has been presenting numerous beneficial results when in contact with the adipose-derived stem cells. Advanced techniques for the growth, differentiation and proliferation of adipose-derived stem cells were used, as well as the characterization of graphene oxide sheets. For this study, it was prepared the graphene oxide/6 ARM-Poly (ethylene glycol) amine films with poly (ε-caprolactone). The graphene suspension in organic solvent was prepared by using an ultrasonicator bath and subsequently, the film was formed by solvent evaporation. Total characterization of graphene oxide/6 ARM-PEG-amine/ poly (ε-caprolactone) film was carried out. It was tested growth and adhesion of adipose-derived stem cells on the film, as well as, were verified the histopathological effects of this scaffold when implanted in the urinary bladder to repair the lesion. Our results demonstrated that this scaffold with adipose-derived stem cells enhanced the repair in rat urinary bladder defect model, resulting in a regular bladder. Improved organized muscle bundles and urothelial layer were observed in animals treated with this scaffold with adipose-derived stem cells compared with those treated only suture thread or scaffold. Thus, our biomaterial could be suitable for tissue engineered urinary tract reconstruction.

1. Introduction

Within the symptomatology that affects urologic patients; voiding dysfunction is one of the major problems with direct impact on quality of life. These symptoms can be directly related to urethral stricture, which occurs in the urethra. Urethral stricture can result from inflammatory, ischemic or traumatic reactions [1]. These processes lead to the formation of scar tissue; this decreases the calibre of the urethral lumen, causing resistance to antegrade flow of urine. Treatment reports of urethral stenosis date back to the Egyptians 4000 years ago and very little have changed until the middle of last century. Treatment today is still a challenge, which directed many researchers from different centres, to study not only the use of stromal mesenchymal cells (SMCs) for reconstitution of urological tissue by promoting healing acceleration, as well as demonstrate the possibility of differentiation of these cells stromal mesenchymal cells in particular applied for each region [2, 3]. Studies have demonstrated the differentiation of these in urothelial cell lines [4]. In 2012 a survey conducted by Frennette et al. [2] showed a number of published papers related to stromal mesenchymal cells were divided into 4 groups: immunomodulation, tissue protection, regenerative medicine and improvement grafts.

Bioactive molecules secreted by stromal mesenchymal cells exert trophic effects on the surrounding tissue, causing: (a) anti-apoptotic effects on cell tissue specific, (b) immunomodulatory effects on cells of the immune system, (c) increasing angiogenesis, and (d) chemotaxis to other young cells committed to the repair process [5]. Thus, reducing the initial inflammatory response may occur in some ischemic diseases, restoration of blood supply and adequate tissue repair, depending on the extent and severity of tissue damage.

Thus, the last two years extensive research in the area of nanotechnology presents and pointed out the possibility of having a biocompatible support using graphene oxide (GO) in the construction of scaffolds that allows the growth and stromal mesenchymal cells differentiation [6-13].

One of these studies examined the interference of the graphene oxide and reduced graphene oxide (R) on the differentiation of skeletal myoblasts and morphology of mice. The authors found that graphene oxide and its reduced form did not affect negatively this differentiation, but acted as stimulators of the same, without change morphologically these cells. The graphene oxide was able to increase in the viability 1.53 times in two days and 1.24 fold after 4 days, compared to control cells, is better than the reduced graphene oxide [12].

But to develop studies with these materials should be considered the risks to animal and humans health, and therefore should be studied toxicological tests for minimize risk before moving in the medical area [14, 15].

These studies can be said the promising future of graphene oxide as material to be used in the medical field.

2. Methods

*2.1. Conjugate synthesis GO/*6 ARM-PEG-amine

In order to obtain the conjugate GO/6 ARM-PEG-amine, 10 mg of graphene oxide (Single-layer graphene oxide, purity 99%, thickness 0.7-1.2 nm (AFM); ~300-800nm X&Y dimensions is the standard size <450 nm & 1-20 μm lateral dimensions from Cheap Tubes Inc., Bratleboro, USA) used which was dispersed in 18 mL deionized water. The stirred dispersion to homogenization and held in an ultrasound bath for 30 min. Then added 10 mg of EDC (hydrochloride 1-ethyl-3- (3-dimethylaminopropyl) carbodiimide)) previously dispersed in 1.0 ml of water and kept at ultrasonic dispersion bath for 30 min with subsequent agitation for mis 30 min. Thereafter, was added 10 mg of PEG (6 ARM-Poly (ethylene glycol) amine (6 ARM-PEG-amine) (15kD) previously dispersed in 1.0 ml of deionized water. The reaction medium was maintained under stirring for a period of 24 h in the absence of light.

The next day, the mixture was centrifuged at 14,000 rpm for 20 minutes. Thus, the conjugate was resuspended in deionized water and centrifuged under the same conditions described above. This procedure was repeated 3 times to remove unreacted 6 ARM-PEG-amine with graphene oxide.

Preliminary studies have shown that it is possible to obtain stable suspensions of graphene oxide in deionized water (1mg / ml) for up to 10 days. The suspensions of graphene oxide (1.0 mg/mL) will be prepared in sterile deionized water and conditioned in 15 mL Falcon tubes (sterile) when placed in ultrasonic bath for 30 minutes.

IOP Conf. Series: Journal of Physics: Conf. Series **838** (2017) 012035 doi:10.1088/1742-6596/838/1/012035

*2.2. Obtaining the scaffold GO/*6 ARM-PEG-amine */PCL*

First, 1.5 g of poly (ε-caprolactone) (PCL) (80 kD) were dissolved in 20 mL of chloroform (Merck, Germany) and kept under stirring for 2 h for complete dissolution of the polymer. At the same time 0.5 mg of graphene oxide were added to 10 mL of chloroform and placed in ultrasonic bath Unique, Model USC 700 for 30 min to disperse graphene oxide. This step was reduced to 5 min when used GO/6 ARM-PEG-amine, showing the significant improvement in dispersibility of the new graphene oxide conjugate. Then dispersed solution of graphene oxide was dissolved and aggregated to the PCL solution and placed on a magnetic stirrer for homogenization for 10 min. The pre-test concentrations to define the optimum ratio were applied by Lee et al. [16] and Kim et al. [13]. The final solution obtained was poured into glass mold (glass vessel 10 cm X 15 cm) to complete evaporation of the solvent to room temperature without controlled atmosphere to obtain the scaffolds.

*2.3. Characterization of the conjugate GO/*6 ARM-PEG-amine */PCL scaffold*

2.3.1. Scanning Electron Microscopy (SEM) of the membrane scaffold PCL and GO/6 ARM-PEG-amine/PCL

As done with samples of graphene oxide and graphene oxide conjugate, the morphologies of scaffolds GO/6 ARM-PEG-amine/PCL were studied in the scanning electron microscope for issuing FEI QUANTA FEG250 field, operated at 5 kV. Samples scaffolds were placed under the adhesive tape and the covered sample port with a thin layer Iridium, using a high vacuum sputter modular Baltec 020 MED.

2.3.2. X-Ray Diffraction (XRD) GO/6 ARM-PEG-amine, PCL membrane and scaffold GO/6 ARM-PEG-amine/PCL

The comparative analysis was made to be able to compare the peaks of different compounds and verify that there was indeed an interaction between them, using XRD equipment by Philips X'Perta.

2.4. Obtaining adipose mesenchymal stromal cells (AMSCs)

The AMSCs will be obtained from human adipose tissue undergoing liposuction surgery under general anesthesia. All procedures were approved by the Ethics Committee of the UNICAMP (process CEP-UNICAMP- No. 839/2008). The Term of Free and Informed Consent (process CEP-UNICAMP- No. 839/2008) was applied to all patients undergoing liposuction which have agreed to the donation. The selected patients were between 25 and 50 years and were excluded diabetic and hypertensive patients. All the procedures performed involving adipose mesenchymal stromal cells as extracting mesenchymal stromal cells and cell culture; immunophenotype of the mesenchymal stromal cells by flow cytometry; mesodermal differentiation in vitro; chondrogenic differentiation; osteogenic differentiation; adipogenic differentiation; cytochemical analysis of differentiated mesenchymal stem cells; Confocal immunofluorescence microscopy; activity of the enzyme telomerase and cytogenetic were adapted from the techniques used by different authors by the team of Prof. Dr. Angela C.M. Luzo, molecular biology laboratory. - Blood Center - UNICAMP, Campinas [17].

*2.5. Viability of cells on the scaffold AMSCs GO/*6 ARM-PEG-amine*/PCL*

Inoculum was prepared from adipose mesenchymal stromal cells laboratory standards obtained from human adipose tissue in DMEM (Dulbeco's Modified Eagle Medium), which were incubated for 72 h to attain the maximum log growth for the time of incubation in an incubator at 37°C in 5% CO_2 with 10% bovine serum (mother inoculum) [8].

Thus, from the mother inoculum (initial) was prepared one MSCs solution through detachment of these adherent cells in disposable petri dish (diameter 150 mm) using 1 mL of trypsin (0.005%) for 3 min at 37^0C, made washing with PBS (phosphate buffered saline (0.1 M)). Cell viability was verified by trypan blue staining technique and counting in a Neubauer chamber [18]. This cell solution was used to inoculate scaffolds. After placement of the scaffold, aseptically, each well received 125 uL of a solution (6x105 cells / ml) which was added 1.0 ml of DMEM with 100 of 2% bovine serum and and they were inoculated to 37^0C for 24 h. The LIVE / DEAD® technique was performed to check the cell viability and the images were acquired in microplate reader Cytation 5 (Biotek, Vermont, USA) using the GFP filter (ex/ at ~ 469 nm / ~ 525 nm) for reading Calcein-AM and RFP (ex / at ~ 531 nm / ~ 593 nm) for reading the ethidium homodimer.

IOP Conf. Series: Journal of Physics: Conf. Series **838** (2017) 012035 doi:10.1088/1742-6596/838/1/012035

2.6. Obtaining the bladder injury and implantation of the scaffold
A total of 12 rats of the Fischer 344 strain were divided into 4 groups (3 animals per group) and with the animal placed, there were aseptic in the lower third of the region of the abdomen then the animals were anesthetized i.p. using 0.2 mL of xylazine / ketamine (1: 1) and the incision performed. Experimental group (n = 3 animals per group) was divided at Control group, Group suture thread attached with adipose mesenchymal stromal cells, Scaffold Group and Scaffold group adhered to adipose mesenchymal stromal cells which the latter two were also sutured with suture thread from adipose mesenchymal stromal cells. The scaffold was positioned so that the side joined with the adipose mesenchymal stromal cells stay facing the lumen of the organ and the suture was performed. To check if there escape of urine, after suturing performed filling the same using 0.9% saline stained with blue pigment. A urinary catheter (22 gauge) was introduced into the animal's urethra and a small amount of serum was given slowly and observing whether there was leakage or not. After 30 days of follow up, the rats were sacrificed and the bladder was collected using mooring technique where after deep anesthesia procedure and opening of the abdomen, the bladder was emptied completely by introducing catheter and then was completed solution Bouin (fixer) and tied at the bottom portion to be fixed with total volume. The histopathology was made to assess healing and inflammatory processes.
2.7. Histopathological analysis
After fixed with Bouin liquid for 12 h the bladder follow standard fixation procedures, dehydration, and inclusion diafanization using Paraplast Plus (ST. Louis, MO, USA). The materials was sectioned on rotary microtome CUT5062 (SLEE MAINZ, Munich, Germany) with a thickness of 5 microns, stained with hematoxylin and eosin (HE) and Masson, and photographed the light microscope Leica DM2500 (Leica, Munich, Germany) equipped with DFC295 camera (Leica, Munich, Germany) [19]. The following criteria were observed: angiogenesis, urothelial hyperplasia, inflammatory infiltration, foreign body giant cell reaction around the suture and scaffolds and collagenous.

3. Results and Discussion

3.1. Characterization of the scaffold:

3.1.1. Scanning Electron Microscopy (SEM) of the membrane scaffold PCL and GO/6 ARM-PEG-amine/PCL
The images of SEM from GO-PCL and GO/6 ARM-PEG-amine/PCL scaffolds showed the surface characteristics of both sides of the scaffolds: B-side (bright) was defined as the side in contact with the forming surface; F-side (front) was defined as the side in contact through evaporation environment (e.g., oxygen environment contact). Due to the solvent evaporation rate used to dilute the graphene oxide, and the surface appeared more homogenous and soft (Fig. 1, samples A,C,E,G) that the F-side (Fig. 1, samples B,D,F,H), the B-side was chosen to inoculate the adipose mesenchymal stromal cells in the scaffolds at the cell viability test, to facilitate the adhesion and growth of adipose mesenchymal stromal cells.

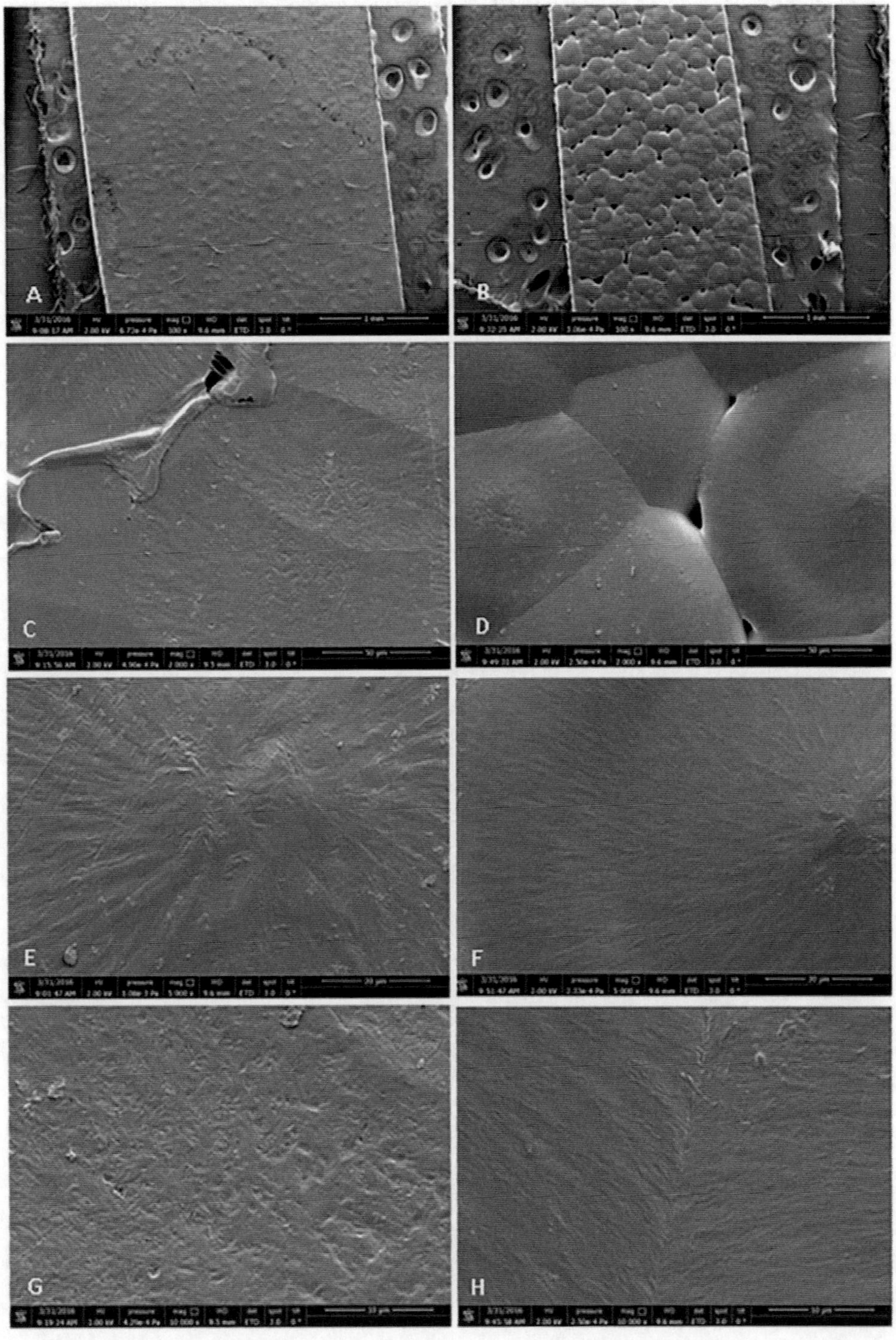

Figure 1. SEM images of the B-side (A – 100X, C – 2.000X, E – 5.000X, G – 10.000X) and F-side (B– 100X, D – 2.000X , F – 5.000X, H – 10.000X) from GO/6 ARM-PEG-amine/PCL scaffold.

IOP Conf. Series: Journal of Physics: Conf. Series **838** (2017) 012035 doi:10.1088/1742-6596/838/1/012035

3.1.2. X-Ray Diffraction (XRD) GO /6 ARM-PEG-amin, PCL membrane and scaffold GO/6 ARM-PEG-amine /PCL

The diffraction peak of pure graphene oxide was located at 11.5° (2θ), corresponding to distance of graphene oxide interlayer with approximately 0.8 nm, which is an important indicator of interleaving between graphene oxide sheets, due to its aromatic groups. In the presence of 6 ARM-PEG-amine, 11.5° diffraction peak was shifted to 6°, which corresponds to basal spacing of approximately 1.4 nm, due to incorporation of 6 ARM-PEG-amine (6 arms with angled structures capable of distancing layers graphene oxide, increasing spaces of the interlayer). By comparing the membrane PCL and the scaffold itself, there was no observed diffraction peak of GO in the scaffold, due to the high concentration of the PCL (Fig. 2).

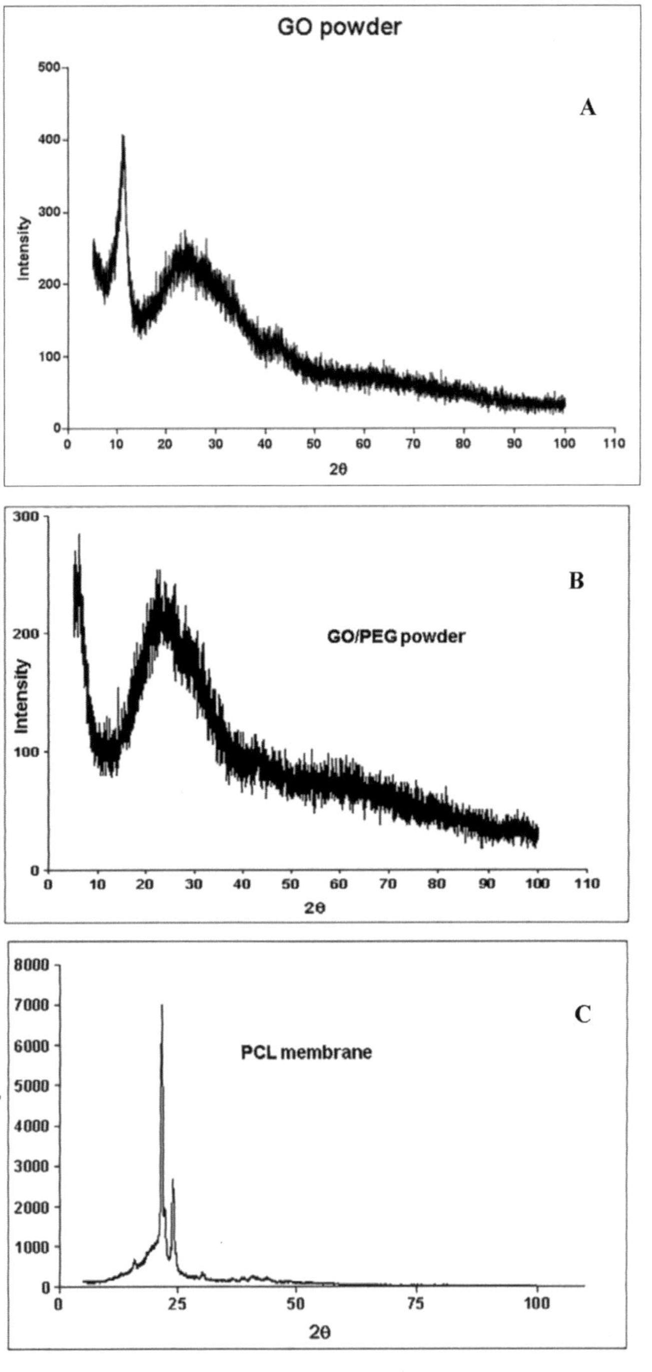

IOP Conf. Series: Journal of Physics: Conf. Series **838** (2017) 012035 doi:10.1088/1742-6596/838/1/012035

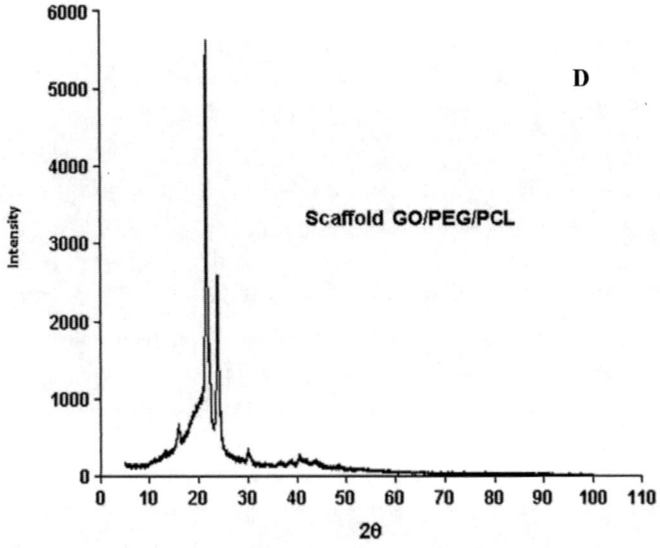

Figure 2. DRX showing displacement graphene oxide peaks powder (A) when incorporated into the PEG in GO/ powder 6 ARM-PEG-amine (B), and PCL membrane scaffold (C) versus GO/6 ARM-PEG-amine/PCL (D).

3.2. Viability of cells on the scaffold AMSCs GO/6 ARM-PEG-amine/ PCL

When compared with controls (fig. 3A and B), the incubation after 24 h of adipose mesenchymal stromal cells with GO/6 ARM-PEG-amine/PCL scaffold, in relation to cell viability, there were no morphological changes in adipose mesenchymal stromal cells (Fig 3C and D), but it was possible to detect that the GO/6 ARM-PEG-amine/PCL scaffold increased adipose mesenchymal stromal cells proliferation, indicating synergistic interaction (Fig. 3).

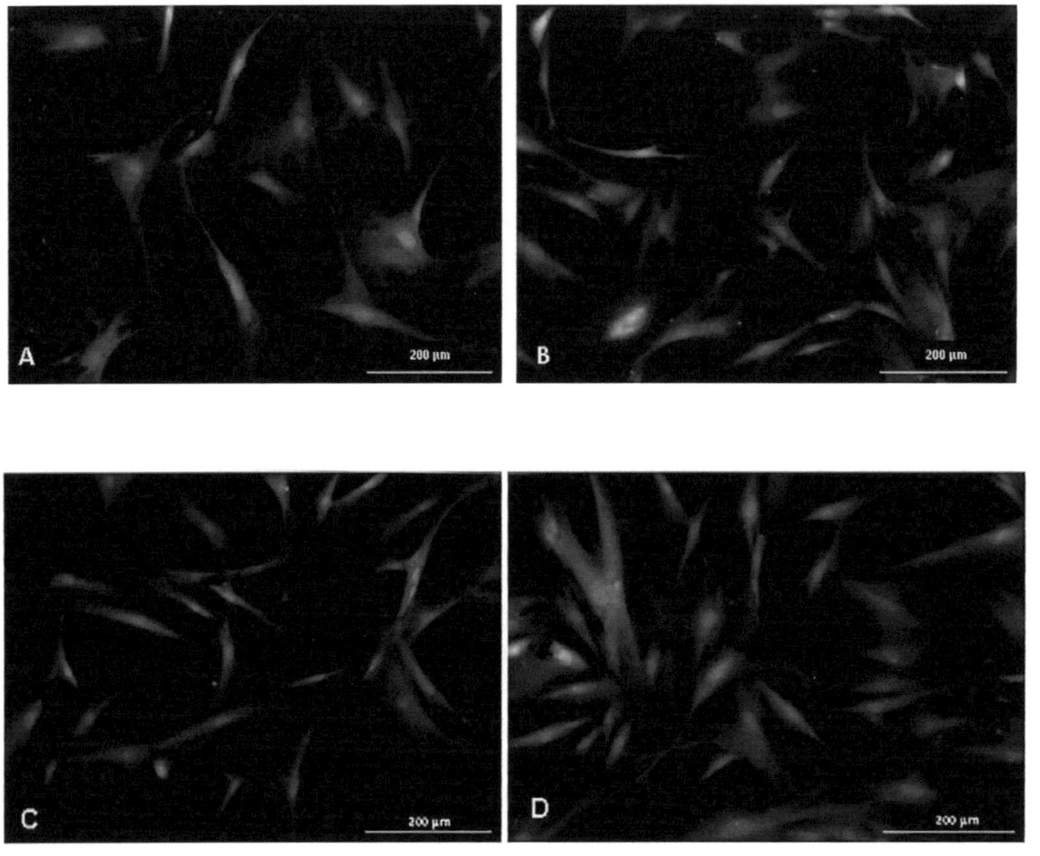

Figure 3. Photomicrographs A and B showing the adipose mesenchymal stromal cells grow control. The photomicrographs C and D show proliferation of the adipose mesenchymal stromal cells in culture incubated with GO/6 ARM-PEG-amine/ PCL scaffold.

3.3. Histological analysis

The histological results from 12 rats of the Fischer 344 strain, showed that angiogenesis process was more intense in the Scaffold+ADMSCs group in relation to Scaffold and Suture thread groups, in which the angiogenic process was predominantly moderate (Fig.4). In contrast, inflammatory process and giant cell reaction were numerically greater in the Suture thread in relation to Scaffold and Scaffold+ADMScs groups, which showed moderate tissue reaction (Fig. 4). The collagenization was intense in the Scaffold+ADMScs group, moderate in Scaffold group and weak in the Suture thread group (Fig. 4).

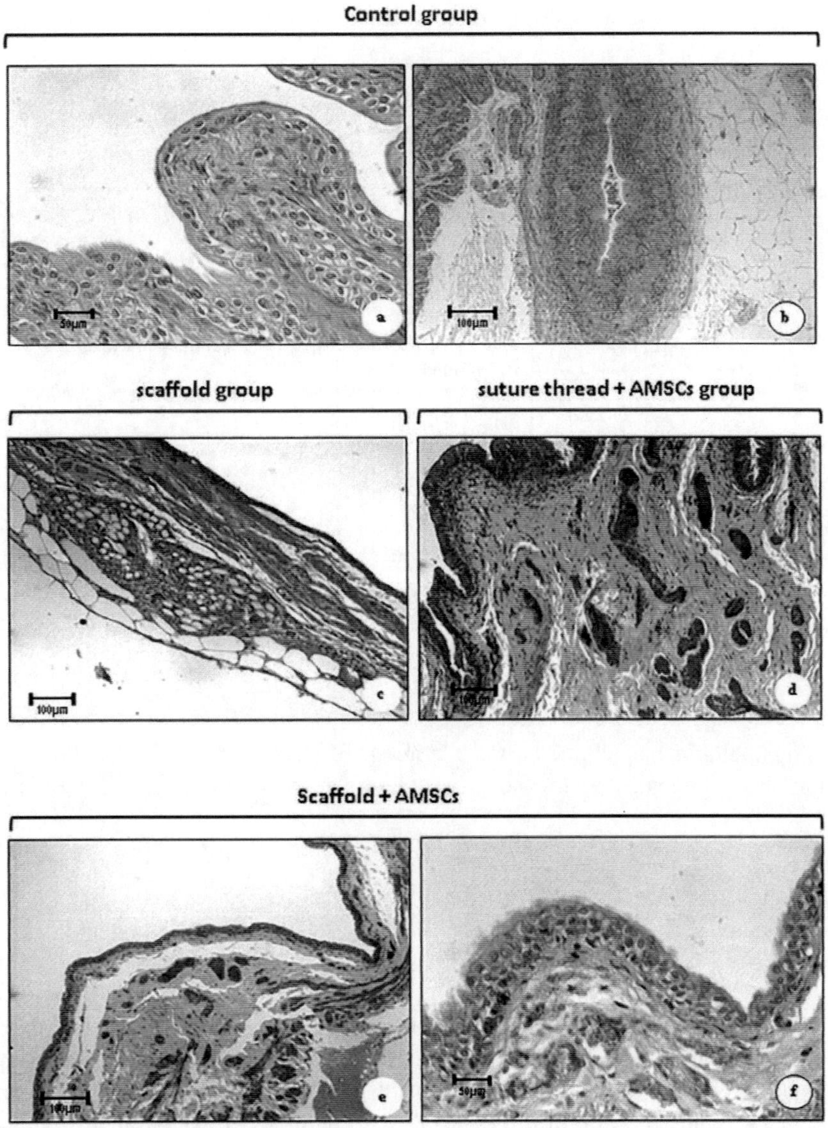

Figure 4. Photomicrographs of Control (a, b), Suture thread+ADMSCs (c), Scaffold (d) and Scaffold+ADMSCs (e, f) groups. (a), (b) Normal urothelium composed of 2–3 layers: a basal cell layer, an intermediate cell layer, and a superficial or apical layer composed of umbrella cells; lamina propria layer composed by collagen fibers and smooth muscle; detrusor layer (muscle layer). (c) Moderate angiogenesis, inflammatory infiltrate and giant cell reaction processes, and weak collagenization in the urinary bladder. (d) Moderate angiogenesis and collagenization processes, and weak inflammatory infiltrate and giant cell reaction in the urinary bladder. (e), (f) Normal urothelium; intense angiogenesis and collagenization processes, and weak inflammatory infiltrate and giant cell reaction in the urinary bladder. Hematoxylin-Eosin: a, c, f. Masson`s trichrome: b, d, e.

All of these data are summarized in Table 1. In *in vivo* trials the GO/6 ARM-PEG-amine/PCL and adipose mesenchymal stromal cells sewed on bladder showed complete healing after of surgical bladder injury treatment. The histopathological analyses showed the presence of healing cells, as compared with the previous surgery injuries. After cell adhesion it was found a high bladder healing area.

Groups (n=3)	Angiogenisis	Inflamatory infiltrade	Giant cell reaction	Collagenization
Control	All presented grade 1	All presented grade 1	3 presented grade 1	3 presented grade 1
Suture thread	1 presented grade 1 2 presented grade 3	1 presented grade 2 2 presented grade 3	2 presented grade 3 1 presented grade 4	3 presented grade 2
Scaffold	1 presented grade 2 2 presented grade 3	2 presented grade 2 1 presented grade 3	2 presented grade 2 1 presented grade 3	1 presented grade 1 2 presented grade 3
Scaffold + AMSCs	1 presented grade 3 2 presented grade 4	1 presented grade 1 2 presented grade 2	1 presented grade 1 2 presented grade 2	3 presented grade 4

Histological grades for angiogenesis, inflammatory infiltrate, giant cell reaction and collagenization processes; absent - 0; weak - 1; moderate - 2; intense - 3.

Table 1. Histopathological evaluation of the angiogenesis, presence of inflammatory infiltrate, giant cell reaction and collagenization in the urinary bladder of the rats from different experimental groups.

4. Conclusions

Total characterization of GO/6 ARM-PEG-amine film was carried out. This optimized platform exhibited growth and adhesion of adipose-derived stem cells (ADMSC) and excellent properties for the wound healing processes. The trials to made the adhesion of adipose mesenchymal stromal cells was confirmed by LIVE and DEAD®. In *in vivo* trials the GO/6 ARM-PEG-amine/PCL and adipose-derived stem cells sewed on bladder showed complete healing after of surgical bladder injury treatment. The histopathological analyses showed the presence of healing cells, as compared with the previous surgery injuries. After cell adhesion it was found a high bladder healing area.
Data suggested a great potential of GO/6 ARM-PEG-amine/PCL films as platform to adipose-derived stem cells, as well as, new material for treatment several urological diseases.
Improved organized muscle bundles and urothelial layer were also observed in animals treated with GO/6 ARM-PEG-amine/PCL scaffold with adipose-derived stem cells compared with those treated only wit Suture thread or Scaffold. Thus, our biomaterial could be suitable for tissue engineered urinary tract reconstruction.

Acknowledgements: FAPESP (Proc. 2014/08349-5), the Laboratory of Nanostructure Synthesis and Biosystem Interactions-NANOBIOSS (MCTI), Brazilian Network on Nanotoxicology and Laboratory of Urogenital Carcinogenesis and Immunotherapy, UNICAMP.

IOP Conf. Series: Journal of Physics: Conf. Series **838** (2017) 012035 doi:10.1088/1742-6596/838/1/012035

References

[1] Chapple C, Barbagli G, Jordan G, Mundy AR, Rodrigues-Netto N, Pansadoro V and McAninch JW 2004 *BJU Int.* **93** 1195
[2] Frenette PS, Pinho S, Lucas D and Scheiermann C 2012 *Med. Annu. Rev. Immunol.* **12** 285
[3] Moad M, Pal D, Hepburn AC, Williamson SC, Wilson L, Lako M, Armstrong L, Hayward SW, Franco OE, Cates JM, Fordham SE, Przyborski S, Carr-Wilkinson J, Robson CN and Heer R 2013 *Eur. Urol.* **64** 753
[4] Shi JG, Fu WJ, Wang XX, Xu YD, Li G, Hong BF, Wang Y, Du ZY and Zhang X.J 2012 *Biomed. Mater. Res.* **Part A 100A** 2612
[5] Meirelles LS, Chagastelles PC and Nardi NB 2006 *J. Cell Sci.* **119** 2204
[6] Lee WC, Lim CH, Shi H, Tang LA, Wang Y, Lim CT and Loh KP 2011a *ASC Nano* **5** 7334
[7] Paul W and Sharma CP 2011 *Trends Biomater. Artif. Organs*. **25** 91
[8] Akhavan O, Ghaderi E and Akhavan A 2012 *Biomaterials* **33** 8017
[9] Burande AM and Burande AR. 2012 *Inter. J. Recent Trends Sci. Technol.* **4** 50
[10] Chen GY, Pang DWP, Hwang SM,Tuan HY and Hu YC 2012 *Biomaterials* **33** 418.
[11] Kim J, Kim YR, Kim Y, Lim KT, Seonwoo H, Park S, Cho SP, Hong BH, Choung PH, Chung TD, Choung YH and Chung JH 2013 *J. Mater. Chem. B.* **1** 933
[12] Ku SH and Park CB. 2013 *Biomaterials* **34** 2017
[13] Zuo PP, Feng HF, Xu ZZ, Zhang LF, Zhang YL, Xia W and Zhang WQ 2013 *Chem. Cent.* **7** 1
[14] Durán N, Guterres SS and Alves OL 2014 (Eds) *Springer* **pp 412** (ISBN 978-1-4614-8992-4)
[15] Seabra AB, Paula AJ, de Lima R, Alves OL and Durán N 2014 *Chem. Res. Toxicol.* **27** 159
[16] Lee H, Shin US, Lee J, Kim H 2011b *J. Biomed. Mat. Res. B: Appl. Biomat.* **98B** 246
[17] Volpe BB, Santos Duarte AS, Ribeiro TB, Stocchero I, Kharmandayan P, Olalla Saad ST, Bustorff-Silva JM and Luzo ACM 2014 *Cytotherapy* **16** 1709
[18] Eça LP, Ramalho RB, Oliveira IS, Gomes PO, Pontes P, Ferreira AT and Mazzetti MP 2009 *Acta Cir. Bras*. **24** 400
[19] Fávaro W, Billis A, Nunes IS and Durán N 2012 *J. Urol.* **187** E231

IOP Conf. Series: Journal of Physics: Conf. Series **838** (2017) 012036 doi:10.1088/1742-6596/838/1/012036

Developing parameters for multi-mode ambient air models including the nanometer mode

Paolo Tronville[1] and Richard Rivers[2]

[1]Department of Energy, Politecnico di Torino, 10129 Turin, Italy
[2]Environmental Quality Sciences, Louisville, KY 40204-2106, USA

E-mail: paolo.tronville@polito.it

Abstract. The particle count, surface and mass in an occupied space can be modeled when the HVAC system airflows are known, along with the particle-size distribution for outdoor air, internal generation rates as a function of particle size, and the efficiency as a function of particle size for filters present. Outdoor air particle-size distribution is rarely available, but measures of particle mass concentration, $PM_{2.5}$ and PM_{10}, are often available for building locations. Outdoor air aerosol size distributions are well modeled by sums of two or three log-normal distributions, with essentially all mass in two larger modes. Studies have also shown that some mode parameters are, in general, related by simple functions. This paper shows how these relationships can be combined with known characteristics of $PM_{2.5}$ and PM_{10} samplers to create reasonable inclusive models of outdoor air aerosol-size distributions. This information plus knowledge of indoor particle generation allows calculation of aerosol mass in occupied spaces. Estimation of parameters of aerosol modes with sizes below100 nm and measurement of filter efficiencies in that range are described.

1. Introduction

To calculate the aerosol mass downstream of an air filter, two sets of data are needed. The first is the particulate mass and particle size distribution upstream of the filter. The second is the efficiency of the filter as a function of particle size. The range of sizes for both sets must include all sizes for which there is any significant quantity of particles in the upstream flow. Neither of these two data sets is easy to obtain if the calculation is made for a filter in a real-world building HVAC ventilation system. The outdoor air size distribution changes hour by hour. Even its average pattern will not be known unless some agency samples the air near the system intake, measures the size distribution more or less continuously, and publishes its data.

The particulate concentration and size distribution inside a ventilated space depends on many factors. Among these are the magnitude of the ventilation flows into and out of the space, including recirculation flow; aerosol generation in the space; leaks into and from the space; and the existence of additional filters in flow paths. Some of these factors are easy to quantify, others quite difficult. For ambient aerosols, even the concept of "size" is ambiguous. Only a spherical aerosol particle has a definite, single size. Filtration behavior of cylindrical particles or agglomerates must be expressed in terms of "aerodynamic equivalent diameter" spheres. In addition, filter efficiency also depends on particle density, which can have a wide range of values. For these reasons, standardized filter efficiency tests [1]-[4] make use of spherical liquid aerosols and nearly spherical solid aerosols of specified density.

Similarly, studies of the distribution of ambient and indoor particle sizes generally ignore density variations, and accept the sizes defined by whatever principle the distribution-measuring instrument uses. If the distribution is measured by a differential mobility analyzer (DMA) or scanning mobility spectrometer (SMPS), the diameters measured are equivalent to the size measured for a spherical particle with a specified density and a single electron charge. Such diameters are called "mobility diameters". Instruments for optical measurements of size distributions are usually calibrated in terms of the light scattered by spherical polystyrene latex particles; ambient particles are assigned the diameter of a latex particle which scatters the same amount of light. This is then called an "optical particle size". If the size-measurement device is a cascade impactor, the measured mean size on each impactor stage is a function of particle aerodynamic drag and density. Sizes thus measured are expressed as "aerodynamic equivalent diameters", usually for an assumed density of 1000 kg/m^3.

All of these instruments are widely used in studies of indoor air quality (IAQ) and outdoor air quality (OAQ), and to measure the local values of particle mass termed PM_1, $PM_{2.5}$ and PM_{10}. The exact meaning of these three measures of OAQ and IAQ and the observed relations between their values are essential to our discussion. We define them in the next section. References [5] and [6] provide summaries of the operating principles of the various instruments used to monitor ambient aerosol concentration and obtain PM_1, $PM_{2.5}$ and PM_{10}.

Data on filter efficiency as a function of particle size is subject to less ambiguity than ambient particle-size distribution data, provided one is willing to accept that the results of efficiency tests by standard procedures are representative of the results that would be obtained with typical ambient aerosols. Efficiency-vs-particle size data is also likely to be more easily obtained, since all internationally recognized standards require such data as part of the qualification of a filter.

However, obtaining this efficiency-vs-size data may require some persuasion of the filter manufacturer, for filter advertising material is almost always limited to filter "ratings". Ratings are single numbers which place a filter in a category, based on tests performed following the standard. Under EN 779:2012 [1] the rating is based on a single particle diameter, 0.4 µm. ASHRAE Standard 52.2-2 [2] has MERV ratings, Minimum Efficiency Reporting Values. A MERV rating is the minimum efficiency for particles of potassium chloride measured for the clean filter and after three loadings of the ASHRAE standardized loading dust. ISO 16890 [3] produces three rating numbers for filters, ePM_1, $ePM_{2.5}$ and ePM_{10}, based on calculating the particle mass removed by the filter from two particle-size distributions typical of urban and rural locations. The subscripted numbers represent the approximate upper limits of particle diameter (in µm) for the ISO calculation procedure.

Filters do not always operate at the "rated capacity" set by the manufacturer. The efficiency of a filter is to some degree dependent on its flow rate. All the above-mentioned test standards ignore this fact, and require only that filter efficiency be measured at a single flow rate. If efficiency-vs-particle-size data were made available for low, medium, and high operating flow rates for a filter, efficiency values at other flow rates throughout the tested size range could be calculated by an interpolation algorithm.

The changing efficiency of a filter as it accumulates particles in actual operation is quite complicated, and not well defined in filtration literature. OAQ varies with season, day of the week, and hour of the day. Internal particle generation in many cases is time-dependent. These effects could be incorporated into a computer model to calculate particulate matter concentration, including its variation with time. In this discussion, we limit our model for particulate matter concentration to steady-state conditions, and accept filter efficiency as that measured at a single flow rate.

2. How PM_1, $PM_{2.5}$ and PM_{10} are defined and measured

Many journal articles and websites refer to PM_{10} as "the mass of particles less than 10 µm diameter". PM_1 and $PM_{2.5}$ are often similarly defined, as "…less than 2.5 µm…" and "…less than 1 µm…". These definitions are only approximations. PM_{10} is in fact the mass concentration of particles that would be

IOP Conf. Series: Journal of Physics: Conf. Series **838** (2017) 012036 doi:10.1088/1742-6596/838/1/012036

collected from an air flow by a sampler which has a size-selective device ahead of its filter medium. The size-selective inlet is designed to pass most particles with aerodynamic diameters less than 10 µm, and to capture most particles greater than 10 µm. The percentage passing, called inlet penetration, for an actual $PM_{2.5}$ inlet is shown in Figure 1.

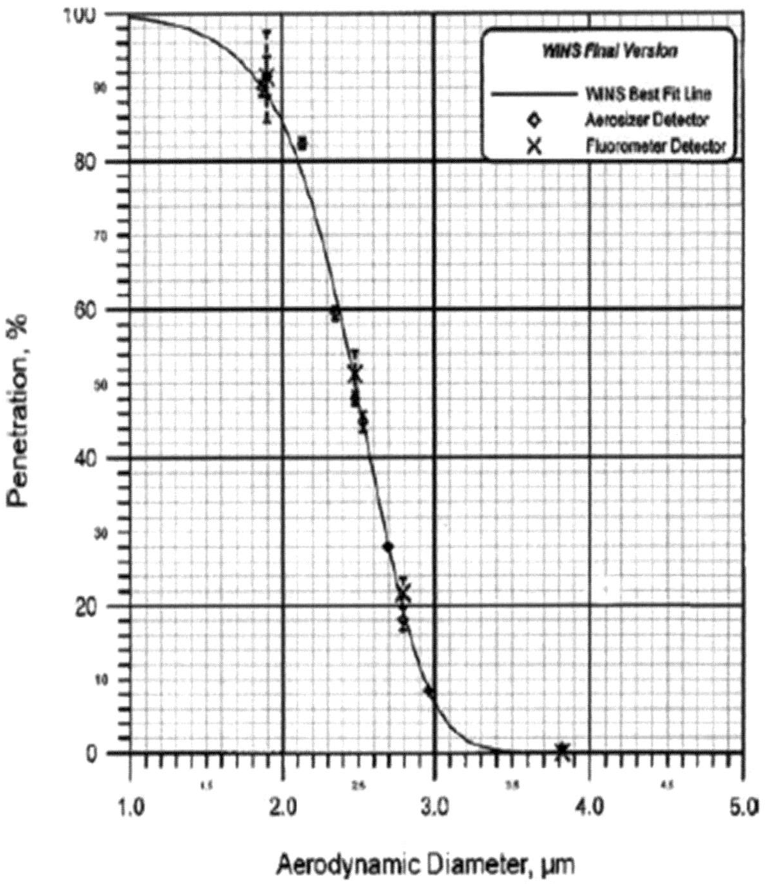

Figure 1. Particle penetration versus particle diameter for WINS $PM_{2.5}$ size-selective sampler inlet. (from reference [7]).

Note that the curve is tangent to the graph X-axes at penetration = 0% and 100%. It passes through diameter = 2.5 µm at penetration = 50 %, and transitions smoothly between these points. The curves for PM_1 and PM_{10} inlets are similar, but with 50 % values of 1 µm and 10 µm. The characteristics of PM_X sampler inlets are not, however, defined by the US Environmental Protection Agency and similar government agencies elsewhere as "ideal" for PM_1, $PM_{2.5}$ and PM_{10} by curves like Figure 1. The "ideal" curve for PM_{10} given in Ref. [8] is apparently a list of actual test data; and has irregularities which are not generally observed for impactors. The basic definition of $PM_{2.5}$ and PM_{10} in the United States, the Federal Reference Method (FRM) [8], is not data or a curve, but a set of drawings of the holder for the sampling media and the size-selective inlet that precedes it.

We have developed expressions which can be coded in computer programs for the PM_X inlet penetrations as functions of particle diameter. Our penetration function - equation (1) - fits the measured penetration data for PM_1 and $PM_{2.5}$ samplers quite well. It is a slight modification of an expression for the PM_X inlet penetrations given in Ref. [9]. Our modified expression is:

$$P_{in} = 100 \cdot \left[1 - \mathrm{K} \cdot \left(1 + \tanh\left(\mathrm{A} \cdot \left(\mathrm{B} \cdot d - \mathrm{PM_X}\right)\right)\right)\right] \quad (1)$$

IOP Conf. Series: Journal of Physics: Conf. Series **838** (2017) 012036 doi:10.1088/1742-6596/838/1/012036

Here P_{in} is the inlet penetration in %; PM_x is 1.0 or 2.5, depending on the PM_x inlet simulated; and *d* is particle diameter in micrometers. For PM_1 we developed best-fit parameters for the penetration curve given in [10]. For $PM_{2.5}$, data were taken from Figure 1 above, from Ref.[7].

The values for K, A, and B are as follows for the PM_1 and $PM_{2.5}$ inlets:

- PM_1: K = 0.500; A = 6.3; B = 1.0; Diameter range: all
- $PM_{2.5}$: K = 0.52453; A = 2.08; B = 0.991 Diameter range: 0.0 – 3.256 μm

Equation (1) provided a rather poor fit to the PM_{10} inlet penetration curve. We smoothed the tabulated data in Ref. [8] using two polynomials, a cubic for the diameter range from 0.0 to 15 μm, and a parabola for the range from 15 to 16 μm, where the penetration reaches zero. These expressions for PM_{10} are:

- Over the diameter range 0 - 15 μm:

$$P_{in} = 96.6444 - 0.098681081 \cdot d - 0.56324808 \cdot d^2 + 0.01066718 \cdot d^3 \quad (2)$$

- Over the diameter range 15 - 16 μm:

$$P_{in} = 593.6 - 72.3 \cdot d + 2.20 \cdot d^2 \quad (3)$$

For all three inlets, penetration is 100 % for *d* below the lower end of the lower diameter range, and 0 % for *d* above the upper end of the upper range.

One thing to be kept in mind in all analyses related to PM_x values is that they are measured or simulated only for the specific instantaneous particle-size distribution of Total Suspended Particulates (TSP) present, or its average over the sampling period. One should never speak of "a $PM_{2.5}$ dust" or "the PM_{10} efficiency of a filter" as values unconnected to the aerosol size distribution in the air of concern.

3. Particulate effects

HVAC systems designers are often faced with the problem of predicting the indoor air quality (IAQ) in buildings. The impetus for this is often the health of occupants, but it may be the effect of pollutants on materials or objects in the ventilated spaces [11]. In this discussion, we are concerned only with particulate pollution.

Many studies have been made of the impact of particulate pollution on human health. Indeed, concerns about outdoor pollution led to the establishment of PM-monitoring networks, and regulations based on PM data. The size-dependent toxicity of particles gave rise to the concepts of PM_1, $PM_{2.5}$ and PM_{10}. Many epidemiological studies provide correlations between PM_1, $PM_{2.5}$ and PM_{10} values and health effects.

Since humans spend much of their lives indoors, it is important to be able to determine the PM_x levels that would be measured by a PM_x sampler inside an occupied, ventilated space. This calculation might include the pattern of blower-driven flows for the spaces, leakages into and from the spaces, internal particle generation, particle deposition, air exchange between spaces, and the effects of filters in the system.

Characteristics of various system types are described in Refs. [12] through [15], and Ref. [13] and [16] provide equations for duct system design, including data on pressure drop in air distribution elements such as duct elbows. To calculate the flows in all system branches we need functions for the pressure drop dependence on air flow volume for all system elements. Both commercial and open-source computer program packages are available to carry out calculations of ventilation systems of considerable complexity.

In this discussion, we limit ourselves to one relatively simple configuration (Figure 2) and the choice of parameters usable in calculating its internal PM_x values.

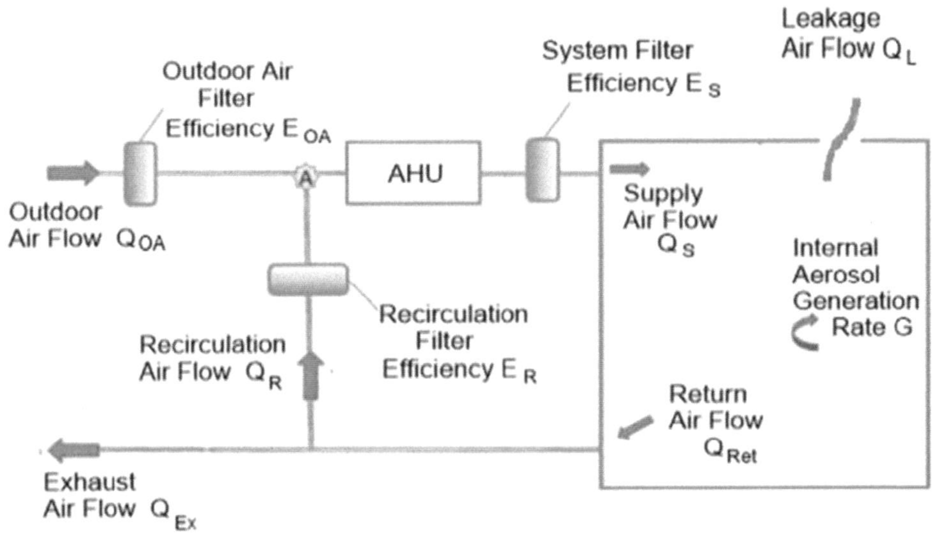

Figure 2. Schematic model of a ventilation system with recirculation and filtration for a single space.

4. Ambient particle size distributions

Many investigations have shown that outdoor aerosol distributions world-wide have three more or less distinct modes, each of which has an approximately log-normal distribution. The design of PM_1, $PM_{2.5}$ and PM_{10} size-selective inlets is not entirely arbitrary; they are intended to remove entire modes from TSP in a way that allows the mass in each mode to be determined. The entire outdoor mass size distribution in a locality can be described with fair accuracy if the three PM_x values for the locality are known, and some added assumptions are made.

One must first accept some values for the density of each mode. In addition, one must have the geometric mean diameter and standard deviation of the modal log-normal distributions. The overall mass distribution as a function of particle size can then be calculated. If the mass distribution is obtained, one further assumption – that all particles are spheres – allows one to calculate the surface and count distributions. Alternatively, if we know the overall count distribution, and accept the assumed density and sphericity of particles and the modal parameters, we can calculate the details of the surface and mass distributions.

5. Calculating the effect of a filter on PM_x

The log-normal function is one form of "probability density", which gives the relative numbers of particles of a specific particle size in an aerosol cloud. The most probable size has the geometric mean diameter d_{50}; probabilities for particles below and above this size have values given by the familiar bell-shaped "Gaussian" curve. The equation for this curve is:

$$f(d,\sigma_g,d_{50})=\frac{1}{\ln\sigma_g}\exp\left[-\frac{(\ln d-\ln d_{50})^2}{2(\ln\sigma_g)^2}\right] \quad (4)$$

Here d is the geometric mean diameter of a particle diameter range from d_1 to d_2, $(d_1 \cdot d_2)^{0.5}$, and σ_g the standard deviation of the modal distribution. The probability density of a multi-mode distribution is simply the sum of the probability densities of each mode, multiplied by the number-weighted fraction y for each mode. For three modes, the probability density is:

IOP Conf. Series: Journal of Physics: Conf. Series **838** (2017) 012036 doi:10.1088/1742-6596/838/1/012036

$$P(d) = y_A \cdot f(d, \sigma_{gA}, d_{50A}) + y_B \cdot f(d, \sigma_{gB}, d_{50B}) + y_C \cdot f(d, \sigma_{gC}, d_{50C}) \quad (5)$$

The sum of y values must always equal 1; in (5), $y_A + y_B + y_C = 1$. These y-values must be specified for the distribution of particle count, surface, or volume, depending on the parameter studied, or the method used obtain the distribution.

Equation (5) can be used for interior PM_X calculations only if the density of individual particles is assumed to be the same throughout the entire size spectrum. This is a rather drastic simplification. The bulk of the mass in the PM_{10} mode, especially in rural locations, is often soil or sand particles with density about 2600 kg/m^3. The carbon and organic condensate particles which dominate in the $PM_{2.5}$ and PM_1 modes have densities nearer 1000 kg/m^3. Computer programs can accommodate such differences quite easily. We will therefore assume that the particles in a mode have uniform density, but that the density values for the modes are not necessarily equal. This means that for mass calculations, the masses in each narrow size range must be calculated separately for all modes present, and the values summed. The steps for a single mode follow here below.

If we take the area under the curve under the function given by (5) to represent the total number of particles in the aerosol sample, the number of particles present between two sizes d_1 and d_2 is given by two equations:

$$d = \frac{\ln d_2 - \ln d_1}{2} \quad (6)$$

and

$$\Delta N = P(d) \cdot (\ln d_2 - \ln d_1) \quad (7)$$

With the assumption that the particles in the mode are all spheres, with density ρ_m, the mass between d_1 and d_2 is:

$$\Delta m(d) = \rho_m \cdot \frac{\pi \cdot d^3}{6} \cdot P(d) \cdot (\ln d_2 - \ln d_1) \quad (8)$$

If this little mass packet is passed through a filter with an efficiency $E(d)$ for particles with diameter d, the mass $\Delta m_2(d)$ attributable to this mode between sizes d_1 and d_2 immediately downstream of the filter is:

$$\Delta m_2(d) = [1 - E(d)] \cdot \Delta m(d) \quad (9)$$

where $E(d)$ is the efficiency of the filter at size d.

The PM_X concentration downstream of the filter including all particle sizes is:

$$PM_X = \sum_D \rho_m \cdot \frac{\pi \cdot d^3}{6} \cdot P(d) \cdot (\ln d_2 - \ln d_1) \cdot [1 - E(d)] \cdot P_{in}(d) \quad (10)$$

Here, $P_{in}(d)$ is given by Equation (1), (2) or (3), depending on which PM_X is being calculated. Summation is performed over all D ranges. In effect, we have simulated what particles would arrive at the sampling filter of a PMX monitor.

Air filter efficiency as a function of particle size is generally available, at least for one "rated" operating velocity, and usually for the particle size range from about 300 nm to 10 µm. Existing international efficiency testing standards for general ventilation filters are limited to this range, but instrumentation needed to extend efficiency measurement to low nanometer particle sizes is increasingly available in testing laboratories. Filtration theory predicts, and tests confirm, that fibrous filters efficiency will generally increase below 100 nm, because of diffusion effects. Nevertheless, manufacturers need to extend filter testing to low nanometer sizes, so that indoor concentrations of nanoparticles can be calculated as accurately as possible. Such calculations require numerical values for efficiency over the entire range of interest, i.e. also below 300 nm size.

The European Committee for Standardization has recognized this need, and draft standards (EN ISO 21083-1 and -2) are being written defining test methods for measuring efficiencies on nanoparticles in two ranges, 20-500 nm and 3-30 nm.

IOP Conf. Series: Journal of Physics: Conf. Series **838** (2017) 012036 doi:10.1088/1742-6596/838/1/012036

6. Sources of data for ventilation system parameters

6.1. Outdoor size distributions

Particle size distributions outdoors have been measured and reported in numerous published papers, but the data are usually presented graphically, making them difficult to average or to incorporate into computer programs. Many distributions are presented in the cumulative, or "percentage greater than stated size" form, that is the integral of equation (4) from $d = 0$ to the stated value of d. It is possible, but difficult, to extract the modes from such plots. Often, the distribution data are based on measurements made with a cascade impactor having 4 or 5 stages. The parameters of mode probabilities that can be extracted from such measurements have very low accuracies. Our search for ambient mode data was limited to higher-resolution methods, such as the Optical Particle Counter (OPC), Tapered Element Oscillating Microbalance (TEOM), Differential Mobility Analyzer (DMA) and Scanning Mobility Particle Sizer (SMPS). Ref. [17] is a survey of these devices for measuring aerosol distributions. Each device has limitations on its useful size range. Table 1 gives the averaged modal parameters found in a group of these distribution references. The mode names shown are common in aerosol literature, though not very precisely defined. The last two rows of Table 1 give parameter values for the four modes suitable for indoor PM_X calculations.

Table 1. Averages of parameters measured for outdoor aerosol modes and recommended for simulations

	Mode:	Nucleation		Aitken		Accumulation		Coarse	
Ref.	Location	d_{50}	σ	d_{50}	σ	d_{50}	σ	d_{50}	σ
20	Finland	11.7	1.71	37.3	1.78	151	1.60	-	-
20	Finland	13.8	1.63	42.5	1.71	152	1.54	-	-
21	Central Europe	-	-	67.4	1.68	265	1.49	-	-
22	United States	20.0	1.60	-	-	350	1.9	5000	2.2
23	Arctic Canada	-	-	-	-	300	2.2	10000	3.1
24	Europe	-	-	50.8	1.98	159	1.57	-	-
25	Pacific Ocean	21.8	1.16	36.0	1.19	155	1.18	-	-
26	Many	14.2	-	64.0	-	269	-	-	-
27	Beijing	-	-	-	-	305	1.67	6800	2.43
Recommended Values									
	d_{50}, σ :	15	1.50	50	1.70	230	1.60	7300	2.60
	m, kg/m^3:	1380		1000		1000		2660	

6.2. Outdoor PM concentrations

The World Health Organization provides a list of annual average $PM_{2.5}$ and PM_{10} values for 2971 cities world-wide [28]. The US EPA reports these data for thousands of urban and rural PM monitor sites [29]. Similar data is available for sites in other countries, for example [30]. There is essentially no data from government agencies for PM_1 concentrations. We list some values appearing in air-pollution literature in Table 2. The ratio (PM_1 / $PM_{2.5}$) for this list is reasonably constant. Multiplying its average value (0.75) times the local $PM_{2.5}$ might be used for PM_1 where no measured PM_1 is available.

IOP Conf. Series: Journal of Physics: Conf. Series **838** (2017) 012036 doi:10.1088/1742-6596/838/1/012036

Table 2. PM concentrations and $PM_1/PM_{2.5}$ ratios.

Ref.	Locale	Sampling Season, Days	Mean PM Values, μg/m3			Ratio,
			PM_1	$PM_{2.5}$	PM_{10}	$PM_1/PM_{2.5}$
31	urban, Vienna Austria	year 365	14.9	18.9	26.5	0.873
31	urban, Vienna Austria	year 365	14.7	18.8	29.1	0.782
31	suburban, Vienna Austria	year 365	17.5	21.1	31.0	0.829
32	urban, Athens Greece	year 365	18.5	23.7	51.3	0.781
32	urban, Athens Greece	year 365	20.1	29.3	52.2	0.686
32	university in hills near sea, Crete	year 365	10.3	17.9	32.5	0.575
33	urban, Taipei, Taiwan	spring 91	14.0	20.2	35.1	0.693
33	urban, Taipei, Taiwan	winter 90	9.7	12.7	26.4	0.764
33	urban, Taipei, Taiwan	spring 91	19.2	29.9	51.3	0.642
33	urban, Taipei, Taiwan	autumn 92	29.5	34.4	46.0	0.858
34	urban, Xi'an China	year 365	127.3	182.2*	-	0.699
35	urban, arid, Phoenix Arizona USA	spring 91	4.4	18.4	25.8	0.239
35	urban, arid, Phoenix Arizona USA	summer 92	5.9	8.4	81.6	0.702
35	urban, arid, Phoenix Arizona USA	autumn 92	9.9	14.2	57.8	0.697
36	urban rooftop, Chengdu China	spring 91	49	56	76	0.875
36	urban rooftop, Chengdu China	summer 92	40	43	49	0.930
36	urban rooftop, Chengdu China	autumn 92	54	56	60	0.964
36	urban rooftop, Chengdu China	winter 90	76	83	92	0.916
37	urban rooftop, normal year, Delhi	winter 90	204	236	338	0.864
37	urban rooftop, normal year, Delhi	summer 90	43	69	178	0.623
37	urban, monsoon season, Delhi	Aug./Sept 61	37	54	132	0.685
37	urban post-monsoon, Delhi	Oct./Nov 61	337	389	548	0.866
38	urban, highway traffic, Barcelona	July/Nov 150	17	25	38	0.680

6.3. Indoor particle generation

Extensive studies of indoor particle generation by industrial processes, such as welding, exist. The generation of particles in non-industrial situations, such as offices and concert halls, has received less attention, because the hazards are far less.

Aerosol generation by ink-jet and 3-D printers has been studied [39], [40], and there is abundant literature on the generation of particles and toxic gaseous contaminants by tobacco smoking. These studies have led to widespread prohibition of indoor smoking, but reference to the literature [41], may be necessary in some situations.

Nanoparticle generation in buildings is studied in [42] and [43]. The particulate progeny of radon and thoron are peculiarly hazardous because they are radioactive

6.4. System flows and leakage

The blower-driven flows Q_{OA} and Q_R in Figure 2 are of course determined by the heating and cooling loads of the space, and the number of occupants. Leakage into or out of the space or its ventilation ducts is dependent on the pressure differential between the space or ducts and outdoors, and can be measured the "blower-door" technique (Ref. [44]). Predicting leakage is difficult, but both new construction and remodeling usually seek to minimize leakage. Ref. [45] and [46] provide guidance on the effects of leakage-reduction. Ref. [47] provides actual measurements of leakage in buildings in several countries, with different wall constructions, where one may need to estimate leakage parameters for cases where no measurement of leakage is attempted.

IOP Conf. Series: Journal of Physics: Conf. Series **838** (2017) 012036 doi:10.1088/1742-6596/838/1/012036

7. Conclusions

Reasonable estimates of the total suspended particulates in ventilated spaces can be calculated, including the values of PM_1, $PM_{2.5}$ and PM_{10} mass concentrations that would be measured in these spaces by monitors designed to measure these indices outdoors. Such calculations allow evaluation of the impact of particle filters on the health and comfort of building occupants, and on potential damage to surfaces and objects in the ventilated space.

Some of the parameters needed for the calculation are well-defined and available for specific locations, but some are can only be assigned typical values. The calculations are complex, and well-organized user-friendly computer programs and databases are needed to make them useful to HVAC system designers.

The effect of filter loading with particles has not been quantified, and filter efficiencies-versus particle size at different operating flows are not usually available. This data must be provided if the cost and energy impact of filters are to be calculated.

References

[1] EN 779 "Particulate Air Filters for General Ventilation – Determination of the Filtration Performance", 2012, European Committee for Standardisation (CEN), Brussels.

[2] ANSI/ASHRAE Standard 52.2 "Method of Testing General Ventilation Air-Cleaning Devices for Removal Efficiency by Particle Size", 2012, American Society of Heating, Ventilation and Air-Conditioning Engineers, Atlanta GA, USA.

[3] ISO 16890-1 "Air Filters for General Ventilation – Part 1: Technical specifications, requirements and classification system based on particulate matter efficiency (ePM)", 2016, International Standards Organization, Geneva.

[4] Tronville P., Rivers R., 2016, "Air filter performance", ASHRAE J., 58: 14 – 25.

[5] Watson J. G., Chow J. C., Moosmüller H., Green M., Frank N., Pitchford M., 1998, Guidance for using continuous monitors in $PM_{2.5}$ monitoring networks (EPA-154/R-98-012), US EPA OAQPS, Research Triangle Park NC, USA.

[6] Amaral S. S., Andrade de Carvallo J., Martins Costa M. A., Pinheiro C., 2015, An overview of particulate matter measurement instruments, Atmosphere, 6:1327-1345.

[7] Peters T M, Vanderpool R W, Weiner R W 2001 Design and calibration of the EPA $PM_{2.5}$ well impactor ninety-six (WINS), Aerosol Sci. & Tech. 34: 389-397.

[8] US Government Printing Office 2015 Federal Register, CFR Title 40 Part 50, Appendix L: Reference Method for the Determination of Fine Particulate Matter as $PM_{2.5}$ in the Atmosphere.

[9] Lawless P. A., Rodes C. E., Evans G., Sheldon L., Creason J., 2001, Aerosol Concentrations during the 1999 Fresno exposure studies as functions of size, season and meteorology, Aerosol Sci. & Tech. 34: 66-74.

[10] Gussman R. A., Kenny L. C., Labickas M., Norton P., 2002, Design, calibration and field test of a cyclone for PM_1 ambient air sampling, Aerosol Sci. & Tech., 36:361-365.

[11] Nazaroff W. W., Ligocki M. P., Salmon L. G., Cass G. R., Fall T., Jones V. C., Liu H. I. H., Ma T., 1993, Airborne Particles in Museums. Getty Conservation Institute, Marina del Rey CA, USA.

[12] Bhatia A., 2012, Design Options for HVAC Distribution Systems, PDH Center, Fairfax VA USA.

[13] Burdick A., 2011, Advanced Strategy Guideline: Air Distribution Basics and Duct Design, US Department of Energy Building Technologies Program.

[14] ASHRAE Guideline 24 "Ventilation and Indoor Air Quality in Low-Rise Residential Buildings" American Society of Heating, Ventilation and Air-Conditioning Engineers, Atlanta GA USA.

[15] ASHRAE 2016 Handbook: Systems and Equipment, American Society of Heating, Ventilation and Air-Conditioning Engineers, Atlanta GA USA.

[16] ASHRAE 2013 Handbook: Fundamentals, American Society of Heating, Ventilation and Air-Conditioning Engineers, Atlanta GA USA.
[17] Ionel I., Popescu F., 2010, Methods for online monitoring of air pollution concentrations, Air Quality, ISBN 978-953-307-131-2.
[18] De Carlo P. F., Slovik J. G., Worsnop D. R., Davidovits P., Jimenez J. L., 2004, Particle morphology and density characterization by combined mobility and aerodynamic diameter measurements. Part 1: Theory, Aerosol Science & Tech., 38: 1185-1205.
[19] Sarangi B., Aggarwal S. G., Sinha D., Gupta P. K., 2016, Aerosol effective density measurement using scanning mobility particle sizer and quartz crystal microbalance, with estimation of involved uncertainty, Atmos. Meas. Tech., 9: 859-875.
[20] Hussein T., Puustinen A., Aalto P. P., Mäkelä J. M., Hämeri K., Kulmala M., 2004, Urban aerosol number size distributions, Atmos. Chem. Phys., 3: 391-411.
[21] Birmili W., Wiedensohler A., Heintzenberg J., Lehmann K., 2001, Atmospheric particle number size distribution in central Europe: statistical relations to air masses and meteorology, J. Geophys. Res. 106: 32005-32018.
[22] Lundgren D. A., Harris F. S., Marlow W. H., Lippmann M., Clark W. E., Durham MD 1979, Aerosol Measurement (p170), University Presses of Florida, Gainesville, FL, USA.
[23] Covert D. S., Wiedensohler A., Aalto P., Heintzenberg J., McMurry P., Leck C., 1996, Aerosol number size from 3 to 500 nm diameter in the arctic marine layer during Summer and Autumn, Tellus, 48B: 197-212.
[24] Asmi A. et. al., 2011, Number size distribution and seasonality of submicron particles in Europe 2008-2009, Atmos. Chem. Phys., 11: 5505-5508.
[25] Ueda S, Miura K, Kawata R, Furutani H, Uematsu M, Omori Y, Taniomoto H 2016 Number size distribution of aerosol particles and new particle formation events in tropical and subtropical Pacific Ocean, Atmos. Envir., 142: 324-339.
[26] Costabile F. et. al., 2009 Spatio-temporal variability and principal components of the particle number size distribution in an urban atmosphere, Atmos. Chem. Phys., 9: 3163-3195.
[27] Wang S. P., Fang L., Zhang X. Y., Wang W. H., 2015, Retrieval of aerosol properties from fine/coarse mode aerosol mixtures over Beijing from PARASOL measurements, Remote Sensing, 7: 9311-9324.
[28] WHO, 2016, WHO Global Urban Ambient Air Pollution Database, www.who.int/phe/health_topics/outdoorair/databases/cities/en, World Health Organization, Geneva.
[29] EPA 2016 Tables of Daily and Daily Summary Data, aqsdr1.epa.gov/aqsweb/aqstmp/airdata/download_files.html, US Environmental Protection Agency, Washington, USA.
[30] Beccaceci S. et. al., 2014, Airborne Particulate Concentrations and Numbers in the UK- Annual Report 2014, https://uk-air.defra.gov.uk/library.
[31] Gomiscek B., Hauck H., Stopper S., Preining O., 2004, Spatial and temporal variations of PM1, PM2.5, PM10 and particle number concentrations during AUPHEP project, Atmos. Envir., 38: 3917 – 3934.
[32] Koulouri E., Grivas G., Kerasopoulos E., Chaloulakou A., Mihalopoulos N., Spyrellis N., 2008, Study of size-segregated particulatee (PM_1, $PM_{2.5}$, PM_{10}) concentrations over Greece, Global NEST Journal, 10: 132-139.
[33] Li C-S, Lin C-H, 2002 PM1 /PM2.5 /PM10 characteristics in the urban atmosphere of Taipei, Aerosol Sci. Tech. 36 469 – 473.
[34] Shen Z. X., Cao J. J., Arimato R., Han Y. M., Zhu C. S., Tian J., Liu S. X., 2010, Chemical characteristics of fine particles (PM1) from Xi'an China, Aerosol Sci. Tech., 44: 461 – 472.
[35] Lundgren D., Hlang D. N., Rich T. A., Marple V. A., 1996, PM_{10}/ $PM_{2.5}$/ PM_1 data from a trichotomous sampler, Aerosol Sci. Tech, 25: 353 – 357.
[36] Cao J. J., Xu H. M., Xu Q., Chen B. H., Kan H. D., 2012, Fine particle matter constituents and

cardiopulmonary mortality in a polluted Chinese city, Int'l J. Envir. Res. Public Health, 120: 373 – 378.
[37] Tiwari S., Chate D. M., Srivastava A. K., Bisht D. S., Padmanabhamurty B., 2012, Assessment of PM_1, $PM_{2.5}$, and PM_{10} concentrations in Delhi at different mean cycles, Geofizika, 29: 125 – 141.
[38] Pérez N., Pey J., Cusack M., Reche C., Querol X., Alastuey A., Viana M., 2010, Variability of particle number, black carbon and PM_{10}, $PM_{2.5}$, and PM_1 levels and speciation: influence of road traffic emissions, Aerosol Sci. Tech., 44: 487 – 499.
[39] He C. G., Morawska L., Taplin L., 2007, Particle emission characteristics of office printers, Envir. Sci. Tech., 41: 6039-6045.
[40] Stephens B., Azimi P., El Orch Z., Ramos T., 2013, Ultrafine particle emissions from desktop 3D printers, Atmos. Envir., 79: 334-339.
[41] Russo E. T., Hulse T. E., Adamkiewicz G., Levy D. E., Bethune L., Kane J., Reid M., Shah S. N., 2015, Comparison of indoor air quality in smoke-permitted and smoke-free multiunit housing: findings from the Boston Housing Authority, Nicotine and Tobacco Res., 2015: 316-322
[42] Wallace L., 2006, Indoor sources of ultrafine and accumulation mode particles: size-resolved concentrations and source strengths, Aerosol Sci. Tech., 40: 348-360.
[43] Tu K.-W., Knutson E. O., George A. C., 1994, Thoron versus radon: comparison of measured progeny aerosol size distributions, Aerosol Sci. Tech., 20: 266-274.
[44] Southface Energy Institute, 2013, Blower Door and Duct Pressure Testing for Duct and Envelope Tightness and Verification, www.southface.org.
[45] Carrié F. R., Jobert R., Leprince V., 2012, Methods and Techniques for Airtight Buildings, Air Infiltration and Ventilation Centre, Coventry, UK.
[46] U.S. Department of Energy EERE, 2011, Building Technolgies Program Building Air Leakage Guide, www.eere.energy.gov/informationcenter.
[47] Orme M., Liddament M. W., Wilson A., 1998, Numerical Data for Air Infiltration and Natural Ventilation Calculations, Air Infiltration and Ventilation Centre, Coventry, UK.